WAVE PROPAGATION – MOVING LOAD – VIBRATION REDUCTION

PROCEEDINGS OF THE INTERNATIONAL WORKSHOP WAVE 2002
OKAYAMA, JAPAN, 18 – 20 SEPTEMBER 2002

Wave propagation
Moving load – Vibration Reduction

Edited by

Nawawi Chouw
Okayama University, Japan

Günther Schmid
Ruhr University Bochum, Germany

A.A. BALKEMA PUBLISHERS LISSE / ABINGDON / EXTON (PA) / TOKYO

Published by: A.A. Balkema, a member of Swets & Zeitlinger Publishers
www.balkema.nl and www.szp.swets.nl

ISBN 90 5809 559 2

Printed in the Netherlands

Wave propagation – Moving load – Vibration reduction, Chouw & Schmid (eds.)
© 2003 Swets & Zeitlinger, Lisse, ISBN 90 5809 559 2

Table of Contents

Vibration reduction

Track-soil analysis

Numerical and experimental investigation

Wave propagation – Moving load – Vibration reduction, Chouw & Schmid (eds.)
© 2003 Swets & Zeitlinger, Lisse, ISBN 90 5809 559 2

Foreword

During the workshop WAVE 2002 researchers from universities and research institutes and practitioners came together to discuss the state of the knowledge in man-made induced disturbances and their mitigation. Special interest was given to ground vibration and noise caused by ground compaction, high-speed trains and explosion in soil, and also to noise and vibration reduction. Most of the colleagues contributed works related to high-speed trains, since speed plays a significant role in the life of many people. For many, speed is important, because they believe it will significantly improve their life effectiveness. People wish, if it is possible, to arrive at their destination as fast as possible. This ability can indeed increase our life effectiveness, however, we should remember that this does not automatically mean our life quality will increase. Often not *not to have but the desire to have* is the true source of many people's dissatisfaction with their situation. The speed can indeed calm their desire. The speed can, however, only contribute to our life quality if we can distinguish the essential from the unessential needs. It does not do us much good, to arrive at our destination very fast, if we do not know what we really want there. Many people arrive at a certain location, and soon have the desire to be at another location. Sometimes it is significant to rest and to be in silence, since it makes us able to recognize our essential needs.

The workshop WAVE 2002 is the result of the co-operation between the Okayama University, the Ruhr University Bochum, Germany, and the Tokyo Institute of Technology. As in the previous two workshops at the Ruhr University Bochum in 1994 and 2000 we are pleased that colleagues from our institutions and also experts from other institutions are present in the workshop. Besides the considered wave propagation and vibration reduction, noise pollution due to a moving source is considered for the first time in this workshop. The Scientific Committee invited 12 specialists and selected additional contributions from a call for papers. The proceedings contain the state of the art contributions. Each contribution is peer reviewed and revised. The addresses of the speakers in the appendix may be used for establishing personal contacts, and for future co-operation.

The workshop was supported by the Japanese Ministry of Education, Science, Sports, and Culture; the Electric Technology Research Foundation of Chugoku; The Yakumo Foundation for Environmental Science; Wesco Scientific Promotion Foundation; Japan Society of Civil Engineering; Okayama Foundation for Science and Technology. Support was also received from the Dean of the Faculty of Environmental Science and Technology, and the President of the Okayama University.

To all members of the Organizing and Scientific Committee we would like to express our sincere thanks for their excellent co-operation. We also would like to thank the reviewers for their splendid work.

Special thanks are due to Dr. Kohei Yamamoto of the Kobayashi Institute of Physical Research in Tokyo for his significant support in organizing the session on noise pollution.

Our thanks to all speakers and participants for their contribution to the discussion and to the proceedings, for the exchange of their knowledge and for creating the nice atmosphere during the meeting.

We hope that WAVE 2002 is one of many sources of knowledge that will continue to propagate in the future.

Nawawi Chouw and Günther Schmid

Foreword of the Dean of the Faculty of Environmental Science and Technology, Okayama University

The third international workshop WAVE 2002 on *Wave Propagation, Moving Load and Vibration Reduction* represents the fifth international symposium annually hosted by the Faculty of Environmental Science and Technology at the Okayama University. The purpose of these annual meetings is to present a forum for the presentation and discussion of current activities and progress in various environmental issues. We also hope these meetings will enhance the interdisciplinary efforts of mathematicians, civil engineers, earthquake engineers, other researchers, and engineers in practice in remedying environmental problems.

This year the meeting aims at a better understanding of wave propagation in the soil and in the air due to a stationary as well as moving source through exchanging the knowledge and experience of the participants.

We would like to thank the Ministry of Education, Science, Sports and Culture, the Electric Technology Research Foundation of Chugoku, the Yakumo Foundation for Environmental Science, Wesco Scientific Promotion Foundation, Japan Society of Civil Engineering and Okayama Foundation for Science and Technology for their kindness to support this meeting.

It is our hope and belief that this meeting will contribute towards the understanding of phenomena to improve our life quality by reducing vibration as well as noise pollution.

Professor Dr. Kenji Sakata

Foreword of the President of the Okayama University

Development in the society is often accompanied by environmental problems. We therefore have to deal with the complexity of environmental issues. In order to solve environmental problems and to improve the standards of living, our Faculty of Environmental Science and Technology was established in 1994. Up to now it is the only kind of faculty established at a National University in Japan. It has four departments: Environmental and Mathematical Sciences, Environmental and Civil Engineering, *Environmental Management Engineering as well as Environmental Chemistry and Materials.*

The topics of this workshop, *Wave Propagation, Moving Load and Vibration Reduction*, are of strong interest to us. Since the 80's we have strong co-operation with Ruhr University Bochum not only in exchange of teaching staff and researchers but also students. Since the first international workshop WAVE'94, hosted at the Ruhr University Bochum, the meetings were already jointly organized by Ruhr and Okayama Universities. This time the Tokyo Institute of Technology also joins us in organizing the meeting.

I hope that the lectures, discussions, and exchange of ideas during this meeting will be useful for your work in the future. I also hope that the established international co-operation will improve the transfer of knowledge not only in education but also in practice.

Professor Dr. Iichiro Kono

Foreword of the Dean of the Faculty of Civil Engineering, Ruhr University Boohum

As the Dean of the Faculty of Civil Engineering of the Ruhr University Bochum I convey the best greetings to the participants of the workshop and to the Faculty of Environmental Science and Technology of Okayama University to which we are related through an Agreement for Academic Exchange and through the organization of now the third WAVE-workshop.

The workshop WAVE 2002 addresses man-made waves which are transmitted through the soil or the air inducing vibrations to buildings or creating the sensation of noise in our ears. In contrast to the traditional conception of waves, which expressed so beautifully the Japanese painter Hokusai, man-made waves create nuisances which we as engineers or geologists have to reduce or eliminate to conserve an environment where we can live well and enjoyably.

I wish WAVE 2002 to be a successful workshop which helps to combine the needed technical inventions with an environment free of unpleasant vibration and noise.

Professor Dr. Harro Stolpe

Foreword of the Rector of the Ruhr University Bochum

With the international workshop WAVE 2002 held at the Okayama University in Japan common goals and improved relations with the Ruhr University Bochum in Germany, where two years ago WAVE 2000 took place, is documented. Whereas two years ago the relations were related to common interest in research projects related to prediction of waves created by earthquakes or by fast moving trains currently exchange of engineering students between our institutions show the intensified cooperation and the interest to learn from each other. And indeed we are much closer connected to each other than most people believe. The seismological centre of the Department of Geology of the Ruhr University detects within short delay the waves which in Japan are excited by the movements of the Catfish which, as we know from our Japanese colleagues, creates earthquakes.

I hope that we continue to be related through man made waves which are not only created by fast moving trains but also by new explosive ideas. Ideas that can also reduce the stress which we induce in our environment, another topic of this WAVE-workshop.

I wish that all participants in this conference have an exciting meeting which radiates new ideas into the international scientific community and that research, teaching and application in practice will interact with these waves.

Professor Dr. Dietmar Petzina

Wave propagation – Moving load – Vibration reduction, Chouw & Schmid (eds.)
© 2003 Swets & Zeitlinger, Lisse, ISBN 90 5809 559 2

Organization

SCIENTIFIC COMMITTEE

Nawawi Chouw (Okayama University, Japan)
Sohichi Hirose (Tokyo Institute of Technology, Japan)
Günther Schmid (Ruhr University Bochum, Germany)
Takeo Taniguchi (Okayama University, Japan)

ORGANIZING COMMITTEE

Nawawi Chouw *
Talal A. Mohamed Etri * ***
Kayoko Hashimoto *
Sohichi Hirose **
Günther Schmid ***
Takeo Taniguchi *
Kenji Tomioka *
Jun Quan * ***

* Okayama University, Japan
** Tokyo Institute of Technology, Japan
*** Ruhr University Bochum, Germany

SPONSORS

Ministry of Education, Science, Sports and Culture, Japan
Electric Technology Research Foundation of Chugoku, Japan
Yakumo Foundation for Environmental Science, Japan
Wesco Scientific Promotion Foundation, Japan
Japan Society of Civil Engineering, Japan
Okayama Foundation for Science and Technology, Japan

REVIEWERS

Nawawi Chouw
Okayama University, Japan
chouw@cc.okayama-u.ac.jp

Uwe E. Dorka
University of Kassel, Germany
uwe.dorka@uni-kassel.de

Otto von Estorff
University Hamburg-Harburg, Germany
estorff@tuhh.de

Hong Hao
University of Western Australia, Australia
hao@civil.uwa.edu.au

Sohichi Hirose
Tokyo Institute of Technology, Japan
shirose@cv.titech.ac.jp

Günther Schmid
Ruhr University Bochum, Germany
Guenther.Schmid@ruhr-uni-bochum.de

Ground vibrations and blast induced vibrations

long periods waves become dominant.

Figure 6 shows the acceleration obtained from the numerical derivation of velocity records using sampling interval of the velocity-meter (denoted Vel-3) next to the accelerometer Acc-3 (see Figure 2 for location). Although records are very similar to each other, the accelerations computed from velocity-meters are larger in amplitude as compared with those from accelerometers. For example, for the vertical component, the maximum amplitude of acceleration wave was 525 gal from the accelerometer as compared with 609 gal computed from the velocity record of the velocity-meter. The amplitude obtained from the velocity-meter records is about 1.16 times that of the true acceleration records. The difference may arise from one or all of the following reasons: slight variation of fixation of instruments, the errors inherent in numerical derivation arising from digital wave forms or the frequency dependence of directly measured acceleration value. The validity of the first and second reasons must be checked by further measurements and analysis. If, especially, the second reason holds true, the empirical damage criteria based on the peak particle velocity of the ground, could not be employed straightforward in the case integration of directly measured acceleration values, or vice versa. Hence a different damage criteria should be developed for the case of direct monitoring of acceleration.

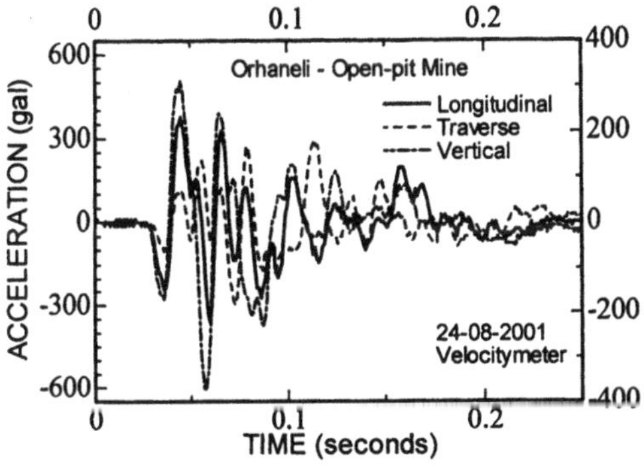

Figure 6: Acceleration components obtained from the numerical derivation of velocity records

Next Fourier spectra of longitudinal, traverse and vertical components of acceleration records of accelerometer Acc-1 and Vel-3 are computed and shown in Figures 7 and 8. As expected the vertical component consist of waves with greater amplitude as compared with the other components. While dominant frequencies of the waves observed at 8-10 Hz and 30-40 Hz, it may be safe to say that blasting results mainly in high-frequency waves due to very close monitoring distance. This implies that the low-rise buildings may be influenced by the vibrations caused by blasting operations if one takes into account the fundamental vibration mode. However, Hao (2002) pointed out that the high frequency blast motion might also excite high vibration modes of high-rise buildings.

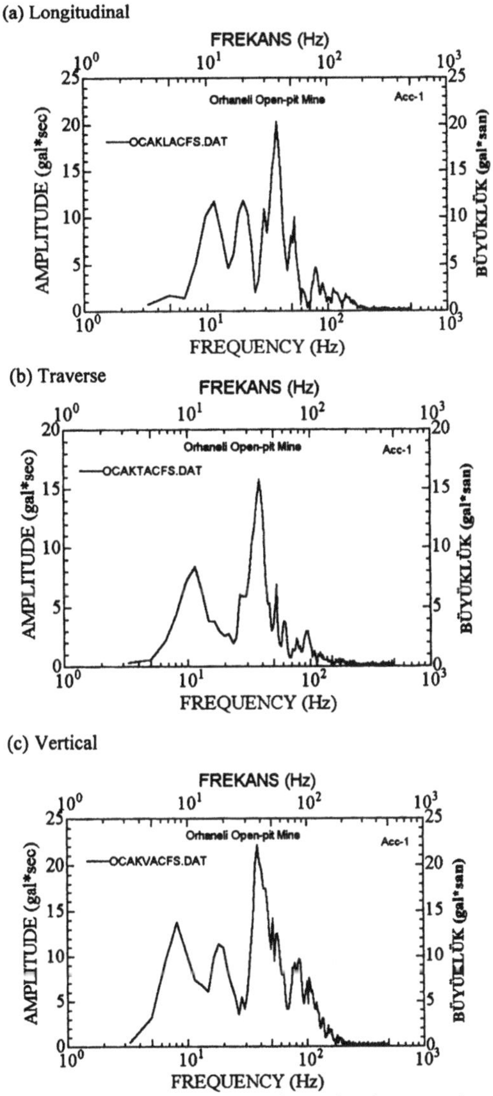

Figure 7: Fourier spectra of acceleration records of accelerometer Acc-1

Using the acceleration records of accelerometer Acc-1 and accelerations computed from the records of velocity-meter Vel-3, a series of response analysis is carried out. Figures 9 and 10 shows the normalized acceleration response spectra of each component of acceleration records, respectively with damping coefficient (*h*) values of 0.000,0.025,0.050.

(a) Longitudinal

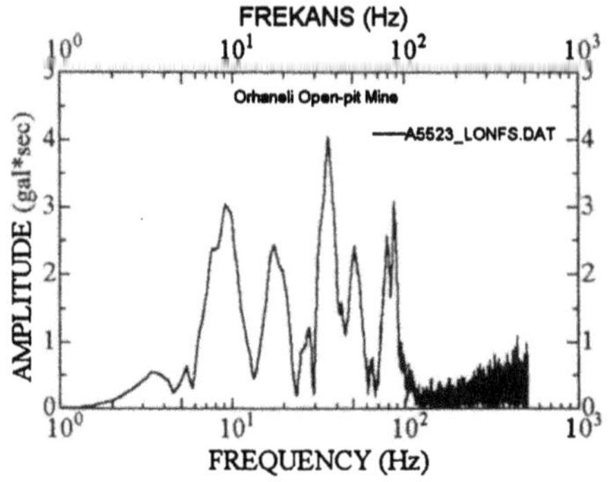

(b) Traverse

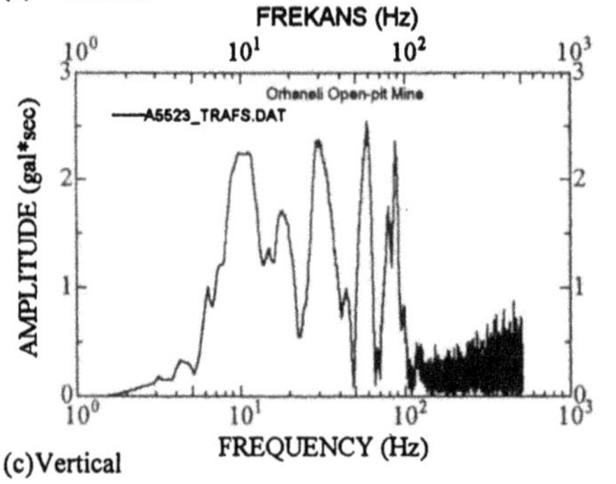

(c)Vertical

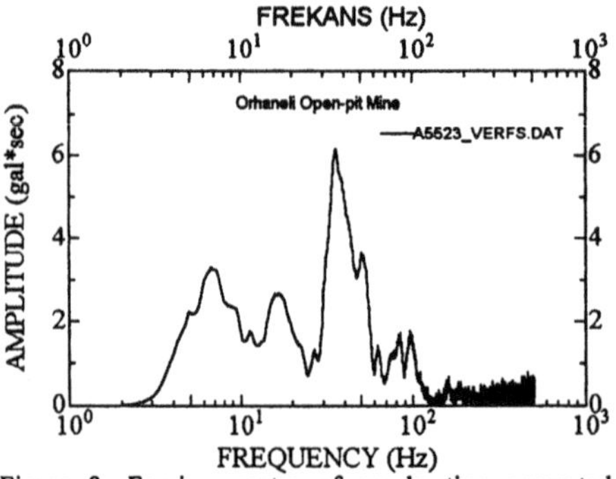

Figure 8: Fourier spectra of acceleration computed from velocity records of velocity-meter Vel-3

The results indicate that structures having a natural period less than 0.05s could be very much influenced. The effect of blasting should be smaller for structures having natural periods greater than 0.1s. Figure 11 and 12 show the responses of a structure with a natural period of less than 0.05 second and damping coefficient of 0.05 (H=0.059) for each acceleration components from the accelerometer measurements and velocity-meter measurements. As understood from Figure 9 and 10, the absolute acceleration acting on structures should be less than the accelerations of input motion. It would be safe to assume that the induced accelerations caused by blasting should act on the ground-structure system without any reduction.

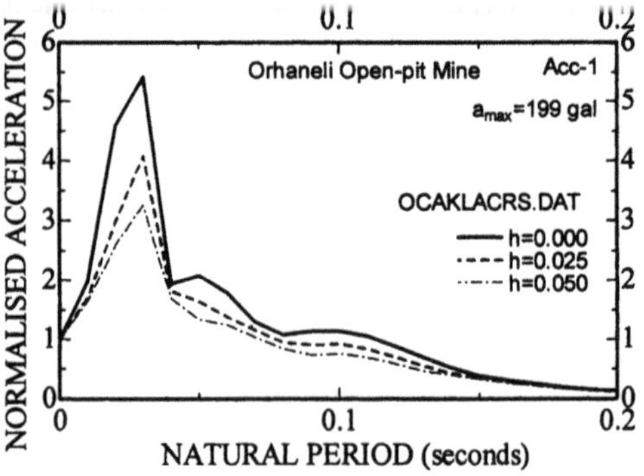

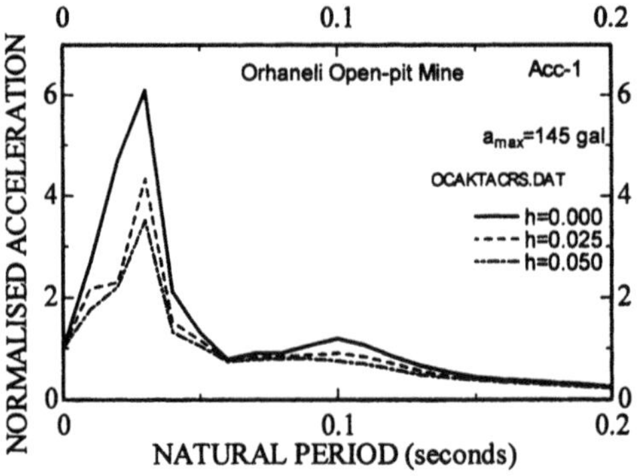

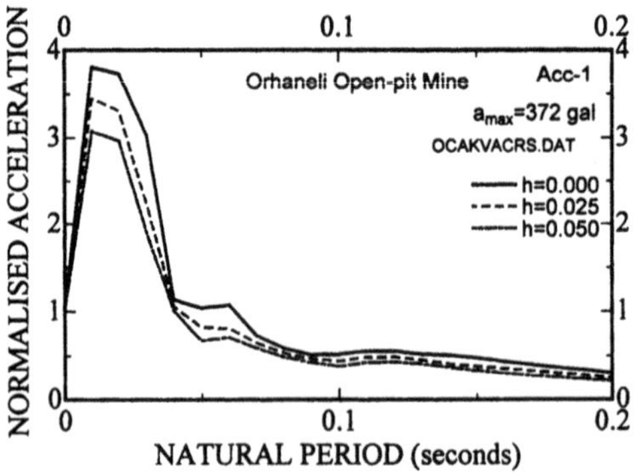

Figure 9 Normalized acceleration response spectra of the records of accelerometer Acc-1

4 NATURAL PERIODS OF STRUCTURES IN TURKEY

Aykut (1973) was first to initiate some experimental studies on the natural periods of structures in Turkey. He also developed an instrument for measuring the natural periods of structures in Turkey. Bayülke (1978) who made new additional measurements on the natural periods of structures throughout Turkey followed Aykut's work.

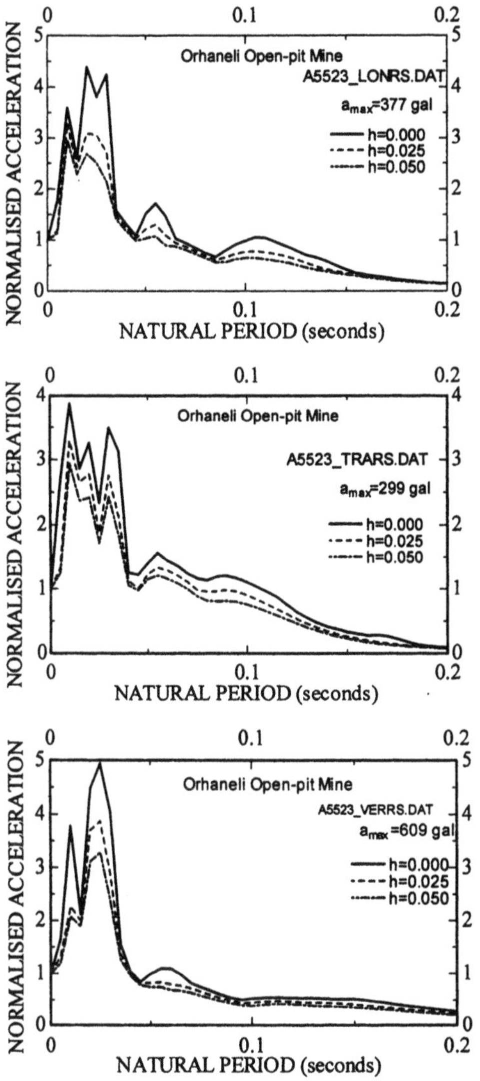

Figure 10 Normalized acceleration response spectra from records of velocity-meter Vel-3

Recently, the data was supplemented with some additional data with the use of better instruments. The data are plotted as a function of number of stories (N) in Figure 13. This experimental data were fit to a linear function by considering the number of stories up to 8 to obtain an empirical relation between the natural period of the building (T) and the number of stories (N):

$$T = 0.05N \qquad (1)$$

The 1997 Turkish Seismic Code requires the use of the following function for the natural period of buildings for the first fundamental mode

$$T = C_t H_t^{3/4} \qquad (2)$$

Where H_t is total height of structure and C_t is a coefficient, which may be selected using three different procedures. The simple recommendations are as follows:

$C_t = 0.07$ for reinforced concrete framed structures

$C_t = 0.08$ for steel framed structures

$C_t = 0.05$ for all kinds of buildings

Equation above may be written in terms of story number N as

$$T = C_t (N * H_f)^{3/4} \qquad (3)$$

Where H_f is story height.

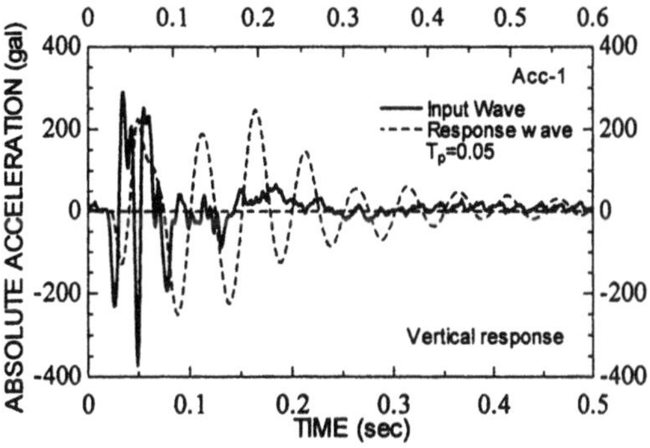

Figure 11 Absolute vertical acceleration response of ground-structure system (Acc-1)

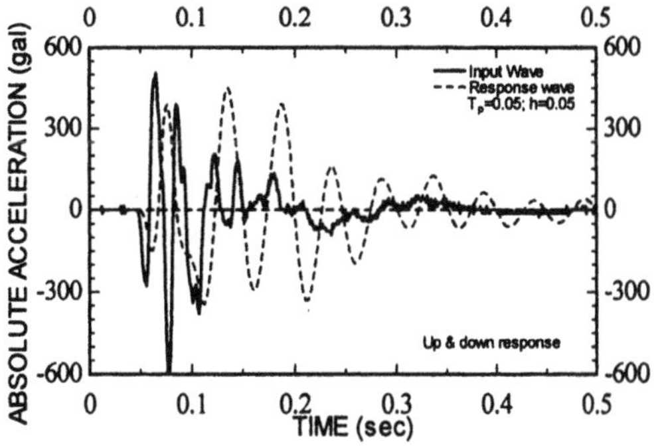

Figure 12 Absolute vertical acceleration response of ground-structure system using the acceleration computed from the velocity records of velocity-meter Vel-3.

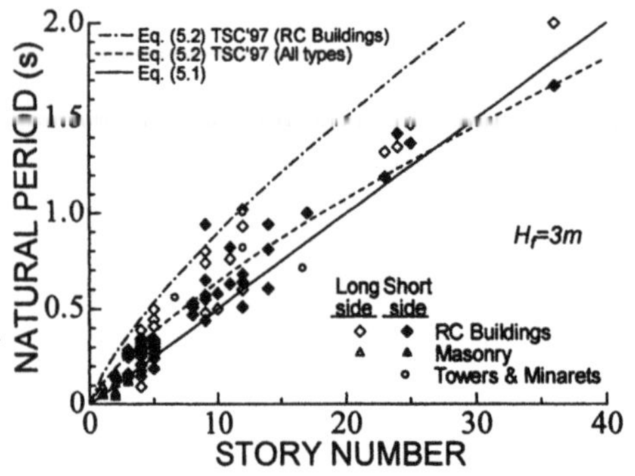

Figure 13: Natural periods of structures in Turkey

These equations are also plotted in Figure 13. Although Eq. (1) acts as a lower bound for a great range of story number, this simple formula (1) can also be used for estimating the natural periods of buildings up to 8 stories which covers most of the building stocks in Turkey. The natural periods of buildings with 4,5,6,7,8 stories are obtained from Eq. (1) are 0.20s, 0.25s, 0.30s, 0.35s and 0.40s. The natural period of minarets also ranges between 0.30-0.40s.

5 THE EFFECT OF THE 1999 KOCAELİ EARTHQUAKE

The nearest strong motion station of the national strong motion network of Turkey is in Bursa, which is about 40km north of Orhaneli open-pit mine. The peak ground accelerations for SN, EW and UD components are 54.32,45.81,42.96 gals, respectively. The hypo-central distances of the Bursa strong motion and Orhaneli open-pit mine are 93 km and 130km, respectively. Figures 14 and 15 show the attenuation relations together with observational data and the shaking response at Bursa station.

Fourier and response spectra of acceleration records are shown in Figure 16 and Figure 17, respectively. The computed response spectra results particularly indicate that the 1999 Kocaeli earthquake should have much larger amplification effects compared to that on structures with a few stories. In other words, no amplification effects should be expected from this earthquake on buildings having one and two stories.

Figure 18 shows the responses of a building with a natural frequency of 0.05 second, which corresponds to single story buildings. Since the amplification effect is almost negligible, the acceleration response of the structure is almost the same as that of ground. Unless the structures are weak against these acceleration waves, it is expected that the 1999 Kocaeli earthquake should have negligible effect on the low story buildings in the area.

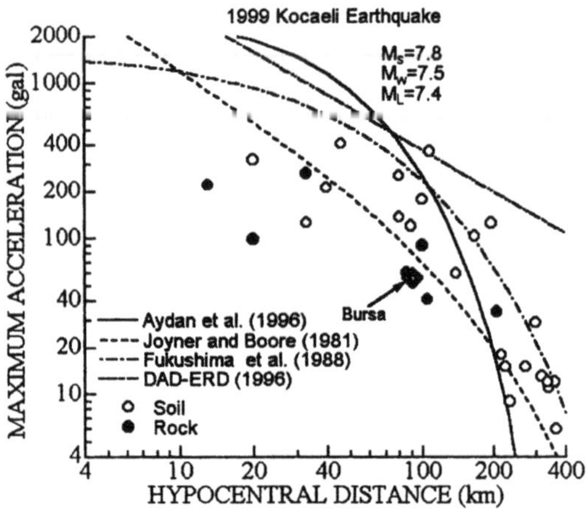

Figure 14 Comparison of attenuation relations for peak ground acceleration with observations

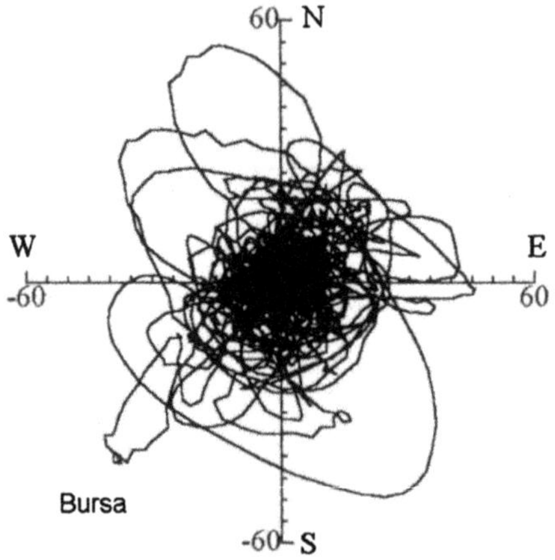

Figure 15 Trace of ground shaking at Bursa station

6 CONCLUSIONS

In this study, the vibrations caused by blasting operations at Orhaneli open-pit mine in Bursa province of Turkey are measured and analyzed. Although this study is at its preliminary phase, the following conclusions may be drawn from this study:

- The amplitude of vertical component of acceleration waves caused by blasting is larger than that of other components. The amplitude of the acceleration waves are in the order of vertical, longitudinal (radial), traverse (tangential). However, the response spectra imply that amplifications are in the reverse order.
- The acceleration obtained from the numerical derivation of velocity records is very similar to those from accelerometers. Nevertheless the amplitude and frequency characteristics of

acceleration waves computed from velocity-meters are different from those from accelerometers. For example, the amplitude obtained from the velocity-meter records was about 1.16 times that of the true acceleration records at the same location.

- Fourier spectra of accelerations indicated that waves caused by blasting consist of high frequency components. There are two dominant frequency ranges 30-40 Hz and 8-10 Hz.
- Acceleration response spectra indicated that there should be no amplification of acceleration waves on structures having a natural period greater than 0.05s (natural frequency less than 20Hz). Therefore, the acting absolute acceleration on structures will be the same as that of ground.

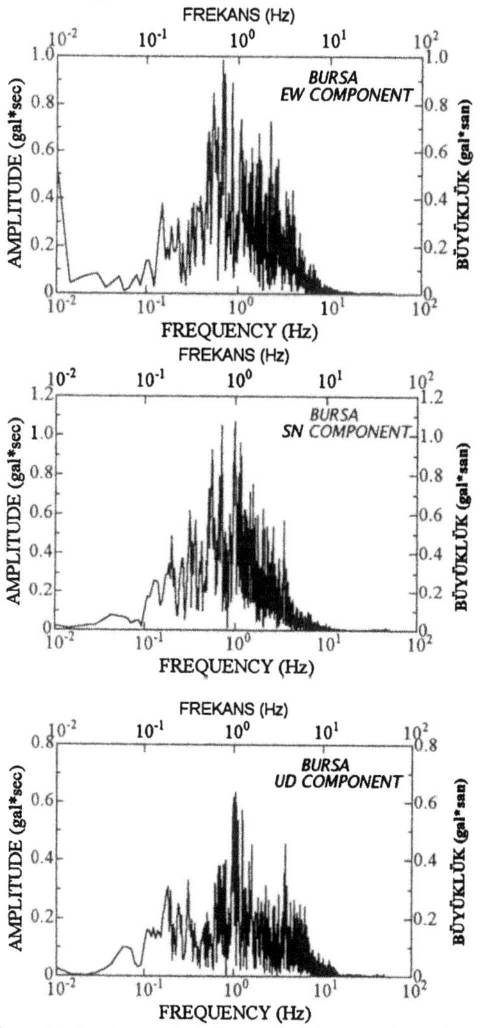

Figure 16 Fourier spectra of strong motion records at Bursa station

- The effect of the 1999 Kocaeli earthquake on structures in the vicinity of Orhaneli open-pit mine could be similar to that observed in Bursa with some reduction in the amplitude of acceleration waves. Since the acceleration response spectra of the strong motion records measured at Bursa station indicated that the amplification of accelerations acting on buildings with single or double stories should be very small. Therefore, the vibration caused by this earthquake in the area will be equivalent to the ground motion imposed on the structure. Unless structures are weak, the effect of the earthquake should therefore be negligible.

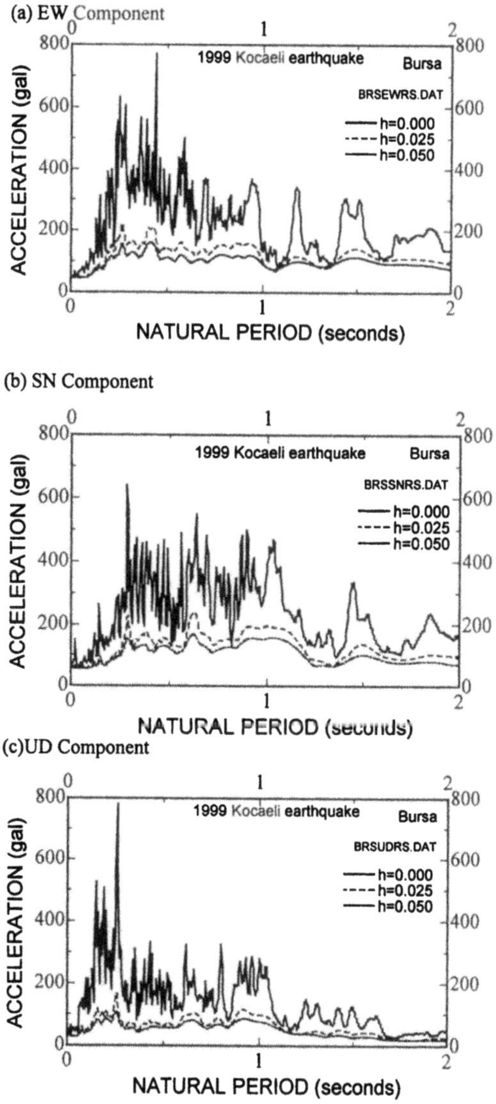

Figure 17 Acceleration response spectra for strong motion records at Bursa

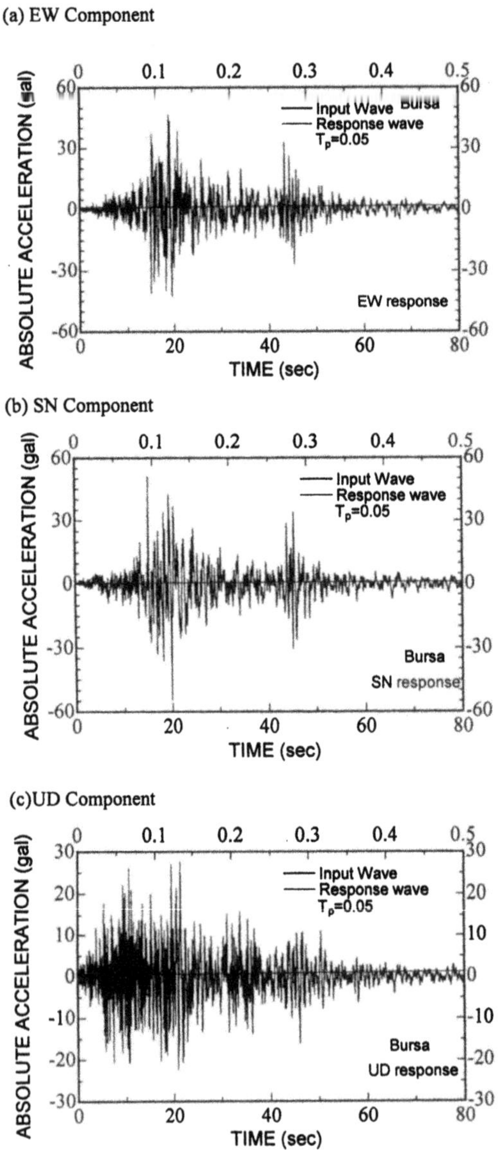

Figure 18 Acceleration responses of a single story building for strong motion records at Bursa

Further experiments are necessary to study the effect of blasting on structures. These studies may be as follows:

- To establish the attenuation relations for peak ground accelerations as a function of distance and energy released by blasting. Furthermore, the effect of structural weakness zones in rock masses should also be considered in such relations.
- To check if the frequency characteristics of acceleration waves vary with distance, amount of explosives and the geometry of weakness zones in rock masses.
- Since most damage assessment criteria associated with blasting are based on the peak particle velocity (velocity approach), the interrelations between velocity type and acceleration type approaches must be established.

REFERENCES

Aykut, A. (1973). Yapıların doğal titreşim periyotlarının deneysel yolla çözümü. Türkiye'de Deprem Sorunu ve Deprem Mühendisliği Sempozyumu, Ankara, Bildiri No. 18, TÜBİTAK.

Aydan, Ö., Sezaki, M., and Yarar, R., 1996. The seismic characteristics of Turkish Earthquakes. 11th World Conference on Earthquake Engineering., Acapulco, Mexico,1-8.

Aydan, Ö., Ulusay, R., Hasgür, Z., and Taşkın, B., 1999. A site investigation of Kocaeli Earthquake of August 17, 1999. Turkish Earthquake Foundation, 180pp.

Bayülke, N. 1978. Tuğla yığma yapıların depremlerdeki davranışı. Deprem Araştırma Enstitüsü Bülteni, 6(22), 26-41.

Fukushima, Y., Tanaka, T. & Kataoka, S. 1988. A new attenuation relationship for peak ground acceleration derived from strong motion accelerograms. IX WCEE, Tokyo.

Hao, H. 2002. Characteristics of non-linear response of structures and damage of RC structures to high frequency blast ground motion, Wave2002, Okayama (this issue).

İnan, E., Çolakoğlu, Z., Koç, N., Bayülke, N., & Çoruh, E. 1996. Catalogue of earthquakes between 1976-1996 with acceleration records. Pub. Of Earthquake Research Department. General Directorate of Disaster Affairs, Ministry of Public Works and Settlement., Ankara, July.

Joyner, W.B. & Boore, D.M. 1981. Peak horizontal acceleration and velocity from strong motion records, including records from the 1979 Imperial Valley, California Earthquake, Bull. Seis. Soc. Am., Vol.71, No.6, 2011-2038.

Wave propagation – Moving load – Vibration reduction, Chouw & Schmid (eds.)
© 2003 Swets & Zeitlinger, Lisse, ISBN 90 5809 559 2

Characteristics of dynamic response and damage of RC structures to blast ground motion

H. Hao
Department of Civil and Resource Engineering, University of Western Australia, Crawley, Australia

ABSTRACT: The dynamic response and damage mechanism of structures to blast ground motion is not well understood yet. Compared to seismic motion, blast motion has higher frequency contents, larger amplitude and shorter duration. Moreover, it displays more significant spatial variation than seismic motion over the same separation distance. This paper discusses the characteristics of dynamic responses and damage of RC frame structures to blast ground motions. The attenuation relations of peak ground acceleration, peak particle velocity, principal frequency, as well as the spatial variation properties of blast ground motion will be presented. Numerical calculations are then carried out to investigate the accuracy of using discrete beam-column element model in estimating the structural responses to high frequency motions, and the effects of blast motion spatial variations on structures. The acceptable model is used in parametric calculations of RC frame damage to blast motions of different amplitudes and principal frequencies. Numerical results of structural damage are discussed with respect to those obtained by continuum damage model and with the current code specifications.

1 INTRODUCTION

Characteristics of dynamic response and damage of structures to seismic motions have been a subject of investigation for the last few decades by many researchers and are quite well understood. It is commonly accepted that structures normally respond at their fundamental and low vibration modes and damage of a structure to seismic motion is mainly caused by excessive structural displacement. Thus, the displacement-related quantities such as storey drift and ductility ratios are usually employed in analysis and design in earthquake engineering to quantify structural damage.

On the other hand, structural performance and damage to blast ground motion is also a concern of mining, construction and defence engineers. Much research effort has been spent in this area as well (Dowding 1996). But most of this work is experimental based, rather than analytical. Structural damage criteria were obtained primarily from field blast tests on old low-rise masonry-type residential buildings. Many empirical relations were derived from those observations and they are also adopted in design codes (NATO 1997, DoD 1992, German 1984, OSM 1977). They correlate structural damage directly with ground motion amplitude, mainly peak particle velocity (PPV), or PPV together with ground motion principal frequency (PF). It is commonly accepted that those criteria are quite conservative to modern reinforced concrete structures (Zhou, et al. 2000). Other drawbacks of those criteria are that they do not take into consideration of the structural properties and they ignore the physical laws that govern the structural response and damage process.

Recently, some research effort has been spent on studying the dynamic response and damage of RC structures to blast ground motion. Some researchers directly employed the methods developed in earthquake engineering in their analyses of structural responses and predictions of structural damage (Quek 2002, Mendis, et al. 2002, Lovholt and Madshus 2001). The characteristics of blast ground motions are very different from earthquake motions. Blast motions have much higher frequency contents, larger amplitudes and short duration as compared with seismic motion. Its effect on structures is more like an impulsive ground movement on structures rather than a cyclic excitation. Some preliminary numerical (Ma, et al 2002) and experimental study (Lu, et al. 2001) of the problem revealed that the dynamic response and damage mechanism of structures to blast ground motion are very different to those under earthquake motion. Unlike that to earthquake excitation, structures usually do not respond at their fundamental modes under high frequency blast motion. Thus, the structural displacement and storey-drift are rather small. Structural damage is usually caused by large shear forces associated with the ground impact. A flexible structure will behaviour or damage as a rigid one under high frequency ground excitations. The numerical model developed in the study (Ma, et al. 2002) is based on continuum damage mechanics. It was proven giving good predictions of structural responses and damage to blast motion, as compared with the experimental results. However, its practical application is limited because it is very time consuming and computationally not feasible for a practical frame structure owing to the continuum damage mechanics modelling.

This paper will first discuss the characteristics of blast motions on ground surface. Surface blast ground motions are obtained either in field blast tests or numerical simulation of underground explosion with a validated computer code. Ground motion amplitude (peak acceleration and peak particle velocity) attenuation, principal frequency as a function of blast weight and stress wave propagation distance, as well as ground motion spatial variation properties will be presented. Then structural responses and damage to blast ground motion of various amplitudes and frequency contents will be estimated. The accuracy of using discrete beam-column elements in simulation of structural responses to high frequency ground excitations will be discussed. The effects of spatial variations of blast ground motion on structural response and damage will also be studied. Based on those analyses, the acceptable structural model that gives accurate structural response and damage estimation will be identified. Using the identified structural model, parametric simulations are carried out to estimate structural damage to blast ground motions of various amplitude and frequency contents. The numerical results will be compared with those obtained from continuum model and various code criteria. The accuracy of the code specifications on allowable ground vibrations will be discussed.

2 CHARACTERISTICS OF BLAST MOTION ON GROUND SURFACE

Many researchers have studied characteristics of blast ground motion, in particular its peak value attenuation relations as a function of scaled distance. Owing to intrinsic difficulties in numerical modelling of this highly nonlinear and fast strain rate problem, and many uncertainties in the geological medium, most available formulae that predict blast motion attenuation are empirical (Dowding 1996, Hendron 1977, Odello 1998, Hao, et al. 2001). Because stress wave propagation is highly site dependent, those empirical formulae obtained by different authors based on test data on different sites might differ by more than 100 times (Dowding 1996).

Recently, much effort has been devoted to simulate blasting stress wave propagation (Yang, et al. 1996, Liu and Katsabanis 1997, Hao, et al. 1998, Hao, et al. 2002). As compared with the measured data from field blast tests, the numerical simulation of stress wave propagation based on anisotropic rock damage model (Hao, et al 2002) gives much improved predictions of stress wave propagation than the previous models with isotropic assumption. In this section, the calibrated model is used to simulate underground explosion induced stress waves on ground surface. The stress wave characteristics, such as the attenuation relations, principal frequency changes, and spatial variations, will be presented and discussed. They will also be compared with the measured data at the same site in field blast tests.

2.1 *Site description*

The site under consideration is a granite quarry site. Because of construction of an underground facility at the site, intensive site investigations, including field blast tests, were carried out. Both field and laboratory investigations revealed that the rock mass at the site has the following properties: density 2610 kg/m^3, uniaxial compressive strength 186 MPa, tensile strength 16.1 MPa, Young's modulus 73.9 GPa, Poisson's ratio 0.25 and seismic wave velocity 5790 m/s. Site investigation results also indicated four major and two minor joint sets in the rock mass. The predominant joint set is sub-vertical and in the NNW-SSE direction with a spacing ranging from 39 cm to 62 cm. The secondary joint sets are in the NNE-SSW, NE-SW, and NW-SE directions and are widely spaced.

Owing to continuous quarry activities for the last three decades, a flat terrain of about 200×200 m in the area and 30 m below the original ground surface with unweathered granite is available. Field blast tests were carried out at this site. Before the test, a charge hole of 11 m deep and 1.5 m in diameter for its upper 6 meters and 0.8 m in diameter for the bottom 5 m was drilled as shown in Figure 1. A total of 8 blast tests with different charge weight were performed. Before each test, the upper 6 m of the charge hole was filled with concrete blocks of 15 ton on a steel plates covering the bottom 5 m charge chamber to simulate a fully contained explosion. The bottom 5 m charge chamber is used to simulated decoupled detonations. More than 100 sensors were placed below ground and on ground surface. Only the recorded motions on ground surface will be discussed here. Figure 1 also illustrates the 9 measurement points on granite surface. Five of them are on a line parallel with the direction of the predominant joint set. Two of them are placed on a line with a 45° degree to the predominant joint direction, and another two on a 90° line. At each measurement point, a steel plate is cemented onto the granite surface.

Two accelerometers, one in radial and one in vertical direction, are mounted onto the steel plate with their magnetic mounting base.

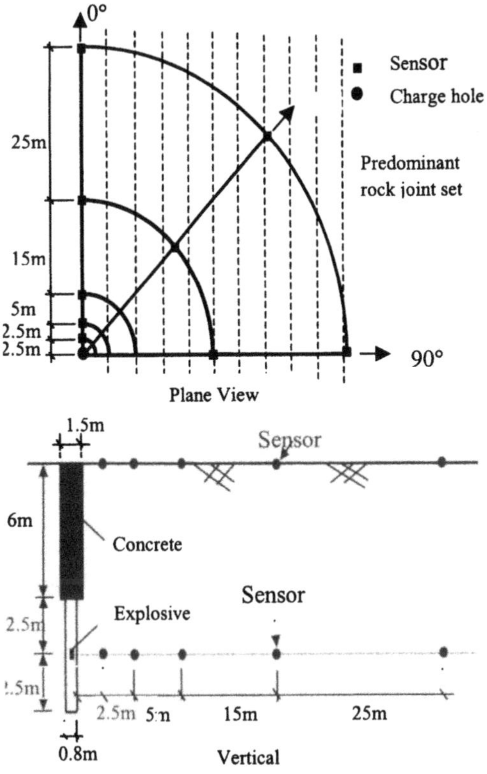

Figure 1. Configuration of the test set-up (not to scale)

Field measured data indicated that stress wave attenuates the fastest if it propagates in the direction perpendicular to the rock joint set, while it attenuates the slowest if it propagates in the direction parallel to the rock joint set. Radial and vertical components of the motion at the same point are highly correlated, but motions at different locations separated by 25 m are only weakly correlated. More detailed discussions of the field measured data can be found in reference (Hao, et al. 2001).

2.2 *Numerical simulation of underground explosion induced stress wave propagation*

A numerical model had been developed to simulate underground explosion induced stress wave propagation and rock mass damage (Hao, et al. 2002). The model is developed based on the theories of continuum damage mechanics and stress wave propagation. It includes a modified Drucker-Prager strength model allowing strength degradation with damage, a modified linear equation of state and an anisotropic damage model. The joints and microcracks in the rock mass are modelled as initial damage of the rock mass before loading. Cumulative damage is assumed to evolve from the initial damage once the equivalent tensile strain is larger than the threshold strain and is irreversible. Numerical model is validated by using the above field test data. More detail description of the numerical model, its implementation and calibration, can be found in the reference (Hao, et al. 2002). Here, the model is used to simulate ground vibrations on rock surface.

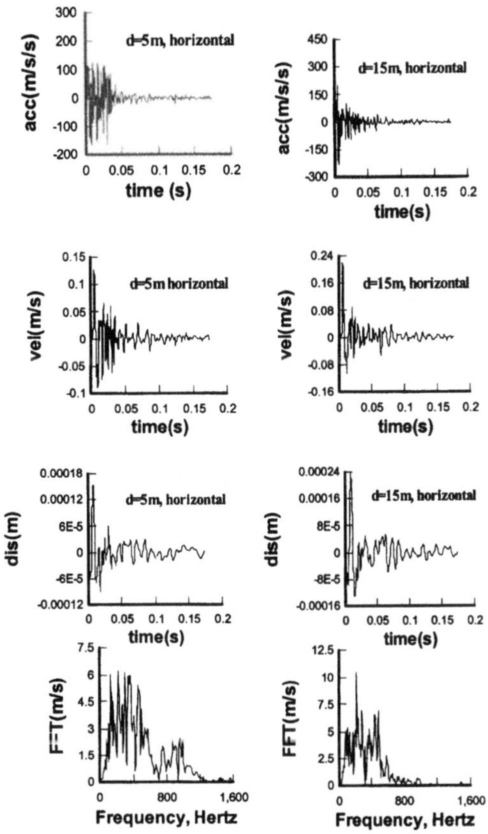

Figure 2. Numerically simulated radial motion on ground surface and their Fourier spectra

Figure 2 shows the numerically simulated horizontal motion on ground surface at 5 m and 15 m from the epicenter of the explosion. The explosive weight used in the simulation is 10 ton TNT equivalent, detonated in a charge chamber 22 m below ground surface. The dimensions of the charge chamber is 10m×10m×10m, with a loading density 10 kg/m^3. As shown, compared to seismic motion, blast motion has very large amplitude, short duration and high frequency contents and wide frequency band.

Because of very high frequency contents, ground velocity and displacement are rather small, although ground acceleration is very large. The FFT spectra of the motions at 5 m and 15 m epicenter distances show rapid attenuation of the high frequency energy as wave propagates. The secondary peak in the FFT spectrum of the motion at 5 m distance at frequencies higher than 800 Hz disappear in the spectrum of the motion at 15 m distance. The wave form of the motions separated by 10 m distance are quite different, indicating substantial spatial variations. This can be attributed to high frequency contents of the motion. High frequency signals not only attenuate fast when wave propagates, they are also very sensitive to initial damages such as joints and micro cracks in the propagation path.

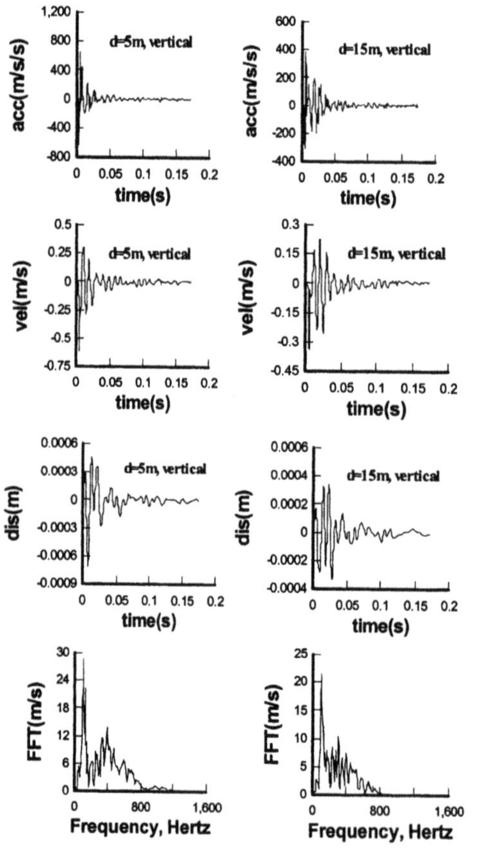

Figure 3. Numerically simulated vertical motion on ground surface and their Fourier spectra

Figure 3 shows the vertical component of the simulated motions at epicenter distances 5 m and 15 m. As shown, the observations made above still can be drawn here. It can also be noted that the vertical component of the motion is actually larger than the horizontal component. This is because the blast motion consists of mainly P-wave and the focal depth is larger than the epicenter distance. As the epicenter distance increases, the horizontal component will eventually becomes larger than the vertical one.

Motions at other points on ground surface are also calculated. Simulations of motions with other explosive weights and loading densities were also carried out. The loading densities considered in the simulations vary from 10 kg/m^3 to 1630 kg/m^3 (fully coupled). Using the simulated data, the attenuation relations of peak ground acceleration (PPA) and peak particle velocities (PPV) of surface motions are derived as

$$PPV = 2.981 f_V (R/Q^{1/3})^{-1.3375} \text{(m/s)} \quad (1)$$

$$PPA = 351 f_A (R/Q^{1/3})^{-2.0975} \text{ (g)} \quad (2)$$

in which R is the distance in meter measured from explosion center, Q is the equivalent TNT charge weight in kilogram, and f_V and f_A are decoupling factors defined as

$$f_V = 0.121(Q/V)^{0.2872} \quad (3)$$

$$f_A = 2.996(Q/V)^{-0.1514} \quad (4)$$

where V is the volume of the charge chamber in cubic meter.

As shown in Figures 2 and 3, the frequency of the motion also attenuates rapidly with the epicenter distance. Since ground motion frequency is also an important parameter that affects structural responses, the principal frequency attenuation relation is also derived here. The principal frequency of a time history is taken as the center frequency of FFT spectrum. The principal frequency attenuation relation is

$$PF = 115 f_F (R/Q^{1/3})^{-0.5483} \text{(Hz)} \quad (5)$$

in which the decoupling factor is

$$f_F = 5.49(Q/V)^{-0.2354} \quad (6)$$

2.3 *Estimation of spatial variation*

Many researchers have investigated the characteristics of seismic motion spatial variation and their effects on structures (Bolt, et al. 1982, Hao 1989). It was found that the spatial variations of seismic motion affect structural responses significantly. Ground motion spatial variation effects can be divided into the wave passage effect and coherency loss effect. Wave passage effect is more critical to relatively flexible structures, whereas the coherency loss effect is more important to stiff structures (Hao 1993). As discussed above, the spatial variation, mainly the coherency loss, of blast motion is more substantial than seismic motion over a short separation distance. It is essential to study the blast motion spatial variation characteristics in order to investigate its effect on structures.

A few authors had studied the spatial variations of blast motions recorded in field blast tests (Mclaughlin, et al. 1983, Reinke and Stump 1988, Hao et al. 2001). Usually, ground motion spatial variations in the frequency domain is modelled by a coherency function defined as

$$\gamma_{kl}(i\omega) = \frac{S_{kl}(i\omega)}{[S_{kk}(\omega)S_{ll}(\omega)]^{1/2}} \qquad (7)$$

where $S_{kk}(\omega)$ is the auto-power spectral density function of the motion at point k, and $S_{kl}(i\omega)$ is the cross-power spectral density function of the motions at points k and l, and i is the imaginary constant and ω is the circular frequency. The coherency loss function, which is the amplitude of the coherency function, of the numerically simulated motions are estimated and shown in the following figures.

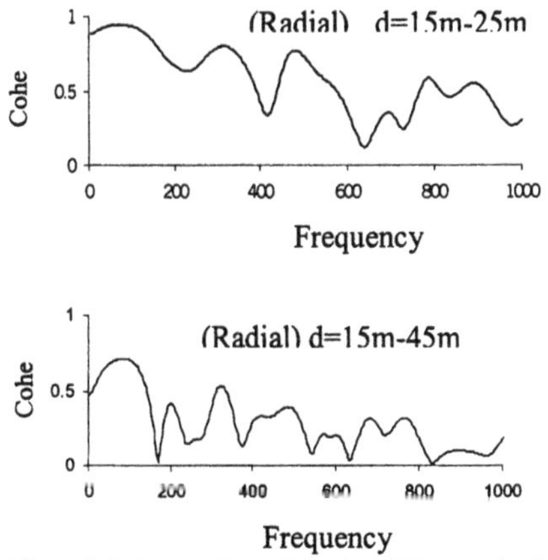

Figure 4. Coherency loss of simulated blast motion in radial direction

As shown the radial motions at epicenter distances of 15 m and 25 m with a separation distance of 10 m are highly correlated when frequency is less than 100 Hz, but are only intermediately correlated at higher frequencies. When the separation distance is 30 m, the motions are weakly correlated at low frequencies, and are almost not correlated when frequency is higher than 200 Hz. This observation indicates significant spatial variations of blast motions over a short separation distance as compared with seismic motion.

Figure 5 shows the coherency loss functions of the vertical component of the blast motion. As shown, vertical component of the motion seems more correlated than the horizontal component, although they are still not as correlated as the seismic motion over the same separation distance.

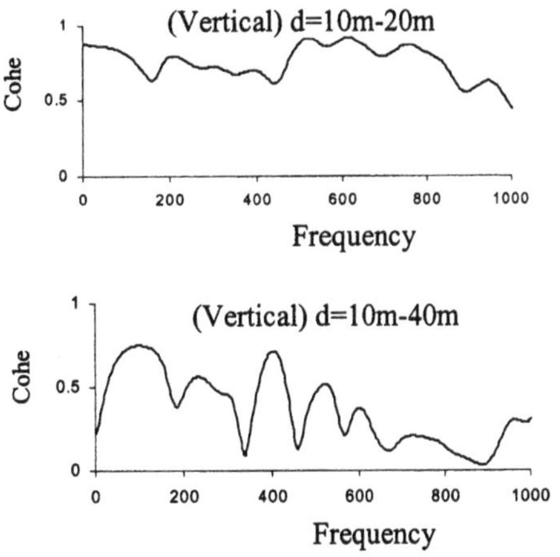

Figure 5. Coherency loss of simulated blast motion in vertical direction

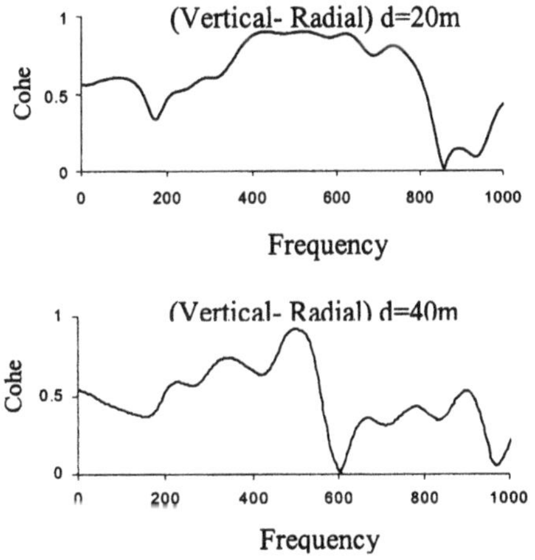

Figure 6. Coherency loss between radial and vertical component of the simulated motion

Figure 6 shows the coherency loss between radial and vertical component of the simulated motions at the epicenter distances of 20 m and 40 m. As shown, when the epicenter distance is 20 m, the two components are highly correlated in the frequency range of 400 Hz to 800 Hz. But they are only weakly correlated at frequencies lower than 400 Hz, and not correlated at high frequencies. This observation indicates that the primary propagating blast wave energy concentrates in the frequency range between 400 Hz and 800 Hz. It is different from the seismic wave where the coherency loss usually decreases with the frequency because wave energy is normally distributed over the low frequency range. At epicenter distance 40 m, the two components are less correlated, especially at frequencies higher than 500 Hz, indicating the wave propagation path effect on high frequency signals.

The above numerically simulated ground motion time histories will be used in the following structural response analyses.

3 CHRACTERISTICS OF STRUCTURAL RESPONSES

Recent numerical (Ma, et al 2002) and experimental (Lu, et al. 2001) studies revealed that the response characteristics and damage mechanism of structures to high frequency ground motions are very different to those to low frequency seismic motions. However, because of the limitations of the testing equipment, the experimental studies are only limited to small scale structures and the force power of the testing equipment can only cause damage to structural model at reduced vibration frequencies. On the other hand, the numerical model is developed based on the theories of continuum damage mechanics and wave propagation. It is computationally very demanding and not suitable for practical application. Nevertheless, these two studies give good indications that directly employing the methods developed in seismic engineering to estimate structural responses and damage to blast motions is not suitable, and the necessity for further studies of this unique problem. In the following, the response characteristics of structures to high frequency ground excitations are further investigated, the possibility of using structural dynamics method and beam-column element model, which is computationally much more efficient and suitable for practical application, is explored.

3.1 *Continuous model*

To investigate the relative contributions of various vibration modes to overall structural responses, a continuous structural model is analyzed first. The model is shown in Figure 7. It consists of a rigid mass M supported by a RC column of length L, rigidity EI, density ρ and cross sectional area A.

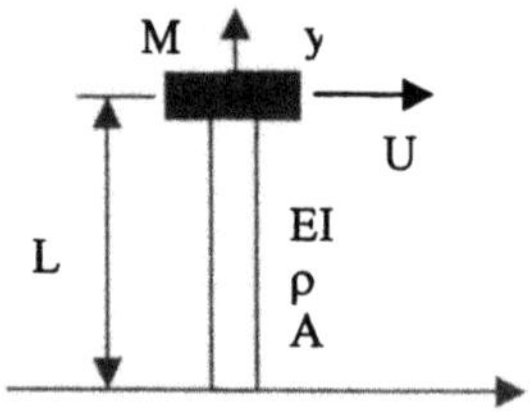

Figure 7. Continuous structural model

The free vibration equation is

$$EI\frac{\partial^4 U}{\partial y^4}+\rho A\frac{\partial^2 U}{\partial t^2}=0 \tag{8}$$

Where U is the lateral displacement of the column. Let $U=\sum\phi_i(y)w_i(t)$, and sub into Eq. (8), the following equations can be derived

$$\phi_i''''(y)-a_i^4\phi_i(y)=0 \tag{9}$$

$$\ddot{w}_i(t)+\omega_i^2 w_i(t)=0 \tag{10}$$

in which $\omega_i^2=\frac{EI}{\rho A}a_i^4$ is the circular vibration frequency of the structure. The boundary conditions are

$$\begin{aligned}&U(0,t)=U'(0,t)=0, \quad \text{or} \quad \phi(0)=\phi'(0)=0\\&U'(L,t)=0 \quad \text{or} \quad \phi'(L)=0\\&EIU'''(L,t)=M\ddot{U}(L,t), \text{ or}\\&EI\phi'''(L)=-\omega^2 w(t)\end{aligned} \tag{11}$$

Solving the differential equation with the above boundary conditions, the transcendental equation

$$\sin\lambda\cosh\lambda+\sinh\lambda\cos\lambda=\beta\lambda(\cos\lambda\cosh\lambda-1) \tag{12}$$

can be derived; in which $\lambda=aL$ and $\beta=\frac{M}{\rho AL}$ is the ratio of the lumped mass to the column mass. For a typical low-rise RC building, the floor mass M supported by each column is usually in a range of 10 to 30 times of the column mass. For a high-rise building, the mass supported by each first storey column is much higher than this. Without losing generality, in the present study, β is taken as 20. Using the numerical method, λ values are obtained from Eq. (12) and those corresponding to the first five modes are listed in Table 1.

Table 1 Parameters of the first five modes

Mode	1	2	3	4	5
λ	0.876	4.720	7.847	10.991	14.134
f(Hz)	3.87	112.3	310.32	608.84	1006.8

Assuming a RC column has dimensions 300 mm× 300 mm, Young's modulus of RC E_c=26 GPa, density ρ=2.4 kg/m^3, it has $\omega=\sqrt{\frac{EI}{\rho A}}a^2=285\frac{\lambda^2}{L^2}$ rad/s. If the column length is L=3 m, it has natural vibration frequency $f=\omega/2\pi=5.04\lambda^2$ Hz. The first five vibration frequencies of the structural model are also given in Table 1.

The mode shape of the structural model is

$$\phi_i(x)=\cos\frac{\lambda_i}{L}x-\cosh\frac{\lambda_i}{L}x+c_i(\sin\frac{\lambda_i}{L}x-\sinh\frac{\lambda_i}{L}x) \quad (13)$$

in which $c_i=\frac{\sin\lambda_i+\sinh\lambda_i}{\cos\lambda_i-\cosh\lambda_i}$. Figure 8 shows the first four mode shapes.

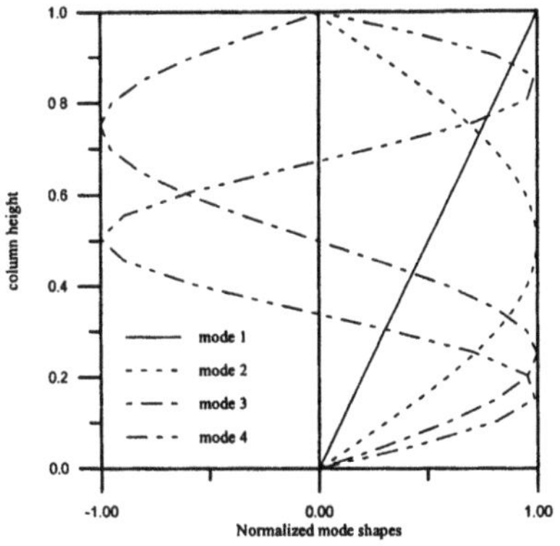

Figure 8. Mode shapes

From Figures 1 and 2, the spectrum of blast motion can be approximately represented by

$$S_g(\omega)=\frac{\omega^4}{(\omega_1^2-\omega^2)^2+4\xi_1^2\omega_1^2\omega^2}\bullet\frac{1+4\xi_g^2\omega^2/\omega_g^2}{(1-\omega^2/\omega_g^2)^2+4\xi_g^2\omega^2/\omega_g^2}S_0 \quad (14)$$

in which ω_1 and ξ_1 are the central frequency and damping ratio of the high-pass filter; ω_g and ξ_g are those of the Tajimi-Kanai power spectral density function; and S_0 is a scale factor depending on the intensity of the ground motion. In the present study, the principal (central) frequency of the motion is varied from 2.5 Hz to 300 Hz in numerical calculations of the structural responses.

When the structure is under forced vibration, its equilibrium equation becomes

$$\rho\frac{\partial^2U}{\partial t^2}+c\frac{\partial U}{\partial t}+EI\frac{\partial^4U}{\partial x^4}=-\rho\ddot{v}_g \quad (15)$$

Sub $U=\sum\phi_i(y)w_i(t)$ into the equation, the decoupled ith mode vibration equation is

$$\ddot{w}_i(t)+2\xi_i\omega_i\dot{w}_i+\omega_i^2w_i=\gamma_i\ddot{v}_g \quad (16)$$

where ξ_i is the ith mode damping ratio, $\ddot{v}_g$ is the ground acceleration and $\gamma_i=-\frac{\int_0^L\phi_i(x)dx}{\int_0^L\phi_i^2(x)dx}$ is the participation coefficient of the ith mode. For the structure model under consideration, the first five participation coefficients are 1.2863, -0.1896, -1.974×10^{-5}, -2.699×10^{-4}, and 4.027×10^{-8}.

Transfer Eq. (16) into the frequency domain, it has

$$w_i=H_i(i\omega)\gamma_i\bar{\ddot{v}}_g(i\omega) \quad (17)$$

in which $H_i(i\omega)$ is the complex transfer function of the ith mode. It has the form

$$H_i(i\omega)=\frac{1}{(\omega^2-\omega_i^2)+2i\xi_i\omega\omega_i} \quad (18)$$

The spectral density functions of the displacement response can be derived as

$$S_U(\omega)=\sum_{j=1}^{n}\sum_{k=1}^{n}\phi_j(x)\phi_k(x)H_j(i\omega)H_k^*(i\omega)\gamma_j\gamma_kS_g(\omega) \quad (19)$$

That of the acceleration response is simply

$$S_{\ddot{U}}(\omega)=\omega^4S_U(\omega) \quad (20)$$

The spectral density functions of bending moment and shear force can be similarly derived as

$$S_M(\omega)=(EI)^2\sum_{j=1}^{n}\sum_{k=1}^{n}\phi_j''(x)\phi_k''(x)H_j(i\omega)H_k^*(i\omega)\gamma_j\gamma_kS_g(\omega) \quad (21)$$

and

$$S_V(\omega)=(EI)^2\sum_{j=1}^{n}\sum_{k=1}^{n}\phi_j'''(x)\phi_k'''(x)H_j(i\omega)H_k^*(i\omega)\gamma_j\gamma_kS_g(\omega) \quad (22)$$

in which n is the number of modes included in the calculation.

Numerical results of the spectral responses are calculated by normalizing ground motion PPV=20 cm/s. Results indicate that displacement responses is always dominated by the first mode vibration, even when the principal excitation frequency of ground motion is 300 Hz. The displacement response decreases with the increase of the ground motion vibration frequency. Even the ground motion principal vibration frequency coincides with the second vibration mode of the structure, the displacement response is still primarily associated with the first vibration mode. Figure 9 shows the displacement response spectral density functions along the column height obtained with ground motion principal frequency ω_g=16 Hz and ω_g=100 Hz, respectively.

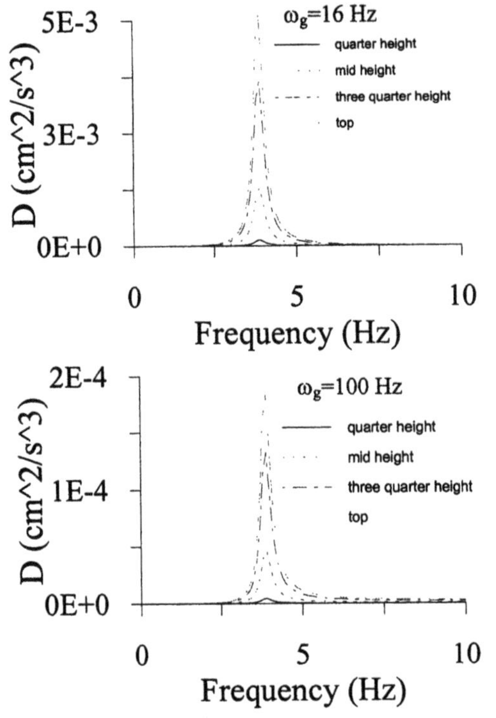

Figure 9. Displacement response spectra

The other response quantities, namely, acceleration, bending moment and shear force, are dominated by the first mode vibration only when the ground motion vibration frequency is small. When ground motion vibration frequency is large, the responses are mainly caused by second vibration mode. The contributions from the third or higher vibration modes are rather small even ground motion resonates with the vibration mode. This is because the participation coefficients corresponding to the higher modes are very small, It requires much higher ground motion energy to excite those modes. The following figures show the response spectra obtained when the principal vibration frequencies of the ground motions are 16 Hz and 300 Hz, respectively.

From the above observations, it can be concluded that different structural vibration modes contribute to different structural response quantities under high frequency ground excitations. Fundamental mode always dominates displacement responses, whereas the first structural elemental mode (second mode here) could dominate acceleration, bending moment and shear force responses if the ground motion frequency is very high. However, it is unlikely that the second or higher structural elemental mode (third mode in the model) could become the dominant response mode even the ground motion resonates with them because it requires much higher energy to excite such high modes. Thus, in discrete modelling, the highest mode that needs be captured in the calculation is probably the first elemental mode. This will be verified in the next section.

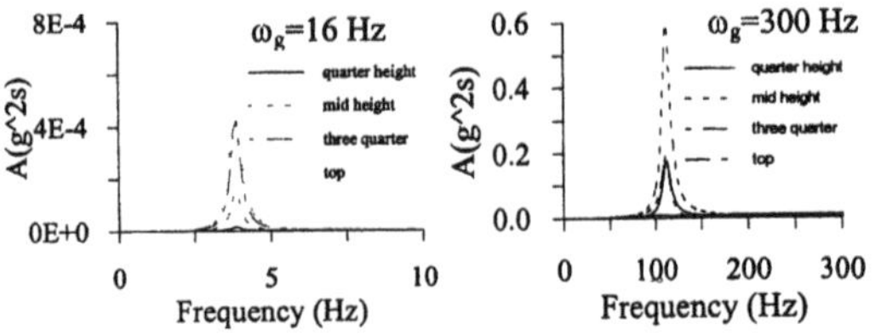

Figure 10. Acceleration response spectra

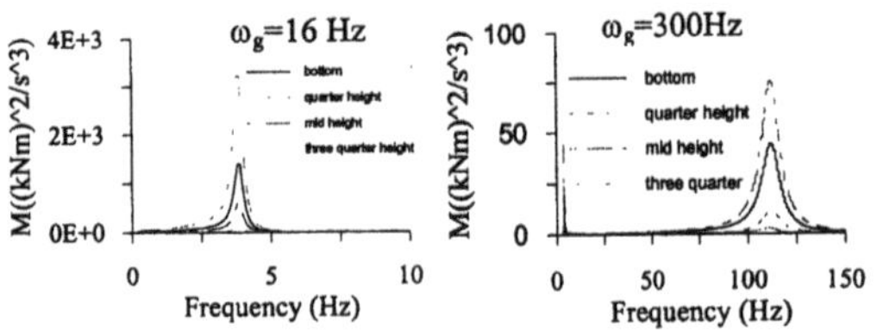

Figure 11. Bending moment response spectra

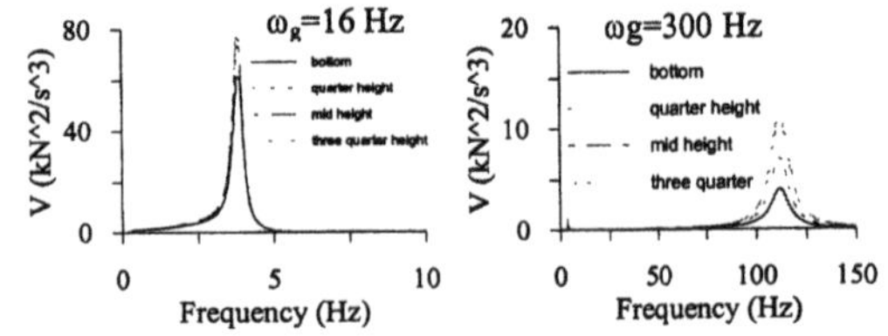

Figure 12. Shear force response spectra

3.2 *Discrete model*

A previous study developed a numerical model to simulate RC structural responses and damage to high frequency blast ground motion (Ma, et al. 2002). That model was proven giving good predictions. However, it is computationally very demanding and is not suitable for practical application.

The above analysis revealed that only the responses from the first elemental mode contribute significantly to the overall responses of structures under high frequency excitations. It indicates the possibility of discrete modelling of the structures with beam-column elements, instead of the continuous modelling. This section will investigate the suitability of using discrete model in calculating RC frame structural responses. It should be noted that spatially varying ground motions are used in the calculations. The simulated motions at 5 m, 10 m and 15 m epicentral distances on ground surface are used as input. The PPV of the motion, which is calculated by $PPV = \max\sqrt{v_r^2(t)+v_v^2(t)}$, is normalized to 1 m/s, and the PF shifted to 35 Hz with the total ground motion duration of 1.75 sec. Structural models considered are shown in Figure 13. It is a 2-storey 2-span RC frame. The material and geometrical properties of the structure are: unit weight ρ=24 kN/m^3, modulus of elasticity E=27 GPa, uniaxial compressive strength 30 MPa, steel confinement ratio 0.0055, viscous damping ratio 0.05, Poisson's ratio 0.3, floor area 100 m^2 with three frames, slab thickness 150 mm, beam size 230×400 mm, external column size 200×300 mm, internal column 250×300 mm, and reinforcement ratio of 2% for all the beams and columns.

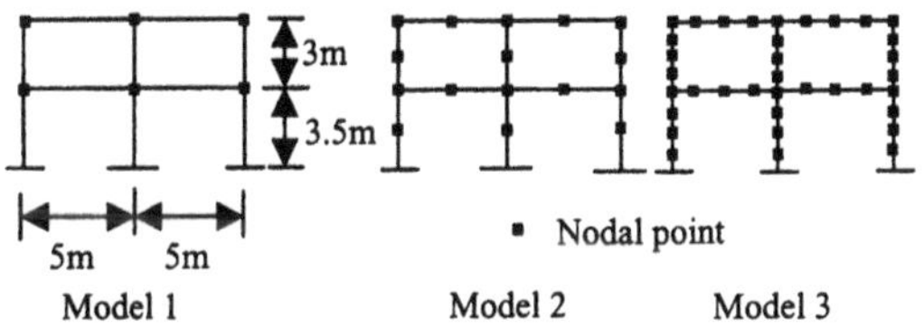

Figure 13 Structure model

To test the accuracy of using discrete element, three models are considered. The first one is the traditional model commonly used in seismic engineering with nodal points at beam-column joints only, the second one discretizes each beam column member into two elements with one nodal point at mid point of each member, and the third one discretizes each member into four elements.

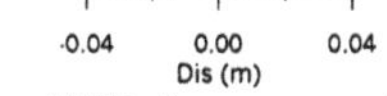

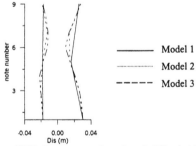

Figure 14 Displacement envelope along building height

Figure 14 shows the displacement envelope along the building height. As shown, structural responses corresponding to the first elemental mode is obvious and it is not captured by model 1. Responses obtained from model 2 and mode 3 are quite similar, indicating again contributions from higher vibration modes than the first elemental mode is minimum and can be neglected. It also shows that the storey drifts are ver small.

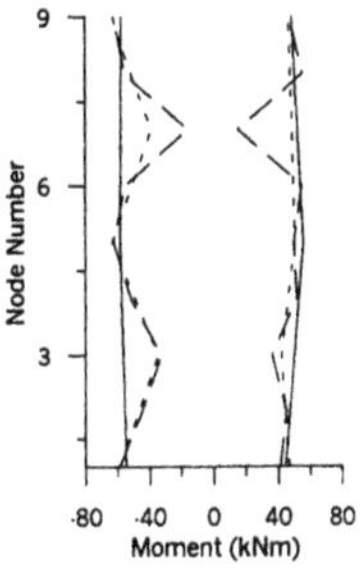

Figure 15. Bending moment envelope along column 1

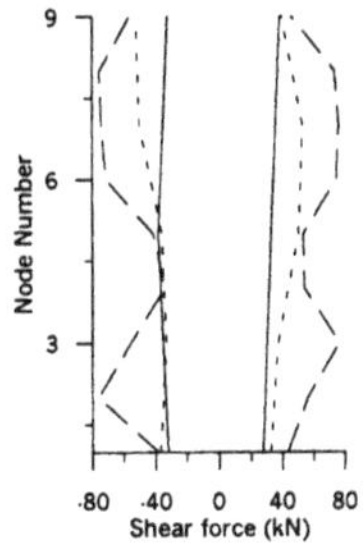

Figure 16. Shear force envelope along column 1

Figures 15 and 16 show bending moment and shear force envelope of column 1. As shown, similar to the displacement response, model 2 and model 3 give quite similar bending moment responses, but model 3 results in significantly larger shear force responses than the first two models. These observations indicate the importance of the first elemental mode contributions to shear force responses. It should be noted that results obtained by further subdividing the structural member to more elements are also calculated, and they give similar estimations of shear forces as those by model 3, indicating the convergence of the numerical results and no further subdividing of structural elements is necessary.

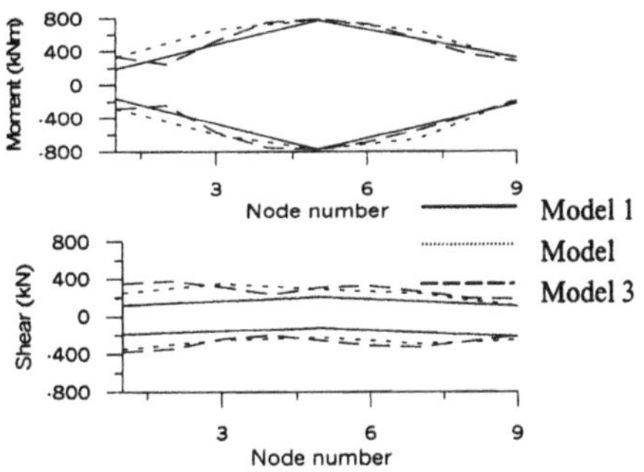

Figure 17. Bending moment and shear force envelope of the first storey bea

Figure 17 illustrates the bending moment and shear force envelopes along the first storey beam. As shown, models 2 and 3 give quite similar results.

It should be noted that the bending moment and shear force responses of the first storey beam are much larger than column responses. This is because of the very large vertical component of ground motion as shown in figure 3. Responses of the second storey beam are very similar to the first storey one, indicating the strong vertical motion excites the entire structure and the whole building vibrates vertically at the same level. This observation is similar to the previous numerical results based on continuous model (Ma. et al. 2002).

Table 2 Accumulated plastic hinge rotations ($\times 10^{-3}$ rad)

Model /Node		1	2	3	4	5	6	7	8	9
C	1	4.1				29.1				20.4
o		-8.1				-27.1				-22.1
l	2	6.8		7.5		18.6		1.9		18.2
u		-8.6		-7.4		-21.2		-2.1		-20.9
m	3	4.3	2.4	2.4	8.4	14.0	8.5	0	8.3	13.7
		-5.7	-1.7	-2.3	-7.8	-15.9	-7.6	0	-7.4	-15.6
B	1	0				3.8				0
e		0				-3.4				0
a	2	0		0.4		2.5		0.4		0
m		0		-0.6		-2.3		-0.4		0
	3	0	0	0.3	0.9	1.7	0.3	0.1	0	0
		0	0	-0.1	-0.5	-1.7	-0.4	-0.2	0	0

Table 2 lists the accumulated plastic hinge rotations of column 1 and the first storey beam. As shown, plastic hinge actually occurs at every node on the column. Model 1 could not capture the distributed plastic hinge rotations/damages along the structural member. It gives concentrated structural damage at the beam column joints, and might results in overestimation of the plastic hinge rotations at the beam column joints.

The above results indicate that model 3 should be used in response analysis of structures to high frequency motions. It also illustrates that the displacement response, hence storey drift, is very small, but shear forces are rather large. Thus structural damage might be caused primarily by shear forces, rather than by insufficient ductility capacities. In such cases, a flexible structure might suffer a brittle damage. The traditional design and damage assessment criteria used in seismic engineering based on structural displacement related quantities, such as ductility ratios and storey drifts, cannot be directly used to assess structural damage to high frequency motions.

3.3 *Effects of ground motion spatial variations*

As discussed above, blast ground motion displays more substantial spatial variations over a short propagation distance than seismic motion. However, its effect on structural responses are not necessarily more significant than seismic motion. This is because spatial ground motion effects on structures consist of coherency loss effect and phase shift (wave passage) effect (Hao 1993). Blast motion coherency loss is very significant and coherency loss effect is pronounced especially to stiff structures. Since under high frequency ground excitation, even a flexible structure will respond at its high modes, in other words, behaves stiffly, the blast motion coherency loss effect is expected to be substantial to structural responses. On the other hand, however, the coherency loss effect also depends on ground displacement. But ground displacement of blast motion is very small, as shown in Figures 2 and 3. Thus the blast motion spatial variation effect on structures is expected to be less significant. The significance of the combined effect, namely coherency loss and phase shift, on structures is not known. This section will investigate the spatial variation effect of blast motion on structures. The above structural model 3 is used. Simulated motions at 5 m, 10 m and 15 m epicentral distances are used as the spatially varying input. As a comparison, simulated motion at 10 m epicentral distance is used as the uniform input at the three structural supports. It should be noted that ground motion frequency contents shown in Figures 2 and 3 are scaled down 10 times again in the calculation by increasing the sampling rate 10 times. The principal frequency of the motion is changed from around 350 Hz to 35 Hz and the total duration to 1.5 sec.

Figure 18 shows the displacement envelope of the structure obtained by uniform and spatially varying input. It shows that uniform input gives slightly larger displacement response, but those obtained by both uniform and spatially varying input are quite similar.

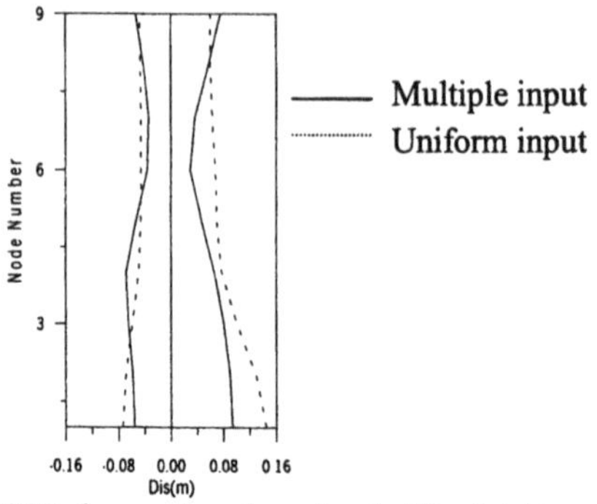

Figure 18 Displacement envelope along building height

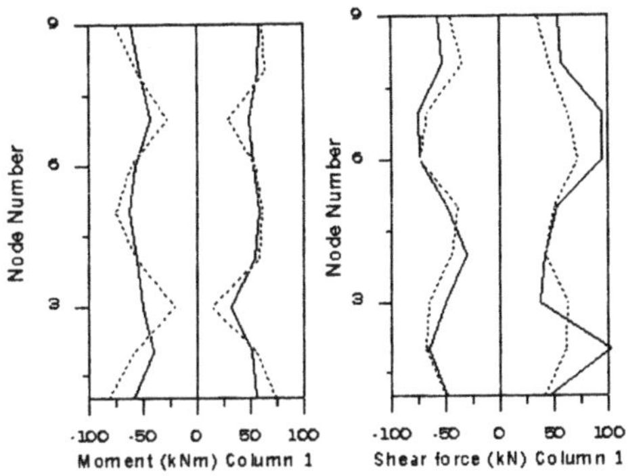

Figure 19 Bending moment and shear force envelope of column 1 (solid line-multiple input, dash line-uniform input)

Bending moment and shear force envelopes shown in Figure 19 indicate that the overall differences between those obtained by uniform and multiple input are not substantial, especially the bending moment. However, neglecting blast motion special variation might cause 50% underestimation of the shear force response at about the quarter height of the first storey column. Hence, ground motion spatial variations should be considered in the analysis.

4 STRUCTURAL DAMAGE ESTIMATION

The above structural model 3 and simulated motion at epicentral distances 5 m, 10 m and 15 m are used in the parametric calculations of structural damage to blast motion. The ground motion PPV and principal frequency is scaled in the calculations. Because motions at the three points have different PPV and PF, the scaling is based on the motion at 10 cm epicentral distance and those at the other two points are scaled with the same scaling factor.

Currently, there are mainly two approaches to assess building structural damage to ground motions. One is based on the storey drift, or ductility ratios together with the hysteretic energy losses (Park and Ang 1985). Another one is based on the stiffness degradation by estimating the change in fundamental vibration frequency of the structure before and after the excitations (DiPasquale 1990). Those methods are developed primarily to assess structural damage to seismic motions. As discussed above, structural damage mechanism to blast ground motions is quite different from that under seismic motion. The structural displacement response is very small and even a flexible structure might respond stiffly. Moreover, because the duration of blast motion is very short and structure will suffer brittle damages, hysteretic energy loss will be minimum. Hence, the damage criteria based on displacement-related responses such as storey drift and ductility ratios and hysteretic energy loss are not applicable here. Structural stiffness degradation is a possible choice to estimate its damage to high frequency motions. It was used in a previous study (Ma, et al. 2002). In that study, the damage index is defined as one minus the squared ratio of the damaged fundamental vibration period to initial period. Such a definition works in earthquake engineering because structures respond primarily in its fundamental mode. As discussed above, structures to high frequency blast motion do not respond primarily at its fundamental vibration mode. Damage to structures usually spreads over entire structure member, instead of limited at the beam-column joints. Although damage will cause an increase on the fundamental vibration period, but it might cause a more significant increase on the periods of higher vibration modes, especially the local elemental vibration modes. The difficulty in using that definition is that it is not clear the change of which vibration period should be used in estimating the damage index, fundamental period, or the period of local vibration modes, or a weighted combination of all of them.

Since structural damage spread over almost the entire structural elements, in this study, a new structural damage index is proposed. It is based on the maximum plastic deformation of all the nodal points. Since the ultimate ductility ratio for a RC structural element damaged primarily by shear is about 2 to 4, in the present study, the damage index is defined as

$$D = \frac{\sum_{i=1}^{n} \frac{\theta_p}{\theta_y}}{4n} \qquad (23)$$

in which θ_p and θ_y is plastic deformation and yielding displacement of the structural element; n is the number of joints. In numerical calculation, when θ_p/θ_y is larger than 4, it is taken as 4. The definition indicates when there is no plastic deformation at a nodal point, its local damage is zero. When the plastic deformation is four times of the yield deformation, the local damage is 1. Local damages at all the nodal points are equally weighted to estimate the global structural damage.

Using the above definition, the damage indices of the structural model to ground motions of different PPV and PF are estimated. It should be noted that motions of different PPV and PF are obtained by scaling the numerically simulated motions at 5 m, 10 m and 15 m epicentral distances.

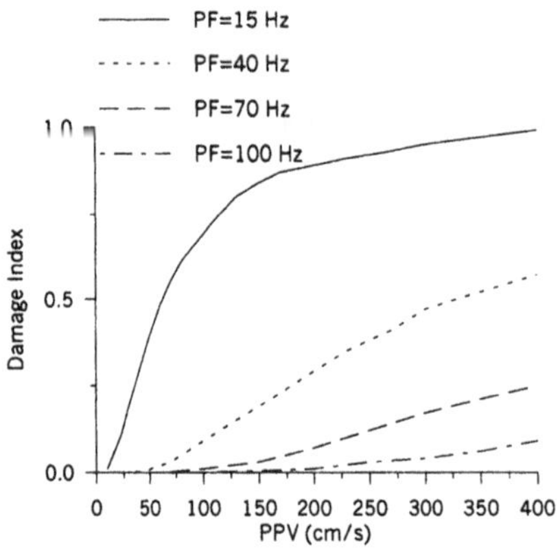

Figure 20 Estimated damage indices of the structure to blast motions of different PPV and PF

Figure 20 shows the estimated structural damage indices. As shown, the damage is highly ground motion frequency dependent. When PF is 15 Hz, structure suffers significant damage when PPV is about 50 cm/s, but it almost remains entirely in elastic range when PPV=50 cm/s and PF=100 Hz. When PF=100 Hz, structural only suffer minor damage even PPV is as large as 400 cm/s. This observation is consistent with previous experimental results on a single storey RC frame (Lu, et al. 2001). It was found that when PF is larger than 120 Hz, only when PPV is larger than 50 cm/s, some minor cracking at structural joints are observed. When PF=10 Hz, minor cracking occurs at PPV=10 cm/s and major cracking at some structural joints appear when PPV=25 cm/s. When PF= 70 Hz, at PPV=50 cm/s, no crack was observed.

Since PF of blast ground motion is usually larger than 30 Hz, the above results indicate that the current specification on allowable vibration limits of 23 cm/s is very conservative for RC structures. When PF is larger than 40 Hz, at PPV=23 cm/s, the current results indicate no plastic deformation at all. Structural response remains in the elastic range. Plastic response starts to occur at some of the nodal points when PPV=40 cm/s and PF=40 Hz, 70 cm/s when PF=70 Hz, 150 cm/s when PF=100 Hz. If PF=350 Hz, structural response remains in the elastic range even when PPV=400 cm/s.

The results are also consistent with the numerical results generated by wave propagation method and damage mechanics theory (Ma, et al. 2002). Previous numerical results indicated that at about PF=100 Hz, PPV=120 cm/s, concrete suffered major damage, but steel reinforcement still remains in elastic range, implying no plastic hinge is formed in the structure under such a blast motion excitation.

The present and the previous results all indicate that the allowable vibration limits of 23 cm/s is very conservative for RC structures. They also indicate that a reasonable criterion should include ground motion frequency into consideration. It is believed that rock mass is likely to be damaged when PPV is larger than 400 cm/s, which occurs at about the scaled distance 0.8. Thus, the present results imply that if PF is larger than 100 Hz, it is very unlikely the blast motion would cause major damage to RC structures if its foundation does not fail, but some cracks in concrete will appear when PPV is larger than 100 cm/s. However, because high frequency motion generates large acceleration responses, it might cause damage to secondary structures and overtopping of furniture and equipment housed in the structure. It might also cause damage to infill masonry wall. Further investigation is deemed necessary.

It should be noted that in the present calculation, effect of strong motion duration is not considered. Studies in seismic engineering revealed that strong motion duration has a pronounced effect on nonlinear structural responses. The duration effect of blast motion, however, needs be investigated. This is because blast motion duration is very short, usually in an order of micro to mini seconds. It is believed that hysteretic loop will not be formed, structure will suffer brittle damage. Effect of blast motion duration might not be important. Nonetheless, further investigation on this need be carried out.

It should also be noted that the current definition of damage index might underestimate structural damage levels, especially when PF is small. This is because some nodal points experience larger plastic deformation than others. For example, usually beam-column joint will have larger plastic deformation than the internal nodes of an element. This is especially true when structure responds at its low vibration modes. If plastic hinge is formed, i.e., local damage equals to 1.0, at both ends of the first storey columns, the structure is actually totally damaged as it loses the stability. However, the current definition gives equal weight to plastic deformations at all the nodes. However, this definition accounts for the spread damage over the entire structural element associated with high mode vibrations. Again, further investigation on a more suitable damage index is necessary.

5 CONCLUSIONS

This paper first discusses characteristics of surface motions induced by underground explosions in a granite mass. The numerically simulated blast motions were used in the analyses. Those motions were then used as input to calculate RC structural responses and damage. It has been found that displacement responses of structures are always dominated by the structural fundamental vibration mode, while acceleration, bending moment and shear force responses are dominated by fundamental vibration mode only when ground motion frequency is low. When ground motion vibration frequency is high, first structural elemental vibration mode contributes most to these responses. Contributions from modes higher than the first elemental modes are insignificant even they resonate with ground motion. Structural responses can be calculated by using beam-column element model. However, each element needs be discretized into at least four sub elements in order to capture high mode vibration and give accurate estimation of shear forces. Blast motion varies significantly over a short separation distance. Neglecting it in calculation might result in under estimation of structural responses.

Structural damage not only depends on ground motion amplitude, but also strongly depends on ground motion frequency. Current code specification of allowable vibration limits of 23 cm/s is very conservative when ground motion principal frequency is larger than 40 Hz. When blast motion principal frequency is higher than 100 Hz, it is unlikely that it will cause significant damage to RC structures, rather it will damage structural foundations first.

6. REFERENCES

Bolt, B. A. Loh, C. H., Penzien, J., Tsai, Y. B. & Yeh, Y. T. 1982, 'Preliminary report on the SMART-1 strong motion array in Taiwan', Report No. UCB/EERC-82-13, Earthquake Engineering Research Center, Univ. of California at Berkeley.

DiPasquale, E., Ju, J. W., Askar, A. & Cakmak, A. S., 1990. Relation between global damage indices and local stiffness degradation, ASCE J. Struct. Eng., Vol 116(5), 1440-1456.

Dowding, C. H. 1996. *Construction Vibrations*, Prentice-Hall, New Jersey.

DoD Ammunition and Explosive Safety Board, 1992. DoD 6055.9-STD, Assistant Secretary of Defense (Production and Logistics), Washington, DC.

German Standards Organization, 1984. DIN 4150, Vibrations in building construction, Berlin.

Hao, H. 1989. Effects of spatial variation of ground motions on large multiply-supported structures, Report No. EERC-89/06, Earthquake Engineering Research Center, University of California at Berkeley.

Hao, H. 1993. Arch response to correlated multiple excitations. Journal of Earthquake Engineering and Structural Dynamics, Vol. 22, 389-404.

Hao, H. Ma, G. & Zhou, Y. X. 1998. Numerical simulation of underground explosions. Fragblast, International Journal of Blasting and Fragment., Vol. 2, 383-395.

Hao, H., Wu, Y., Ma, G. & Zhou, Y.X. 2001. Characteristics of surface ground motions induced by blasts in jointed rock mass. Journal of Soil Dynamics and Earthquake Engineering, Vol. 21, 85-98.

Hao, H., Wu, C. & Zhou, Y. X. 2002. Numerical analysis of blast-induced stress waves in a rock mass with anisotropic continuum damage models: Part I: Equivalent material property approach. Journal of Rock Mechanics and Rock Engineering, Vol. 35 (2), 79-94.

Hendron, A. J. 1977. Engineering of rock blasting on civil projects, in structural and geotecnical mechanics. A Volume Honoring N. M. Newmark, Ed. W. J. Hall, Prentice-Hall, NJ, 242-277.

Liu, L. & Katsabanis, P. D. 1997. Development of a continuum damage model for blasting analysis. Int. J. Rock Mech. Min. Sci., Vol. 34, 217-231.

Lovholt, F. & Madshus, C. 2001. Risk based ground shock lethality model for underground ammunition storage. Norwegian Geotechnical Institute, Report No. 515178-3.

Lu, Y. Hao, H., Ma. G. & Zhou, Y. X. 2001. Simulation of structural response under high-frequency ground shock excitation. Int. J. Earthquake Engineering and Structural Dynamics, Vol. 30, 307-325.

Ma, G. Hao, H., Lu, Y. & Zhou, Y. X. 2002. Localized structural damage generated by high-frequency ground motion. ASCE J. Struct. Eng., Vol. 128(3), 390-399.

Mclaughlin, K., Johnson, L. & McEvily, T. 1983. Two-dimensional array measurements of near source ground accelerations, Bull. Seismo. Soc. Am., Vol. 73, 349-376.

Mendis, O., Ngo, T. & Kusuma, G. 2002. Assessment of tall buildings under blast loading and aircraft impact. Proc. 17th Australaisan Conf. On Mech. Struct. Materials, Gold Coast, Ausrtalia, 12-14 June 2002, 495-500.

NATO, 1997, Manual on NATO safety principles for the storage of ammunition and explosives. In document : AC/258-D258, Brussels, Belgium.

Office of Surface Mining, 1977. Surface mining reclamation and enforcement provisions, Public Law,95-87, Federal Register, Vol.42, No.289

Odello, R. J. 1998. Origins and implications of underground explosives storage regulations. International Symposium of Transient Loading and Response of Structures, Troudhein, Norway, 1998.

Park, Y. J. & Ang, A. M. S. 1985. Mechanistic seismic damage model of reinforced concrete. ASCE J. Struct. Eng., Vol. 111(4), 722-739.

Quek, S. T., Bian, C. Lu, X. Lu, W. & Xiong, X., 2002. Tests of low ductility RC frames under high- and low-frequency excitations. J. Earthquake Engineering and Structural Dynamics, Vol. 31(2), 159-171.

Reinke, R.E. and Stump, B. W. 1988. Stochastic geologic effects on near-field ground motions in alluvium, Bull. Seismo. Soc. Am., Vol. 78, 1037-1058.

Zhou, Y. X, Seah, C. C., Hao, H. & Quek, S. T., (Editors) 1999. Proceedings of the joint Singapore-Norway technical workshop on ground shock. Singapore 31 May – 4 June, 1999, Singapore.

Wave propagation – Moving load – Vibration reduction, Chouw & Schmid (eds.)
© 2003 Swets & Zeitlinger, Lisse, ISBN 90 5809 559 2

Ground Vibrations Caused by Soil Compaction

K. Rainer Massarsch
Geo Engineering AB, Stockholm, Sweden

ABSTRACT: Presently used empirical methods do not account for many important aspects of ground vibrations, for, instance the dynamic interaction between the energy source and the soil. Also the influence of dynamic soil properties on wave propagation is generally neglected. A concept is presented, which describes the dynamic resistance of the compaction source. Two cases are discussed, the impact of a rigid plate on an elastic half-space and the energy transfer to the soil from the shaft of a cylindrical compaction probe. Soil impedance and rate of loading determine the dynamic soil response and thus ground vibrations. There is an upper limit to the vibrations, which can be transmitted to the soil. Based on a closed-form solution it is possible to assess the relationship between static and the dynamic soil resistance. In the vicinity of the compaction source, shear wave velocity and thus soil impedance are strongly affected by strain level. A relationship is proposed for estimating the reduction of shear wave velocity with shear strain. It is possible to estimate the vibration intensity at the compaction source and to predict vibration propagation from the base and the shaft of a compaction probe. The variation of force with time (loading rate) of an impacting plate, and thus the predominant vibration frequency can be estimated from time factors. Resonance effects can occur during compaction, which amplify ground vibrations. A case history is presented, which illustrates the compaction of a deposit of sand using a falling weight. The results of extensive vibration measurements on the impacting plate and in the soil are reported and compared with the proposed analytical solutions.

1 INTRODUCTION

Compaction methods are used increasingly to improve loose, granular soils to great depth more efficiently and at lower cost. In the recent past, new, more powerful compaction machines and construction equipment have been developed, which make the densification process more efficient. However, also the risk of excessive ground vibrations increases. Electronic monitoring systems have been developed, which can control the compaction process and monitor the performance of the equipment, thereby aiding the machine operator during the different phases of project execution. Sensors on the compaction machine and in the ground can be used to measure different parameters, such as compaction frequency, power consumption and ground vibration velocity, which are important for the compaction process, Massarsch (2002). This information can be used to document and verify that the anticipated compaction effect has been actually achieved. Monitoring provides also valuable information about the performance of the equipment, and the effect of soil compaction on the surrounding soil, Massarsch and Fellenius (2002). This increasing database of site information has helped to better understand the compaction process, as well as vibration propagation in the ground, (Massarsch, 1993 and Massarsch & Broms, 2001).

The main objectives of soil compaction are to eliminate total and differential settlements during static loading, and to reduce the susceptibility of water-saturated sands to cyclic loading (liquefaction), for instance during earthquakes.

Soil compaction systems can be classified according to two aspects: the location of the point of energy application (at ground surface or at depth) and the type of applied energy (impact or vibration). These considerations are important for ground vibrations caused by soil compaction.

2 EMPIRICAL RELATIONSHIPS

2.1 *Energy-based vibration prediction*

A commonly used concept to assess ground vibrations from pile driving or soil compaction is based on the energy of the compaction device. The following empirical relationship has been proposed to predict ground vibrations from pile driving, (Attwell and Farmer, 1973)

$$v = K\frac{\sqrt{W}}{D} \qquad (1)$$

where K is an empirical coefficient, W is the energy applied to the pile head (or compaction probe) and D is the lateral distance from the pile measured on the ground surface. A coefficient $K = 0.5$ is often assumed but its actual value depends on many factors and can vary within wide limits (Massarsch, 1992).

Figure 1 illustrates the problems associated with this simplified empirical approach. The compaction energy actually applied to the soil is usually lower than the nominal energy of the compaction device. However, in practice it is difficult to account for these energy losses. Also, as will be shown below, the maximum energy, which can be transferred to the soil and generates ground vibrations, is limited by the dynamic response (impedance) of the soil.

It is also common practice to use in Equation 1 the lateral distance, measured at the ground surface. This causes an error, which can be significant when ground vibrations are measured close to the compaction source, as the depth increases while the lateral distance remains almost unchanged.

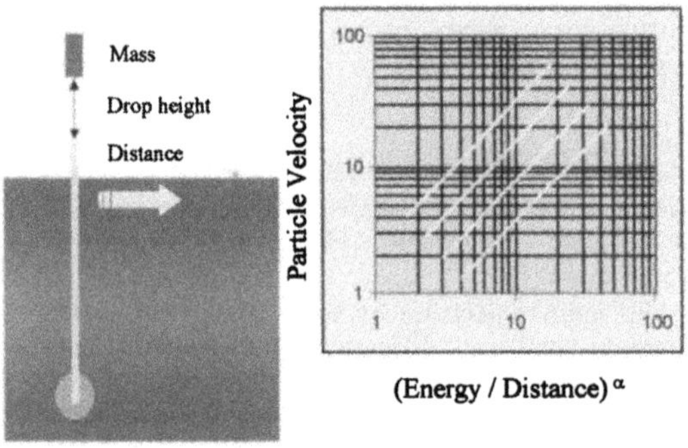

Figure 1. Estimation of ground vibrations caused by soil compaction or pile driving based on applied energy and scaled distance. The exponent α is determined empirically.

Another important aspect is vibration amplification due to the dynamic response of the compaction probe or pile. If the impedance of the compaction probe is low, ground vibrations can be amplified significantly. This effect has been measured by Heckman & Hagerty (1978). The practical implications of the impedance effect on pile driving were discussed by Massarsch (1992, 1993). It is preferable to use a compaction probe with a low impedance, as this will increase vibration transmission to the soil and thus the compaction effect.

Perhaps the most important limitation is that Equation 1 does not consider the effect of ground conditions on vibration propagation. It is not surprising that a large scatter is obtained, even when the measurement results are plotted in a double-logarithmic diagram. Therefore, empirical relationships as proposed in Figure 1 or Equation 1 should be treated with caution and used only for site-specific calibration of vibration measurements. Extrapolation and application of measurements to other site conditions can be misleading.

2.2 *Effect of Compaction Depth and Distance*

Another aspect, which is important when assessing ground vibrations caused by soil compaction, is the depth of the vibration source, in relation to the distance to the point of vibration observations or measurement, Figure 2.

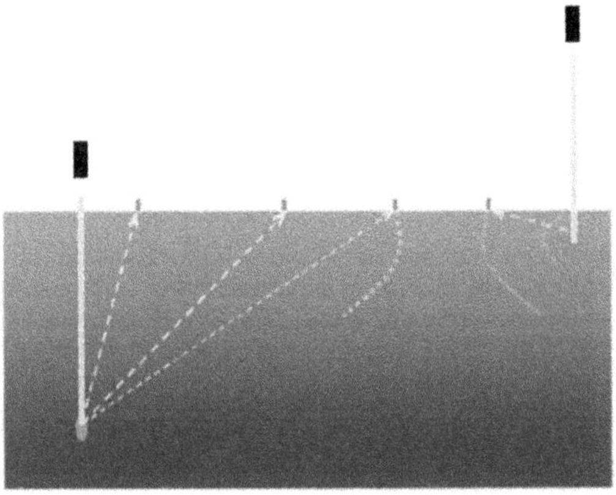

Figure 2. Influence of depth of energy source on vibrations on ground surface.

To the right, surface compaction, or the initial phase of deep compaction is shown, where the energy source is near the ground surface, compared to the distance to the observer. Vibrations can be assumed to emanate from a point source and that waves propagate mainly in the surface layer. In the near-field, vibrations will be caused predominantly by body waves (compression and shear waves, propagating along the ground surface). At larger distances (corresponding to at least two wave lengths), surface waves will start to dominate.

To the left in Figure 2, the case of deep compaction is shown. The energy source has now penetrated into the soil deposit. It is often assumed that vibrations emanate mainly from the base (lower end) of the compaction probe. However, in the case of deep compaction using impact or vibratory probes, a significant amount of compaction energy can also be generated along the shaft, Massarsch (2002). Thus, vibrations can be generated by several energy sources (base and/or shaft) simultaneously.

Even in the case of a deep vibration source, body waves will dominate in the near-field, i.e. at a distance less than the depth of the energy source. Surface waves usually begin to manifest themselves at a critical distance, which can be seen as the location on the ground surface, where body waves emerging at depth are refracted as surface waves. The critical distance depends on several factors, such as the wave length of vibrations (i.e. the vibration frequency and wave velocity in the soil layers), the variation of wave velocity with depth (soil layering), but also on the depth of the energy source. As can be seen from Figure 2, the general assumption that ground vibrations are mainly due to propagation of surface waves, is not generally applicable in the case of deep soil compaction. Only at a distance, which is several times large than the depth of the energy source, surface waves dominate. At large distances, it can also be assumed that vibrations are generated by a point source, even in the case of vibrations emanating from the base and shaft.

3 DYNAMIC SOIL RESISTANCE

In spite of the practical importance of ground vibrations associated with soil compaction, little information is available in the geotechnical literature, describing the fundamental aspects of this problem. There is a need to better understand the factors, which influence the generation of ground vibrations. The problem is complex, compared to conventional vibration problems, such as ground vibrations from industrial activities (machine vibrations) or blasting. During soil compaction, the location and the dynamic characteristics of the energy source change. In addition, as one of the objectives is soil densification, the dynamic and static soil properties will change gradually. Thus, the soil conditions at the end of compaction will not be comparable to those, determined by soil investigations prior to compaction. For instance, the shear wave velocity of loose, uncompacted sand ranges typically between 100 and 150 m/s. After compaction, the shear wave velocity can have increased from 200 to 300 m/s. Also the soil volume decreases and thus the density increase. It is thus necessary to assess how compaction affects the dynamic and static soil properties.

Vibrations can originate either at the base, and/or along the shaft of the compaction probe. While the dynamic response of piles subjected to impact loading is relatively well understood, the mechanism governing pile-soil interaction during vibratory driving or vibratory compaction is still not fully understood. In the following sections, the interaction of a plate and of a shaft with the surrounding soil will be discussed.

3.1 *Dynamic Soil Resistance of Plate on Elastic Half-space*

The action of a compaction probe can be analysed assuming a plate impacting on the ground surface. Figure 3 illustrates the difference between the static and the dynamic soil resistance. When a rigid plate is loaded statically, it can be assumed that the soil resistance increases gradually with deformation. Of course, ground vibrations will not be generated due to the slow loading rate. However, if the plate is dropped on the ground surface, or loaded dynamically, an additional dynamic resistance will occur and the total soil resistance will be higher than in the case of static loading. Further, an observer will notice ground vibration. Thus, the problem of ground vibrations is linked to the dynamic soil resistance. It should be noted that Figure 3 gives a simplified picture of a complex problem, especially, when the applied stress reaches the strength of the soil (non-elastic range).

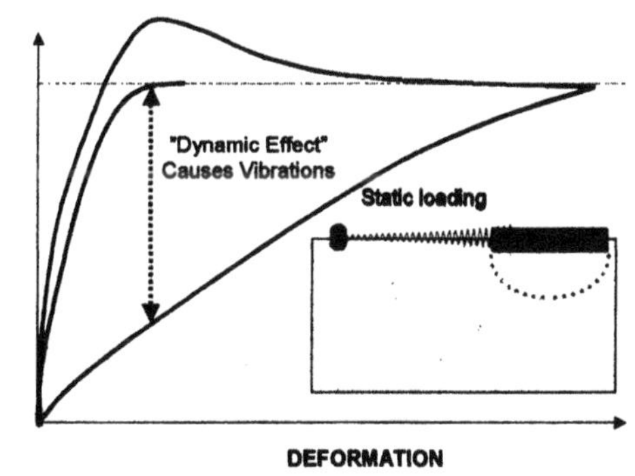

Figure 3. Variation of the total soil resistance with deformation, consisting of a static and a dynamic component. The dynamic soil resistance is responsible for ground vibrations.

A simplified analytical model will be used to assess the parameters, which influence the emission of vibrations from the source. This approach has the advantage over more sophisticated models, that the relative importance of different parameters becomes apparent.

It can be assumed, that a rigid, circular plate impacts on the surface of an elastic half-space. Bodare and Orrje (1988) have developed a closed-form solution for the dynamic resistance at the interface between a rigid plate and the soil, which is based on the "Herlitz equation". This solution is attractive, as the results of a complex problem can be presented in a relatively simple form. Orrje (1996) has shown by comparison with model tests, that this theoretical model, which is strictly applicable only in elastic material, describes many of the significant aspects of dynamic plate-soil interaction.

In the case of an impacting circular plate, the dynamic stress σ_{dyn} acting on the surface can be estimated from the following relationship

$$\sigma_{dyn} = z_0 \, v \tag{2}$$

where the z_0 is the specific impedance and v is the particle velocity. The specific soil impedance z_s can be determined from

$$z_0 = \frac{c_s \, \rho}{s} = \frac{z_s}{s} \tag{3}$$

where ρ is the bulk density, c_S is the shear wave velocity. As will be shown below, the shear wave velocity c_s is not a soil constant but is affected by shear strain. Consequently, also the soil impedance is strain-dependent. The parameter s is the ratio between the shear wave velocity, c_s and the compression wave velocity, c_p and can also be expressed as a function of Poisson's ration, ν

$$s = \sqrt{\frac{1-2\nu}{2(1-\nu)}} = \frac{c_s}{c_p} \tag{4}$$

The dynamic stress σ_{dyn} acting on the interface of a circular plate and the soil can thus be estimated from the following relationship

$$\sigma_{dyn} = v \, c_s \, \rho \sqrt{\frac{2(1-\nu)}{1-2\nu}} \tag{5}$$

The dynamic stress is depends mainly on the shear wave velocity below the plate, the soil density and Poisson's ratio, cf. Figure 3. From Equation 5, the maximum vibration v_{max} during loading can be determined

$$v_{max} = \frac{\sigma_{mob}}{\rho \, c_s^{*}} \, s \tag{6}$$

The soil density ρ and Poisson's ratio ν do not vary significantly. However, the shear wave velocity c^{*}_{s} decreases with increasing shear strain. Thus, there is an upper limit to the vibrations, which can be transferred to the soil due to dynamic plate loading. Equations 2 and 5 confirm that ground vibrations will be higher when the rate of loading (impact velocity) increases. The shear strain level γ d can be estimated if the particle velocity v and the shear wave velocity, c_s are known

$$\gamma = \frac{v}{c_s} \tag{7}$$

3.2 *Evaluation of Dynamic Base Resistance*

As shown in Figure 3, the total soil resistance during dynamic loading consist of a static and a dynamic resistance component. The dynamic resistance is mainly influenced by the loading rate (loading velocity) and the strain-dependent soil impedance (strain-dependent shear wave velocity). Bodare and Orrje (1988) have presented solutions of the Herlitz equation for the elastic case.

Figure 4 defines three parameters, which influence the force-time relationship. The two time factors t_p and t_s express the relationship between the plate radius, r and the compression wave velocity, c_p and the shear wave velocity, c_s, respectively. The time factor t_p is the time required for the compression wave c_p to travel from the centre of the plate to the perimeter and t_S is the time for the shear wave to travel from the centre to the perimeter. The time factor T_0 defines the time at which the dynamic force reaches its maximum.

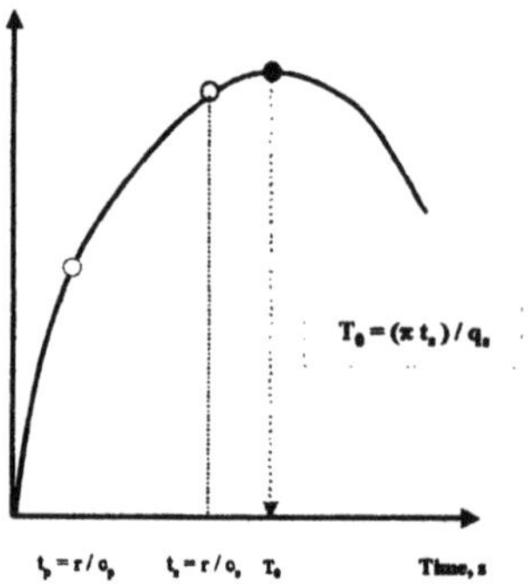

Figure 4. Determination of dynamic soil resistance of a circular plate on in an elastic half-space. The shape is defined by three time factors.

The time factor T_0 can be determined according to Figure 4 if the time factor t_s and the parameter q_s are known. The force-time relationship is thus defined by three "characteristic time factors", t_p, t_s and T_O. It is apparent that the plate radius and the velocity of the compression wave and of the shear wave below the plate influence the duration of loading (rise time). If the plate radius is increased, the predominant frequency of ground vibrations will decrease.

Bodare and Orrje (1988) have published solutions, from which the relationship between the static and the dynamic loading resistance can be calculated. Figure 5 has been developed based on this relationship. The dynamic soil resistance is responsible for ground vibrations to occur. The force ratio P_{dyn}/P_{stat} depends on the factor q_s, which in turn is affected by the two time constants t_s and T_o. The dynamic force increases with increasing q_s In dry sand, it can be as high as 6 times the static soil resistance and many times higher in sand below the ground water. This is due to the high compression wave velocity of water compared to dry soil.

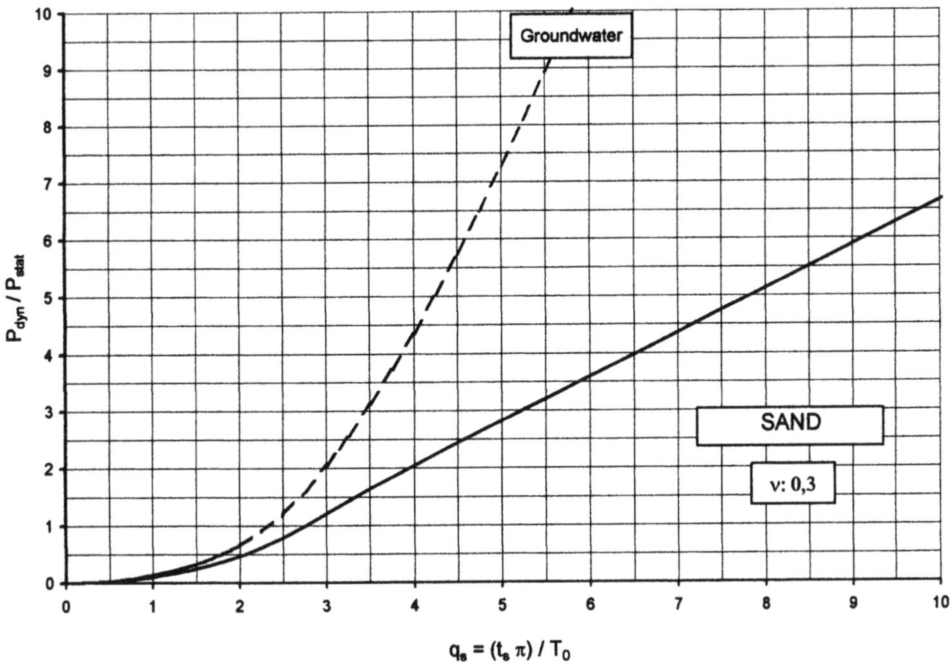

Figure 5. Ratio between the dynamic and the static force in dry sand and in sand below the ground water, cf. Figure 3.

It can be concluded that ground vibrations will be significantly higher in water-saturated soils. Also, with increasing soil compaction, the shear wave velocity and thus t_s increases and result in higher ground vibrations.

3.3 *Evaluation of the Dynamic Shaft Resistance*

3.3.1 *Vertical Ground Vibrations*

The dynamic resistance acting along the shaft of a vertically vibrating compaction probe can be estimated similar to the concept presented in the previous section. However, dynamic shaft resistance can be estimate more easily if it is assumed that the probe moves as a rigid cylinder. Along the perimeter, the dynamic shaft resistance can be estimated from

$$\tau_{dyn} = v\, z_s \tag{7}$$

where v is the loading velocity and z_s is the soil impedance

$$z_s = c_s\, \rho \tag{8}$$

where c_s is the strain-dependent shear wave velocity and ρ is the soil density. The total shaft resistance results from the combined effect of the static and the dynamic component. Again, there is an upper limit to the dynamic energy, which can be transferred from the compaction probe to the surrounding soil, cf. Equation 8. This depends mainly on the shear wave velocity at large strain (in the transition from the plastic to the elasto-plastic zone). However, with increasing soil densification, the dynamic soil resistance increases, and thus also the level of ground vibrations. From equation 7 and 8 it can be concluded that ground vibrations generated along the shaft of the compaction probe will increase with increasing shear wave velocity. Thus, increasing soil densification will results in higher ground vibration amplitudes. Vibrations will be transmitted from the shaft of the compaction probe to the soil as vertically polarized shear waves. This cylindrical wave front is similar to that of a Rayleigh wave and has the same geometric damping. It can be difficult to predict along which section of the compaction probe (or pile) the maximum amount of vibration energy will be disseminated. This depends on several factors, the most important of which are the rigidity of the compaction probe (probe impedance) and the dynamic probe-soil interaction (impedance ratio between soil and probe).

3.3.2 *Horizontal Ground Vibrations*

In the geotechnical literature is often assumed that in the case of a vertically oscillating probe or pile, only vertical ground vibrations are generated from the shaft, cf. previous section. However, in addition to vertically polarised shear waves, also horizontal vibrations can exist in granular soils, Massarsch (2002). This is due to the friction between the shaft of the compaction probe and the soil. The horizontal stress changes result in a compression wave, which contributes beneficially to the compaction of the soil, as the lateral earth pressure increases. Figure 6 shows the results of field measurements during vibratory compaction, Krogh & Lindgren (1997).

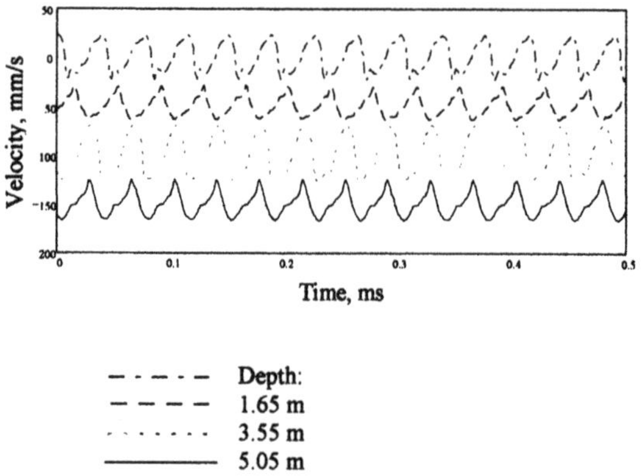

Figure 6. Horizontal vibration amplitude measured at different depth levels during vibratory compaction, from Krogh & Lindgren (1997).

Horizontally oriented vibration sensors (geophones) were installed on, and at different levels below the ground surface, at a distance of 2.9 m from the centre of the compaction probe. At the time of the vibration measurements, the tip of the compaction probe had passed the lowest measuring point. It can thus be confirmed that strong horizontal vibrations are generated in spite of the fact that the compaction probe is oscillating only in the vertical direction, Massarsch (2002). The horizontal ground vibrations amplitudes were of the same magnitude as the vertical vibration amplitudes. It has also been shown that as a result of vibratory compaction, the horizontal stresses increase in the soil. This compaction effect is of practical importance as it changes permanently the stress conditions after compaction.

4 EFFECT OF STRAIN LEVEL ON SHEAR WAVE VELOCITY AND IMPEDANCE

The soil impedance (and thus the shear wave velocity) and Poisson's ratio are of importance when assessing the dynamic soil resistance and ground vibrations during soil compaction. As has been emphasised, the soil impedance must be determined using the strain-adjusted shear wave velocity. Otherwise, too high impedance values will be obtained, which over-predict the dynamic soil resistance and thus vibration levels in the vicinity of the energy source.

The shear wave velocity can be measured in the field or in the laboratory, or estimated from empirical relationships. However, it is not generally recognized that the shear wave velocity decreases even at relatively low strain levels. Figure 7 shows the result of a resonant column test on a reconstituted sample of medium dense sand. The shear modulus and the shear wave velocity decrease when a critical shear strain level (approximately 10^{-3} %) is exceeded. At a strain level of 0,1 %, the shear modulus and the shear wave velocity have decreased from 76 MPa to 25 MPa, and from 205 m/s to 120 m/s, respectively. This effect is significant, especially in sandy soils and is so large that it can not be neglected. The shear wave velocity at 1 % shear strain (which corresponds to strain levels in the vicinity of the compaction source) is about 50 % for plastic soils (clays) and as low as 20 % for sands, when compared to the shear wave velocity at small strain (elastic wave). The shear wave velocity can be readily determined from the shear modulus if the soil density ρ is known

$$c_s = \sqrt{\frac{G}{\rho}} \tag{9}$$

The decrease of the shear modulus and of the shear wave velocity can be described by reduction factors R_G and R_c, respectively. The shear modulus, G and the shear wave velocity, c_s at a given shear strain level can be determined if the respective maximum values and the reduction factors R_G and R_c are known

$$G = R_G\, G_{max} \tag{10a}$$

$$c = R_c\, c_{max} \tag{10b}$$

According to Equation 9, the following relationship exists between the reduction factors of the shear modulus and the shear wave velocity, respectively

$$R_c = \sqrt{R_G} \tag{11}$$

Extensive investigations have been carried out to study the effect of shear strain on the shear modulus, for instance Massarsch (1985), Vucetic & Dobry (1985), Rollins et al. (1998). Döringer (1997) has performed an extensive literature survey of resonant column tests. Based on this compilation of numerous tests in cohesive and low-plastic soils (clays, silts and silty sands), a relationship was developed between the shear wave velocity reduction factor and shear wave velocity, which can be expressed in the following expression

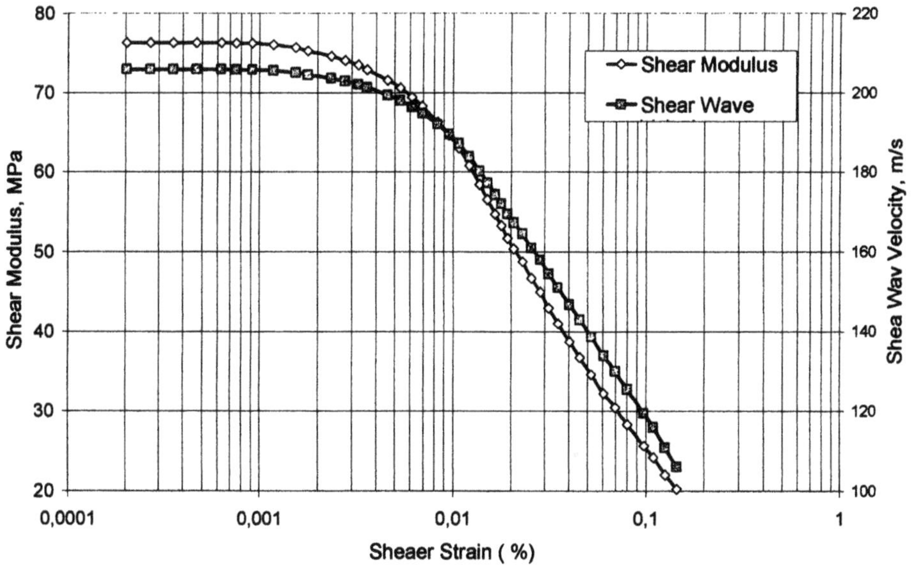

Figure 7. Resonant column test on a reconstituted sample of medium dense sand. Shear modulus (left axis) and shear wave velocity (right axis) as a function of shear strain level (semi-logarithmic scale).

$$\frac{G}{G_{max}} = \frac{1}{\left[1+\alpha\,\gamma\left(1+10^{-\beta\gamma}\right)\right]} \qquad (12a)$$

$$\alpha = \frac{794}{I_P^{\;0.36505}} \qquad (12b)$$

$$\beta = 0{,}046 + 0{,}5475\ \log(I_P) \qquad (12c)$$

where α and β are empirically determined coefficients and I_P is the plasticity index. The relationship of Equation 12 is shown in Figure 8. All soils exhibit strain softening and this effect depends on soil plasticity. The shear wave velocity is reduced when a critical shear strain level is exceeded. The reduction of the shear wave velocity is more pronounced in soils with low plasticity, i.e. sands and silty sands. In soils with higher plasticity, the shear wave velocity reduction is lower but still of importance.

5 VIBRATION PROPAGATION IN SOILS

The propagation of vibrations from a source into the surrounding soil is complex. In the vicinity of the compaction source, the soil is usually in a plastic, or elasto-plastic state.

Three compaction zones can be identified adjacent to the compaction source:

1. Plastic zone: where the soil is in a failure condition and subjected to large strain levels >10^{-1} %. The source-soil interaction in this zone is complex but of less practice significance.
2. Elasto-plastic zone: where the strain level ranges between approximately 10^{-3} and 10^{-1} %.
3. Elastic zone: where the shear strain level is below 10^{-3} %, and no permanent deformations can be expected.

In the plastic zone the vibration velocity is limited by the shear wave velocity at large strain (and thus the strength) of the soil. In the plastic, and the elasto-plastic zone, the wave propagation velocity is strain-dependent. The shear wave velocity increases with increasing distance from the energy source. In the elastic zone, the wave propagation velocity is constant (elastic wave velocity).

Wave propagation in the elastic and elasto-plastic zone can be analyzed using simple or more sophisticated models. However, it is necessary to account for the influence of shear strain, cf. Figure 8.

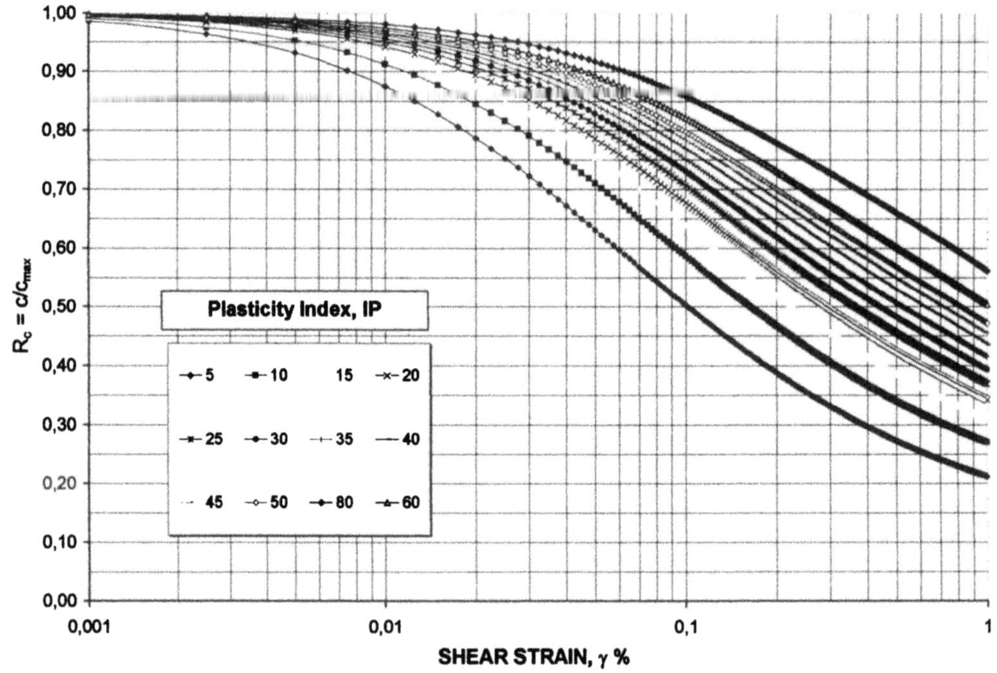

Figure 8. Effect of shear strain on shear wave velocity as a function of soil plasticity.

5.1 *Wave propagation*

The propagation of waves in an elastic or elasto-plastic medium can be analysed using sophisticated methods. However, in many cases, also simplified solutions will be satisfactory, provided that source-soil interaction is accounted for and appropriate soil parameters have been selected. This aspect appears to be more important than the degree of analytical sophistication.

The vibration velocity at the interface of the plastic and elasto-plastic zone can be estimated using the relationships provided in Section 3. The maximum vibration velocity can then be used to predict the vibration propagation from the shaft and from the base of the compaction probe into the surrounding soil.

Also in the case of vibration propagation the effect of strain level on vibration velocity in the near-field (elasto-plastic zone), must be taken into account. For most soil compaction problems, an estimate of vibration velocity using the simple wave attenuation from a point source will yield satisfactory results

$$\frac{A_2}{A_1} = \left(\frac{R_2}{R_1}\right)^{-n} e^{-\alpha(R_2 - R_1)} \qquad (13)$$

where A_1 and A_2 are the vibration amplitudes at distance R_1 and R_2, respectively. The exponent n depends on the wave type as shown in Table 1.

Table 1. Wave exponent n for different wave types

Wave type	Exponent n
Body Wave	1,0
Body Wave at surface	2,0
Surface wave (or cylindrical wave)	0,5

The absorption coefficient α takes into account the effect of soil damping

$$\alpha = \frac{2\pi\, D\, f}{c_s} \qquad (14)$$

where D is material damping, f is the predominant vibration frequency and c_s is the shear wave velocity. Material damping at small strain levels is almost independent of soil type and ranges typically between 2 and 4 %. However, also material damping increases with increases with shear strain, Vucetic and Dobri (1985), Rolling et al (1990). At a shear strain levels between 0.5 and 1 %, the damping ratio can be as high as 20 %.

5.2 *Predominant Frequency*

The predominant vibration frequency can be determined by field measurements. In the case of vibratory compaction, the predominant frequency at some distance from the compaction point corresponds usually to the operating frequency of the vibrator. However, in the case of impact loading, it is more difficult to predict the predominant vibration frequency, which depends on the dynamic interaction between the source and the surrounding soil. The information provided in Section 3 can be used to estimate the predominant frequency in the case of impact loading.

The predominant period of an impulse, for instance due to the impact of a plate on the ground surface, can be estimated according to Figure 4. The time factors t_p and t_s can be calculated from the plate radius and the respective wave velocity. The parameter q_s can be estimated, based on Figure 5, and is for soil compaction typically in the range of 2 to 4. With this information it is possible to estimate the rise time to the maximum force. The predominant vibration frequency can then be estimated from

$$f = \frac{1}{4\,T_0} \qquad (15)$$

It is interesting to note that the predominant vibration frequency is mainly dependent on the plate geometry and the wave velocity below the plate. The frequency increases with decreasing time factor. Thus a smaller plate gives a higher vibration frequency than a larger plate at the same ground conditions.

5.3 *Vibration Amplification*

An important aspect of ground vibration prediction in connection with soil compaction is vibration amplification. Vibration amplification can be used in connection with vibratory compaction to enhance the compaction efficiency, Massarsch (2002). This compaction method is known as "resonance compaction" (MRC). A compaction probe is inserted in the ground, using a vibrator with variable frequency. During the probe penetration phase, the optimal penetration speed is achieved when the probe is vibrated at a high frequency (around 30 Hz), cf. Figure 9. During the compaction phase, the vibrator frequency is lowered to achieve vibration amplification (resonance of the vibrator-probe-soil system). Figure 9 shows how the vertical vibration velocity, measured on the ground surface, varies as a function of the vibration frequency. During probe penetration and extraction, a high vibration frequency (around 30 Hz) is used, which does not cause significant ground vibrations. During the compaction phase, the operating frequency of the vibrator is reduced to the resonance frequency of the probe-soil system. At resonance, ground vibrations are strongly amplified. The probe and the surrounding soil vibrate in phase, resulting in a more efficient transfer of compaction energy.

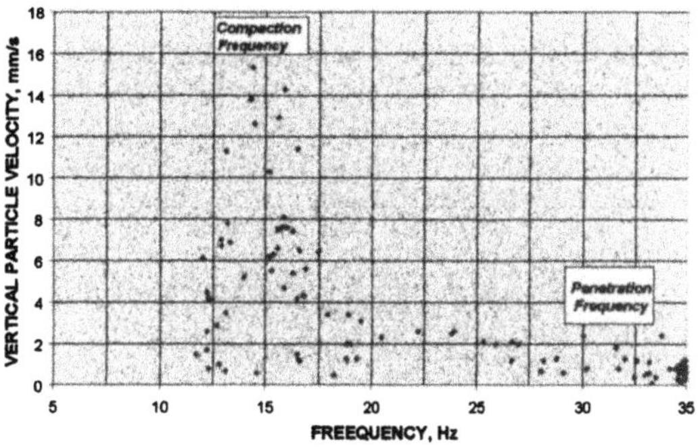

Figure 9. Vertical ground vibration velocity at a distance of 4 m from the compaction probe during probe penetration and resonance compaction

The dynamic response of the soil deposit during compaction can also be used to monitor the compaction effect. With increasing densification of the soil layers, the resonance compaction frequency rises. Also the ground vibration velocity increases and soil damping is reduced.

It can be concluded that ground vibrations are significantly amplified if the compaction frequency is close to the resonance frequency of the compaction system. Ground vibration amplification can be as high as 10 to 20 times and should therefore be taken into consideration.

6 CASE HISTORY – DYNAMIC COMPACTION

Changi airport was initially constructed during the 1970' s when 700 ha of land were reclaimed from the sea. This ambitious land reclamation project was continued to provide two additional landing strips and terminal buildings. The land reclamation comprised 1 550 ha (210 million m^3). After removal of a soft marine clay layer from the sea bottom the extension of the airport was constructed on a 5 to 15 m thick sand fill. The sand was loose below a dense surface layer. The average cone penetration resistance prior to compaction was 4 to 8 MPa. The fill consisted mainly of sand with an effective particle diameter d_{10} = 0.5 mm. The ground water level was located about 3 m below the formation level.

Extensive field investigations were carried out by the author in connection with the compaction project. Two different soil compaction methods were compared: dynamic compaction, using a heavy falling weight, Figure 10, and a vibratory compaction probe, Figure 11.

Dynamic compaction is probably the oldest ground improvement method. In spite of its simplicity, the efficient execution requires careful planning and extensive field testing. Compaction is usually performed in several passes. The compaction energy (mass of pounder times drop height) is usually chosen, based on past experience, taking into consideration the soil conditions and the required degree of densification. However, at large projects trial compaction is usually carried out to determine the required number of compaction passes, the optimal pounder mass and drop height.

Figure 10. Deep soil compaction ("dynamic compaction") at change airport.

Figure 11. Vibratory compaction (MRC), using vibrator with variable frequency in order to achieve vibration amplification. Note the geophone for resonance measurements in the picture to the right.

In the following, the results of field measurements during dynamic compaction will be described. As a result of extensive field tests, the following dynamic compaction procedure was chosen in order to achieve a cone penetration resistance of at least 10 MPa. The mass of the pounder was 21 ton and the drop height was varied between 15 and 20 m. The degree of compaction increased from the ground surface and reached a maximum at about 4 m depth (15 to 29 MPa). Below this level, the compaction effect decreased again.

During the initial compaction tests, comprehensive dynamic measurements were performed, Krogh & Lindgren (1997). The pounder was instrumented with an accelerometer and ground vibrations were measured on and below the ground surface with accelerometers (in the near-field) and with geophones. Figure 12 shows vibration measurements during the impact of the pounder.

Figure 12. Vibration measurements during dynamic compaction at Changi airport, Singapore.

Different types of seismic measurements were performed in order to determine the shear wave velocity prior to, during and after compaction. Geophones were also installed below the ground surface, cf. Figure 6.

The shear wave velocity varied within the stratified soil deposit, also prior to compaction. The shear wave velocity was relatively high (around 200 m/s) in the upper surface layer. In the uncompacted sand at the ground water level, the shear wave velocity decreased to 160 m/s.

The results of the vibration measurements and the interpretation of the measurements are presented in the following figures. Figure 13 shows the acceleration (retardation) of the pounder during impact. The time factor T_0 can be estimated from the acceleration record. Due to practical reasons and because of the air trapped below the pounder at impact, the rise time is not as sharp. This aspect was taken into account when selecting the time factor as 0.025 seconds.

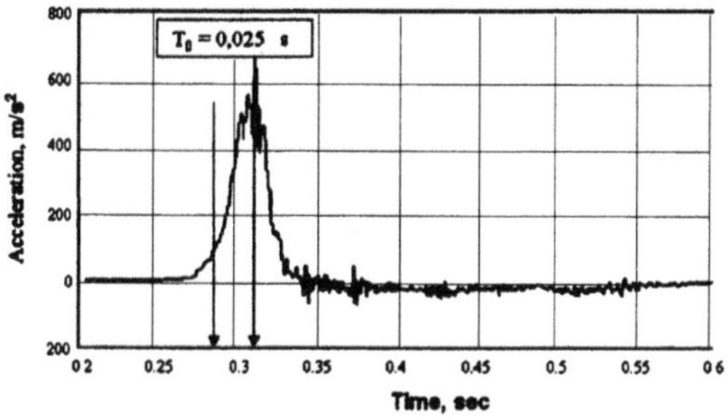

Figure 13. Acceleration measured on the pounder during impact. Indicated is the estimate time factor, T_0.

From the acceleration measurement the impact velocity can be determined by integration. From the measured acceleration and the pounder mass the force can be calculated. If the applied stress is plotted against the velocity, the specific soil impedance can be readily determined, Figure 14.

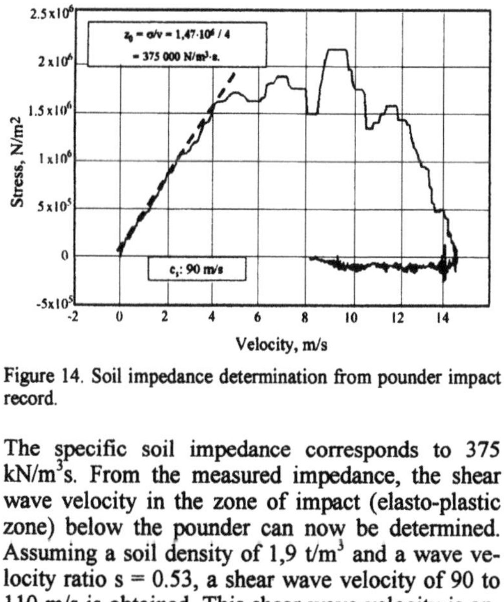

Figure 14. Soil impedance determination from pounder impact record.

The specific soil impedance corresponds to 375 kN/m^{3}s. From the measured impedance, the shear wave velocity in the zone of impact (elasto-plastic zone) below the pounder can now be determined. Assuming a soil density of 1,9 t/m^3 and a wave velocity ratio s = 0.53, a shear wave velocity of 90 to 110 m/s is obtained. This shear wave velocity is approximately 50 % of the shear wave velocity measured at small strain.

From the acceleration record, T_0 and the shear wave velocity, the parameter q_s can be determined. It is then possible to estimate the dynamic soil resistance and the vibration amplification, Figure 15. In the present case, $P_{dyn} / P_{stat} = 0.52$, which is a relatively low value, which is probably due to the large plate diameter and the low shear wave velocity at impact.

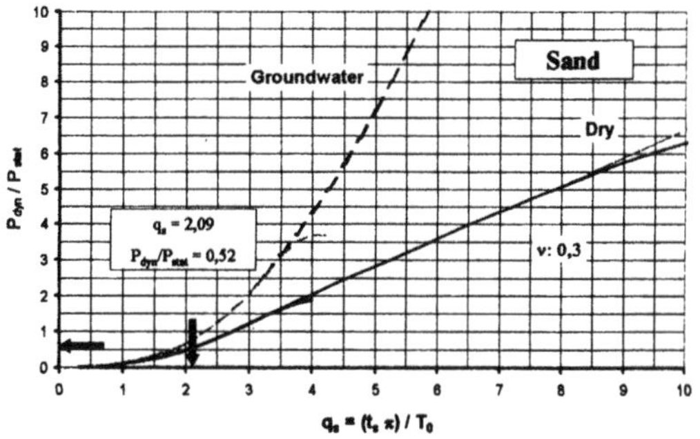

Figure 15. Estimation of dynamic soil resistance and vibration amplification, cf. Figure 5.

The penetration of the pounder can be calculated by double integration of the acceleration record. Figure 16 shows the movement of the pounder during impact, and the calculated penetration depth was about 0.35 m. This value is in good agreement with field observations and confirms the accuracy of the acceleration measurements.

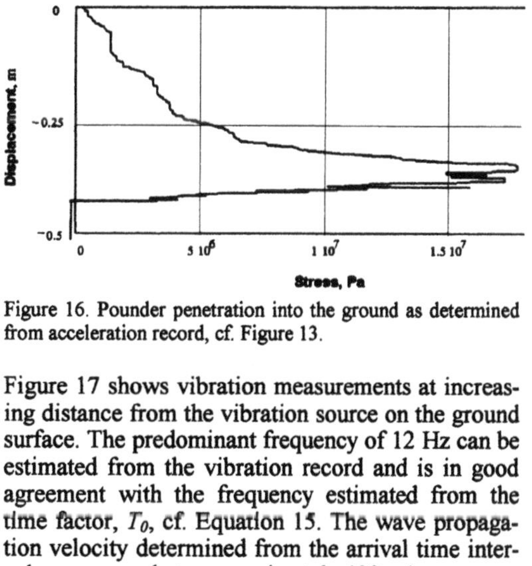

Figure 16. Pounder penetration into the ground as determined from acceleration record, cf. Figure 13.

Figure 17 shows vibration measurements at increasing distance from the vibration source on the ground surface. The predominant frequency of 12 Hz can be estimated from the vibration record and is in good agreement with the frequency estimated from the time factor, T_0, cf. Equation 15. The wave propagation velocity determined from the arrival time intervals corresponds to approximately 400 m/s.

7 SUMMARY AND CONCLUSION

Vibrations caused by soil compaction are important when planning projects in urbanized areas or in the vicinity of vibration-sensitive structures or installations. In spite of its practical importance, little quantitative information has been reported in the literature about this problem. Most correlations are of an empirical nature and neglect important aspect, such as interaction between the energy source and the ground, and wave propagation in the ground.

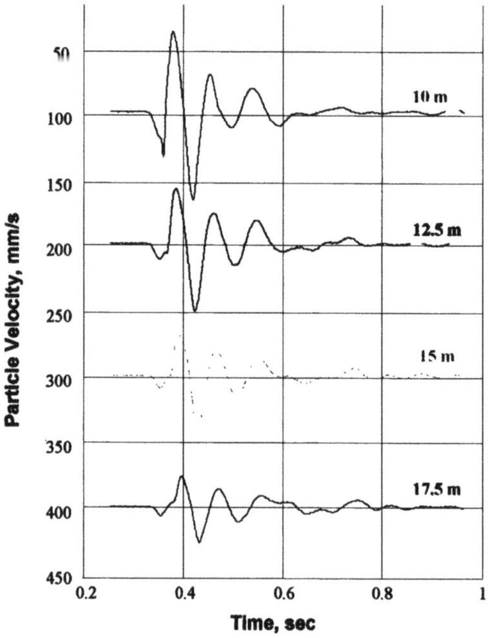

Figure 17. Ground vibration measurements on the ground surface caused by pounder impact.

An attempt has been made to describe quantitatively the interaction between the compaction source and the surrounding soil. The interaction of a plate and of a cylindrical shaft with the soil have been analysed and solutions are proposed for estimating the dynamic soil resistance. The most important parameter is soil impedance, which determines the dynamic resistance of the soil during dynamic loading. However, the shear wave velocity and thus the soil impedance decrease with shear strain and this effect must be taken into consideration.

The soil impedance limits the dynamic energy and thus the vibrations that can be transmitted from a compaction source.

A simple chart is presented, which makes it possible to estimate the ratio between the dynamic and the static soil resistance. The dynamic resistance, and thus ground vibrations, increase with increasing shear wave velocity and increasing loading rate.

A semi-empirical relationship is proposed for the assessment of the reduction of shear wave velocity as a result of shear strain. The strain-softening effect is most pronounced in granular soils.

The propagation of ground vibrations from the shaft and from the base of the compaction probe have been described. A vertically oscillating compaction probe generates along the shaft vertically polarized shear waves. However, in granular soils, due to soil friction, also horizontal vibrations are created.

Ground vibrations increase during soil compaction and also the vibration frequency increases. Another important factor is vibration amplification, which occurs during vibratory compaction when the operating frequency approaches the frequency of the vibrator-probe-soil system.

Results of extensive field tests during compaction trials at Changi airport are described. The energy transfer from a plate impacting on the ground surface was measured. From the acceleration measurements on the pounder the specific soil impedance and the predominant vibration frequency can be determined. The field tests confirm that the soil impedance is lower in the vicinity below the compaction plate. The measured predominant vibration frequency is in agreement with the predicted value, based on the time factor T_0.

8 ACKNOWLEDGEMENTS

The research on soil compaction was supported by a grant from the Lished Foundation (Lished Stiftelsen), which is gratefully acknowledged. Anders Lindgren participated in the field tests at Changi and performed part of the data evaluation. Dr. Anders Bodare has developed the theory of dynamic plate loading, based on the Herlitz equation. His comments and suggestions have been most valuable and are acknowledged. Krupp TBT, Germany supported the field tests and their support and assistance is highly appreciated.

The astute comments by the reviewer of the paper are acknowledged.

REFERENCES

Attwell, P, B. & Farmer, I. W., (1973), Attenuation of Ground Vibrations from Pile Driving, *Ground Engineering, Vol. 3, No. 7, pp. 26 - 29.*

Bodare, A., & Orrje., 1988. Impulse load in a circular surface in an infinite elastic medium. (Closed solution according o the theory of elasticity). *Extended version. KTH jord och bergmekanik, JoB Report No. 23, Stockholm.*

Döringer, H., 1997. Verformungseigenschaften von bindigen Böden bei kleinen Deformationen (Deformation properties of cohesive soils at small strain). *Royal Institute of Technology, Stockholm, Soil and Rock Mechanics. Examensarbete 97/8, 54 p.*

Heckman, W. S. & Hagerty, D. J., 1978. Vibrations associated with Pile Driving. *ASCE Journal of the Construction Division. Vol. 104. No. C04, pp. 385 – 394.*

Krogh, P., & Lindgren, A. 1997. Dynamic field measurements during deep compaction at Changi airport, Singapore. Examensarbete, *JoB, Institutionen för Anläggning och Miljö, Royal Institute of Technology (KTH), JOB Report 97/9, 88 s.*

Massarsch, K. R., 1979. Closing remarks, Discussions, Session 1, *Seventh European Conference on Foundation Engineering, Brighton, Proceedings*, *Vol. 4, p. 59.*
Massarsch, K. R., 1985. Stress-Strain Behaviour of Clays. *11th International Conference on Soil Mechanics and Foundation Engineering, San Francisco, Proceedings, Volume 2, pp. 571 - 574.*
Massarsch, K. R., 1992. Static and Dynamic Soil Displacements Caused by Pile Driving, Keynote Lecture. *Fourth International Conference on the Application of Stress-Wave Theory to Piles, the Hague, the Netherlands, September 21-24, 1992, pp.15 - 24.*
Massarsch, K. R., 1993. Man-made Vibrations and Solutions, State-of-the-Art Lecture. *Third International Conference on Case Histories in Geotechnical Engineering, St. Louis, Missouri, June 1 - 6, 1993, Vol. II, pp. 1393 - 1405.*
Massarsch, K. R. & Broms, B. B., 2001. New Aspects of Deep Vibratory Compaction. *Proceedings, Material Science for the 21st Century, JSMS 50th Anniversary Invited Papers, Vol. A, The Society of Material Science, Japan, pp. 172 – 179.*
Massarsch, K. R., 2002. Effects of Vibratory Compaction. *TransVib 2002 – International Conference on Vibratory Pile Driving and Deep Soil Compaction. Louvain-la-Neuve. Balkeama, Keynote Lecture, pp. 33 - 42.*
Massarsch, K.M. & Fellenius, B.H., 2002. Dynamic compaction of granular soils. *Canadian Geotechnical Journal, Vol. 39, No. 3, pp. 695 - 709.*
Orrje, O., 1996. The use of dynamic plate load tests in determining deformation properties of soil. *Doctoral thesis, Division of Soil and Rock Mechanics, Royal Institute of Technology (KTH), Stockholm. 111 p.*
Rollins, K. M., Evans, M. D., Diehl, N. B. and Daily, WD, III., 1998. Shear modulus and damping relationships for gravels. *ASCE. Journal of Geotechnical and Geoenvironmental Engineering, Vol 124, No 5, pp. 396-405*
Vucetic, M. and Dobry, R. 1990. Effect of Soil Plasticity on Cyclic Response. *Journal of the Geotechnical Engineering Division, ASCE Vol. 117, No. 1. Jan, pp. 89 - 107.*

Wave propagation – Moving load – Vibration reduction, Chouw & Schmid (eds.)
© 2003 Swets & Zeitlinger, Lisse, ISBN 90 5809 559 2

Analyses on accumulation of propagating ground surface wave under running train

Y. Sato
Railway Track System Institute, Tokyo, Japan

S. Okamoto, C. Tamura and M. Hakuno
University of Tokyo, Tokyo, Japan

S. Morichi
Tokyo University of Science , Noda-shi, Chiba-ken Japan

ABSTRACT: The study on the accumulation of the propagated ground surface wave under running train with high velocity proceeded by authors from 1971 to 1973 is in report. It was a pilot research in this area, but has never been published except part of the data. The research were (1) Survey on existing studies, (2) Photoelastic study for confirming the phenomena, (3) Study on actual data obtained on Shinkansen (4) Study on the effect of damping in ground. As results, the effect of damping in ground, the experimental confirmation of the accumulation of lateral waves under running load with the velocity approaching to that of ground wave and the characteristics of ground vibration such as their increase with train velocity, their frequency and amplitude depending on the layer structure of ground are presented.

1 INTRODUCTION

The study on the accumulation of the propagating ground surface wave under running train with large velocity was asked from Railway Technical Research Institute to authors, JNR, from 1971 to 1973. It was a pilot research in this area, but the whole figure except part of them has never been published.

2 STUDY ON EXISTING PAPERS

It is easily imagined with the analogy to the supersonic plane surpassing the sonic velocity that some unusual phenomenon will happen if the train velocity approaches to the ground surface wave velocity. For existing train, as the train velocity was not so high, it had not been studied so much up to the age of 1960. In early age of 1942, Sato & Rikitake (1942) calculated the case for train speed of 90 km/h and the longitudinal wave (P-wave) velocity of 5 km/s. The dynamic effect is 0.000014 times of that of static load. In this case, the ratio of wave velocity in the ground to train velocity is 200:1. However, in the case that train velocity is 300 km/h and that of P-wave in soft ground, the ratio of velocities is 3.6:1. In this case the situation will be very different.

In 1960's such problems had been studied by several scholars for the other purpose than the train running (Beitin 1969, Cole et al 1958, Demkin et al 1970, Eason 1965, Gakenheimer et al 1969, Keer 1969, Payton 1964, Scott 1964 & Singh et al 1970). They treated the problem with the models of elastic half space on the surface of which various forms of load run. These are classified as follows depending on the load velocity and by the form of the load.

(1) Types of velocity and variations of load
  (a) Constant velocity (b) Constant acceleration
  (a) Constant pressure (b) Harmonically varied
(2) Distribution of load
  (a) Line load running perpendicularly to the direction of load distribution
  (b)Pointed, circular or rectangular load
(3) Domain of velocity
  (a) Load velocity of smaller than that of elastic wave
  (2) Load velocity of larger than that of elastic wave

With these studies, elastic solutions for the moving load with a constant velocity were obtained in either case in (3). As these are elastic solutions, if the load velocity is coincided with any velocity of lateral wave, longitudinal wave or surface wave, the displacement and strain become infinity. Actually the plasticity existing in ground prevents such dispersion. In another word, the interactive phenomena between train running and ground could not be analyzed just with completely elastic body.

As an example the stresses in the ground are demonstrated for the running velocity in Figure 1 by the theoretical study of Eason (1965). The ratio of load running velocity to that of lateral wave, $\alpha_2 = V/c_2$, is given in abscissa and the ratio of stress

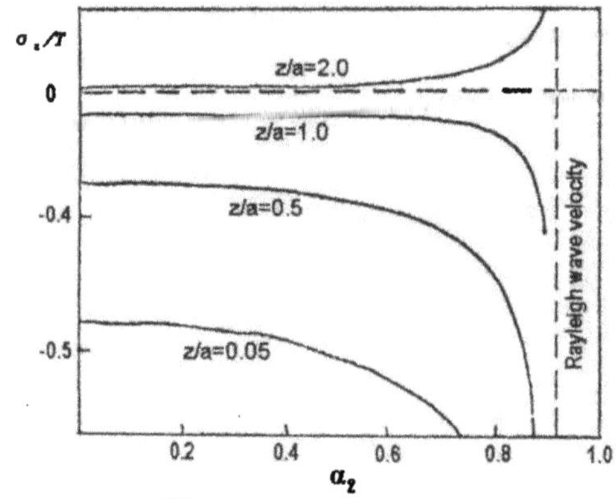

Figure 1. σ_z/T in ground to α_z

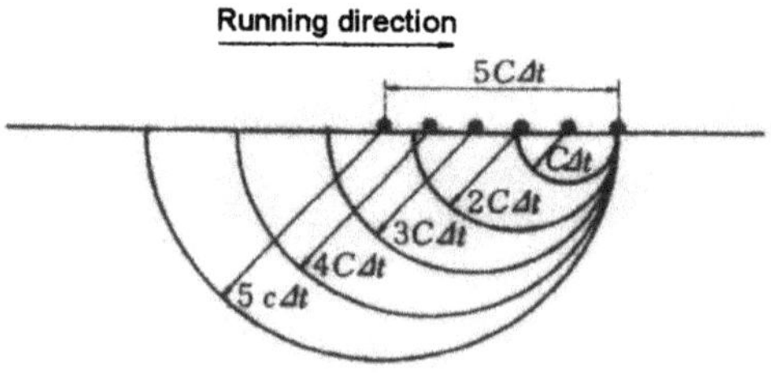

Figure 2. Accumulation of propagating waves

under the load in propagating direction to load pressure, σ_z/T, in ordinate. The ratio of depth to the side length of square load, z/a, is given as a parameter. Here, α_z is a coefficient like Mach in aerodynamics. The stress becomes smaller with the depth and the sign is changed from minus to plus at a certain depth. It grows with the velocity and becomes infinity at the velocity of Rayleigh wave.

3 EXPERIMENT WITH GELATIN GEL (Morichi et al 1974 & 1981)

To confirm the accumulation of displacement and stress caused by moving load, the experiment by using the gelatin gel with very slow elastic lateral wave (S-wave) of 2-3 m/s was performed. The gelatin gel specimen with the size of 10 × 10 × 50 cm (Figure 4) was impacted laterally with pendulums at 4 locations. The distance between these locations is 5cm. In order to simulate a running train, impact loads were induced at the different four locations successively at different time.

The impaction of test piece with pendulums is shown in Figure 3. In the figure, the rectangular bar of 3 × 1 × 10 cm set at the top of gelatin gel laterally is the part to generate the lateral displacement. The pendulum of 800g swung from the height of 20 cm and impacted circular bar of

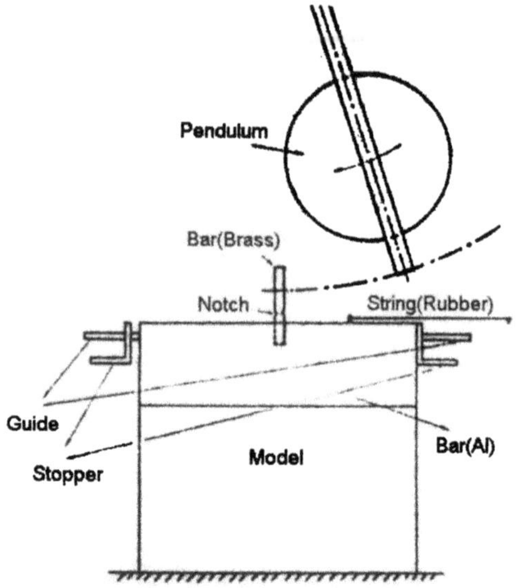

Figure 3. Impaction of gelatin test piece

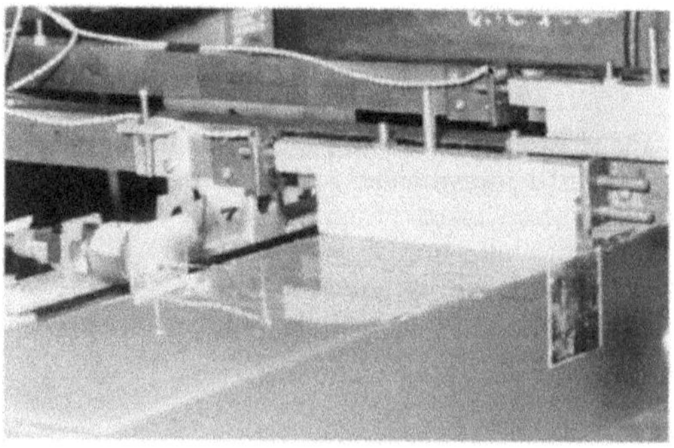

Figure 4. Test speciemen of gelatin gel

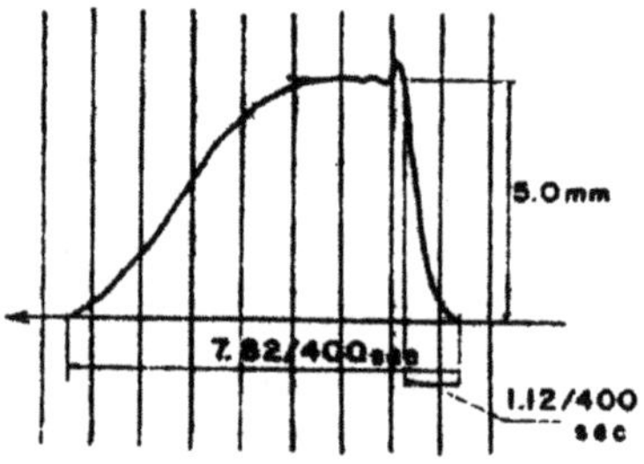

Figure 5. Generated impulsive displacement

brass (4 mm ϕ). The stopper placed in front of the rectangular bar stopped the displacement generated by the impact. At this moment the circular bar of brass was broken at a notched part and pulled back with a rubber string.

With the equipment, an impulsive displacement as shown in Figure 5 was generated.

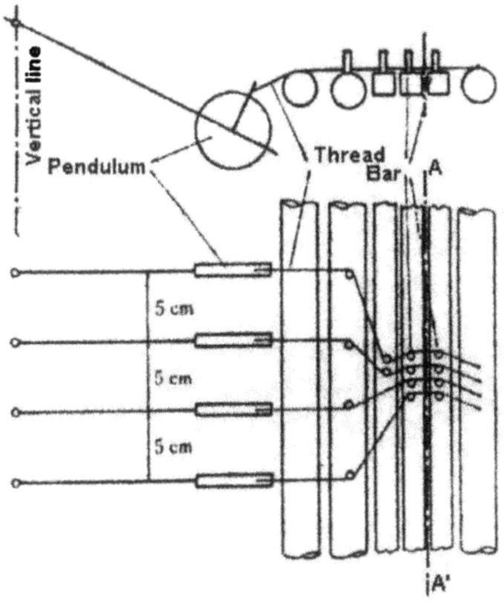

Figure 6. Mechanism giving time difference for giving impulsive displacement

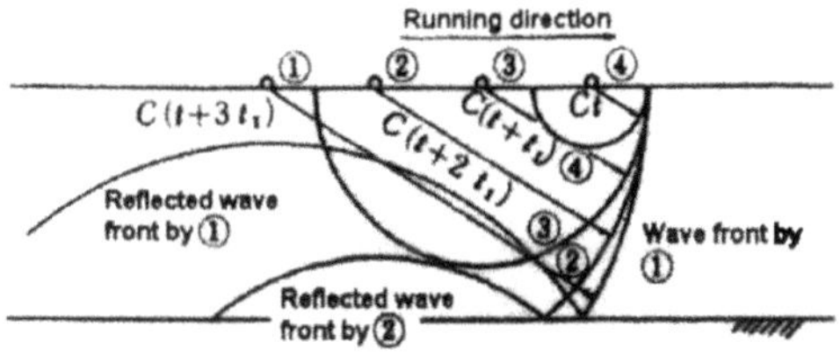

Figure 8. Accumulation of propagated waves

Figure 6 shows the mechanism for giving time difference to four impulsive displacements. The strings which supported the pendulums and concentrated in a small section were cut by a razor set at the bottom of another pendulum (2.4 kg) which run along A-A' line. The running speed of the pendulum with razor was controlled with the height of holding the pendulum.

The accumulation of propagating lateral waves originated from four origins is given in Figure 7. The clear accumulation of lateral waves at the wave front is observed. The fringes given in the figure is not mostly photoelastic ones, but shadows due to the surface deformation of gelatin gel being soft enough. The accumulation of movements is explained in Figure 8 for the case (10) in Figure 7.

To confirm the accumulation of the displacements due to propagating wave numerically, the lateral movement of test specimen was measured with gap sensors of induction type as shown in Figure 9. At the end of embedded bar of brass the tin plate of 3 × 3 × 0.035 mm was set for the measurement.

The impacted places (S1 and S2) and measuring places (M1 and M2) are given in Figure 10.

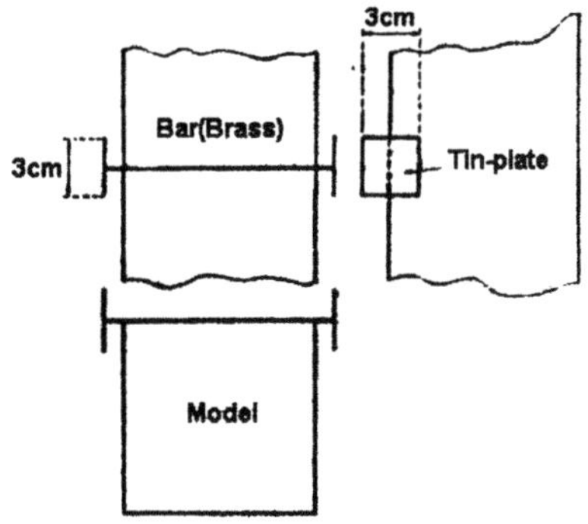

Figure 9. Measurement of displacement

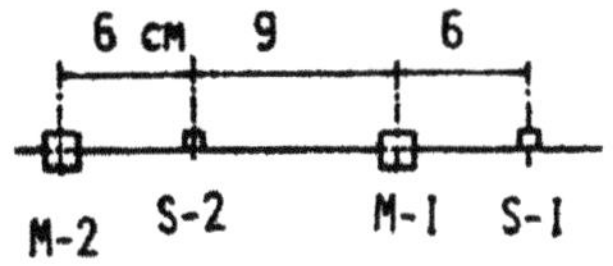

Figure 10. Arrangement of impacting and measuring points

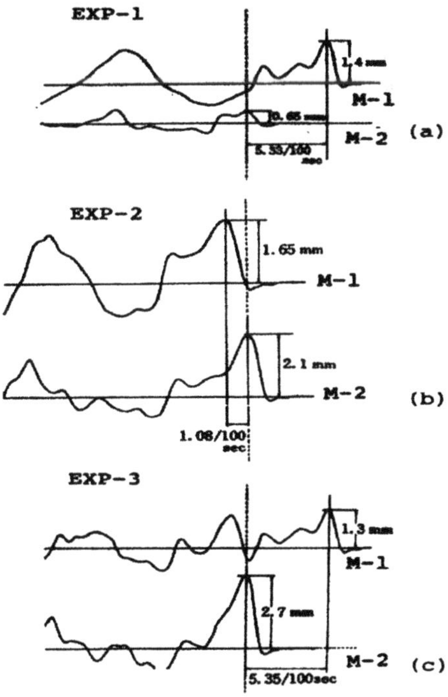

Figure 11. Measured results

In experimental results in Figure 11 (a) is for the impact at S1 (Experiment 1), (b), for that at S2 (Experiment 2) and (c), the case in which S1 and S2 are impacted with the time difference of accumulating major waves from S1 and S2.

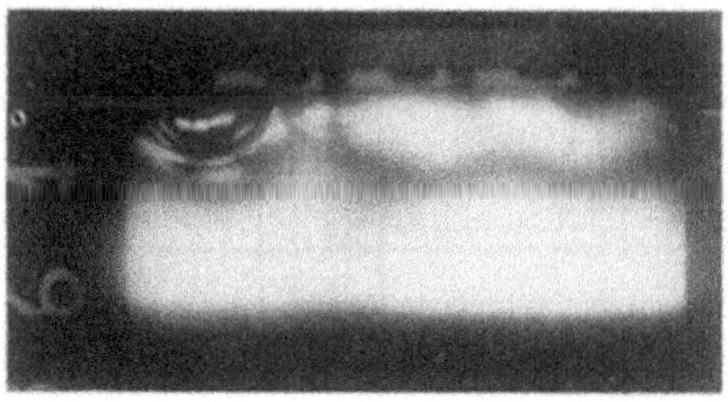

(1) Starting from ①

(2)

(3) Starting from ②

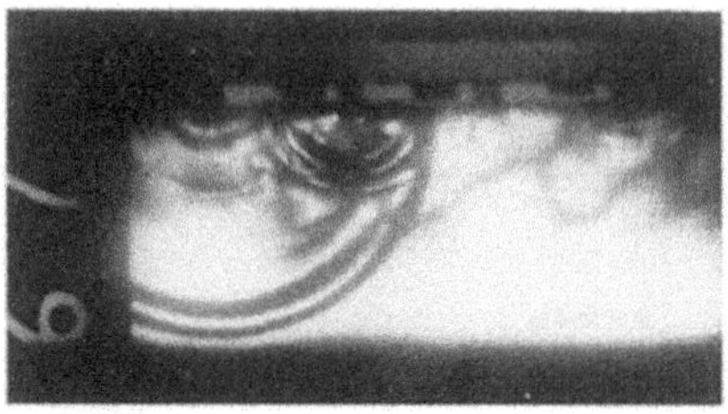

(4)

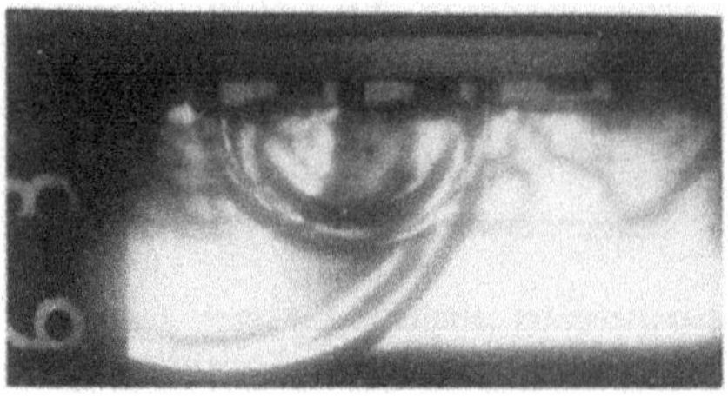

(5)

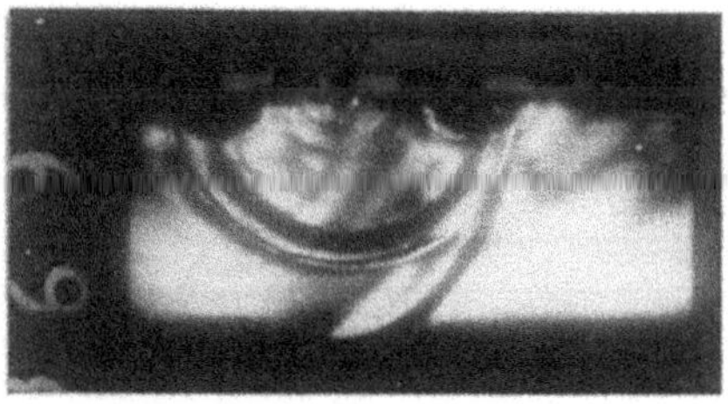

(6) Starting from ③

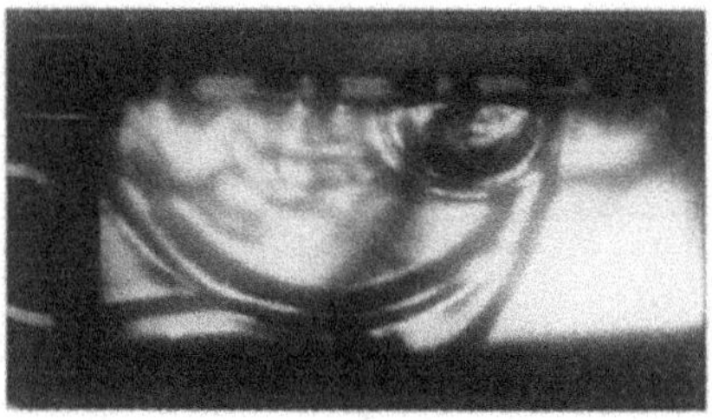

(7)

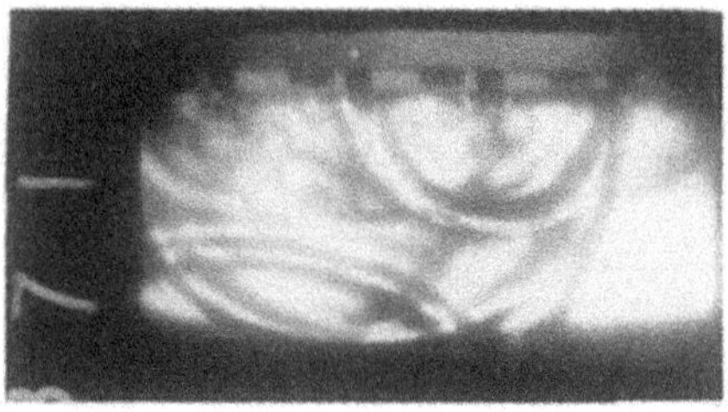

(8)

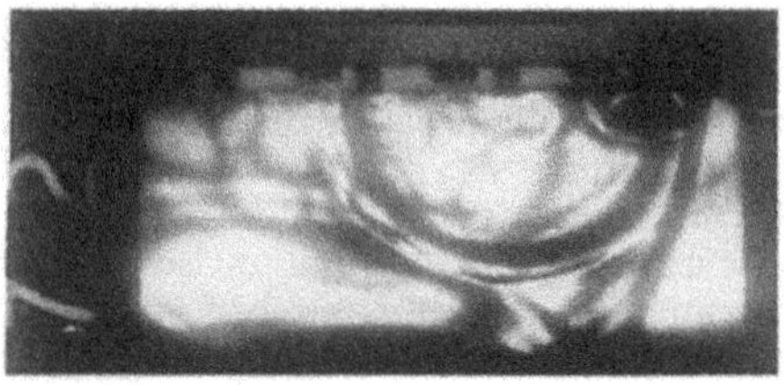

(9) Starting from ④

(10)

Figure 7. Accumulation of propagating waves

Table 1 Superposition of waves

Exp.	Position of impact	Time difference (s)	Distance from impact (cm)		Mxm. Displacement (mm)		Time difference at M1 & M2 (s)	Wave velocity (m/s)
			M1	M2	M1	M2		
1	S1	-	6	21	1.4	0.65(0.75)	5.33/100	2.8
2	S2	-	9	6	1.65(1.7)	2.1	1.08/100	2.78
3	S1 & S2	5.2/100	-	-	1.3	2.7	5.35/100	-

The maximum values in Figure 11 are given in Table 1. Comparing the measurements in Figure 11 (a) and (b), the decrease of amplitude in displacement depending on the distance from the impacted point is observed. Considering the propagation is cylindrical, the amplitude is assumed to decrease reciprocally to the square of distance. They are given in () in Table 1. They show good coincidences with the measured values.

The velocity calculated from the time difference of maximum values and from the distance of measuring points are also given in Table 1. They are also in good coincidence with the value of 2.98 m/s obtained from the shear vibration test performed independently.

As shown in Table 1, the maximum displacement of M2 is 2.7 mm. It corresponds to the sum of maximum displacements of 0.65 mm in Experiment 1 and 2.1 mm in Experiment 2. Further the displacement in Figure 11 (c) corresponds well to the sum of Figures 11(a) and (b) which are set to show the maximum displacements of M2 at the same time. It demonstrates experimentally that, if the moving velocity of load coincides that of wave propagation, the very large displacement could be caused under the moving load.

Further it is demonstrated that such a phenomenon as for complicated boundary conditions could be analyzed with such an experimental work.

4 STUDY ON MEASURED GROUND VIBRATION BY SHINKANSEN

To solve the problem in the title, it is necessary to clarify the propagation of waves in the ground. As it had not been discussed up to this moment in the area of civil engineering, it was considered important to formulate the matter theoretically.

For this purpose, the semi-infinite elastic body with *n* layers was assumed and theoretically analyzed. It was shocked with a point source and waves were propagated from it symmetrically. Here the downward displacement in radial direction was in concern. As results, the characteristics of Rayleigh's and Love's waves were given.

It is necessary to check the actual data in parallel to the theoretical analysis. As the data in such a high speed as coincided with the wave velocity did not exist at this moment, the data obtained in Nishi-Akashi area for the test train of 951 type consisted of two cars on Shinkansen were investigated. Those were the ground vibrations (velocities) in the vertical and lateral directions at the places of 10 m and 66.3 m from the center of the viaduct in the area with a soft surface layer (paddy field) of 1 m deep. The running velocity of train was more than 200 km/h. It was fairly slow compared with the wave velocity in the sub-layer of ground, but seemed to be semi-supersonic for the surface layer.

From the data obtained followings were noticed through precise studies

(1) The sub-layer of ground has the P-wave velocity of 1360 m/s under the surface layer with that of 140 m/s. As the velocity of S-wave 180 m/s and 350 m/s were noticed for the sub-layer, but that for surface layer was estimated as 50-70 m/s.
(2) In the vibrational velocities of data the period of 0.04 s was noticed especially in the wavelets at 66.3 m.
(3) At 10m location from the center of viaduct the amplitude in vertical direction was the double of the horizontal one. On the contrary, the horizontal one was 2-4 times of the vertical at 66.3 m. At any places and at any components the amplitude increased with train velocity.
(4) The frequency at 10 m location from the center of viaduct was eminent at 40 Hz in horizontal and at 25 Hz in vertical. That at 66.3 m was eminent at 25 Hz in both directions. The frequency was not affected by train velocity. It seemed to be determined by the railway structure or the structure of ground.
(5) Data in the horizontal direction at 66.3 m show several wavelets of about 25 Hz before and after (more before) the train approaching nearest place to the measurement. Further it was noticed that the arriving time of wavelet before the train passing at the nearest place was shortened and that after the train arrival was retarded with the increase of train velocity. It could be the Doppler's effect. The frequency of about 25Hz was also noticed to be higher with the increase of velocity.

These observations were discussed from the viewpoints of Love's wave, multiple reflections and the arrival time of wavelets. Results were as follows:

(1) The amplitude of ground vibration was increased with the train velocity.
(2) The frequency could be determined depending on the ground structure.
(3) At the distance of about 50 m from the center of viaduct, the vibration due to the surface wave affected by the layer structure of ground or the multiple reflections in surface layer due to S-wave transmitted through the foundation might give large amplitude.
(4) As the origin of vibration, the irregularities of track, load series and natural frequencies of structure and ground are to be noticed.

5 CALUCULATION WITH VASICOUS DAMPING

As given in Chapter 2, the existing theoretical analyses assumed the semi-infinite elastic body and show the infinitely increasing displacement and stresses of the ground under the load with the velocity approaching to that of wave propagation. However, the actual ground consisting of earth and sand dissipates the wave propagating energy with its viscous or plastic components. The ground responses under the load running with the velocity approaching to the critical wave velocity of the ground were calculated for the ground with the viscosity of Voigt type.

The fundamental equations and boundary conditions were based on the paper by Sato & Rikitake (1942). The coefficients were changed from elastic ones to Voigt's ones. Calculations were performed by integrating introduced formula. Under the load in Figure 12 the calculations were performed for following coefficients.

Uniformly distributed load, $Q=10^4$ kgf/m^2, Lame's constants, $\lambda=8\times10^7$ and $\mu=2\times10^7$, Density, $\rho=2\times10^3$ kg/m^3 and Poisson's ratio, $\nu=0.4$

The displacement at the front end of the load is given in Figure 13. Figure 13(a) shows the case with loading area of 2 × 2m. It increases at the train velocity larger than 90 m/s and shows the disturbance in the range larger than 95m/s. However, the theoretical analysis does not show such a disturbance by smoothly increasing to infinity. As it is estimated due to errors of numerical integration, the displacement was defined as approaching infinity with the cause of the disturbance.

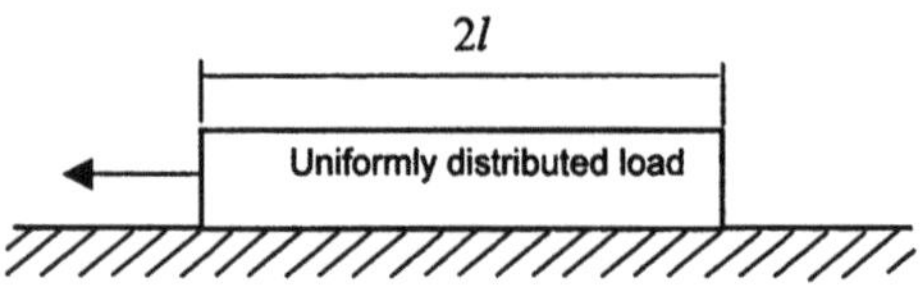

Figure 12. Load acting on ground

Here, increasing the viscosity ratio λ'/λ and μ'/μ from $1/10^5$ to $1/10^3$, the disturbance is observed for the first, but not for the latter. In Figure 13(b) with the load of 2 × 20m the displacement shows the disturbance without viscosity. However, at the case with the viscosity of 5% the tendency of disturbance is disappeared, but with some smooth

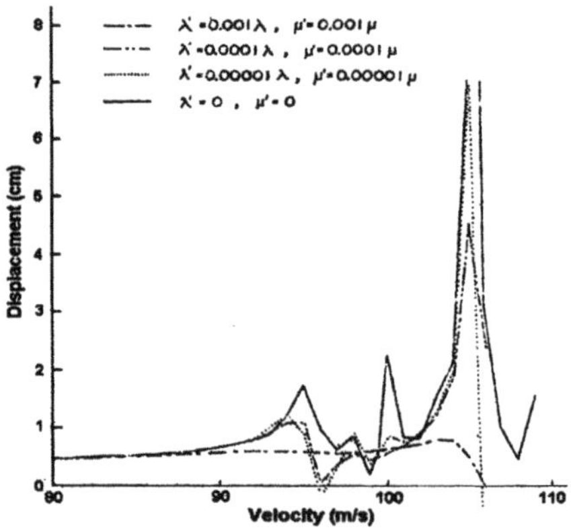

(a) Load of 2 × 2 m

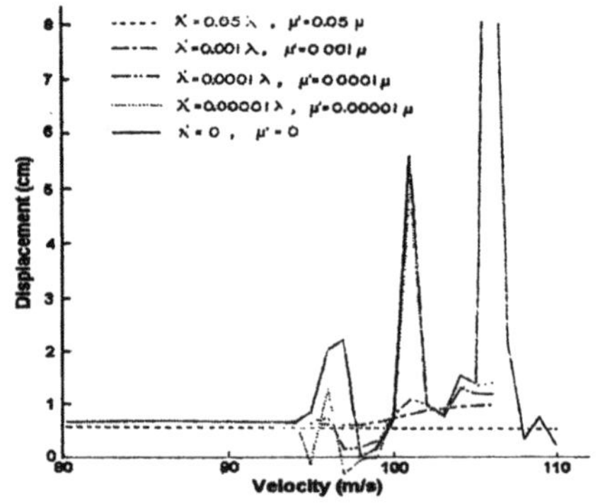

(b) Load of 2 × 10 m

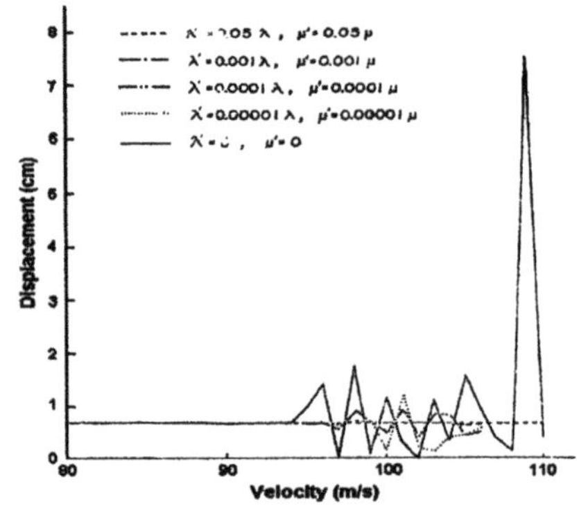

(c) Load of 2 × 100 m

Figure 13. Displacement at front end of load

increase. In Figure 13(c) with the load of 2 × 100 m the disturbance is disappeared with the viscosity over $1/10^3$.

Next, the forms of displacement depending on the velocity and the effect of damping on the form of displacement were checked. They are shown in Figures 14 and 15. Figure 14 shows the form of displacement near the load of 2 × 2 m on complete elastic body ($\lambda' = \mu' = 0$). The abscissa is fixed to the running load by placing it on *x-vt*. The displacement is changed with the increase of the velocity of the load. Figure 15 shows the forms of displacement at the surface of the complete elastic body and at that of viscoelastic body. It shows that the displacement decreases with the viscosity. On the elastic body the increase of displacement is symmetric to the abscissa fixed to the running load, but it becomes asymmetric and looks like the wave due to a running boat on water with high speed. With a further increase of viscosity, the form of displacement becomes smoother. It suggests that the strain in ground may also be smaller.

Additionally, the forms of displacement in the longitudinal direction were also calculated for the coordinate fixed to the load for different load length changing velocity as shown in Figure 16. Figure 16(a) shows how the form of displacement changes with the velocity on the ground with viscosity. That is, the wave head is higher with the velocity. Figures 16(b) and (c) show the forms of displacement to velocity for the load length of 20m and 100m, respectively. The same tendency are given as in case (a), but the changes of .the form of displacement are much calmer than in Figure 16(a).

The results reveals:

(1) Assuming that the ground is a viscous elastic body, the displacement at the critical velocity was decreased with an increase of viscosity and disappeared at the viscosity of 5 percent.
(2) Assuming that the ground has some viscosity,

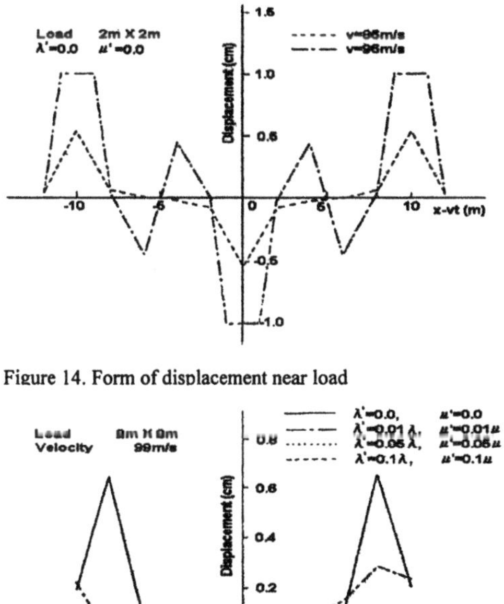

Figure 14. Form of displacement near load

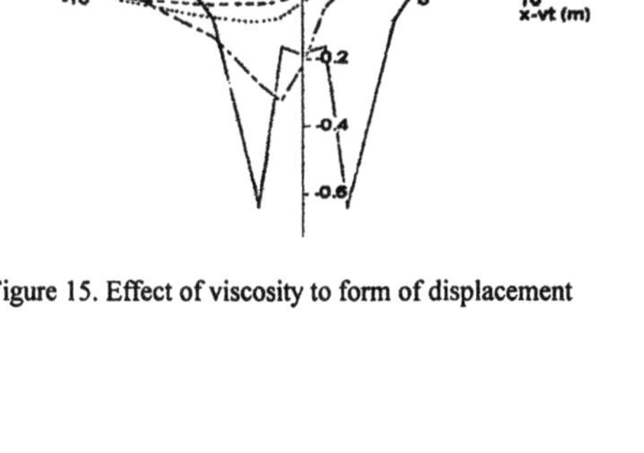

Figure 15. Effect of viscosity to form of displacement

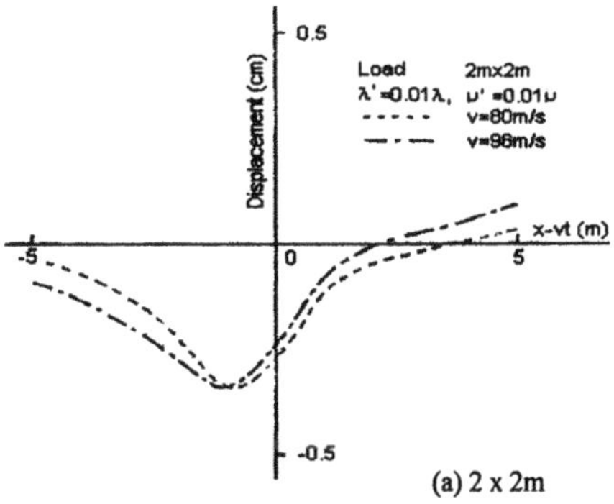

(a) 2 x 2m

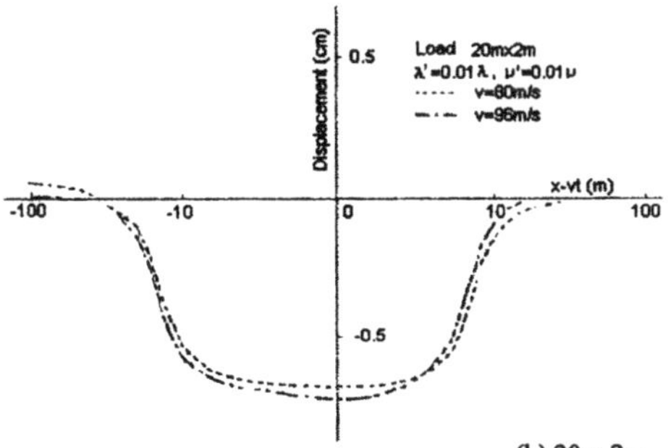

(b) 20 x 2m

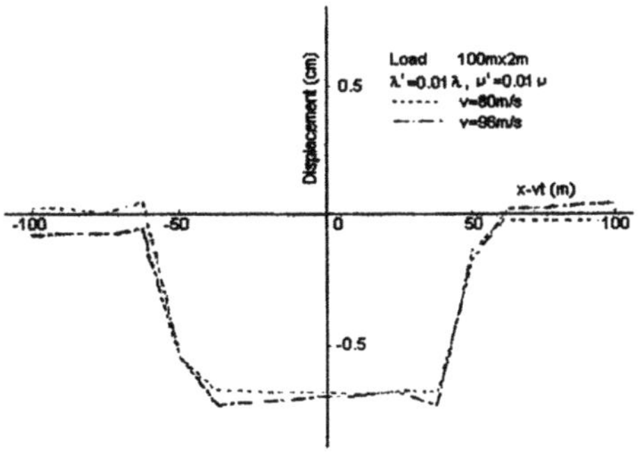

Figure 16. Form of displacement under long load

the deformation of ground was greatly changed.

(3) The deformation of ground had positive angle in the running direction like the wave caused by a boat running on the water with high velocity.

(5) Following photoelastic study, the scale model train running with high velocity on scale model track might be useful.

6 CONCLUSIONS

From the study following conclusions are obtained.

(1) The study on the existing papers on elastic half space at these ages gave that the displacement and stresses under the running train grew infinitely when the train velocity approaches the wave propagation velocity in the ground. However, assuming that the ground has a viscosity, the approach of train velocity to that of wave propagation velocity does not cause an increase of the ground displacement.

(2) The experiment was performed by laterally impacting a gelatin gel specimen at discrete places with the time interval corresponding to the train running velocity. It proved that the accumulation of displacements and gave a clear image of accumulation of propagating waves. It established the photoelastic method for such problems.

(3) The precise study on the actual data revealed the characteristics of ground vibration such as its increase with train velocity. The frequency and amplitude of vibration depend on the layer structure of ground.

Further discussions are on following matters.

(1) As the increase of vibration was observed in contrary to the theoretical analysis with damping which refrains the increase of ground displacement, the studies for the actual damping of ground including structures are important.

(2) The actual value of ground damping is unknown.

(3) What are the effects of structures (Rail, viaduct, piers etc.)?

(4) Though just the introduction of damping diminished the sharp accumulation of vibration, the effects of unknown factors must be considered.

ACKNOWLEDGEMENT

Authors are grateful for the preparation of data on ground vibration by Shinkansen to Dr. Yoshimasa Kobayashi, Honorary Professor of Kyoto University (Formerly, Railway Technical Research Institute, JNR) and for the work accepting the research commission to Professor Narioki Akiyama of Nihon University (Formerly, Saitama University).

REFERENCES

Beitin, K.I. 1969.12. Response of an Elastic Half Space to a Decelerating Surface Point Load. In *Jour. of App. Mech.* (Abbreviated as "*JAM*"): pp.819-826.

Cole, J.D. & Huth, J.H. 1958.:Stress Produced in a Half Plane by Moving Loads. In *JAM* 25: pp. 433-436. ASME 80.

Demkin, J.W. & Corbin, D.G. 1970. Deformation of Layered Elastic Half Space by Uniformly Moving Loads. In *Bull. of Seism. Soc. Am.*: 60: pp. 167-191.

Eason, G. 1965. The Stresses Produced in a Semi-Infinite Solid by a Moving Surface Force. In Interm. *J. Eng. Sci.*: 2 pp. 581/609.

Gakenheimer, D.C. & Miklowitz J. 1969.9. Transient Excitation of an Elastic Half Space by a Point Load Traveling on the Surface. In *JAM*: pp.515-515.

Keer, L.M. 1969. Moving and Simultaneously Fluctuating Loads on an Elastic Half Plane. In *Jour, of Acoustical Soc. of America*: 47-5 (part 2).

Morichi, S. & Tamura, C. 1974. An Experimental Study on Propagation of Elastic Wave due to a Wave Source Moving on the Surface. In *Seisan Kenkyu*: 26-7. (in Japanese)

Morichi, S & Tamura, C. 1981. On Dynamic Experimental Analyses by Testing of Models Made from Gel-like Materials, *Trans. of JSCE*: 310. (in Japanese)

Payton, R.G. 1964. An Application of the Dynamic Betti - Rayleigh Reciprocal Theorem to Moving-Point Loads in Elastic Media. In *Quat. of App. Math.*: Vol. 21: pp.299-313.

Sato, Y. & Rikitake, T. 1942. Deformation of Semi-infinite Elastic Surface under Moving Rectangular Load. In *Earthquake*: 14-5. (in Japanese)

Scott, R.A. 1969.12. Line Loads Moving on the Surface of an Inhomogeneous Elastic Half-space. In *App. Sci. Rev.*: 21: pp. 356-365.

Singh, S.K. & Kuo, J.T. 1970.3. Response of an Elastic Half Space to Uniformly Moving Circular Surface Load. In *JAM*: pp.109-115.

Wave propagation – Moving load – Vibration reduction, Chouw & Schmid (eds.)
© 2003 Swets & Zeitlinger, Lisse, ISBN 90 5809 559 2

Soil-structure interaction due to moving loads

S.A. Savidis & C. Bode
Institute of Soil Mechanics and Foundation Engineering, Technical University Berlin, Germany

ABSTRACT: The paper summarises some of the authors' activities concerning a three-dimensional numerical approach to analyse dynamic soil-structure interaction (SSI) of railway tracks on layered soil under transient load (moving loads) in the time domain. The approach can also be used to calculate vibration propagation in the soil and its effect on nearby buildings. However, the approach is by no means restricted to railway structures, but can be applied to any other SSI-problem. A basic feature is the application of the substructure method, where the track or any other structure is analysed by means of finite elements, whereas the soil is analysed by means of boundary elements with fundamental solutions for a layered halfspace. The SSI-model has been implemented in ANSYS accounting for nonlinear properties of the track. The method is applied to analyse the dynamic response of railway tracks due to a moving wheel set. In particular, the effect of the travelling speed on the rail deflections and the influence of 'through-the-soil-coupling' are investigated.

1 INTRODUCTION

Vibrations generated by trains and transmitted through the ground are giving rise to increasing environmental nuisance such as annoyance to people living next to railway lines and sometimes even damage to nearby buildings. The reasons are, among others, the world-wide increasing passenger transport and heavy-axle freight traffic, the increasing density of traffic networks and especially the rapid growing of high-speed lines all over the world.

Hence, reliable and applicable methods are required to calculate dynamic soil-track interaction and to predict vibration level away from the line, and also to design tracks and constructional measures to reduce vibration level. Various approaches have been developed which have to be classified as either numerical methods, analytical methods or (semi-) empirical methods.

Semi-empirical methods are usually based on a large number of measurements to come to a statistical prediction model for railway induced vibrations (Madshus et al. 1996). Further experimental data especially for high-speed train induced vibrations have been presented for example by Auersch (1988), Degrande & Lombaert (2000), Kaynia et al. (2000), Madshus & Kaynia (2000), Degrande & Schillemans (2001) and Ditzel et al. (2001).

Analytical or semi-analytical methods are usually applied to get physical insight into specific phenomena by avoiding complicating factors and accounting for more or less complex (or simple) excitation mechanisms, track models and soil representations. Typical aspects that have been investigated are the influence of soil stratification and train speed on traffic-induced free field vibrations. Representative works are those of Auersch (1988), Krylov (1994, 1995), Dietermann & Metrikine (1996, 1997), Jones et al. (1996, 2000), Knothe & Wu (1998), Grundmann et al. (1999), Metrikine & Popp (2000) and Lombaert et al. (2001).

Numerical methods allow a close-to-reality modelling of the track system usually by applying the Finite Element Method (FEM). If finite elements are employed to model the "complete" system, i.e. the track and also a soil island near the track, this is called *direct method*. However, an important consideration is the simulation of the infinite medium around the island and the fulfilment of the radiation condition, i.e. avoiding reflections from artificial model boundaries. This can be done approximately by incorporating transmitting boundaries (for a brief comparative review see e.g. Kausel 1988), or by introducing an artificial material damping that is later removed (Song & Wolf 1994).

If, on the other hand, the Boundary Element Method (BEM) is applied for the soil, this leads to a combined FEM-BEM approach (Adam et al. 2000, Mohammadi & Karabalis 1995). Since different analysis techniques are applied for different parts of the entire system (substructures) this approach is based on the *substructure method*.

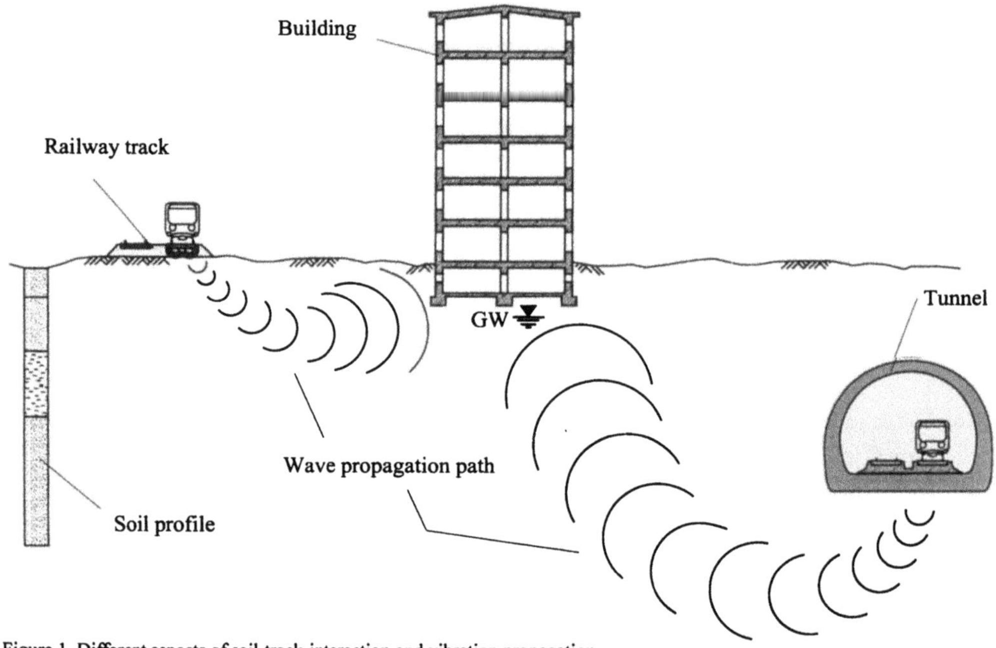

Figure 1. Different aspects of soil-track-interaction and vibration propagation.

The present paper gives an overview of the authors' activities concerning a three-dimensional numerical model to analyse soil-structure interaction of railway tracks on layered soil under dynamic load (moving loads). The approach can also be used to calculate vibration propagation in the soil and its effect on nearby buildings (Fig. 1). However, the model is by no means restricted to railway structures, but can be applied to any other dynamic SSI-problem.

2 NUMERICAL APPROACH – CONCEPT

The numerical model is based on the application of the substructure method. By using this method a complex system, for example a more or less big portion of the system shown in Figure 1, is divided into several parts. One part is kept as a kind of "main structure" while the substructures are cut off. All substructures can be analysed independently with respect to the degrees of freedom (DOF) at the common interface with the main structure. By satisfying equilibrium and compatibility conditions at all interfaces the whole structure is reassembled and a solution for the complete system is obtained (Fig. 2).

Particularly for substructures with different (physical) properties specific techniques can be applied and combined to retain the advantages of one method and to reduce or even eliminate the disadvantages of the respective others. Another advantage is the reduction of memory requirements since small systems are analysed. However, to obtain the displacements and stresses at arbitrary points of a substructure, a post-solution step must be performed with the displacements and stresses at the interface of the substructure as input.

In particular finite elements are applied for the track (section 3), which may exhibit nonlinear and inhomogeneous properties. Since finite elements are well established in practice, very powerful FEM packages are commercially available. They provide the full range of mechanical features like nonlinear material properties, large strains, and multi-physics, as well as the full range of numerical algorithms for nonlinearities, dynamic analysis in frequency and time domain, sparse solvers and iterative solvers,

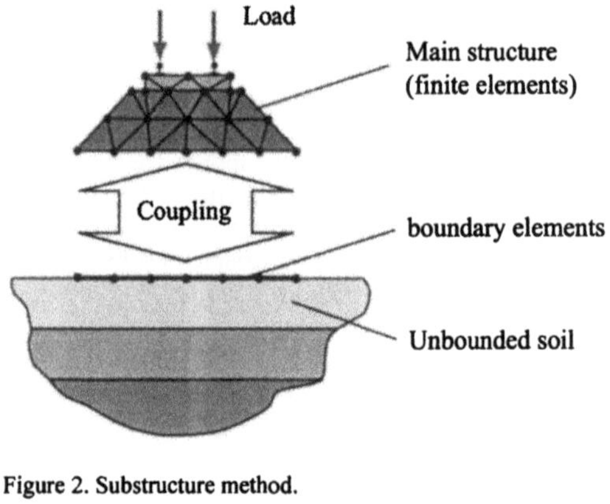

Figure 2. Substructure method.

and also substructuring. Additionally they provide a software interface for user-supplied routines in nearly any stage of the solution process.

To analyse the unbounded soil the Boundary Element Method is used to fully satisfy the radiation condition towards infinity. In this work boundary element routines have been developed based on half-space fundamental solutions (section 4). Hence, the spatial discretisation can be confined to the soil-structure interface, i.e. no auxiliary grid beyond the interface as for conventional boundary element formulations is required. Finally it remains the task to couple both the boundary elements and the finite elements (section 5).

3 FINITE ELEMENT ANALYSIS OF TRACK

The application of the Finite Element Method for modelling a finite structure – in this case the track – leads to a discrete equation of motion which is given by

$$\boldsymbol{M}\cdot\ddot{\boldsymbol{u}}(t)+\boldsymbol{C}\cdot\dot{\boldsymbol{u}}(t)+\boldsymbol{K}\cdot\boldsymbol{u}(t)=\boldsymbol{P}(t)-\boldsymbol{Q}(t). \qquad (1)$$

The matrices $\boldsymbol{M},\boldsymbol{C}$ and $\boldsymbol{K}$ are the mass matrix, the damping matrix, and the stiffness matrix of the structure, respectively, the vector $\boldsymbol{u}$ denotes the nodal displacements, $\boldsymbol{P}$ and $\boldsymbol{Q}$ are the external forces and the resultant nodal forces of the contact stresses at the interface of track and soil (interaction forces), respectively.

The time-discretised equation of motion (1) expressed in terms of the nodal displacements $\boldsymbol{u}^{i+1}$ at time t^{i+1} will be of the form

$$\boldsymbol{M}\cdot\ddot{\boldsymbol{u}}^{i+1}+\boldsymbol{C}\cdot\dot{\boldsymbol{u}}^{i+1}+\boldsymbol{K}\cdot\boldsymbol{u}^{i+1}=\boldsymbol{P}^{i+1}-\boldsymbol{Q}^{i+1}\left(\boldsymbol{u}^{i+1}\right). \qquad (2)$$

It is worth emphasizing that the interaction forces $\boldsymbol{Q}^{i+1}$ depend on the unknown displacements $\boldsymbol{u}^{i+1}$.

In the following the DOFs of Equation (2) are sorted by DOFs at the interface (subscript I) and the remaining DOFs outside the interface (subscript R). This leads to

$$\begin{bmatrix}\boldsymbol{M}_{RR} & \boldsymbol{M}_{RI}\\ \boldsymbol{M}_{IR} & \boldsymbol{M}_{II}\end{bmatrix}\cdot\begin{Bmatrix}\ddot{\boldsymbol{u}}_R\\ \ddot{\boldsymbol{u}}_I\end{Bmatrix}^{i+1}+\begin{bmatrix}\boldsymbol{C}_{RR} & \boldsymbol{C}_{RI}\\ \boldsymbol{C}_{IR} & \boldsymbol{C}_{II}\end{bmatrix}\cdot\begin{Bmatrix}\dot{\boldsymbol{u}}_R\\ \dot{\boldsymbol{u}}_I\end{Bmatrix}^{i+1}+ \qquad (3)$$
$$\begin{bmatrix}\boldsymbol{K}_{RR} & \boldsymbol{K}_{RI}\\ \boldsymbol{K}_{IR} & \boldsymbol{K}_{II}\end{bmatrix}\cdot\begin{Bmatrix}\boldsymbol{u}_R\\ \boldsymbol{u}_I\end{Bmatrix}^{i+1}=\begin{Bmatrix}\boldsymbol{P}_R\\ \boldsymbol{P}_I-\boldsymbol{Q}_I\end{Bmatrix}^{i+1}.$$

For simplicity the symbols $\boldsymbol{Q}$ and $\boldsymbol{Q}_I$ are used interchangeable, because by definition the interaction forces only exist at the interface. The calculation of the matrices $\boldsymbol{M},\boldsymbol{C}$ and $\boldsymbol{K}$ in (1) and (2) is subject to standard finite element procedures and is therefore skipped in the following. The remaining task is the computation of $\boldsymbol{Q}$, which is described in the following section.

4 BOUNDARY ELEMENT ANALYSIS OF SOIL

Standard boundary element schemes yield to an integral equation (Manolis & Beskos 1988)

$$c_{ij}(\xi)w_j(\xi,t)+\int_\Gamma T_{ij}(x,t;\xi)*w_j(x,t)d\Gamma= \int_\Gamma G_{ij}(x,t;\xi)*q_j(x,t)d\Gamma \qquad (4)$$

where w_j and q_j denote the unknown displacements and stresses of the soil, G_{ij} and T_{ij} are the fundamental solutions for the displacements and stresses at point $\boldsymbol{x}$ in the j-direction due to a concentrated pulse at point $\boldsymbol{\xi}$ in the i-direction, Γ denotes the boundary, c_{ij} is the discontinuity jump term which depends on the location of point $\boldsymbol{\xi}$, the operation $*$ indicates the time convolution, and t is the time variable. Body forces and initial conditions are assumed to be zero.

Usually fundamental solutions for a fullspace are used. Since they do not satisfy the boundary conditions at the free surface, they require – strictly speaking – placement of an *infinite* number of boundary elements on the surface. Therefore one must carefully choose a distance from the interface where to truncate the discretisation such that the influence of the violated boundary conditions of the fundamental solutions becomes negligible. However, this auxiliary grid around the track model often entails a substantial additional numerical effort.

These obstacles can be avoided by choosing fundamental solutions which satisfy exactly the free traction boundary condition at the surface of a half-space (Green's functions). Thus, it is sufficient to discretise the soil solely at the soil-structure interface.

Using the halfspace Green's functions, the integral on the left hand side of (4) vanishes completely at all non-singular points. Furthermore, in this case it can be proven that the left hand side is equal to $\delta_{ij}w_j(\boldsymbol{\xi})=w_i(\boldsymbol{\xi})$ leading to

$$w_i(\xi,t)=\int_\Gamma G_{ij}(x,t;\xi)*q_j(x,t)d\Gamma. \qquad (5)$$

In the foregoing equations $G_{ij}(\boldsymbol{x},t;\boldsymbol{\xi})$ are the time domain Green's functions for the halfspace displacements due to a concentrated force with a Dirac delta time-dependency. Usually they are derived from the Green's functions $H_{ij}(\boldsymbol{x},t;\boldsymbol{\xi})$ for the displacements due to a concentrated force with a Heaviside time-dependency by means of a numerical differentation with respect to time

$$G_{ij}(x,t;\xi)=\frac{\partial}{\partial t}H_{ij}(x,t;\xi). \qquad (6)$$

However, the numerical accuracy and stability of this approach is often unsatisfactory (Johnson 1974, Triantafyllidis 1989). But, using (6), the convolution integral (5) can also be represented in the form

$$w_i(\xi,t)=\int_{\Gamma} H_{ij}(\boldsymbol{x},t;\boldsymbol{\xi})*\frac{\partial}{\partial t}q_j(\boldsymbol{x},t)d\Gamma \qquad (7)$$

where the Green's functions $H_{ij}(\boldsymbol{x},t;\boldsymbol{\xi})$ are required in their original form while the time derivative now applies on the stresses $q_j(\boldsymbol{x},t)$. An approach which is based on (7) is used e.g. by Bode (2000).

The main disadvantage of this approach is, whatever Greens's functions are used, H_{ij} or G_{ij}, they are available only for an homogeneous halfspace. But, in most cases the soil must be considered at least as a layered halfspace.

However, Green's functions which also consider an horizontal stratification (layered halfspace Green's functions) are by now restricted to the frequency domain. They can be calculated by means of the Thin Layer Method (Kausel 1986, 1994), a powerful semi-analytical solution, which includes material damping and all effects due to the layering like reflection and refraction at layer boundaries, dispersion, and geometric damping.

Hence, a corresponding approach starts from the boundary integral equation in frequency domain similar to (5) but without convolution

$$w_i(\xi,\omega)=\int_{\Gamma} G_{ij}(\boldsymbol{x},\omega;\boldsymbol{\xi})q_j(\boldsymbol{x},\omega)d\Gamma \qquad (8)$$

and with $G_{ij}(\boldsymbol{x},\omega;\boldsymbol{\xi})$ the corresponding frequency domain Green's functions, i.e. the halfspace displacements due to a concentrated harmonic force with circular frequency ω.

After spatially discretizing with boundary elements and applying (8) to each nodal point $\boldsymbol{\xi}$ – the so-called interaction points – a linear equation system is obtained

$$\boldsymbol{w}(\omega)=\boldsymbol{F}(\omega)\cdot\boldsymbol{q}(\omega) \qquad (9)$$

with $\boldsymbol{F}(\omega)$ the so-called flexibility matrix of soil, and the vectors $\boldsymbol{w}$ and $\boldsymbol{q}$ expressing the soil displacements and contact stresses, respectively.

It is worth mentioning that the matrix $\boldsymbol{F}$ in (9) is generally non-quadratic and non-symmetric. But, for the coupling with the finite element structure the flexibility matrix must be invertible, thus it must be quadratic. This can be accomplished by choosing identical shape functions for displacements and stresses. However, for the coupling the shape functions for displacements and stresses of finite elements should match the respective ones of boundary elements, but in commercial finite element software often elements with same order shape functions for displacements and stresses are not available.

Hence, a compromise must be accepted, for example by using finite elements with linear displacements and constant stresses. By choosing boundary elements with constant stresses and displacements, the equilibrium is satisfied, but the compatibility can be enforced only at the midpoints (interaction points, Fig. 3).

The transition from the frequency domain formulation (9) to a time domain formulation can be accomplished by means of a simple Inverse Fourier Transform (IFT)

$$\boldsymbol{w}(t)=\frac{1}{2\pi}\int_{-\infty}^{\infty}\boldsymbol{F}(\omega)\cdot\boldsymbol{q}(\omega)e^{i\omega t}d\omega \qquad (10)$$

or equivalently by means of a convolution integral

$$\boldsymbol{w}(t)=\int_{0}^{t}\boldsymbol{F}(t-\tau)\cdot\boldsymbol{q}(\tau)d\tau \qquad (11)$$

with $\boldsymbol{F}(t)$ the soil flexibility given as the (Dirac) impulse response matrix. $\boldsymbol{F}(t)$ can be derived from its counterpart in the frequency domain $\boldsymbol{F}(\omega)$, as introduced in Equation (9), again by applying an Inverse Fourier Transform

$$\boldsymbol{F}(t)=\frac{1}{2\pi}\int_{-\infty}^{\infty}\boldsymbol{F}(\omega)e^{i\omega t}d\omega . \qquad (12)$$

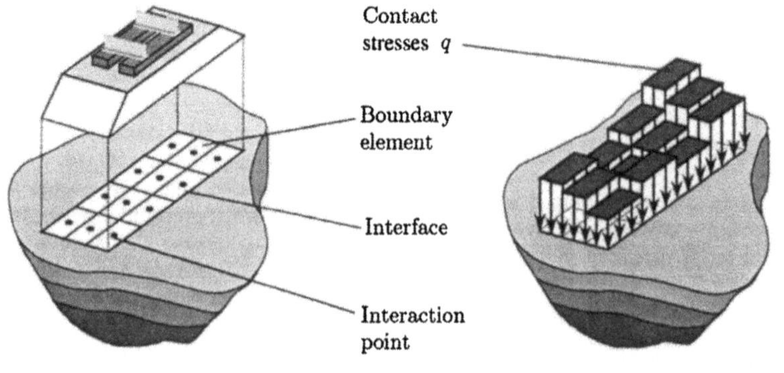

Figure 3. Discretisation of the interface between track and subsoil assuming constant contact stresses within each element.

However, the numerical evaluation of (12) causes some trouble due to the high frequency contents of the flexibility matrix $F(\omega)$. This can be circumvented by using a modified impulse response matrix

$$F(t)=\frac{1}{2\pi}\int_{-\infty}^{\infty}F(\omega)g(\omega)e^{i\omega t}d\omega \qquad (13)$$

with $g(\omega)$ a Gaussian distribution (bell shape) as suggested by Wolf (1988)

$$g(\omega)=\exp\left(-\sigma^2\omega^2\right) \qquad (14)$$

which is the Fourier transform of

$$g(t)=\frac{1}{2\sigma\sqrt{\pi}}\exp\left(-\frac{t^2}{4\sigma^2}\right). \qquad (15)$$

Due to the exponential decay of $g(\omega)$, Equation (13) is much more suited for the application of standard IFFT algorithms. Notice that if the parameter σ tends to zero, the bell shape impulse $g(t)$ converges to the Dirac impulse $\delta(t)$.

Using the modified impulse response to evaluate the convolution (11) in time discretised form leads to an approximation of the time-continuous contact stresses by a sequence of bell shaped impulses as illustrated in Fig. 4. The discretisation of (11) yields

$$w^{i+1}=\underbrace{\Delta t\,F^{i}\cdot q^{1}+\ldots+}_{w^{hist}}\underbrace{\Delta t\,F^{0}}_{F^{cur}}\cdot q^{i+1}$$

$$=w^{hist}+F^{cur}\cdot q^{i+1} \qquad (16)$$

or in reversed order

$$q^{i+1}=\left[F^{cur}\right]^{-1}\cdot\left\{w^{i+1}-w^{hist}\right\}. \qquad (17)$$

Here w^{i+1} and q^{i+1} are the displacement vector and contact stress vector, respectively, at time step $i+1$, F^{cur} is the current soil flexibility matrix, and w^{hist} is the displacement vector due to contact stress history. For efficiency, it is convenient to choose time steps of equal length Δt. In this case F^{cur} does not change during calculation.

The time domain procedure described above is called a 'hybrid domain approach', since frequency domain Green's functions have been used. Beyond that, approaches starting from the integral equations (5) or (7) using time domain Green's functions are called 'pure time domain approaches' (Bode 2000, Bode et al. 2002). Although they are substantially more efficient, they are by now limited to a homogeneous, linear elastic and isotropic halfspace.

5 FEM – BEM COUPLING

The coupling of boundary elements and finite elements requires that both equilibrium and compatibility conditions must be satisfied at the interface between track and subsoil.

As mentioned above the compatibility of displacements is satisfied only at the interaction points. This relationship is given as

$$w^{i+1}=T_u\cdot u^{i+1} \qquad (18)$$

with T_u a transformation matrix. Satisfying equilibrium, the contact stresses q^{i+1} as obtained from Equation (17) are transformed into nodal forces by a simple linear transformation

$$Q^{i+1}=T_q\cdot q^{i+1} \qquad (19)$$

with T_q the corresponding transformation matrix. Combining (17), (18) and (19) leads to

$$Q^{i+1}=\underbrace{T_q\cdot\left[F^{cur}\right]^{-1}\cdot T_u}_{K^{cur}_{Soil}}\cdot u^{i+1}-\underbrace{T_q\cdot\left[F^{cur}\right]^{-1}\cdot w^{hist}}_{Q^{hist}}$$

$$=K^{cur}_{Soil}\cdot u^{i+1}-Q^{hist} \qquad (20)$$

with K^{cur}_{Soil} the current stiffness matrix of the soil. With Equation (20) the equation of motion of the structure (3) will be of the final form

$$\begin{bmatrix}M_{RR} & M_{RI}\\ M_{IR} & M_{II}\end{bmatrix}\cdot\begin{Bmatrix}\ddot{u}_R\\ \ddot{u}_I\end{Bmatrix}^{i+1}+\begin{bmatrix}C_{RR} & C_{RI}\\ C_{IR} & C_{II}\end{bmatrix}\cdot\begin{Bmatrix}\dot{u}_R\\ \dot{u}_I\end{Bmatrix}^{i+1}+$$
$$\begin{bmatrix}K_{RR} & K_{RI}\\ K_{IR} & K_{II}+K^{cur}_{Soil}\end{bmatrix}\cdot\begin{Bmatrix}u_R\\ u_I\end{Bmatrix}^{i+1}=\begin{Bmatrix}P_R\\ P_I+Q^{hist}\end{Bmatrix}^{i+1}. \qquad (21)$$

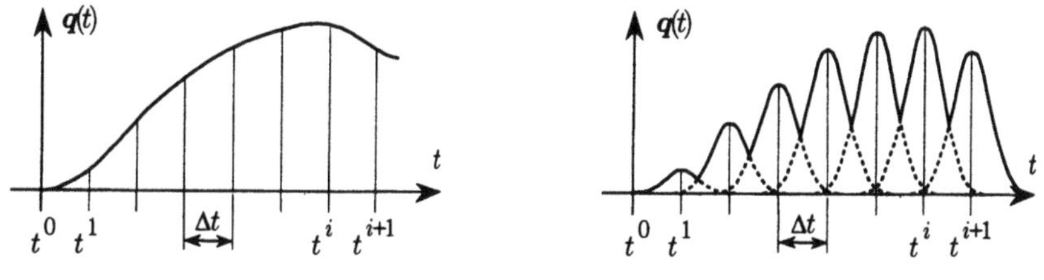

Figure 4. Approximation of the time-continuous contact stresses by a sequence of bell shape impulses.

For any initial value problem, the equation of motion (21) can be solved by a time step integration scheme, such as Newmark's β method (Bathe & Wilson 1976). Finally, after the interaction problem has been solved, it is also possible to calculate the soil displacements at any point in the subsoil.

5.1 *Implementation in ANSYS*

In order to provide the full power of a commercial Finite Element Software for the calculation of the track, the boundary element routines for the soil have been implemented in ANSYS. This section provides a short summary of the methodology of the coupling procedure. Essentially the so-called UPF – User Programmable Features – interface and the substructuring features are used. The procedure is divided into the following steps:

1. The finite element model of the track is built using the ANSYS preprocessor.
2. The soil is attached as a so-called "superelement" to the interface nodes (MDOFs = Master Degrees of Freedom) by using the ANSYS element type MATRIX50.
3. The boundary element routines are activated to compute the transformation matrices $\boldsymbol{T}_u$ and $\boldsymbol{T}_q$, the flexibility matrix of soil $\boldsymbol{F}^{cur}$, and finally, the stiffness matrix of soil $\boldsymbol{K}^{cur}_{Soil}$.
4. Once $\boldsymbol{K}^{cur}_{Soil}$ has been calculated, it is plugged into the stiffness matrix of the ANSYS model by replacing the stiffness matrix of the previously defined superelement.
5. Next, the load vector $\boldsymbol{Q}^{hist}$ is computed by corresponding routines. Since $\boldsymbol{Q}^{hist}$ is time-dependent, it is computed and incorporated into the ANSYS model before each time step.
6. Finally, the soil-structure interaction is solved by means of the Newmark time integration method used by ANSYS. The default Newmark parameters $\beta = 1/4$ and $\gamma = 1/2$ are used, expressing the constant average acceleration scheme.

A detailed description of the coupling procedure is given by Hirschauer (2001).

5.2 *Partial Uplift*

The coupling described above implies that the track is "welded" with the subsoil, i.e. no uplift is allowed to occur at the interaction points (full contact). However, this does not necessarily hold. Therefore, to include a non-cohesive tensionless connection between track and soil, an iteration scheme has been implemented that can take into account a potential partial uplift. In this nonlinear case, the stiffness matrix $\boldsymbol{K}^{cur}_{Soil}$ of the superelement must be updated and replaced before each time or iteration step. Details can be found in Savidis et al. (1999a, b, 2000).

6 SIMULATION OF MOVING LOADS

6.1 *Simplified track model*

The first example deals with the dynamic response of a simplified railway track of 30 m length subjected to a moving wheel set. The track model shown in Figure 5 consists of 2 rails according to type UIC60, each of which being modelled by means of 600 Timoshenko-beam elements (ANSYS BEAM4), and 50 concrete railroad ties similar to German type B70, each of which being modelled by means of 20 solid elements (ANSYS SOLID45). The ties are directly attached to a homogeneous soft soil without any base or ballast in-between. The track parameters are summarised in Figure 5.

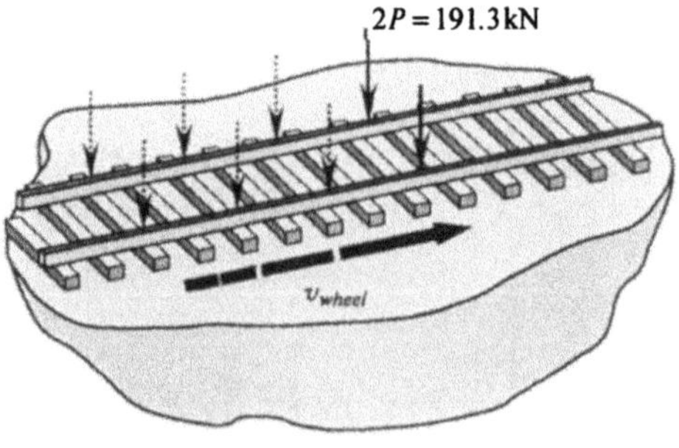

Soil (homogeneous):

Shear modulus:	20×10^6 N/m^2
Shear wave velocity:	100 m/s
Poisson's ratio:	1/3

Rails (according to type UIC60):

Young's modulus:	2.1×10^{11} N/m^2
Moment of inertia:	3.055×10^{-5} m^4
Cross section:	7.686×10^{-3} m^2
Mass density:	7850 kg/m^3
Poisson's ratio:	0.3

Ties (similar to German type B70):

Young's modulus:	3.0×10^{10} N/m^2
Length:	2.6 m
Width:	0.26 m
Height:	0.20 m
Distance:	0.6 m
Total Mass:	290 kg
Poisson's ratio:	0.16

Figure 5. Simplified railway track.

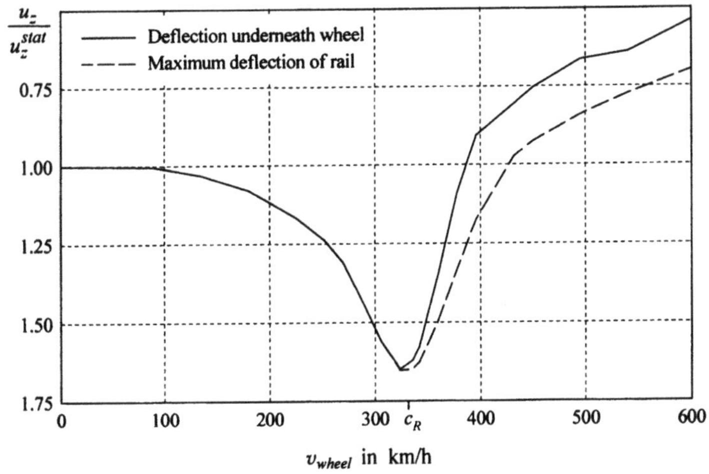

Figure 6. Vertical rail deflection u_z as a function of wheel speed v_{wheel} (scaled with the static deflection u_z^{stat}).

The wheel set is modelled by two downwards directed constant loads, travelling at a constant speed v_{wheel} along the track. The total axle load is assumed to be $2P = 191.3\,\text{kN}$.

To calculate the interaction with the underlying soil, the interface underneath each tie is discretised into 4×20 boundary elements, for a total of 4000 elements. Only vertical dynamics of the track are considered, and it is assumed that the interaction forces only act in vertical direction. The time step is selected as $\Delta t = 1.0 \times 10^{-3}\,\text{s}$.

Figure 6 displays as solid line the vertical rail deflection underneath the wheel as a function of wheel speed for the moment when the wheel passes the 35th tie. The result is scaled with the static deflection. As can be seen, the largest deformations occur when the wheel speed coincides with the Rayleigh wave velocity, which is in the present case $c_R = 335\,\text{km/h}$. For load speeds $v_{wheel} > 390\,\text{km/h}$ even deflections smaller than the static deflection emerge (solid line).

However, the rail deflection underneath the wheel does not necessarily coincide with the maximum deflection of the rail (see dashed line in Fig. 6). Especially when the wheel speed exceeds the Rayleigh wave velocity (supersonic speed), the maximum rail deflection appears behind the wheel and is always larger than underneath the wheel.

This is demonstrated in Figure 7 by comparing the deflection line of the rail for two load speeds:

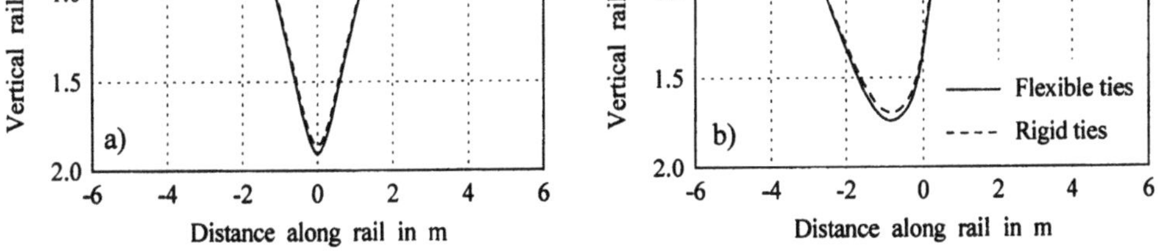

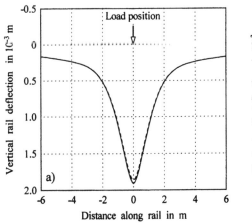

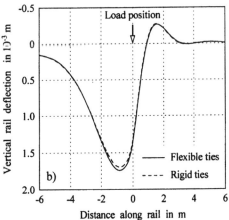

Figure 7. Deflection line of the rail due to loads travelling at constant speeds along a simplified railway track. (a) Subsonic speed $v_{wheel} = 180$ km/h $< c_R = 335$ km/h and (b) supersonic speed $v_{wheel} = 450$ km/h $> c_R = 335$ km/h.

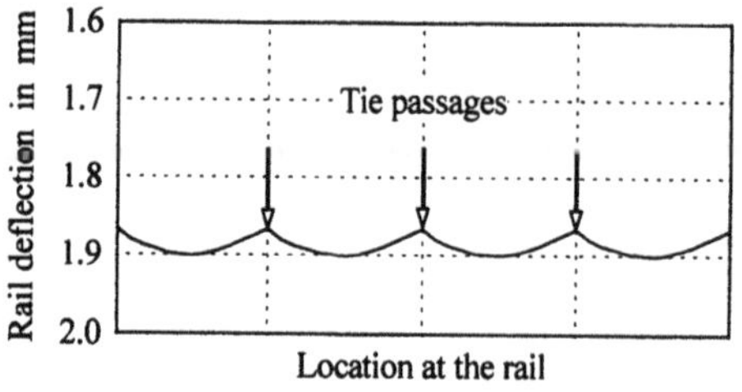

Figure 8. Rail deflection underneath the wheel while passing the track with $v_{wheel} = 180$ km/h.

first $v_{wheel} = 180$ km/h, which is about half of the Rayleigh wave velocity c_R (subsonic case), and second $v_{wheel} = 450$ km/h, which is 34 % larger than c_R (supersonic case). In the situation shown, the load is passing the 20th tie from left to right. A distance ahead of the load is shown as positive.

As can be seen, the shape of the deflection line obtained for the subsonic case (Fig. 7a) is (almost) symmetric with respect to the load position and reminds to that of a stationary load, i.e. $v_{wheel} = 0$. By contrast, in the supersonic case (Fig. 7b) the deflection line ceases to be symmetric. On the one hand the maximum deflection appears far behind the load, and on the other hand an uplift ahead of the load has emerged. In the present case, this is related to the fact that the load speed exceeds the Rayleigh wave velocity, but in general, that depends also on the various stiffnesses (rails, ties, soil).

Furthermore, the deflection line does not change significantly neither if the load acts between two ties rather than above a tie (Triantafyllidis & Prange 1994), nor if the ties are assumed to be rigid (see dashed lines in Figure 7). However, that depends also on the stiffness of the soil. For a much stiffer soil with a shear modulus $G = 95.7 \times 10^6$ N/m² and a shear wave velocity $c_S = 248.37$ m/s, as investigated by Bode et al. (2002), the maximum deflection is about 10 % larger in the case of flexible ties compared to that of rigid ties.

Nevertheless, the periodically varying stiffness along the track (Fig. 8) leads to a parametric excitation of the running train which has to be considered for the interaction of track and vehicle. In the present case the vertical rail deflection underneath the wheel changes periodically from 1.87 mm when the load acts above a tie to 1.90 mm when it acts between two consecutive ties.

A very important consideration in soil-structure interaction analyses is the so-called 'through-the-soil-coupling'. All waves emanated from any tie and propagating through the soil in either direction have, in general, an effect on all its neighbours. Especially in the case of large scaled problems, this aspect often entails a substantial numerical effort. However, due to the geometrical damping and, if present, the material damping, this effect decreases with increasing distance. Therefore, sometimes a reduced through-the-soil-coupling is considered, i.e. the influence of the wave propagation is truncated at an artificial distance. But, this has to be done carefully, since an incomplete 'through-the-soil-coupling' would not be adequate in any case to produce correct results, see Bode et al. 2002.

The effect of various coupling schemes on the rail deflection is demonstrated in the Figures 9 and 10 for both the subsonic and the supersonic case. As far as the subsonic case is concerned (Fig. 9), the reduced coupling leads to a loss of symmetry which

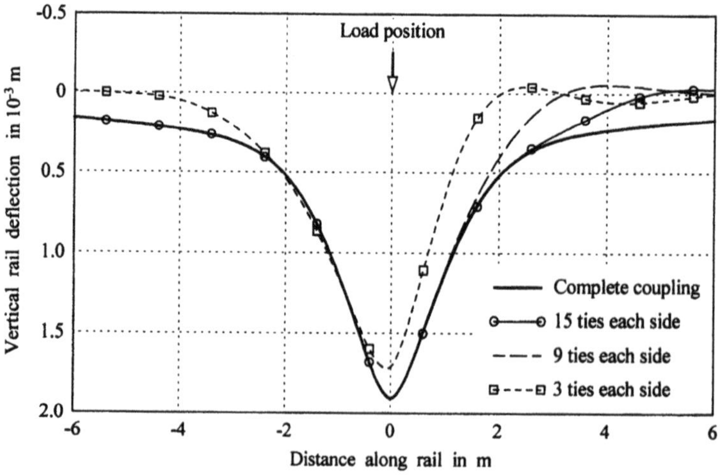

Figure 9. Effect of the coupling scheme on the deflection line of the rail due to a load travelling at the subsonic speed $v_{wheel} = 180$ km/h $< c_R = 335$ km/h along the track.

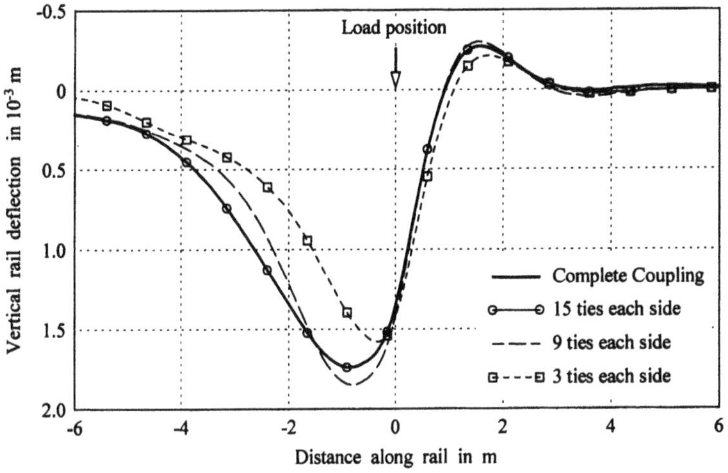

Figure 10. Effect of the coupling scheme on the deflection line due to a load travelling at the supersonic speed $v_{wheel} = 450$ km/h $> c_R = 335$ km/h along the track.

increases as the degree of coupling decreases. In the same way the incorrect uplift ahead of the load shifts towards the contact point. However, except for the most reduced coupling scheme, i.e. 3 ties each side, an excellent agreement of the maximum deflection can be observed. The results for the supersonic case in Figure 10 reveals almost the same: the deviation from the "true" deflection line increases as the degree of coupling decreases. But, to reproduce the "true" maximum a larger degree of coupling is required. In general, the shape of the deflection line depends on both, the ratio of track stiffness (rails and ties) to soil stiffness and the ratio of load speed to Rayleigh wave speed (Auersch 1988). Actually, the stiffness ratio is changed when reducing the tie-soil-tie coupling.

Finally, Figure 11 shows a snapshot of the displacements of track and surrounding soil for a wheel

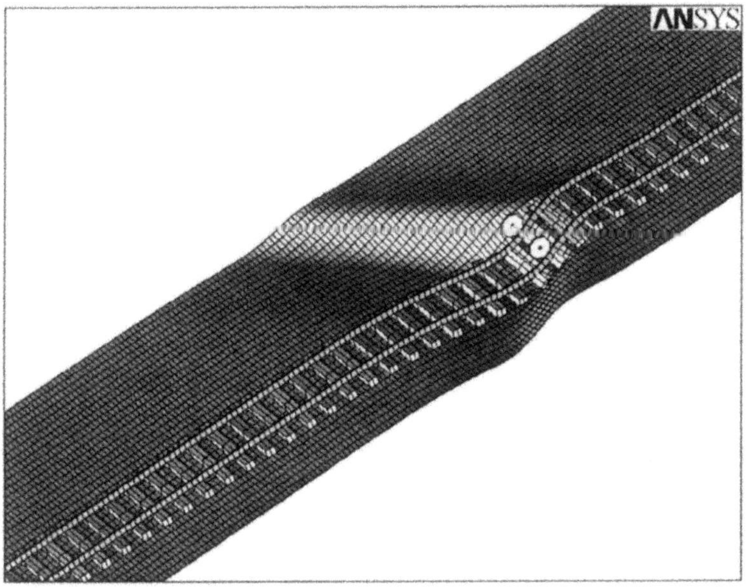

Figure 11. Snapshot of the displacement field of the railway track and the free field (exaggerated) caused by a load travelling at a supersonic speed $v_{wheel} = 400$ km/h $> c_R = 335$ km/h .

set moving at supersonic speed $v_{wheel} = 400\,\text{km/h}$. As expected, the wave pattern generated at the soil surface results in a so-called 'Mach cone', where the slope of its wedge shaped wave front is determined by the ratio of load speed to Rayleigh wave velocity. The Mach cone also appears in the soil below the track (see Bode et al. 2000). Furthermore, although the flexibility of the ties has been taken into account, the ties seem to behave almost rigid. As far as the rails are concerned, the uplift ahead of the load can be seen.

6.2 *Ballast track model*

Although a simplified track model is generally useful to get insight into basic phenomena, in the following a more realistic and common railway track is investigated, namely a ballast supported track of 30 m length (50 ties). The ballast layer has a thickness $h = 0.35$ m and is modelled by means of linear elastic 3-D solids (SOLID45) with a Young's modulus $E = 3.8\times10^8\,\text{N/m}^2$, a mass density $\rho = 1700\,\text{kg/m}^3$, and a shear wave velocity $c_S = 300\,\text{m/s}$. The parameters of rails and ties are the same as before (see Fig. 5). In addition, pads between rails and ties are included, which are modelled as linear spring-damper elements (LINK11) with a stiffness coefficient $k = 6.0\times10^8\,\text{N/m}$ and a damping coefficient $c = 2.0\times10^4\,\text{Ns/m}$.

The soil conditions are changed from soft to medium, with a shear modulus $G = 68\times10^6\,\text{N/m}^2$, a shear wave velocity $c_S = 200\,\text{m/s}$ and a Poisson's ratio $\nu = 0.25$. The track parameters are taken from a benchmark test performed within a priority programme of the German Research Foundation (see Rücker et al. 2002), hence here only the essential parameters are summarised.

The finite element mesh used for the track is shown in Figure 12, whereas the unbounded soil is not shown, but included as described in the sections above. The complete model consists of 5900 finite elements and 2400 boundary elements. Again, only vertical dynamics of the track and complete coupling are considered, and it is assumed that the interaction forces only act in vertical direction. The time step used for the calculation is $\Delta t = 1.0\times10^{-3}\,\text{s}$.

In Figure 13 the deflection of the rail is compared for three load speeds: first $v_{wheel} = 100\,\text{m/s}$, which is far below the Rayleigh wave velocity of the soil ($c_R = 184\,\text{m/s}$), then $v_{wheel} = 225\,\text{m/s}$ which is above c_R but below the shear wave velocity of the ballast ($c_S = 300\,\text{m/s}$), and finally $v_{wheel} = 350\,\text{m/s}$ which is above both. In the situation shown, the load is passing the 25th tie from left to right.

As can be seen, the situation for the load speed $v_{wheel} = 100\,\text{m/s}$ is quite similar to the subsonic case of the preceding example (Fig. 7a). However, things change when the load speed increases, since the dynamic behaviour of the track is getting more and more affected by the wave propagation in the ballast.

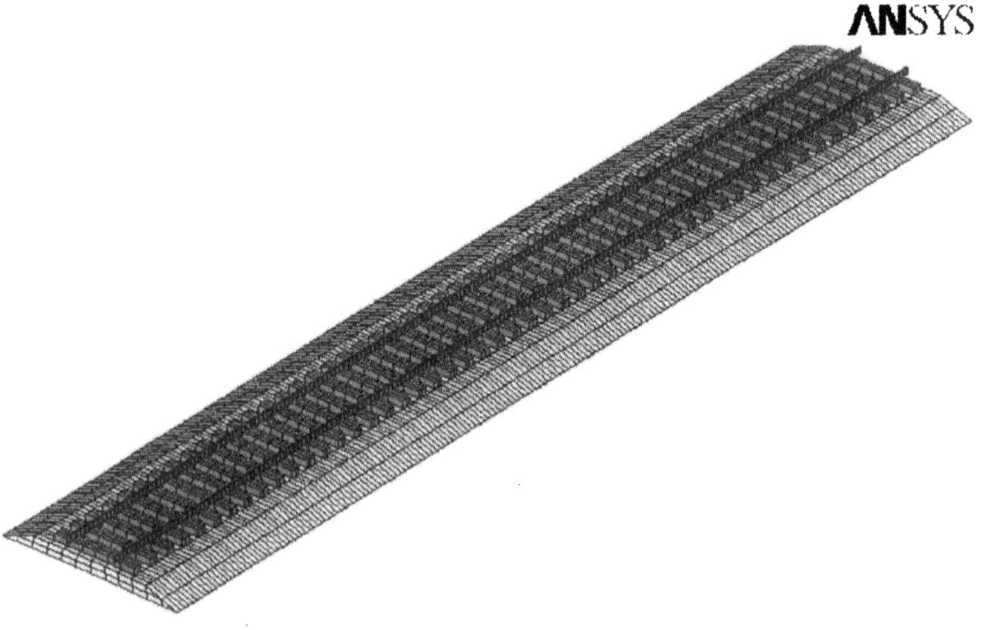

Figure 12. Finite element mesh used for the calculation of the ballast track (50 ties). The unbounded soil is not shown, but included by means of boundary elements as described in the sections above.

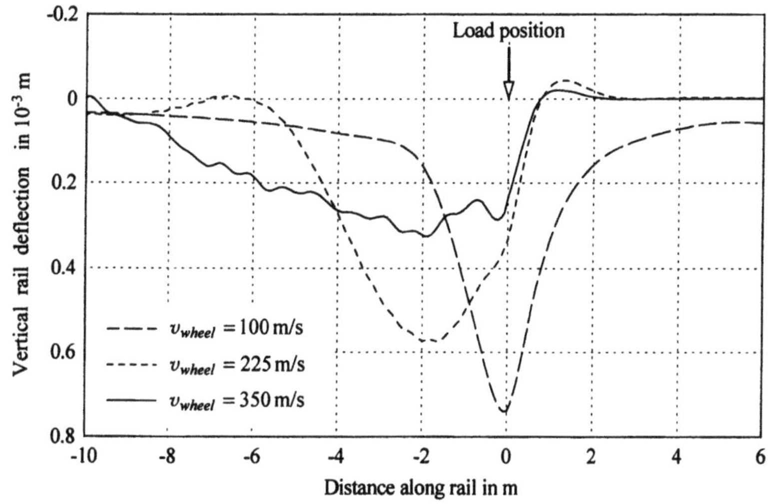

Figure 13. Rail deflection of the ballast supported track subjected to loads travelling at constant speeds.

This can be seen already for the second load speed $v_{wheel} = 225\,\mathrm{m/s}$. Since this speed is still far below the shear wave velocity of the ballast, the influence on the deflection line should be rather small. Nevertheless, in contrast to the supersonic case shown in Figure 7b, first a small change in curvature appears immediately behind the load, while farther behind (at about – 7 m) a local maximum can be observed. If the load speed also exceeds the shear wave velocity of the ballast, i.e. if $v_{wheel} = 350\,\mathrm{m/s}$, the situation is completely different. The deflection behind the load remains almost constant within a large range of about 4 m, while the slight wave-like tendency behind the load obviously results from the wave propagation in the ballast.

It is worth pointing out that the amount of through-the-soil-coupling has a significant effect on the rail deflection even for the ballast supported track under consideration. This is demonstrated in Figure 14 exemplarily for $v_{wheel} = 350\mathrm{m/s}$. As can be seen, the effect is quite similar to that shown in Figure 10. Again, the most reduced coupling scheme, i.e. 3 ties each side, does not approach the "true" deflection line quite well.

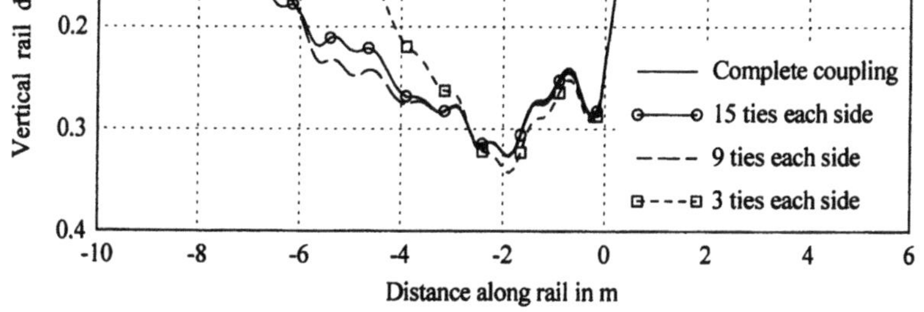

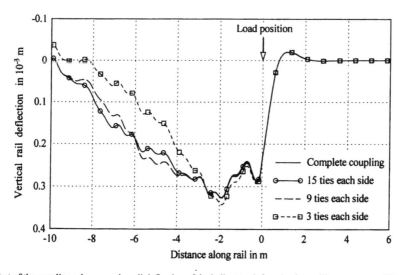

Figure 14. Effect of the coupling scheme on the rail deflection of the ballast track for a load travelling at $v_{wheel} = 350$ m/s.

At the end it is worth pointing out that the strong influence of the load speed on the deflection line shown in Figure 13 is greatly affected by the stiffness of the rail pads. For much softer pads, as they are actually used e.g. in Germany for slab track systems (Savidis et al. 2002, Rücker et al. 2002), the influence of load speed is less strongly pronounced. In those cases, the dynamic behaviour of all components below the pads, including the soil, has a relatively low impact on the deflection line.

7 SUMMARY

A numerical time domain approach for three-dimensional analyses of dynamic soil-structure interaction of railway tracks on layered soil is presented in this paper. The approach is applied to simulate loads travelling at various speeds along different railway tracks.

For a simplified track model without any base or ballast in-between and a ballast supported track model, the vertical rail deflection is calculated as a function of load speed. It turned out, that the largest deformations occur when the load speed coincides with the Rayleigh wave velocity of the soil. But, as the load speed exceeds the Rayleigh wave velocity (supersonic case), the maximum rail deflection appears behind the load and is always larger than at the load position.

Furthermore, it has been demonstrated that the through-the-soil-coupling may have a significant effect on the shape of the deflection line of the rail. An incomplete 'through-the-soil-coupling' would not be adequate in any case to produce correct results.

The approach can also be used in the process of design and refitting of railway tracks. In particular the possibility to analyse the propagation of vibrations excited by moving trains and their reduction by constructional measures is substantial for railway lines in populated areas.

However, the approach is by no means restricted to rail-bound traffic. Since finite elements are used to analyse the structure, also vibrations induced by street-bound traffic and other man-made vibrations can be investigated, as well as vibrations caused by seismic excitation (Savidis et al. 1999b, 2000). In addition, a similar approach has been developed for frequency domain analyses (see Savidis et al. 2002).

REFERENCES

Adam, M., Pflanz, G. & Schmid, G. 2000. Two- and three-dimensional modelling of half-space and train embankment under dynamic loading. *Soil Dynamics and Earthquake Engineering* 19(8): 559-573.

Auersch, L. 1988. Zur Entstehung und Ausbreitung von Schienenverkehrserschütterungen – theoretische Untersuchungen und Messungen am Hochgeschwindigkeitszug Intercity Experimental. *Forschungsbericht 155*. Bundesanstalt für Materialforschung und –prüfung, Berlin.

Bathe, K.J. & Wilson, E.L. 1976. *Numerical Methods in Finite Element Analysis*. Englewood Cliffs, New Jersey: Prentice Hall.

Bode, C. 2000. Numerische Verfahren zur Berechnung von Baugrund-Bauwerk-Interaktionen im Zeitbereich mittels Greenscher Funktionen für den Halbraum. *Veröffentlichungen des Grundbauinstitutes der Technischen Universität Berlin*, Heft 28.

Bode, C., Hirschauer, R. & Savidis, S.A. 2000. Three-dimensional time domain analysis of moving loads on railway tracks on layered soils. In N. Chouw & G. Schmid (eds.), *Wave2000*: 3-12. Rotterdam: Balkema.

Bode, C., Hirschauer, R. & Savidis, S.A. 2002. Soil-structure interaction in the time domain using halfspace Green's functions. *Soil Dynamics and Earthquake Engineering* 22(4): 283-295.

Degrande, G. & Lombaert, G. 2000. High-speed train induced free field vibrations: In situ measurements and numerical modelling. In N. Chouw & G. Schmid (eds.), *Wave2000*: 29-41. Rotterdam: Balkema.

Degrande, G. & Schillemans, L. 2001. Free-field vibrations during the passage of a Thalys high-speed train at variable speed. *Journal of Sound and Vibration* 247(1): 131-144.

Dietermann, H.A. & Metrikine, A. 1996. The equivalent stiffness of a half-space interacting with a beam. Critical velocities of a moving load along the beam. *European Journal of Mechanics, A/Solids* 15(1): 67-90.

Dietermann, H.A. & Metrikine, A. 1997. Steady-state displacements of a beam on an elastic half-space due to a uniformly moving constant load. *European Journal of Mechanics, A/Solids* 16(2): 295-306.

Ditzel, A., Herman, G.C. & Drijkoningen, G.G. 2001. Seismograms of moving trains: Comparison of theory and measurements. *Journal of Sound and Vibration* 248(4): 635-652.

Grundmann, H., Lieb, M. & Trommer, E. 1999. The response of a layered halfspace to traffic loads moving along its surface. *Archive of Applied Mechanics* 69(1): 55-67.

Hirschauer, R. 2001. Kopplung von Finiten Elementen mit Rand-Elementen zur Berechnung der dynamischen Baugrund-Bauwerk-Interaktion. *Veröffentlichungen des Grundbauinstitutes der Technischen Universität Berlin*, Heft 31.

Johnson, L.R. 1974. Green's Functions for Lamb's Problem. *Geophysical Journal of the Royal Astronomical Society* 37: 99-131.

Jones, C.J.C. & Block, J.R. 1996. Prediction of ground vibration from freight trains. *Journal of Sound and Vibration* 193(1): 205-213.

Jones, C.J.C., Sheng, X. & Thompson, D.J. 2000. Ground vibration from dynamic and quasi-static loads moving along a railway track on layered ground. In N. Chouw & G. Schmid (eds.), *Wave2000*: 83-97. Rotterdam: Balkema.

Kaynia, A.M., Madshus, C. & Zackrisson, P. 2000. Ground vibration from high-speed trains: Prediction and Countermeasure. *Journal of Geotechnical and Geoenvironmental Engineering* 126(6): 531-537.

Kausel, E. 1986. Wave propagation in anisotrpoic layered media. *International Journal for Numerical Methods in Engineering* 23: 1567-1578.

Kausel, E. 1988. Local Transmitting Boundaries. *Journal of Engineering Mechanics* 114(6): 1011-1027.

Kausel, E. 1994. Thin-Layer Method. *International Journal for Numerical Methods in Engineering* 37: 927-941.

Knothe, K. & Wu, Y. 1998. Receptance behaviour of railway track and subgrade. *Archive of Applied Mechanics* 68(7/8): 457-470.

Krylov, V.V. 1994. On the theory of railway-induced ground vibrations. *Journal de Physique IV* 4(C5): 769-772.

Krylov, V.V. 1995. Generation of ground vibrations by superfast trains. *Applied Acoustics* 44(2): 149-164.

Lombaert, G., Degrande, G. & Clouteau, D. 2001. The influence of the soil stratification on free field traffic-induced vibrations. *Archive of Applied Mechanics* 71(10): 661-678.

Madshus, C., Bessason, B. & Hårvik, L. 1996. Prediction model for low frequency vibration from high speed railways on soft grounds. *Journal of Sound and Vibration* 193(1): 195-203.

Madshus, C., & Kaynia, A.M. 2000. High-speed railway lines on soft ground: Dynamic behaviour at critical train speed. *Journal of Sound and Vibration* 231(3): 689-701.

Manolis, G.D., Beskos, D.E. 1988. *Boundary Element Methods in Elastodynamics*. London: Unwin Hyman Ltd.

Metrikine, A.V. & Popp, K. 2000. Steady-state vibrations of an elastic beam on a visco-elastic layer under moving load. *Archive of Applied Mechanics* 70(6): 399-408.

Mohammadi, M. & Karabalis, D.L. 1995. Dynamic 3-D soil-railway track interaction by FEM-BEM. *Earthquake Engineering and Structural Dynamics* 24: 1177-1193.

Rücker, W. et al. 2002. A comparative study of results from numerical track-subsoil calculations. In K. Popp & W. Schiehlen (eds.), *System Dynamics and Long-Term Behaviour of Railway Vehicles, Track and Subgrade*. Berlin, Heidelberg, New York: Springer. 2002.

Savidis, S.A., Bode, C., Hirschauer, R. & Hornig, J. 1999a. Dynamic Soil-Structure Interaction with Partial Uplift. In L. Frýba & J. Náprstek (eds.), *Structural Dynamics Eurodyn' 99*: 957-962. Rotterdam: Balkema.

Savidis, S.A., Bode, C. & Hirschauer, R. 1999b. 3-Dimensionale dynamische Baugrund-Bauwerk Wechselwirkung infolge seismischer Erregung unter Berücksichtigung klaffender Fugen. In S.A. Savidis (ed.), *Entwicklungsstand in Forschung und Praxis auf den Gebieten des Erdbebeningenieurwesens, der Boden- und Baudynamik, DGEB-Publikation* Nr. 10: 135-146. Berlin: DGEB.

Savidis, S.A., Bode, C. & Hirschauer, R. 2000. Three-Dimensional Structure-Soil-Structure Interaction under Seismic Excitation with Partial Uplift. In *Proceedings of the 12th World Conference on Earthuake Engineering*, Auckland, New Zealand 2000.

Savidis, S.A., Hirschauer, R., Bode, C. & Schepers, W. 2002. 3D-Simulation of Dynamic Interaction Between Track and Layered Subground. In K. Popp & W. Schiehlen (eds.), *System Dynamics and Long-Term Behaviour of Railway Vehicles, Track and Subgrade*. Berlin, Heidelberg, New York: Springer. 2002.

Song, C. & Wolf, J.P. 1994. Dynamic Stiffness of Unbounded Medium Based on Damping-Solvent Extraction. *Earthquake Engineering and Structural Dynamics* 23: 169-181.

Triantafyllidis, T. 1989. Halbraumlösungen zur Behandlung bodendynamischer Probleme mit der Randelementmethode. *Veröffentlichungen des Institutes für Bodenmechanik und Felsmechanik der Universität Karlsruhe*, Heft 116.

Triantafyllidis, T. & Prange, W. 1994. Mitgeführte Biegelinie beim Hochgeschwindigkeitszug "ICE" – Teil I: Theoretische Grundlagen. *Archive of Applied Mechanics* 64: 154-168.

Wolf, J.P. 1988. *Soil-Structure-Interaction Analysis in Time Domain*. Englewood Cliffs, New Jersey: Prentice Hall.

Wave propagation – Moving load – Vibration reduction, Chouw & Schmid (eds.)
© 2003 Swets & Zeitlinger, Lisse, ISBN 90 5809 559 2

Numerical simulation of blast wave propagation in soil mass

Zhonqi Wang
School of Civil and Environmental Engineering, Nanyang Technological University, Singapore

Hong Hao
Department of Civil and Resource Engineering, the University of Western Australia, Australia

Yong Lu
School of Civil and Environmental Engineering, Nanyang Technological University, Singapore

ABSTRACT: In this paper, a three-phase soil model under shock loading is developed to simulate blast wave propagation in the soil mass. The soil is modeled as a three-phase mass that includes the solid particles, water and air. The solid particles form a skeleton and their void are filled with water and air. The deformation of solid particles and strain rate effect are also considered. The model is implemented into the commercial hydrodynamic software AUTODYN as its user subroutines. The numerical results of the propagation of the blast wave in two kinds of soil mass are presented, one is in saturated soil, and the other is in unsaturated soil. The results appear to be satisfactory and some characteristics of the blast wave propagation are observed.

1 INTRODUCTION

The blast wave propagation in soil mass is a very important problem to engineers in mining, construction and defense engineering. Because of the complexity of the soil mass, it is difficult to predict accurately the blast wave propagation in soil. Current practice in modeling the blast wave propagation in soil is mainly based on some empirical formulae obtained from the field blast tests. Because the properties of the soil mass in each site are significant different, the empirical formulae are highly site dependent (Dowding, 1996). The reliability of the prediction made by the empirical formulae is thus very limited.

With the development of the computer science, many numerical models have been put forward to describe the behavior of the soil mass. Most of the soil models in soil dynamics are based on experiments with static or transient loading (Saleeb, 1981). The load is very small comparing to the load induced by explosion. The stress in the vicinity of a charge can reach several GPa (giga pascal). The mechanism of deformation of soil under such high stress condition is different from that under low stress condition. When the wave generated by the blast propagates in the soil mass, it will attenuate gradually with the distance to the center of the explosion, and the stress will become very low at a certain distance away from the charge. Therefore, the soil model should be capable of covering the whole range of loading conditions in order to simulate the blast wave propagation in soil. Unfortunately there was not a model that can meet such requirement until recently.

Kandaur (Vovk et al. 1968) put forward a conceptual model for the deformation of soil under blast loading. The model considered the three-phase structure of soil. Because of the complexity of the model, it has not been actually used for dynamic problems, even though it describes the behavior of soils in a more reasonable manner. Wang et al (2002) established a three-phase soil model for simulating blast wave propagation based on Kandaur's idea. In this model the soil is considered as an assemblage of solid particles with different sizes and shapes that form a skeleton and their void are filled with water and air or gas. The solid particles, water, gas or air as well as the skeleton formed by the solid particles deform under different laws when the external load acts on the soil mass. In contrast to the soil models that are used in the soil mechanics and soil dynamics, the deformation of the solid particles is also included in the model because the loading caused by the explosion is so large that they should not be neglected.

In this paper the above mentioned three-phase soil model (Wang et al., 2002) is applied to simulate the blast wave propagation in soil mass. The model was implemented into the commercial hydrodynamic software AUTODYN as its user subroutines to allow for the numerical simulation.

Results of the blast wave propagation in two kinds of soil masses are presented, one is in saturated soil, and the other in unsaturated soil. The numerical results are found to agree well with available experimental results, and the effects of varying air content in the soil can be clearly observed from the numerical results.

2 THE THREE-PHASE SOIL MODEL UNDER SHOCK LOADING

Figure 1 illustrates schematically a soil element. The solid mineral particles form a structure whose voids are filled with water and gas. The structure formed by the particles is called skeleton. There are frictions between the solid particles. In many kinds of soils weak bond linkages between the solid particles also exist. Primarily, two deformation mechanisms exist in soil (Henrych, 1979), they are:

a) displacements of the particles and the elastic deformations of bonds on the contact surfaces of particles at low pressure and, at high pressures, a failure in bond.
b) the deformation of all the soil phases, determined by their volume compression.

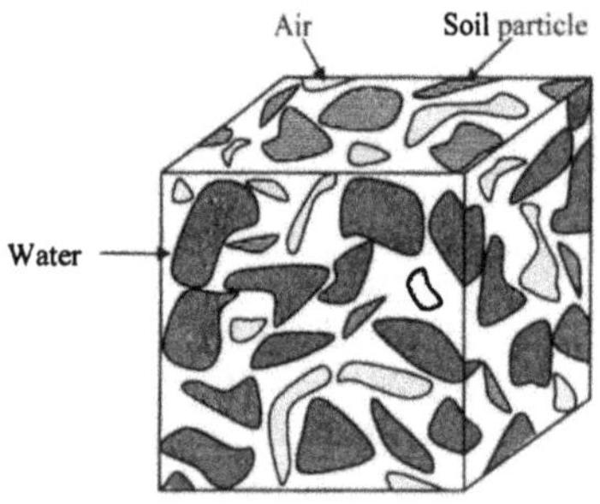

Figure 1 A schematic soil element

When the soil bears a load, both mechanisms always act simultaneously. A dry soil contains air and a small amount of water, whose compressibility considerably exceeds that of the skeleton; therefore, with static and dynamic loading, the first mechanism dominates the deformation at first. With increasing pressure, the soil is compacted so that the second mechanism becomes more and more important. In a water-bearing soil, the voids of the skeleton are filled with water and a little air. With a rapid dynamic loading, the water and air have a higher resistance than the bonds between the solid particles. The deformation and resistance are dominated by the second mechanism, particularly by the water and air deformation and the solid phase only becomes effective at high pressure. Under a slow or static loading, on the other hand, the water and air of the water-bearing soil are pressed out of the void and the compressibility is mainly controlled by the solid skeleton.

For shock loading, such as explosion loading, the duration is not long enough for the gas and water to flow through the soil skeleton (Shamsher, 1981); rather, they will deform with the soil skeleton. Thus, the relative movement between the skeleton and water and air can be neglected. And because of the high pressure of the shock loading, the deformation of the solid phase should also be considered.

Based on the above analysis, a conceptual model can be established, as shown in Figure 2.

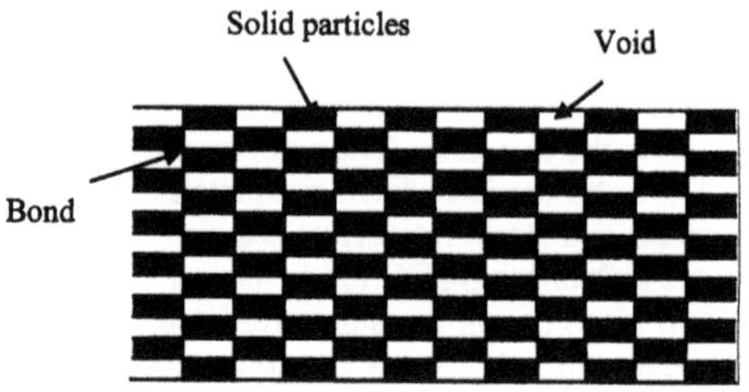

Figure 2 The conceptual model of soil

This conceptual model can be further explained by the diagram shown in Figure 3. In this figure A, B, C correspond to the deformation of the solid particles, water and air, respectively. D is the friction between the solid particles, and E is the elastobrittle bond linkage between the solid particles. The load is created by branches a, b and c. Branch a is the friction force between the particles, branch b is the force borne by the water and air in the void of the skeleton, branch c is the force borne by the bond between the solid particles. The first deformation mechanism corresponds to the elements D and E and the second deformation mechanism corresponds to the elements A, B and C.

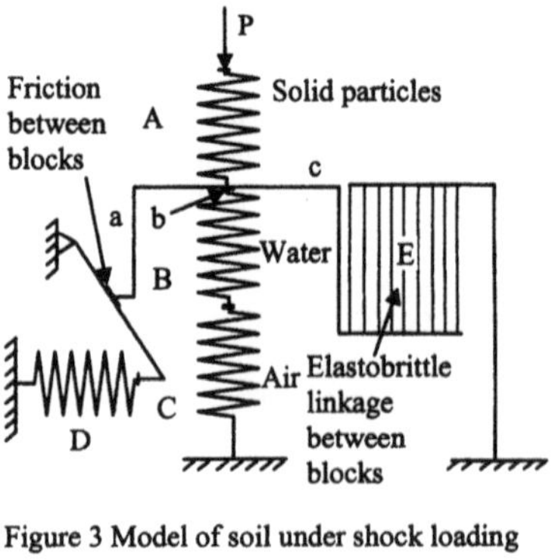

Figure 3 Model of soil under shock loading

3 GENERAL FORMULATION OF MODEL

3.1 *Overall equilibrium*

Under fast shock loading there is not enough time to squeeze the gas and water from the soil skeleton. So they will deform as a unity in a macro view. Overall equilibrium exists between the total stress, body forces and inertia forces.

The equilibrium equation is (Pande & Zienkiewicz, 1982)

$$\sigma_{ij,j} + \rho g_i = \rho \ddot{u}_i \tag{1}$$

where σ_{ij} is the total stress, u_i is the displacements (average) of the solid matrix, g_i is the vector of gravitational accelerations, ρ is the mass density and $\dot{u}_i = \frac{\partial}{\partial t} u_i$.

3.2 *Equations of state of soil*

The deformation of the soil must satisfy the continuity requirements that prevents "gap" between the phases of a deformed multiphase system in order to ensure the conservation of mass. Then, the total volume change must be equal to the sum of volume changes associated with each phase (Fredlund & Rahardjo, 1993)

$$\frac{\Delta V}{V_0} = \frac{\Delta V_w}{V_0} + \frac{\Delta V_g}{V_0} + \frac{\Delta V_s}{V_0} \tag{2}$$

where V_0 is the initial total volume of a soil element, V is the volume of a soil element, V_w is the volume of water, V_g and V_s are those of air and soil particles, respectively.

The load is created by the deformation of each phase, the friction force and the bonds between the solid particles, hence

$$p = p_s \tag{3}$$

$$p = p_a + p_b + p_c \tag{4}$$

$$p_e = p_a + p_c \tag{5}$$

where p is the total hydrostatic pressure, p_s is the pressure exerted on the solid phase. p_a is the pressure borne by the friction between the solid particles, p_b is the pressure borne by the water and gas, or the "pore pressure", p_c is the pressure borne by the bond between the solid particles, p_e is the sum of p_a and p_c, or the pressure carried by the soil skeleton.

The relations of volume of each phase are:

$$V = V_s + V_p \tag{6}$$

$$V_p = V_g + V_w \tag{7}$$

where V, V_w, V_g, V_s are the volume of a soil element, water, air and solid phase, respectively, V_p is the volume of the void

From Equation (4),

$$dp = \frac{\partial p_a}{\partial V_p} dV_p + \frac{\partial p_b}{\partial V_p} dV_p + \frac{\partial p_c}{\partial V_p} dV_p \tag{8}$$

Combining equations (3), (6), (7) into (8) yields

$$dp - \left(dV - \frac{\partial V_s}{\partial p} dp \right) \left[\left(\frac{\partial V_g}{\partial p_b} + \frac{\partial V_w}{\partial p_b} \right)^{-1} + \frac{\partial p_a}{\partial V_p} + \frac{\partial p_c}{\partial V_p} \right] = 0 \tag{9}$$

Equation (9) describes the volumetric deformation under the hydrostatic pressure, where $\frac{\partial V_s}{\partial p}$, $\frac{\partial V_g}{\partial p_b}$, $\frac{\partial V_w}{\partial p_b}$, $\frac{\partial p_a}{\partial V_p}$, $\frac{\partial p_c}{\partial V_p}$ are given by their independent equation of state or stress-strain relation.

In the present study the equation of state or stress-strain relation of each part is as follows:

For water (Cole,1948)

$$p_w = p_{w0} + \frac{\rho_{w0} c_{w0}^2}{k_w} \left[\left(\frac{\rho_w}{\rho_{w0}} \right)^{k_w} - 1 \right] \tag{10}$$

where c_{w0} is the initial sound speed of water, ρ_{w0} is the initial density of water, ρ_w is the density of water; p_{w0}, p_w are the initial pressure and pressure of water; respectively, k_w is a constant.

For solid particles, Lyakhov (1964) recommended the use of equation (10) with the subscripts w replaced by s.

$$p_s = p_{s0} + \frac{\rho_{s0} c_{s0}^2}{k_s} \left[\left(\frac{\rho_s}{\rho_{s0}} \right)^{k_s} - 1 \right] \tag{11}$$

A polytropic gas is thus used to model air in the voids (Henrych, 1979),

$$p_g = p_{g0} \left(\frac{\rho_g}{\rho_{g0}} \right)^{k_g} \tag{12}$$

where p_{g0} is the initial pressure of air; ρ_{w0} is the density of air at initial pressure, ρ_g is the density of air at pressure p_g; k_g is the isentropic exponent.

As for the stress volume deformation relationship of the skeleton, the friction between the particles is dependent on the normal force p' between the particles.

$$p_a = fp' \tag{13}$$

where f is the coefficient of friction of the solid particles. The normal force p' is proportional to deformation of the soil skeleton

$$p' = K_p \Delta V_p \tag{14}$$

where K_p is the coefficient of proportionality, which is a constant for a given soil and moisture, $\Delta V_p = V_p - V_{p0}$, V_p, V_{p0} are the volume of void and the initial volume of void, respectively.

The bonds between the solid particles can be represented by a series of elastic brittle filaments. The resistant force in each filament obeys the Hooke's law until the filament breaks,

$$p_c = E\Delta V_p \tag{15}$$

where E is a variable deformation modulus; ΔV_p is the increment of the volume of voids in the soil. According to the elastic brittle assumption, E may be written as

$$E = E_0(1 - D) \tag{16}$$

where E_0 is the initial modulus of the bonds, D is the damage variable that is decided by the failure of the elasto-brittle bond between the solid particles. Here the detailed deduction of the damage variable is omitted. Only the result is given. Details can be found in Wang et al., (2002).

According the parallel bar damage model (Krajcinovic, 1996), each filament is a two-force structural element endowed only with axial stiffness and strength. During the deformation of the soil some filaments will be destroyed, its ability to bear the load decreases. Assuming that parallel bar system is characterized by the Weibull distribution of filaments rupture strengths (which is commonly used in rupture problems), the damage variable can be defined as

$$D = 1 - exp\left[-\frac{1}{\eta}(B\varepsilon)^{\eta}\right] \tag{17}$$

where B, η are the constant related to the properties of the soil, ε is the strain of the parallel bar system. Then the force-deformation relation is:

$$F = E_0\varepsilon\, exp\left[-\frac{1}{\eta}(B\varepsilon)^{\eta}\right] \tag{18}$$

The parameter η and B can be obtained from fitting the force-deformation curve.

In the triaxial stress state, the definition of damage can be extended by replacing the strain ε by the effective strain ε_{eff}. The definition of the effective strain ε_{eff} of the soil skeleton is

$$\varepsilon_{eff} = \frac{\sqrt{2}}{3}\left[(\varepsilon_1 - \varepsilon_2)^2 + (\varepsilon_2 - \varepsilon_3)^2 + (\varepsilon_3 - \varepsilon_1)^2\right]^{1/2} \tag{19}$$

From equation (17), the damage is defined as,

$$D = 1 - exp\left[-\frac{1}{\eta}(b\varepsilon_{eff})^{\eta}\right] \tag{20}$$

where b is a constant.

With the above definition and the initial condition $p(V_0)=p_0$, the pressure p at any instant can be obtained from Equation (9).

3.3 *The elasto-plastic constitutive relationship of soil skeleton*

The total shear stress is created by the soil skeleton formed by the solid particles. Thus, the deviatoric stress of the soil skeleton is equal to that of the total soil element. The hydrostatic pressure p_e of the soil skeleton is determined from the solution of the equation of state. The stress states of the soil skeleton can be obtained from p_e and the deviatoric stress of the soil skeleton. In the following, the stress and strain are those related to the soil skeleton.

The basic premise of elastic plastic constitutive relation is the assumption that certain materials are capable of undergoing small plastic (permanent) as well as elastic (recoverable) deformation at each loading increment; i.e.: (Chen & Baladi, 1985)

$$d\varepsilon_{ij} = d\varepsilon_{ij}^{e} + d\varepsilon_{ij}^{p} \tag{21}$$

where $d\varepsilon_{ij}$ are the components of the total strain increment tensor, $d\varepsilon_{ij}^{e}$ are the components of the elastic strain increment tensor, and $d\varepsilon_{ij}^{p}$ are those of the plastic strain increment tensor.

Within the elastic range, the behavior of the material can be described by an incremental elastic constitutive relation of the type:

$$d\varepsilon_{ij}^{e} = C_{ijkl}\, d\sigma_{kl} \tag{22}$$

where C_{ijkl} is the material response function, and $d\sigma_{kl}$ are the components of stress increment tensor.

For isotropic elastic materials, the strain increment tensor takes the following form:

$$d\varepsilon_{ij}^{e} = \frac{dI_1}{9K}\delta_{ij} + \frac{1}{2G}ds_{ij} \tag{23}$$

where $I_1=\sigma_{kk}$ is the first invariant of the stress tensor, $s_{ij}=\sigma_{ij}-(I_1/3)\delta_{ij}$ is the deviatoric stress tensor, δ_{ij} is the Kronecker delta, K is the elastic bulk modulus, and G is the elastic shear modulus.

The bulk K and shear moduli G are functions of the invariants of the stress tensor and the damage invariable, $K=K(I_1,D)$, and $G=G(J_2, D)$, where $J_2=s_{ij}s_{ij}/2$ is the second invariant of the deviatoric stress tensor.

For an isotropic material the yield function f is a function of invariants of stress tensor

$$f\left[I_1, \sqrt{J_2}, (J_3)^{1/3}\right] = 0 \tag{24}$$

where I_1 is the first invariant of the stress tensor. J_2, J_3 are respectively the second and third invariants of the deviatoric stress tensor.

For a plastic material the plastic increment tensor can be obtained from the plastic potential function Q. The plastic potential function is also a function of invariants of stress tensor for an isotropic material. The plastic strain increment tensor can be expressed as:

$$d\varepsilon_{ij}^{p} = d\lambda \frac{\partial Q}{\partial \sigma_{ij}} \quad (25)$$

where $d\lambda$ is a positive scalar factor of proportionality. When $Q=f$, it is known as associated flow rule, otherwise it is a non-associated flow rule.

The total strain increment tensor can be obtained by combining equations (21), (23)and(25):

$$d\varepsilon_{ij} = \frac{dI_1}{9K}\delta_{ij} + \frac{ds_{ij}}{2G} + d\lambda\left[\frac{\partial Q}{\partial \sigma_{kk}}\delta_{ij} + \frac{\partial Q}{\partial s_{ij}}\right] \quad (26)$$

Similarly, the stress increment tensor can be written as:

$$d\sigma_{ij} = Kd\varepsilon_{kk}\delta_{ij} + 2Gde_{ij} - d\lambda\left[3K\frac{\partial Q}{\partial \sigma_{kk}}\delta_{ij} + 2G\frac{\partial Q}{\partial s_{ij}}\right] \quad (27)$$

To include the effect of hydrostatic stress on the shearing resistance of the material, such as soil, the von Mises' yield criterion was modified by Drucker and Prager (1952) to include the first invariant of the stress tensor. Thus:

$$f = \sqrt{J_2} - \alpha I_1 - k = 0 \quad (28)$$

in which α and k are material constants, related to the frictional and cohesive strengths of the material, respectively. When f is less than zero, the soil skeleton will undergo elastic deformation only. When equation (28) is satisfied, it wound flow plastically.

Under shock loading, such as an explosion, the high strain rate will affect the strength of the soil. A number of investigators have reported that the undrained shear strength of the soil increases linearly with the increase of the strain rate (Prapaharan et al. 1989). To include the strain rate effect on yield strength, the yield function is modified as:

$$f = \sqrt{J_2} - (\alpha I_1 - k)(1 + \beta \ln\frac{\dot{\varepsilon}_{eff}}{\dot{\varepsilon}_0}) = 0 \quad (29)$$

in which $\dot{\varepsilon}_0$ is the reference effective strain rate, β is the slope of the strength against logarithm of strain rate curve, $\dot{\varepsilon}_{eff}$ is the effective strain rate,

$$\dot{\varepsilon}_{eff} = \sqrt{\frac{2}{3}d\dot{\varepsilon}_{ij}d\dot{\varepsilon}_{ij}} \quad (30)$$

For soils, the direction of the plastic strain increment $d\varepsilon_{ij}^{P}$ is not normal to the yield surface, the plastic potential function is deferent to the yield function. Moreover, the non-associated flow rule can also avoid the problem of resulting in unstable non-unique solutions (Zhang, 1993). Thus, most hydrocodes employ the non-associated flow rule. The plastic potential function employed in AUTODYN (1997) is the Prandtl-Reuss type, expressed as:

$$Q(\sqrt{J_2}) = \sqrt{J_2} - Y = 0 \quad (31)$$

in which Y is the yield limit defined by the yield function.

Using equations (26), (27), the yield function f and the potential function Q, and hence the total increment tensor, can be determined.

4 NUMERICAL ANALYSIS OF THE PROPAGATION OF BLAST WAVE IN SOIL

The propagation of the blast wave in two kinds of soils are studied. One is in a saturated soil, the other is in an unsaturated soil. Figure 4 shows the configuration of the numerical models. It is the same as the arrangement of Lyakhov's experiment (Henrych, 1979).

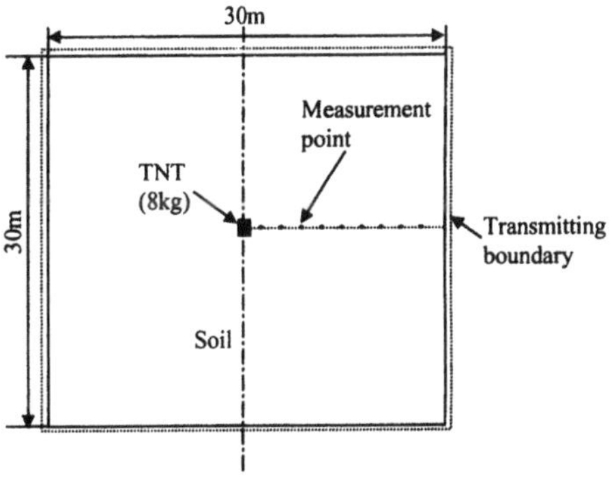

Figure 4 Configuration of the numerical model

The experiments were performed in the water-saturated and water-unsaturated fine-grained sandy loam. Table 1 shows the natural properties of the sandy loam. In which, α_1, α_2, α_3 are the volume ratio of the solid particles, water and air in the soil under consideration. The explosion product of the TNT satisfied the JWL equation of state (AUTODYN, 1997). The weight of the charge is 8 kg.

Table 1 Basic physico-mechanical soil characteristics

α_1	α_2	α_3
0.55 to 0.64	0.10 to 0.29	0.07 to 0.27

In the numerical calculation, the porosity was $\alpha_2+\alpha_3=0.4$. In the saturated soil, the relative air content $\alpha_3=0.0$. In the unsaturated soil, the relative air content is α_3=0.04. Table 2 shows the parameters of the soil in the calculation.

Table 2 Parameters in simulating the stress wave propagation in the soil

Soil	Air phase
$\alpha_1 = 0.6$	$\rho_{g0} = 1.2\ kg/m^3$
$\alpha_2 = 0.4$(Sat.)or0.36(Unsat.)	$c_{g0} = 340\ m/s$
$\alpha_3 = 0.0$(Sat.)or0.04(Unsat.)	$k_g = 1.4$
$\rho_0 = 1.99 \times 10^3\ kg/m^3$(Sat.)	Soil skeleton
$=1.95 \times 10^3\ kg/m^3$(Unsat.)	G = 55MPa
Solid particles phase	K_p = 165MPa
$\rho_{s0} = 2.65 \times 10^3\ kg/m^3$	f = 0.56
$c_{s0} = 4500\ m/s$	α = 0.25
$k_s = 3$	k = 0.2
Water phase	$\dot{\varepsilon}_0$ = 1%/min
$\rho_{w0} = 1.0 \times 10^3\ kg/m^3$	β = 0.1
$c_{w0} = 1500\ m/s$	η = 1.0
$k_w = 7$	E_0 = 20MPa
	b = 5.0

Figure 5 shows the attenuation of the peak pressure of the blast wave in the saturated and unsaturated soil respectively. The straight lines in these figures are the experimental relationships by Lyakhov (Henrych,1979). They are expressed by equation

$$p_m = A(\bar{R})^a \quad (32)$$

where A and a are constant, $\bar{R} = R/\sqrt[3]{W}$ is the scaled distance, R is the distance to the center of the explosion, W is the weight of the charge. The values of the constants A and a are (Henrych, 1979): A =600 and a =1.05 (For saturated soil, α_3=0.0); A =45 and a =2.5 (For unsaturated soil, α_3=0.04).

As shown in Figure 5, the calculation results of the peak pressure attenuation agree well with experimental results. They exhibit almost the same attenuation rate.

Figure 6 shows the attenuation of the peak particle velocity (PPV) in these two kinds of soil mass. Since PPV is not available from the experimental results, only the numerical results are presented here. The following observations on the different trends of propagation in the two kinds of soil mass can be made from the above two figures: a) The peak pressure and PPV in the unsaturated soil reduce more quickly than in the saturated soil. b) In these two figures the attenuation laws of peak pressure and PPV in the saturated soils are more close to a straight line than in the unsaturated soil. It is worth noting that at locations very close to the charge the PPV in the unsaturated soil is larger than in the saturated soil. This phenomenon is probably caused by the deformation mechanism of the unsaturated soil as a three-phase system. Within the range of the small scaled distance the pressure is very high, the soil is compacted seriously, the deformation of the soil is determined by the second mechanism, i.e., all three phases of the soil are deformed. As the scaled distance increases, the pressure decreases, the deformation of the solid phase becomes less important, the first mechanism gradually becomes more effective. The complexity of the change of the deformation mechanism of the three-phase system contributes to the several nonlinear attenuation trends in the unsaturated soil.

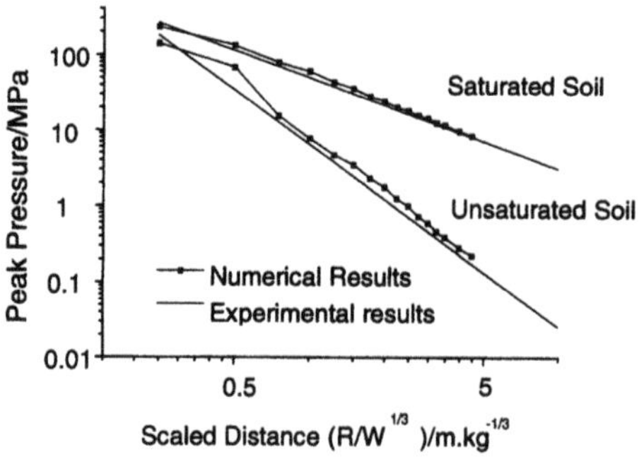

Figure 5 Attenuation of the peak pressure

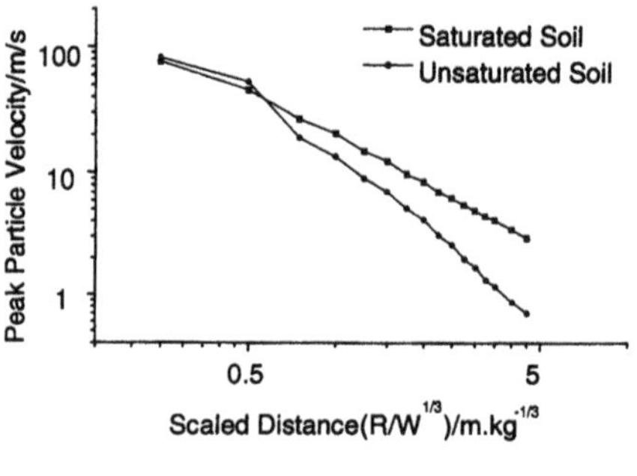

Figure 6 Comparison of the attenuation of the peak particle velocity

Figure 7 shows the variation of the blast wave front velocity with the strength of the wave in the saturated and unsaturated soils. It can be seen that the velocities of the wave front decrease with the reduction of the strength of the wave. The influence of the presence of air in the soil can be clearly observed. In the saturated soil the velocity of wave front varies very slightly; but in the unsaturated soil it decreases rapidly with the decrease of wave strength.

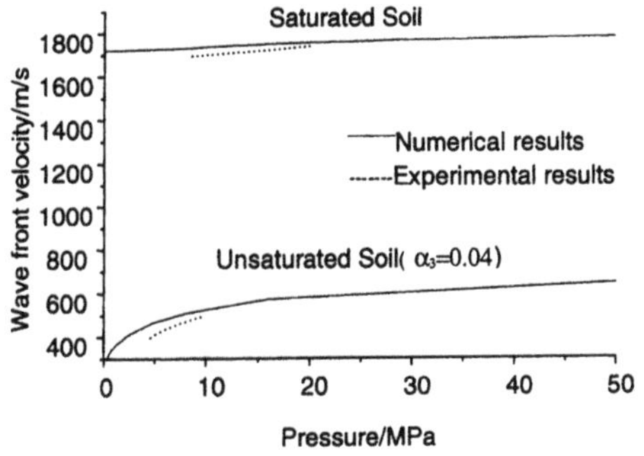

Figure 7 Wave front velocity as a function of its strength

Figure 8 shows the relationship between the particle velocity on the blast wave front and the strength of the wave. The significant effect of air is observed once again. In the unsaturated soil the particle velocity is considerably larger than that in the saturated soil.

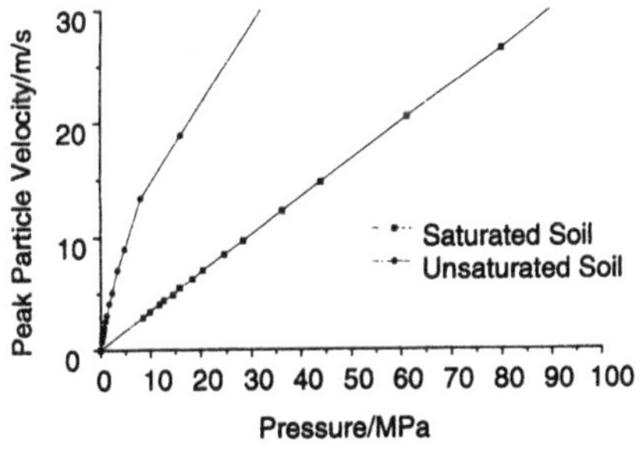

Figure 8 Particle velocity in the wave front as a function of the wave strength

The above results from the numerical simulation provide useful insight into the law of the propagation of the wave in soil mass, and they also reflect the actual deformation mechanism of the soils under blast loading.

5 CONCLUSION

The propagation of the blast wave in soil mass was simulated in this study based on a three-phase soil model for shock loading. The numerical results agree well with the experimental results, and the calculations also revealed some important characteristics of the propagation of the blast wave in soil. The peak pressure and PPV (peak particle velocity) decrease with the increase of the distance to the charge. The containment of air in the soil has significant influence on the attenuation of the blast wave. The peak pressure and PPV in the unsaturated soil attenuate much quicker that in the saturated soil. The wave front velocity in the unsaturated soil appears to be much lower than in the saturated soil. The blast wave front velocity and the peak particle velocity on the wave front all decrease with the reduction of the strength of the wave.

The characteristics of the propagation of the blast wave in the soil mass are closely related to the complex deformation mechanism of soils as a three-phase system. The deformation mechanism of soils under blast loading will be studied further in the subsequent research.

REFERENCE

AUTODYN Theory Manual, revision 3.0. 1997 Century Dynamics, San Ramon, California.

Chen W.F. & Baladi G.Y. 1985 *Soil Plasticity Theory and Implementation*. Elsevier.

Cole R.H. 1948. *Underwater Explosions*. Princeton, New Jersey: Princeton University Press.

Dowding C.H. 1996. *Construction vibrations*. Printice-Hall : Englewood Cliffs.

Drucker D.C.& Prager W. 1952 Soil mechanics and plastic analysis or limit design. *Q. Appl. Math.* 10: 157-165.

Fredlund D.G. & Rahardjo H. 1993. *Soil Mechanics for Unsaturated Soils*. John Wiley&Sons: Chichester.

Henrych J. 1979. *The Dynamics of Explosion and Its Use*. Elsevier: New York.

Krajcinovic D. 1996. *Damage Mechanics*. Elservier: New York

Lyakhov G.M. 1964. *Principles of Explosion Dynamics in Soils and in Liquid media*. Mockba: Henpa. (in Russian).

Pande C.N. & Zienkiewicz O.C. 1982. *Soil Mechanics-Transient and Cyclic Loads*. Chichester: John Wiley&Sons.

Prapaharan S. Chameau J.L. & Holtz R.D. 1989.Effect of Strain Rate on Undrained Strength Derived from Pressuremeter Tests. *Geotechnique* 39(4): 615-624.

Saleeb A.F. 1981 *Constitutive Models for Soils in Landslides*. Thesis of Ph.D, Purdue University.

Shamsher P.1981. *Soil Dynamics*. New York: McGraw-Hill.

Vovк V.V. Cнёrnii G.I. et al. 1968. *Principles of soil Dynamics*., Kneb: Haykoba nymka (in Russia)

Wang Z.Q. Lu Y, & Hao H. 2002. *The three-phase soil model under shock loading*. Singapore: Nanyang Technological University

Zhang X.Y. 1993. *Plastic Mechanics of Rock and Soil*. Beijing: People's Transport Press (in Chinese).

Moving load

Wave propagation – Moving load – Vibration reduction, Chouw & Schmid (eds.)
© 2003 Swets & Zeitlinger, Lisse, ISBN 90 5809 559 2

Investigation of ground vibrations in the vicinity of a train track embankment

M.A. Adam
Department of Civil Engineering, Zagazig University, Shoubra, Cairo, Egypt

G. Schmid
Department of Civil Engineering, Ruhr University, Bochum, Germany

ABSTRACT: A theoretical study of the train induced vibrations to the nearby soil surface is presented. A practical case of railway embankment that includes four existing tracks and a proposed new fifth one at Berlin, Germany is investigated. The analysis is performed utilizing a coupled Boundary Element - Finite Element (BE-FE) method. An impulsive type unit load as well as a harmonic sinusoidal type load is applied on each track, separately. The obtained results show that the proposed new fifth track causes the highest surface response and the surface displacements exhibit large differences in phase and amplitude. Moreover, the resonance at the embankment fundamental frequency strongly amplifies the surface response. The existence of a vertical wall at the right side of the embankment results in strong amplification of the surface response at the right side. On the other hand, the inclined left edge gives longer wave path inside the soil leading to relatively low surface response.

1 INTRODUCTION

Railway operations are considered to be one of the major sources of noise and vibration pollution in urban areas (Chang 1980, Nelson 1987, Flemming 1993). Vibration in the frequency range of 4-80 Hz is transmitted through the track structure to the ground surface as waves of small amplitudes. High levels of vibration in this frequency range are commonly associated with heavy axle-load freight o - erations (Okumura 1991, Jones 1996). The vibration is then transmitted to the building foundations and may excite resonance in the structural components (Jones 1994). If the building vibration has sufficient amplitude it may be felt by the residents and the structural elements may be subjected to distress. Therefore, it is a major issue in the environmental impact assessment of new railways and where new traffic is to be added on existing railways.

Generally, the waves propagate away from the source in the form of surface and body waves. The surface waves propagate parallel to the ground su - face via Rayleigh wave modes with low rates of attenuation with distance. The body waves pass through the soil where geometric dispersion and materials damping in the soil modify the vibration amplitude and frequency. This propagation path is very complicated because of the reflections and r - fraction at the interfaces of soil strata each of which support different shear and compression wave speeds. Therefore, it has been shown that it is important to include the effects of both the track structure and the layered structure of the ground (Jones 1994, Sheng 1999, Petyt 1999). In order to establish these effects and thereby enable the design of a low vibration railway a realistic model of wave generation and propagation is required.

In a previous work (Adam et al. 2000), the author presented two modelling approaches for the dynamic analyses of half-space and train-track embankment. The first is a three dimensional approach and the second is a two dimensional one that employs the coupled Boundary Element - Finite El ment (BE-FE) method. Very good agreement was found between the results of both approaches and it has been concluded that the two-dimensional solution is sufficient for the analysis of such problems under the dynamic loading of normal speed trains.

In this work, the two dimensional approach is applied for the investigation and analysis of a practical case of railway embankment at Berlin, Germany. The embankment includes four existing train tracks and is subjected to a new extension for a fifth track. The aim is to investigate the ground-born vibrations in the embankment vicinity du to the existing tracks as well as the proposed new extension. The applied load is modeled as an impulsive type load and as a harmonic sinusoidal type load. The response time histories at certain selected surface locations as well as the maximum responses along the surface on both sides are obtained and discussed.

2 STATEMENT OF THE PROBLEM

Figure 1 shows the dimensions and configurations of the railway embankment. The original geometry is symmetric as indicated by the dotted inclined edge at the right side. The hatched area in Figure 1 defines the proposed extension of the embankment for construction of the new fifth track. Because of the restricted property line the right side edge should be vertical, therefore, a retaining concrete wall has to be constructed. The embankment is constructed on sandy-soil half space where the ground water level is found to be at a depth of 3.0 m below the ground surface. The rails and the sleeper beneath them are considered to be a rigid unit with equivalent martial properties defined as the track layer hereafter. The mass density ρ, shear wave speed Vs, Poisson's ratio ν and damping ratio ζ for the track layer, embankment and the underlying soil are given in Table 1.

In the analysis, the present model is divided into finite domain and infinite domain as shown in Figure 1. The finite domain is discretized by the finite elements to include all the track structure and a depth of 4.0 m below the ground surface. This part is extended horizontally to a distance of 31.0 m apart from the right vertical side of the embankment and of 26.0 m apart from its left toe. The infinite domain is modeled by the boundary elements and starts from the outer boundary of the finite domain below the ground surface as a common interface and extends to infinity. The discretization of the finite part includes 2119 finite elements and 2283 nodes out of them 152 nodes on the common interface. The boundary element discretization includes 11 surface nodes on each side in addition to the common interface nodes.

Table 1. Material properties and wave speeds for the track structure and the underlying soil.

Description	ρ kg/m³	Vs m/s	ν —	ζ %
Track layer	0930	35,000	0.20	2.0
Ballast	2000	307	0.30	4.0
Sub-ballast	2100	260	0.30	8.0
Embankment	1900	180	0.30	8.0
Dry sand	1800	250	0.33	4.0
Saturated sand	2100	235	0.48	2.0
Concrete wall	2400	1955	0.20	2.0

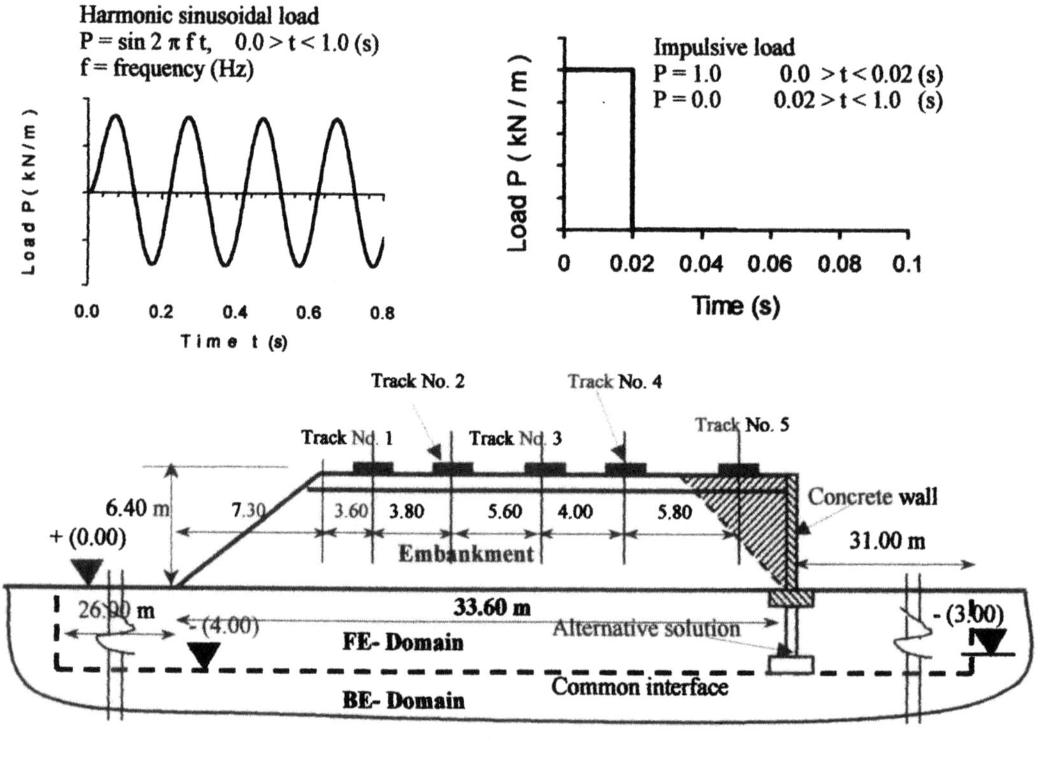

Figure 1. Dimensions and configurations of the railway embankment and the applied load.

The train-load is represented by a single concentrated line load. This load is applied at the center of each rigid track as a separate case of loading. The applied load is designed in the form of an impulse with amplitude of 1 kN/m that lasts for 0.02 seconds (Fig. 1). The frequency contents of this load cover the typical range of frequencies caused by heavy-axle trains (Adam et al. 2000).

Moreover, the train load is represented by a harmonic sinusoidal load with single frequency (Fig. 1) and applied on the new fifth track assuming five different values of frequency f, as f = 2, 5, 10, 25 and 50 Hz, respectively. The computed time range is s - lected to be a complete one second to include all the impulse load duration, enough time for wave propagation and a free vibration range or to contain enough number of load cycles of harmonic load. The step by step time integration scheme is employed a - suming a time increment Δt of 0.001 seconds and a total number of 1000 time steps.

3 RESULTS AND DISCUSSION

3.1 *Response to impulsive load*

The maximum vertical displacement responses due the impulsive type load when applied on each track are depicted in Figure 2. The centerline of the fourth track is chosen as a reference axis (Fig. 1). Figure 2 shows that the maximum peak of vertical response always occurs at the loaded track with very sharp decrease directly close to it. On the ground surface directly to the left or to the right of the embankment sides the response decreases gradually as the distance increases away from the embankment. This is due to the geometric dispersion and materials damping in the soil. When the load applied on the fifth track, the ground surface to the right side of the embankment exhibits the highest vertical response compared to the response to any other track loading. Due to the wave scattering by the vertical side of the embankment and the concrete wall, the ground su - face response on the right side is greater than the response on the left side in all cases of loading. This effect becomes higher when the load is applied closer to the concrete wall, giving rise to a relatively high response even on the left side when the load applied on the fifth track.

The maximum horizontal displacement responses due the impulsive type load are shown in Figure 3. Each track has the lowest horizontal response when it is loaded, except for the new fifth track close to the concrete wall. In addition, the unloaded tracks closer to the loaded one exhibit higher horizontal response than the loaded track itself. Very sharp decrease of the horizontal response is remarked close to the embankment sides, especially at the vertical right side when the load applied on the fifth track. It is also observed that the horizontal response on the ground surface increases to a certain extent as the distance from vertical wall increases. An opposite behavior is observed when the load applied on the other four tracks. The highest response on the right side results due to the load on the fourth or fifth track while the highest response on the left side r - sults due to first track loading. For each load case, the peak response occurs at a different location depending on the distance from the applied load.

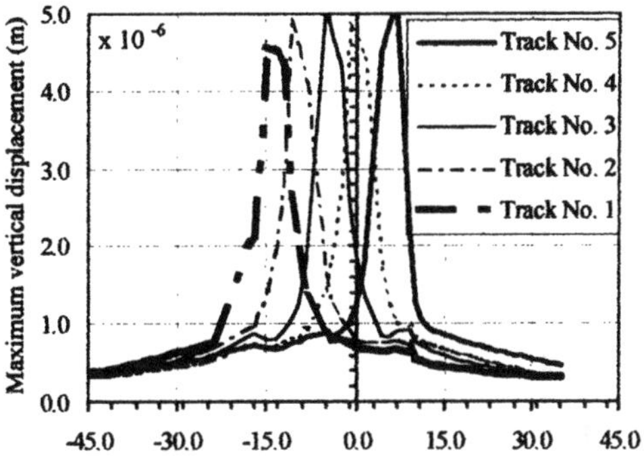

Figure 2. Maximum vertical responses due to impulsive load on each track

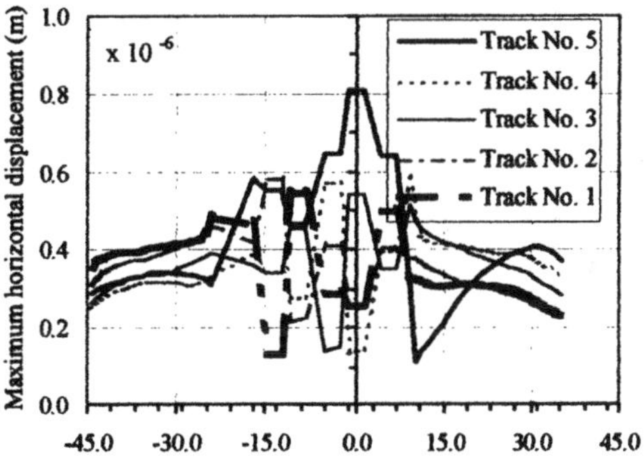

Figure 3. Maximum horizontal responses due to impulsive load on each track.

Based on the these results, the following explanation can be suggested: Since the applied load is vertical, the horizontal response results mainly due to the wave scattering by the effects of soil layering, embankment sides and the concrete wall. As the waves propagate away through the soil, the multiple wave reflections and refraction result in a very co - plicated wave field leading to the amplification of horizontal responses on the unloaded tracks. The existence of the concrete wall close to the loaded fifth track represents a strong wave barrier reflects or scatters the waves leading to the amplification of the track horizontal displacement and the very sharp decrease of the response amplitude at the vertical side. At the same time, the wall emits horizontally propagating surface waves with low attenuation rate leading to the change of the response distribution remarked in Figure 3.

The vertical displacement time histories at different locations on the ground surface at the right side are shown in Figures 4 & 5 when the load is applied on track number 1 and 5, respectively. The distance is measured from the outer face of the vertical concrete wall on the right side. The positive sign of the amplitude indicates the displacement downward in the same direction of the applied load. Only 0.6 seconds time interval from the calculated time is shown in the Figures. Each line represents the response at certain distance as indicted by the number attached to the arrow. The time delay of the first wave arrival at different locations, the amplitude and phase differences of motions are remarked in both Figures. This behavior will result in differential input m - tions at the foundations of the aboveground structures leading to high levels of dynamic stresses in the structure components. As the distance from the applied load increases, the vibration period becomes longer and the amplitude becomes lower. In addition, in all cases of loading the surface response r - store its original zero displacement position slowly.

3.2 *Response to harmonic load*

The maximum vertical and horizontal displacements due to harmonic sinusoidal load with different fr - quencies applied on the new fifth track are shown in Figures 6 & 7, respectively. The maximum response distribution pattern on the right side shows a great difference with the pattern on the left side even for the same applied frequency. For instance, the case of applied frequency of 10 Hz results in higher response than the case of 2 Hz on the right side while the opposite is true on the left side. In addition, the maximum peak occurs at different locations on the ground surface depending on the applied frequency,

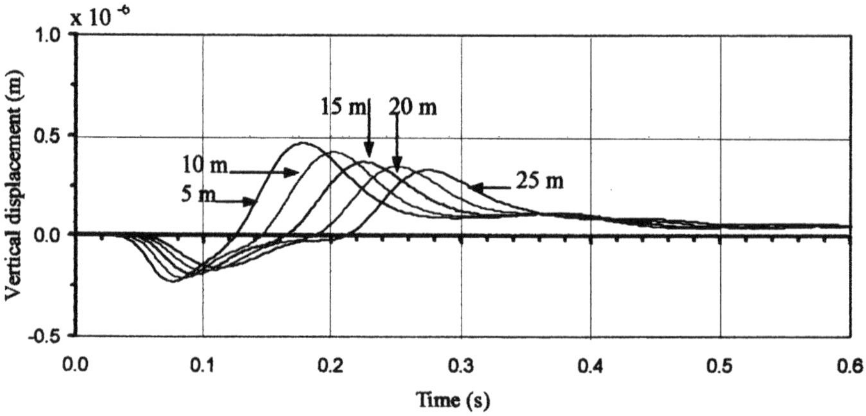

Figure 4. Vertical responses at different distances from the vertical right side due to impulsive load on track number 1.

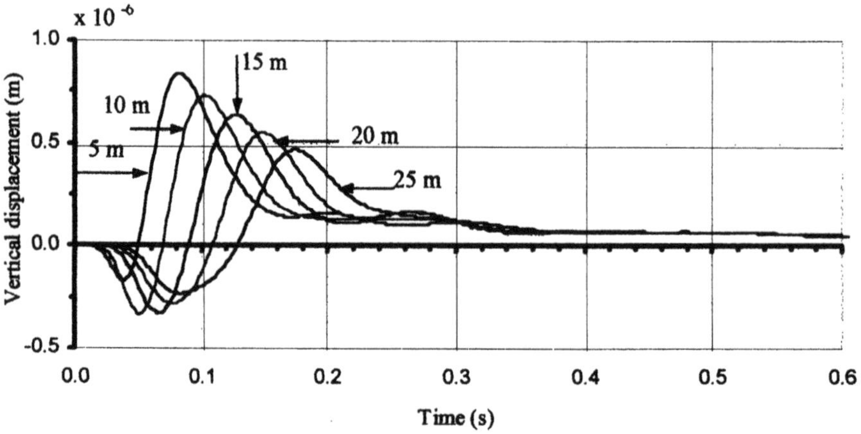

Figure 5. Vertical responses at different distances from the vertical right side due to impulsive load on track number 5.

especially for the horizontal response. For freque - cies of 10, 5 and 2 Hz, the response amplitudes exhibit maximum values as large as double times of the corresponding response to impulsive load on track number 5 (see Figs 2,3). Both figures indicate that the lower the applied frequency, the higher the resulted response. The highest response on each side results due to applied frequency of 5 Hz while the lowest response is caused by frequency of 50 Hz. Moreover, the response amplitudes due to frequency of 5 Hz show maximum values of 15 to 20 times as much as the amplitudes in case of 50 Hz. To explain this behavior, let us assume the embankment as a single layer on rigid base, so we can use the simple formula for the fundamental frequency $f_o = Vs/(4H)$, H is the layer depth, to find that f_o equals to about 7 Hz. Despite the simplicity of this assumption, it gives a reasonable approximation for the fundamental frequency that may lie in the frequency range between 5 and 10Hz. Consequently, we suggest that there is some sort of resonance has been occurred at this range leading to the observed response amplification at frequency of 5 Hz.

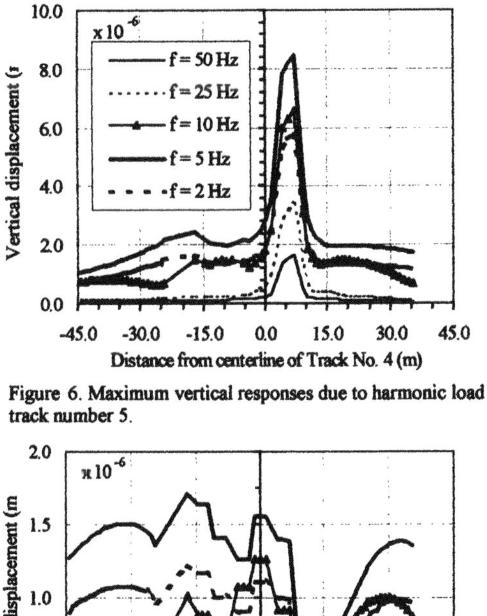

Figure 6. Maximum vertical responses due to harmonic load on track number 5.

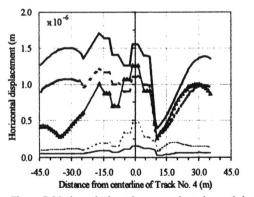

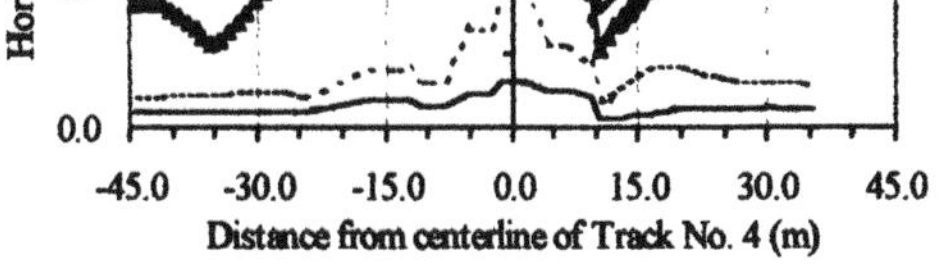

Figure 7. Maximum horizontal responses due to harmonic load on track number 5.

The time histories of the vertical displacement at different distance from the right vertical side due to harmonic type load on track number 5 with frequencies of 50 and 5 Hz are given in Figs 8 & 9, respectively. It can be observed that the peak response at all locations occurred at the second half of the first cycle. The response amplitudes have strong attenuation rate and the vibration dies quickly after the first cycle in case of 50 Hz. Moreover, For distances longer than 10 m the surface vibration does not follow the applied high frequency of 50 Hz. For fr - quencies of 10 and 5 Hz, it is clear that the vibration at all locations follows the frequency of the applied load with high amplitudes. This frequency range is very close the critical range for above ground building vibrations.

For comparison, Figure 10 depicts the vertical response time histories at a distance of 10 m from the right vertical side due to four different applied fr - quencies, f = 50, 25, 10 and 5 Hz. The large difference in the response amplitudes and the low attenuation rate at frequency of 5 Hz can be clearly observed. This comparison confirms the occurrence of resonance at the embankment fundamental frequency and gives the indication that the embankment and underlying soil play an important rule in the wave propagation characteristics and strongly affect the nearby soil surface response and, consequently, the aboveground structure vibrations.

3.3 *Effect of extended concrete wall*

In all previous cases of loading we assumed that the foundation of the concrete wall is constructed near the surface. As an alternative solution, we a - sumed a deeper foundation such that the vertical wall is extended downward to a depth of 3.5 m below the ground surface (see Fig. 1). The maximum vertical and horizontal responses due to impulsive load on the fifth track are shown in Figure 11 compared to the corresponding previous responses. The letters SW indicate the case of near surface found - tion and the letters LW indicate the case of extended wall. It is interesting to observe that a large reduction in the horizontal response of the fifth track and a considerable reduction in the surface horizontal response on the right side are resulted. Moreover, the surface horizontal response on the left side and the surface vertical response on both sides are relatively reduced. This can be owed to a reduction in the surface waves generated by the wall and the incoming waves are forced to travel vertically downward inside the soil.

4 CONCLUSIONS

A theoretical study of the train induced vibrations and the wave propagation through a railway embankment to the nearby soil surface is presented. A practical case of railway embankment that includes five train tracks at Berlin, Germany is investigated. Based on the obtained results, the following conclusions can be drawn out.

- The train track exhibits the maximum peak of vertical response and minimum horizontal response amplitude when it is loaded. The proposed new fifth track causes the highest surface response, especially on the right side of the embankment.
- The existence of the vertical concrete wall at the right side of the embankment results in strong amplification of the surface response, especially at the right side. If the wall is extended vertically downward, a considerable response reduction could be achieved. The inclined side is preferable for such railway embankment since it gives longer wave path inside the soil leading to relatively low surface response and longer vibration period.
- The ground surface response amplitude on both sides of the embankment decrease with the distance increase away from the embankment with different rates of attenuation depending on the loaded track. Moreover, The major part of the high frequency contents is filtered out from the surface vibration as the distance from the e - bankment side increases.
- The surface displacement shows large amplitude and phase differences that should be considered as a differential input motion in the design and analysis of the nearby aboveground structures.

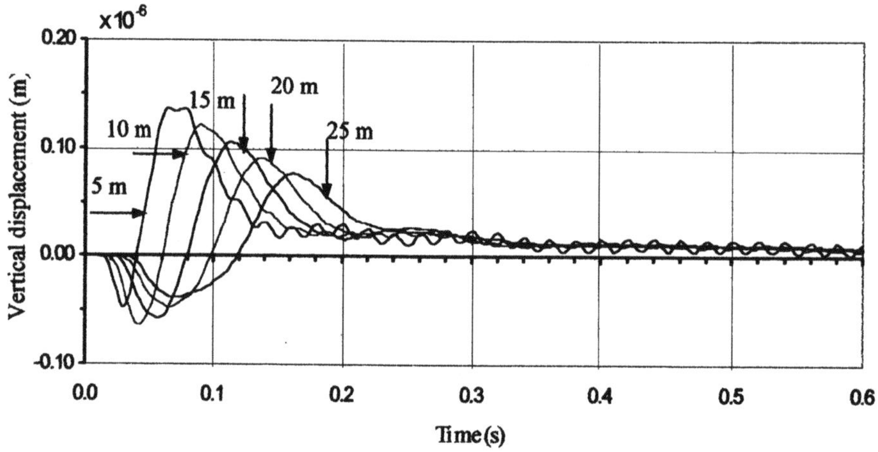

Figure 8. Vertical responses at different distances from the vertical right side due to harmonic load on fifth track, f =50 Hz.

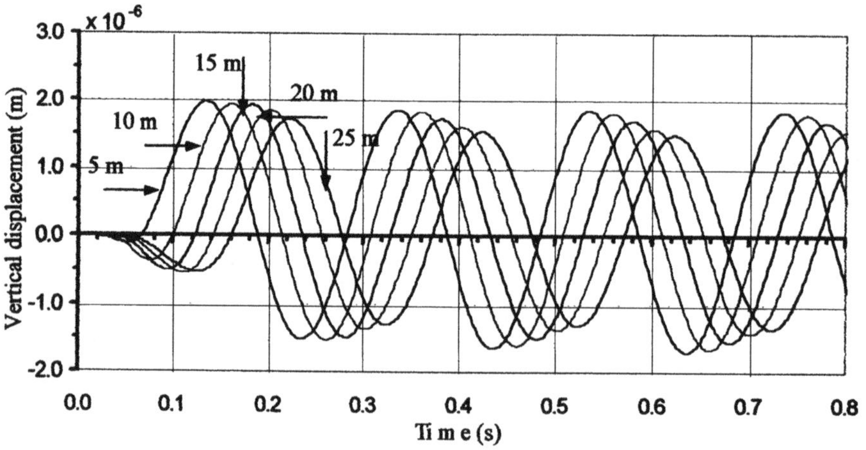

Figure 9. Vertical responses at different distances from the vertical right side due to harmonic load on fifth track, f =5 Hz.

REFERENCES

Adam, M., Pflanz, G. & Schmid, G. 2000a. Two- and three-dimensional modelling of half-space and train track embankment under dynamic loading. *Soil Dynamics and Earthquake Engineering* 19(8): 559-573.

Chang, C.S, Selig, E.T. & Adegoke, C.W. 1980. Geotrack model for railroad track performance. *Journal of Geotechnical Eng. Div.* 106(11): 1201-1218.

Flemming, D.B. 1993. Vibrations from railways. *Proc. Inst. Acoustics* 15(4): 13-22.

Jones, C.J.C. & Block, J.R. 1996. Prediction of ground vibration from freight trains. *Journal of Sound and Vibration* 193(1): 205-213.

Jones, C.J.C. 1994. Use of numerical models to determine the effectiveness of anti-vibration systems for railways. *Proc. Institution of Civil Engineers and Transportation* 105(2): 43-51.

Nelson, P. 1987. *Transportation Noise Reference Book*. London: Butterworths.

Okumura, Y. & Kuno, K. 1991. Statistical analysis of field data of railway noise and vibration collected in an urban area. *Applied Acoustics* (33): 263-280.

Petyt, M. & Jones, C.J.C. 1999. Modelling of ground-borne vibration from railways. In Fryba & Naprstek (eds), *Structural Dynamics – EURODYN '99*: 79-87. Rotterdam: Balkema.

Sheng, X., Jones, C.J.C. & Petyt, M. 1999. Ground vibration generated by a harmonic load acting on a railway track. *Journal of Sound and Vibration* 225(1): 3-28.

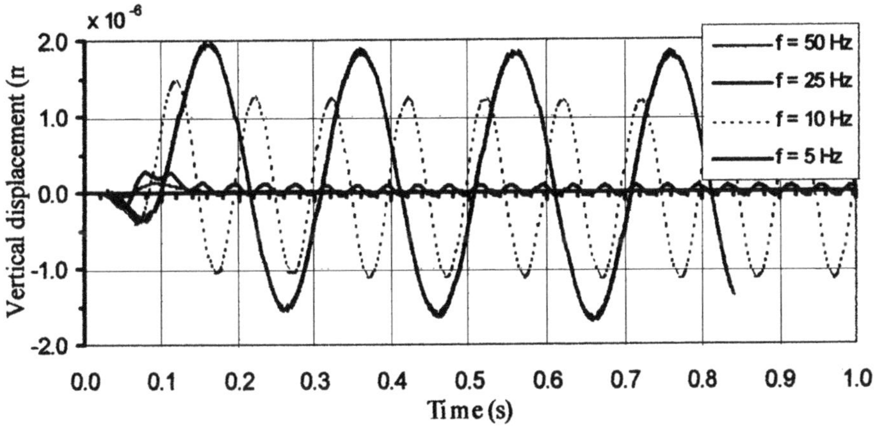

Figure 10. Vertical response time histories at 10m distance from the vertical right side due to harmonic load on track number 5.

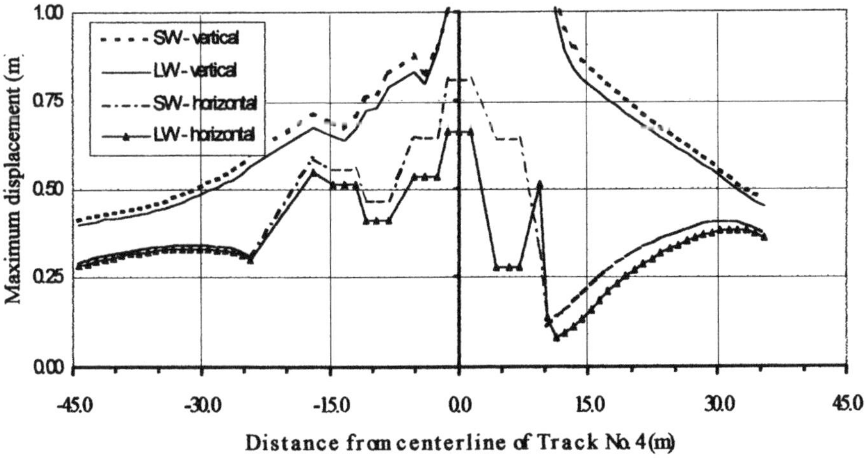

Figure 11. Effect of longer extension of the vertical concrete wall inside the soil.

Wave propagation – Moving load – Vibration reduction, Chouw & Schmid (eds.)
© 2003 Swets & Zeitlinger, Lisse, ISBN 90 5809 559 2

Peculiarities of vibrations of a layered inhomogeneous medium under the action of a load moving on its surface

T.I. Belyankova & V.V. Kalinchuk
Research Institute of Mechanics and Applied Mathematics, University Rostov, Russia

W. Hubert & G. Schmid
Department of Civil Engineering, Ruhr University Bochum, Germany

ABSTRACT: The peculiarities of layered inhomogeneous medium vibrations under the action of a load moving on its surface are investigated. The problem is considered in a coordinate system moving with the load for two inhomogeneous media types: "normal" (more rigid half-space) and "abnormal" (more rigid layer). The results of the investigation are compared with the results for the homogeneous half-space with various load velocities. The integral representation at any point of a structurally non-uniform medium under the action of a moving oscillating load is derived. A detailed analysis of the properties of the Green's function for the medium is carried out. The influence of the load velocity upon the structure of the surface wave field is investigated. Numerical analysis is carried out for the investigation of the medium dispersion properties and of the vertical displacements of the medium surface.

1 INTRODUCTION

Various approaches with different accuracy describe the processes connected with high-speed train dynamics and related wave fields influencing buildings and structures close to the train track. Special interest in real application relates to railway track heterogeneities and to the structure of the subgrade.

Analytical methods are effective only for the investigation of a homogeneous half-space (Barber 1996, Payton 1967).

In case of an inhomogeneous medium the approximate analytical solution can be constructed only in the case when the Green's function of the medium is meromorphic. Accounting for the heterogeneity of the track has to be made at the expense of numerical methods (FEM, BEM). If numerical approaches are used for non-uniform subgrades the modeling is considerably less efficient because of problems related to limitation of computational resources.

The use of BEM is complicated for heterogeneous media. In this connection wide applications were performed combining analytical approaches for the description of dynamic processes in a layered non-uniform medium with numerical methods simulating the dynamics of the non-homogeneous track structure. When the properties of the layers differ greatly the methods based on integral transformations and analytical representation of Green's function of the medium (Jones et al. 2000, Peplow et al. 1999, Sheng et al. 1999) are effective.

In the case when the properties of the medium vary continuously the Thin Layer Method (Kausel 1986) is used, where the layered medium is simulated by a set of layers with constant properties for each of which the Green's function is derived in an analytical way. (Bode et al. 2000) were using the Thin Layer Method in a combination with FEM for modeling a layered medium. (Pflanz et al. 2000) were using the Thin Layer Method/Flexible Volume Method as well as BEM for modeling inhomogeneous medium. (Estorff & Firuziaan 2000) were using a combination of FEM and BEM for modeling the layered inhomogeneous medium.

In a case, when the medium properties change continuously, semi-analytical methods can be used.

In the present work, taking into account a small number of layers, a method based on the use of the analytical representation of Green's function is used.

2 BOUNDARY-VALUE PROBLEM OF INHOMOGENEOUS HALF-SPACE VIBRATIONS

The problem of wave propagation in a layered non-uniform half-space is considered. The coordinate system x_1, x_2, x_3 is introduced. The medium consists of a layer $0 \le x_3 \le h$, resting on a homogeneous half-space surface $x_3 \leq 0$. The vibrations of the medium arise due to a distributed load $\mathbf{q}(x_1,x_2,t)$ in the area Ω on the surface of the medium.

The boundary-value problem is described by the equation of motion ($\Theta^{(n)}$ - stress tensor, $\mathbf{u}^{(n)}$ - displacement vector, $\rho^{(n)},\lambda^{(n)},\mu^{(n)}$ - mechanical parameters density and elastic modules of layer (n=1) and half-space (n=2), respectively):

$$\nabla \cdot \Theta^{(n)} = \rho^{(n)}\ddot{\mathbf{u}}^{(n)} \tag{1}$$

with boundary conditions on the medium surface:

$$x_3 = h: \ \mathbf{n}\cdot\Theta^{(1)} = \begin{cases} \mathbf{q}(x_1,x_2,t), (x_1,x_2)\in\Omega \\ 0, \quad (x_1,x_2)\notin\Omega \end{cases} \tag{2}$$

and with interface conditions between layer and half-space:

$$x_3 = 0: \ \mathbf{u}^{(1)} = \mathbf{u}^{(2)}, \ \Theta^{(1)} = \Theta^{(2)} \tag{3}$$

The problem is properly posed by specifying the radiation condition at infinity.

The tensor $\Theta^{(n)}$ looks like:

$$\Theta^{(n)} = \lambda^{(n)}\,\mathrm{tr}\,\varepsilon(\mathbf{u}^{(n)})\cdot\mathbf{I} + 2\mu^{(n)}\varepsilon(\mathbf{u}^{(n)})$$

where $\mathbf{I}$ - unit tensor, $\varepsilon(\mathbf{u}^{(n)})$ - linear tensor of deformation.

The components of the tensor $\Theta^{(n)}$ in the Cartesian coordinate system are:

$$\Theta_{ks}^{(n)} = \delta_{ks}\,\lambda^{(n)}\frac{\partial u_m^{(n)}}{\partial x_m} + \mu^{(n)}\left[\frac{\partial u_k^{(n)}}{\partial x_s} + \frac{\partial u_s^{(n)}}{\partial x_k}\right] \tag{4}$$

where δ_{ks} - the Kroneker symbol, with summation over repeated indexes.

3 BOUNDARY-VALUE PROBLEM OF VIBRATIONS OF INHOMOGENEOUS MEDIUM UNDER THE ACTION OF A LOAD MOVING ON ITS SURFACE

Let us consider the peculiar features of the vibration of the medium under the action of the load $\mathbf{q}(x_1 - Vt, x_2, t)$ moving on its surface with constant velocity V and oscillating with frequency ω.

Generally, the problem is non-stationary and it is necessary to use the Laplace transformation over time for its solution.

However, in many cases, under the assumption that the motion of the load and its oscillation occur during long time, it is possible to consider the problem in a coordinate system, moving with constant speed V. In this coordinate system we consider the dynamic process to be established.

For further research we shall introduce the coordinate system x_1', x_2', x_3', moving with constant velocity V and connected to the initial (motionless) coordinate system by the relations:

$$x_1' = x_1 - Vt, \ x_2' = x_2, \ x_3' = x_3 \tag{5}$$

The form of the solution of the boundary problem of stationary vibrations of a layered inhomogeneous half-space may be assumed as:

$$\mathbf{u}^{(n)} = \mathbf{u}_0^{(n)}(x_1', x_2', x_3')e^{-i\omega t} \tag{6}$$

Let us notice, that considering (5) in (1) and applying the Fourier transformation on coordinates x_1', x_2' (α_1, α_2 - parameters of the Fourier transformation; in the following the prime at x_3 is omitted) we obtain the system of the equations:

$$\begin{gathered}
\left[-\lambda_1^{(n)}\alpha_1^2 + \rho^{(n)}(V\alpha_1 - \omega)^2 - \mu^{(n)}\alpha_2^2\right]U_1^{(n)} + \\
+\mu^{(n)}\frac{d^2U_1^{(n)}}{dx_3^2} - \lambda_2^{(n)}\alpha_1\alpha_2 U_2^{(n)} - i\alpha_1\lambda_2^{(n)}\frac{dU_3^{(n)}}{dx_3} = 0 \\[2ex]
-\lambda_2^{(n)}\alpha_1\alpha_2 U_1^{(n)} + \mu^{(n)}\frac{d^2U_2^{(n)}}{dx_3^2} - i\alpha_2\lambda_2^{(n)}\frac{dU_3^{(n)}}{dx_3} + \\
+\left[-\lambda_1^{(n)}\alpha_2^2 + \rho^{(n)}(V\alpha_1 - \omega)^2 - \mu^{(n)}\alpha_1^2\right]U_2^{(n)} = 0 \\[2ex]
-i\lambda_2^{(n)}\alpha_1\frac{dU_1^{(n)}}{dx_3} - i\lambda_2^{(n)}\alpha_2\frac{dU_2^{(n)}}{dx_3} + \lambda_1^{(n)}\frac{d^2U_3^{(n)}}{dx_3^2} + \\
+\left[-\mu^{(n)}\alpha_1^2 + \rho^{(n)}(V\alpha_1 - \omega)^2 - \mu^{(n)}\alpha_2^2\right]U_3^{(n)} = 0
\end{gathered} \tag{7}$$

$$\lambda_1^{(n)} = \lambda^{(n)} + 2\mu^{(n)}, \ \lambda_2^{(n)} = \lambda^{(n)} + \mu^{(n)}$$

Thus, for the investigation of the equations in the coordinate system moving with velocity V, it is sufficient to replace ω by $\omega^* = \alpha_1 V - \omega$ in the traditional equation in a motionless coordinate system. It can be seen easily, that the expressions (7) represent an equation system, the type of which depends on the speed of the moving load.

It is meaningful at this juncture to designate: $c_1^{(1)}, c_2^{(1)}$ - velocity of longitudinal and transverse waves in the layer; $c_1^{(2)}, c_2^{(2)}$ - velocity of longitudinal and transverse waves in the half-space.

Depending upon the relation of the elastic properties of the layer and half-space we shall distinguish:

- "normal" medium:
  layer is softer than half-space (to designate this we shall use the index "n");
- "abnormal" medium:
  layer is more rigid than half-space (to designate this we shall use the index "a").

4 SOLUTION OF A BOUNDARY-VALUE PROBLEM OF VIBRATIONS OF INHOMOGENEOUS MEDIUM UNDER THE ACTION OF A LOAD MOVING ON ITS SURFACE

The solution of the boundary-value problem (Equations 1-4) can be given as:

$$\mathbf{u}^{(n)}(x_1,x_2,x_3)=\frac{1}{4\pi^2}\iint_{\Omega}\mathbf{k}^{(n)}(x_1-\xi,x_2-\eta,x_3)\mathbf{q}^{(1)}(\xi,\eta)\,d\xi\, d\eta$$
$$\mathbf{k}^{(n)}(s,t,x_3)=\int_{\Gamma_1}\int_{\Gamma_2}\mathbf{K}^{(n)}(\alpha_1,\alpha_2,x_3)e^{i(\alpha_1 s+\alpha_2 t)}d\alpha_1 d\alpha_2 \qquad (8)$$

The contours Γ_1 and Γ_2 are chosen according to the principle of the "limiting absorption" (Babeshko 1984, Babeshko et al. 2001, equal to the radiation condition) and the behavior of elements of the matrix function $\mathbf{K}^{(n)}(\alpha_1,\alpha_2,x_3)$ on a real axis (Kalinchuk et al. 1989). Expression 8 defines the displacement vector of any point of the layer or the half-space under the influence of the load. The elements of the matrix function $\mathbf{K}^{(n)}(\alpha_1,\alpha_2,x_3)$ are:

– layer (n=1):

$$K_{lp}^{(1)}=\sum_{k=1}^{3}f_{lk}^{(1)}\left[\Delta_{pk}\,\mathrm{sh}\,\sigma_k^{(1)}x_3+\Delta_{p,k+3}\,\mathrm{ch}\,\sigma_k^{(1)}x_3\right]\quad l=1,2$$
$$K_{3p}^{(1)}=\sum_{k=1}^{3}\left[\Delta_{pk}\,\mathrm{ch}\,\sigma_k^{(1)}x_3+\Delta_{p,k+3}\,\mathrm{sh}\,\sigma_k^{(1)}x_3\right] \qquad (9)$$

– half-space (n=2):

$$K_{lp}^{(2)}=\sum_{k=1}^{3}f_{lk}^{(2)}\Delta_{k+6,p}e^{\sigma_k^{(2)}x_3},\quad l=1,2$$
$$K_{3p}^{(2)}=\sum_{k=1}^{3}\Delta_{k+6,p}e^{\sigma_k^{(2)}x_3} \qquad (10)$$

where:

$$f_{m1}^{(n)}=-i\alpha_m\sigma_1^{(n)-1},\ f_{m2}^{(n)}=-i\alpha_m\sigma_2^{(n)}u^{-2},\ m,n=1,2$$

$$f_{31}^{(n)}=f_{32}^{(n)}=1,\ f_{33}^{(n)}=0,\ f_{13}^{(n)}=\alpha_2,\ f_{23}^{(n)}=\alpha_1$$

$$\Delta_{kp}=\Delta_{kp}^0/\Delta^0,\quad k,p=1,2,...,9\ ,\ \Delta^0=\det\left\|T_{kp}\right\|_{k,p=1,2,...,9} \qquad (11)$$

The elements of the matrix $\left\|T_{kp}\right\|_{k,p=1}^{9}$ are given by the expressions:

$$T_{mk}=l_{mk}^{(1)}ch\sigma_k^{(1)}h,\ T_{m,\,k+3}=l_{mk}^{(1)}sh\sigma_k^{(1)}h,\ T_{m,\,k+6}=0\ (m=1,2)$$

$$T_{3k}=l_{3k}^{(1)}sh\sigma_k^{(1)}h,\quad T_{3,k+3}=l_{3k}^{(1)}ch\sigma_k^{(1)}h,\quad T_{3,k+6}=0$$

$$T_{mk}=0,\ T_{m,\,k+3}=f_{m-3,\,k}^{(1)},\ T_{m,\,k+6}=-f_{m-3,\,k}^{(2)}\ (m=4,5)$$

$$T_{6k}=ch\sigma_k^{(1)}h,\ T_{6,\,k+3}=0,\ T_{6,\,k+6}=-1$$

$$T_{m\,k}=l_{m-6,\,k}^{(1)},\ T_{m,\,k+3}=0,\ T_{m,\,k+6}=-l_{m-6,\,k}^{(2)}\ (m=7,8)$$

$$T_{9k}=0,\ T_{9,\,k+3}=l_{3k}^{(1)},\ T_{9,\,k+6}=-l_{3k}^{(2)}$$

with $\Delta_{kp}{}^0(k=1,2,...,9)$ - cofactor of the elements T_{kp}.

$$l_{11}^{(n)}=-2\mu^{(n)}i\alpha_1,\ l_{12}^{(n)}=2\mu^{(n)}i\alpha_1\frac{\theta^{(n)}}{u^2},\ l_{13}^{(n)}=2\mu^{(n)}i\alpha_1\frac{\theta^{(n)}}{u^2}$$

$$l_{21}^{(n)}=2\mu^{(n)}i\alpha_2,\ l_{22}^{(n)}=2\mu^{(n)}i\alpha_2\frac{\theta^{(n)}}{u^2},\ l_{23}^{(n)}=2\mu^{(n)}i\alpha_2\frac{\theta^{(n)}}{u^2}$$

$$l_{31}^{(n)}=2\mu^{(n)}\frac{\theta^{(n)}}{\sigma_1^{(n)}},\quad l_{32}^{(n)}=2\mu^{(n)}\sigma_2^{(n)},\quad l_{33}^{(n)}=2\mu^{(n)}\sigma_2^{(n)}$$

$\sigma_k^{(n)}$ are defined from the characteristic equations:

$$\left(\sigma_1^{(n)}-\frac{\rho^{(n)}\omega_*^2}{\lambda^{(n)}+2\mu^{(n)}}\right)\left(\sigma_k^{(n)}-\frac{\rho^{(n)}\omega_*^2}{\mu^{(n)}}\right)^2=0,\ n=1,2,\ k=2,3$$

5 NUMERICAL EXAMPLES

Let us as an example consider the vibration of a layered inhomogeneous half-space under the influence of the oscillating load $q(x_1,x_2)e^{-i\omega t}$ moving on its surface with constant velocity V. A homogeneous layer of thickness h resting on the surface of a homogeneous half-space simulates an inhomogeneous medium.

When carrying out numerical calculations dimensionless parameters are used:
- geometric linear parameters of the problem are referred to the thickness of layer h;
- stresses are referred to the shear modulus of the half-space $\mu^{(2)}$;
- density is referred to the half-space density $\rho^{(2)}$.

The dimensionless parameter $\kappa_2=\omega h/c_2^{(2)}$ replaces the dimensionless frequency.

The following values are used to characterize the medium in the calculation ($\nu^{(n)}$ - Poisson's coefficient):

– softer layer:
  $c_2^{(1)}$ = 125 m/s, $\rho^{(1)}$ = 1700 kg/m^3, $\nu^{(1)}$ = 0.25, h = 1.0 m;
– stiffer layer:
  $c_2^{(1)}$ = 300 m/s, $\rho^{(1)}$ = 1700 kg/m^3, $\nu^{(1)}$ = 0.25, h = 1.0 m;
– half-space:
  $c_2^{(2)}$ = 200 m/s, $\rho^{(2)}$ = 1700 kg/m^3, $\nu^{(2)}$ = 0.25.

The wave field initiated by the moving load is characterized by an anisotropy of a special kind, i.e. the wave field structure depends on the angle between the direction of wave propagation and the direction of the load motion.

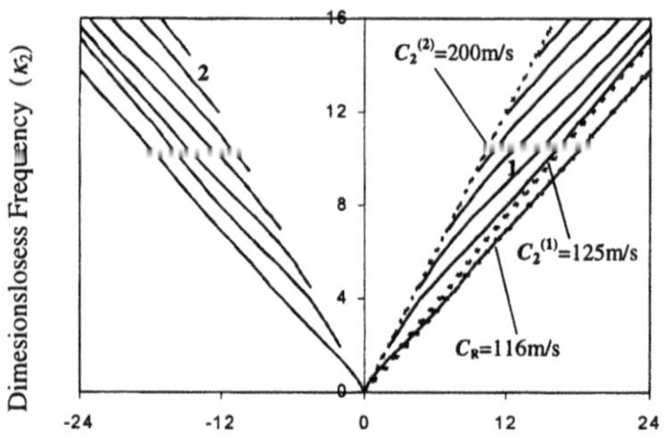

Figure 1. Dispersion diagram for the case of „normal" media. Load fixed. (V=0).

In Figures 1-4 the dispersion diagrams – cross-sections of the dispersion surface of the considered media types with the plane α_2=0 – are shown. In Figure 1 dispersion diagrams (symmetric continuous line sets 1 and 2) referring to the "normal" medium for the fixed load (V=0) are given. Dashed lines refer to both Rayleigh waves (c_R) for the half-space with the same parameters as the layer and transverse waves of the layer ($c_2^{(1)}$) and of the half-space ($c_2^{(2)}$).

In Figures 2, 3 the dispersion diagrams for the "normal" medium for the load moving with velocities of 80 m/s (Fig. 2) and 160 m/s (Fig. 3) are given.

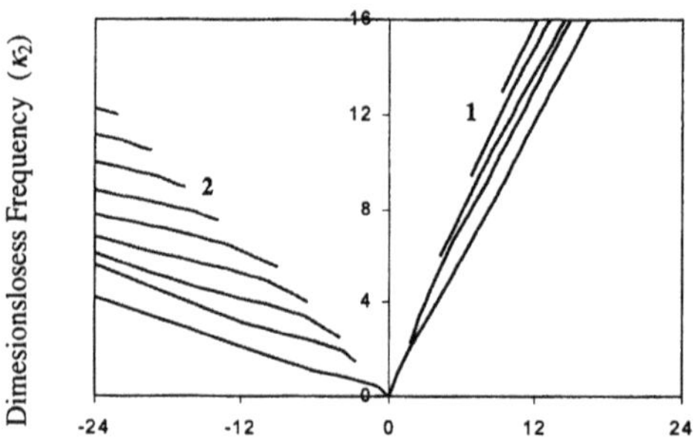

Figure 2. Dispersion diagram for the inhomogeneous half-space („normal" case). Load velocity is 80 m/s.

The load motion leads to a change of the wave field structure. When V = 80 m/s (Fig. 2) the sets 1 and 2 become asymmetric, phase velocities of waves with the positive wave numbers increase. Phase velocities of waves with the negative wave numbers decrease.

When V = 160 m/s (Fig. 3) only one set of "fast" waves with positive wave numbers remains on the medium surface. In the meantime in the domains of positive and negative wave numbers denumerable sets 2 and 3 of "slow" waves appear, which take part

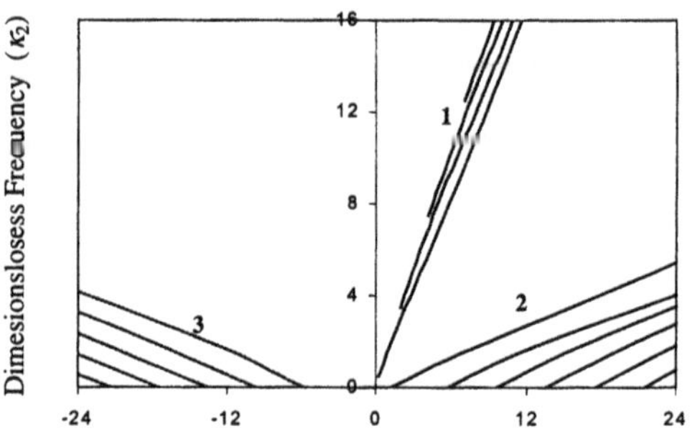

Figure 3. Dispersion diagram for the inhomogeneous half-space („normal" case). Load velocity is 160 m/s.

in the formation of the wave front near the load. It is seen that in this case the motion of the oscillating load essentially influences the quantity, the cut-off frequency and velocity of the propagating surface waves modes.

In the Figure 4 the curves for the motionless load (lines 1) and for the load moving with a velocity of 80 m/s (lines 2, 3) are given for the "abnormal" medium. The Rayleigh wave (c_R) for the half-space with the same parameter as the layer parameters and the transverse waves of the layer ($c_2^{(1)}$) and a half-space ($c_2^{(2)}$) are shown as dashed lines.

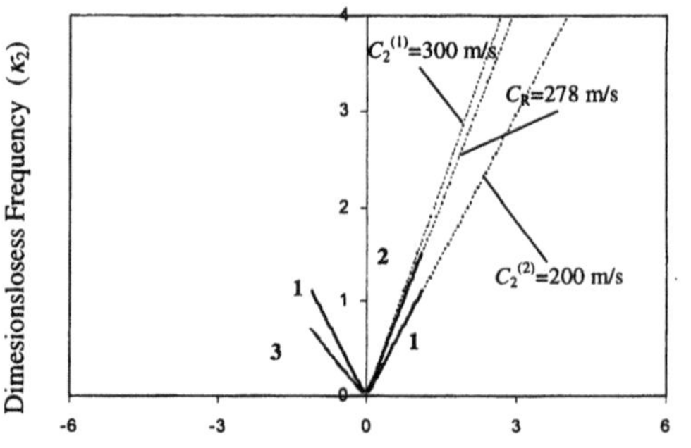

Figure 4. Dispersion diagram for the inhomogeneous half-space („abnormal" case) Load fixed – lines 1; load velocity is 80m/s - line 2 and 3.

The case of an "abnormal" medium is characterized by the presence of so-called cut-off frequencies – the value of the frequencies, above which the surface wave disappears. In this case the motion of the oscillating load influences both the cut-off frequency and the velocity of the propagating surface wave mode. It is seen in the Figure 4 that phase velocities of waves with the positive wave numbers (line 2) and cut-off frequencies increase. Phase velocities of waves with the negative wave numbers (line 3) and cut-off frequencies decrease.

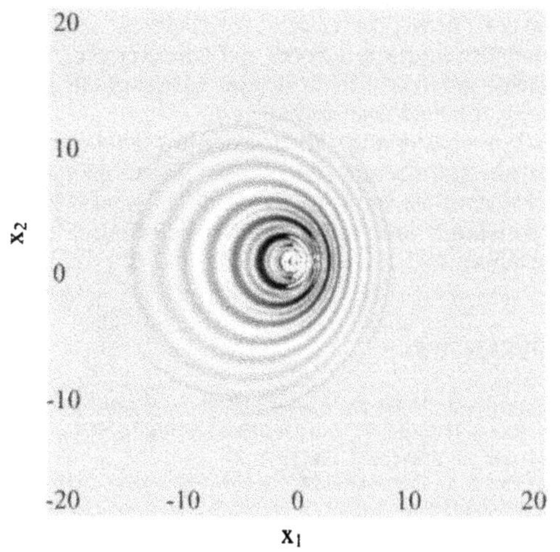

Figure 5. Displacements of homogeneous half-space. Load velocity is 80m/s.

In the Figures 5-10 the displacements of the considered media types are given (bird's view in moving coordinates). The pictures illustrate the influence of the medium properties upon the surface wave field caused by the moving load oscillating with a frequency of 40 Hz. The motion is in the positive direction of the x_1-axis. In the Figures 5-7 the load velocity is 80 m/s. In the Figures 8-10 the load velocity is 160 m/s.

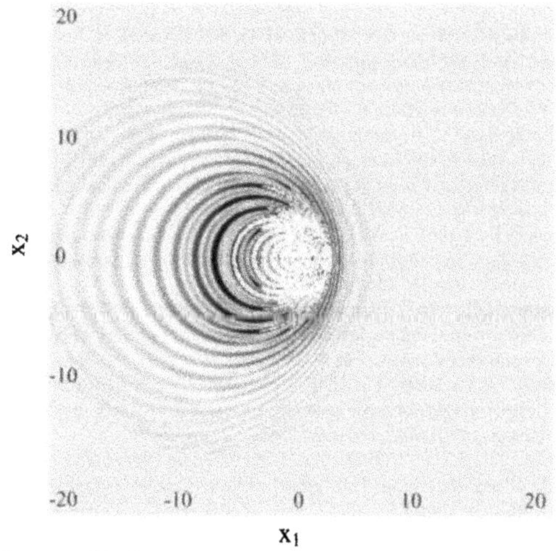

Figure 6. Displacements of inhomogeneous half-space ("normal" case). Load velocity is 80 m/s.

Comparing the Figures 5 and 6 it is noticeable that the presence of the soft layer leads to an essential increase of the anisotropy of the surface wave field. Especially the difference between the velocities of the surface waves propagating to the direction of the load motion and of the surface waves propagating to the opposite direction abruptly increases.

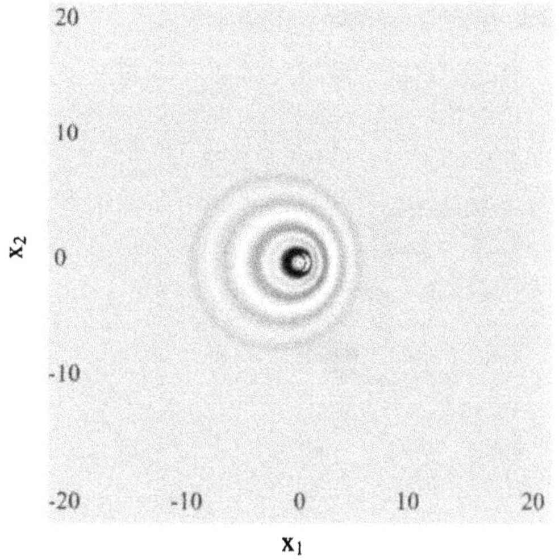

Figure 7. Displacements of inhomogeneous half-space ("abnormal" case). Load velocity is 80 m/s.

Comparing the Figures 6 and 7 it is noticeable that the presence of the rigid layer on the medium surface leads to a decrease of the surface wave field anisotropy.

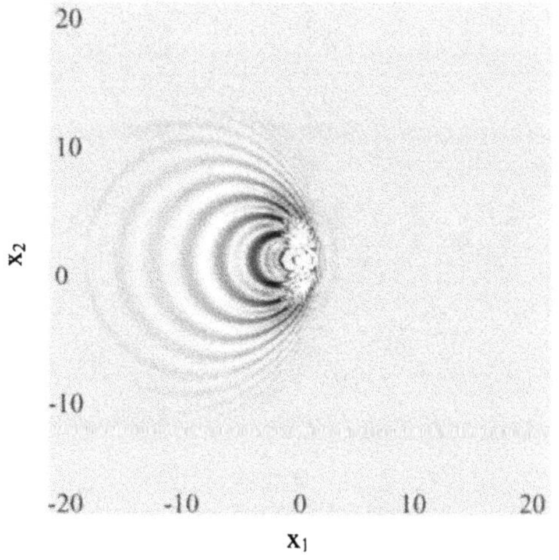

Figure 8. Displacements of homogeneous half-space. Load velocity is 160 m/s.

It is seen in the Figure 8 that the motion with the velocity of 160 m/s leads to an abrupt increase of anisotropy of the surface wave field even for the homogeneous medium.

Comparing the Figures 8 and 9 it can be seen that the presence of the soft layer in this case (load motion velocity is greater than the velocity of the layer transverse wave) leads to a quality change of surface field structure - the 'Mach cone' effect appearance.

From the Figures 8 and 10 it is noticeable that the presence of the stiff layer essentially decreases the anisotropy of the surface wave field.

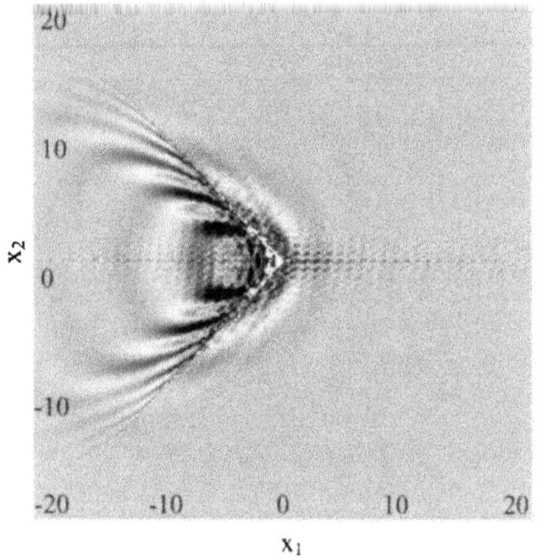

Figure 9. Displacements of inhomogeneous half-space ("normal" case). Load velocity is 160 m/s.

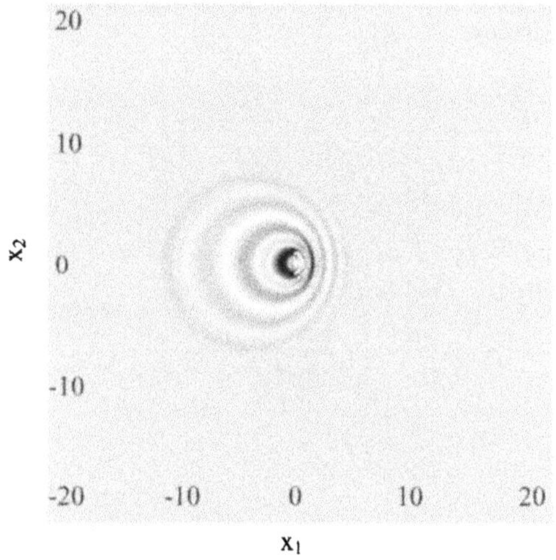

Figure 10. Displacements of inhomogeneous half-space ("abnormal" case). Load velocity is 160 m/s.

6 SUMMARY

The problem of the vibrations of a layered inhomogeneous medium under the action of a load moving on its surface was considered. Three cases of the medium structure were investigated: homogeneous half-space, "normal" medium (softer layer on a more rigid half-space) and "abnormal" medium (more rigid layer on a softer half-space) with subsonic and supersonic load velocities. The research showed that when considering the subsonic load velocity only the medium structure influences the anisotropy of the surface wave field. The soft layer increases the anisotropy, the rigid layer decreases it.

When considering the supersonic load velocity the change of medium structure can lead to the quality change of the surface wave field. The presence of the soft layer leads to the appearance of the 'Mach cone' effect.

REFERENCES

Babeshko, V.A. 1984. *The Generalized Factorization Method and spatial Dynamic Mixed Problems of Elasticity Theory.* Moscow: Nauka (in Russian).

Babeshko, V.A., Belyankova, T.I. & Kalinchuk, V.V. 2001. On the Solution to a Class of the Mixed Problems for a Layered Half-Space. *Doklady Physics* 46(10): 732-735.

Barber, J.R. 1996. Surface displacements due to a steadily moving point force. *Journal of Applied Mechanics* 63: 245-251.

Bode, C., Hirschauer, R. & Savidis, S.A. 2000. Three-dimensional time domain analysis of moving loads on railway tracks on layered soils. In N. Chouw & G. Schmid (eds), *Wave Propagation – Moving Load – Vibration Reduction; Proc. intern. Workshop WAVE2000, Bochum, 13-15 December 2000*: 3-12. Rotterdam: Balkema.

Estorff, O.v. & Firuziaan, M. 2000. FEM and BEM for nonlinear soil-structure interaction analyses. In N. Chouw & G. Schmid (eds), *Wave Propagation – Moving Load – Vibration Reduction; Proc. intern. Workshop WAVE2000, Bochum, 13-15 December 2000*: 357-368. Rotterdam: Balkema.

Jones, C.J.C., Sheng, X. & Thompson, D.J. 2000. Ground vibration from dynamic and quasi-static loads moving along a railway track on layered ground. In N. Chouw & G. Schmid (eds), *Wave Propagation – Moving Load – Vibration Reduction; Proc. intern. Workshop WAVE2000, Bochum, 13-15 December 2000*: 83-93. Rotterdam: Balkema.

Kalinchuk, V.V., Lysenko, I.V. & Polyakova, I.B. 1989. Singularities of the interaction of a vibrating stamp with an inhomogeneous heavy base. *Journal of Applied Mathematics and Mechanics* 53(2): 235-241.

Kausel, E. 1986. Wave Propagation in Anisotropic Layered Media. *International Journal for Numerical Methods in Engineering* 23: 1567-1578.

Payton, R.G. 1967. Transient motion of an elastic half-space due to a moving surface line load. *International Journal of Engineering Sciences* 5: 49-79.

Peplow, A.T., Jones, C.J.C. & Petyt, M. 1999. Surface vibration propagation over a layered elastic half-space with an inclusion. *Applied Acoustics* 56(4): 283-296.

Pflanz, G. 2001. Numerische Untersuchung der elastischen Wellenausbreitung infolge bewegter Lasten mittels der Randelementenmethode im Zeitbereich. *Fortschritt-Berichte VDI Reihe 18 Nr. 265.* Duesseldorf: VDI Verlag.

Pflanz, G., Garcia, J. & Schmid, G. 2000. Vibrations due to loads moving with sub-critical and super-critical velocities on rigid track. In N. Chouw & G. Schmid (eds), *Wave Propagation – Moving Load – Vibration Reduction: Proc. intern. Workshop WAVE2000, Bochum, 13-15 December 2000*: 131-148. Rotterdam: Balkema.

Sheng, X., Jones, C.J.C. & Petyt M. 1999. Ground vibration generated by a harmonic load acting on a railway track. *Journal of Sound and Vibration* 225(1): 3-28.

Wave propagation – Moving load – Vibration reduction, Chouw & Schmid (eds.)
© 2003 Swets & Zeitlinger, Lisse, ISBN 90 5809 559 2

Investigation on soil-viaduct-pile vibrations induced by passing trains

Jong-Dar Yau
Department of Architecture and Building Technology, Tamkang University, Taipei, Taiwan, R.O.C. Member, Taiwan Structural Engineers Association, Taiwan, R.O.C

Hsiao-Hui Hung
China Engineering Consultants, Inc., Taipei, Taiwan, R.O.C.

Yeong-Bin Yang
Department of Civil Engineering, National Taiwan University, Taipei, Taiwan, R.O.C.

ABSTRACT: The soil vibration induced by high speed trains is a problem of increasing importance in engineering practice. Such a problem will be investigated by a finite/infinite element approach in this paper. To simplify the physical simulation of the problem, a moving train is modeled as a vertical harmonic line load and only a two-dimensional profile that contains the cross section of the viaduct, pile foundations, and underlying soils is considered, which is divided into a near-field and a far-field. The near-field including the loads, viaduct, piles, and part of the soils is simulated by finite elements, while the far-field covering the soils extending to infinity by infinite elements. To evaluate the influence of pile foundations in transmitting the train-induced vibrations to the ground afar, the piles are assumed to be sitting on the underlying rocks in this study. The numerical results indicate that the existence of pile foundations can effectively reduce the response of viaducts to trains moving at high speeds, but the ground vibration in the far field will be amplified as the files serve as a connection between the viaduct and underlying rocks, which can enhance the transmission of vibration waves.

1 INTRODUCTION

High-speed railway system has become an important transportation system in modern countries for around four decades. One of the interesting research topics of the high-speed railway system is the ground vibration induced by trains moving at high speeds. The purpose of this paper is to investigate the vibration of a soil-viaduct-pile interaction system that forms the infrastructure of a high-speed railway.

As indicated by the literature on ground-borne vibrations, most of the early researches were conducted by analytical approaches. Whenever an analytical approach was adopted, however, restrictions were inevitably imposed on the geometry and material properties of the problem considered, as closed-form solutions cannot be made available for other complex conditions. Starting from the mid 1970s, enhanced by the advent of high-performance computers, various numerical methods emerged as an effective tool for solving the wave propagation problems, including particularly the finite element method, boundary element method, and their variants.

In the last decade, a great portion of the studies on wave propagation problems were performed by the boundary element method, existing relevant works on such an approach can be found in the review papers by Beskos (1987, 1997). One advantage of the boundary element method is that radiation damping can be accurately taken into account by the use of existing closed-form solutions for some specific problems. However, it is not suitable for simulating the irregularities in geometry and materials of the structure and underlying soils, as may be encountered in practical situations.

On the other hand, the finite element method has the advantage of being able to simulate almost any geometric shapes, allowing us to include the embedded structures and multi layering of the soil deposits. But a major disadvantage of the finite element method is that the soil, which is semi-infinite by nature, can only be approximately represented by elements of finite size. Consequently, the radiation damping relating to the loss of energy by waves traveling away cannot be adequately modeled (Hung, 2000)

To overcome the disadvantages of the boundary element method and the finite element method on wave propagation problems, recourse can be had to a third method called the ***hybrid method.*** By the hybrid method, a soil-structure system is divided into two subsystems, i.e., the near field and the far field (Fig. 1). The near field, including the loads, viaduct, piles, and other geometric/material singularities can be simulated by the finite elements as conventional, while the far-field covering the soils with indefinite boundary by infinite elements.

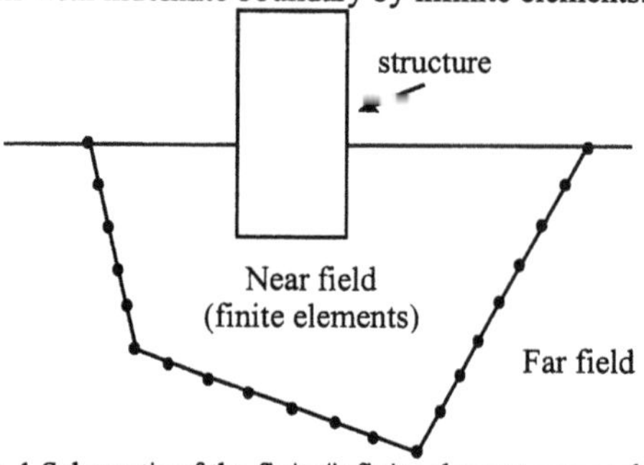

Fig. 1 Schematic of the finite/infinite element approach

Owing to its versatility, the hybrid method has often been used to deal with problems involving various irregularities in geometry and materials, which may include a structure (bridge or building), foundation, layered soils, and the outreaching soils. According to Gutowski and Dym (1976), it is reasonable to simulate the passage of train loads by a moving line load, provided that the receiver from the track is approximately less than $1/\pi$ times the length of the train. As can be seen from the literature, most researches on the ground-borne vibrations were based on the two-dimensional half-space model by assuming the plain strain condition to be valid. In the present study, using the finite/infinite element approach developed previously, a simplified model with two-dimensional profile that contains the cross section of the viaduct, pile foundations and underlying soils will be adopted to investigate the wave propagation behavior of a soil-viaduct-pile interaction system under the passage of high-speed trains.

2 FORMULATION OF THE PROBLEM

2.1 *Basic assumptions*

In high-speed railway engineering, the ground vibration induced by trains moving at high speeds remains a subject of intensive research. To simplify the analysis procedure of this problem, the following assumptions are made:

(1) The material and geometric properties are constant along the moving direction of the loading.
(2) As shown in Fig. 1, a plane strain model with a two-dimensional profile will be used to simulate the dynamic behavior of an elastic body subjected to a series of moving loads.
(3) Since the moving loads acting at a point can always be expressed as a series of harmonic loads, only harmonic line loads are considered.

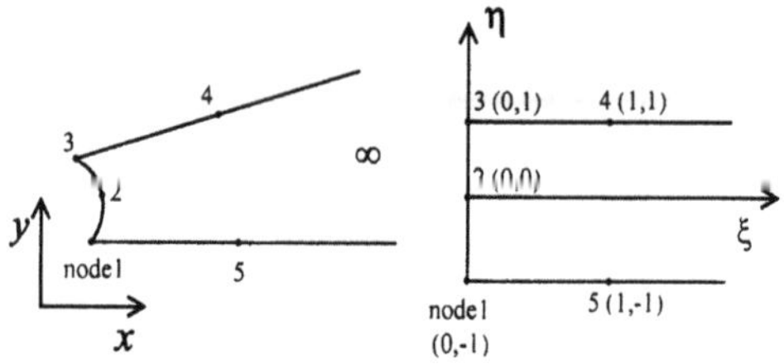

(a) (b)

Fig. 2 Infinite element

2.2 *Mathematical model of soil-structure interaction*

To facilitate simulation of the soil-structure interaction system shown in Fig. 1, the entire domain is divided into two regions, i.e., the near field and the far field. The near field including the structure and nearby soil layers is conveniently represented by the finite elements, and the soils in the far field by infinite elements. Owing to the fact that finite elements are available in standard textbooks, in this paper, only the formulation of infinite elements will be summarized. Considering the infinite element shown in Fig. 2, the transformation for the coordinates x and y are expressed as following:

$$x=\sum_{i=1}^{5} N'_i x_i, \quad y=\sum_{i=1}^{5} N'_i y_i \tag{1}$$

where the shape functions N' are assumed to be linear in ξ and quadratic in η, i.e.,

$$N'_1=-\frac{(\xi-1)(\eta-1)\eta}{2}, \quad N'_2=(\xi-1)(\eta-1)(\eta+1)$$
$$N'_3=-\frac{(\xi-1)(\eta+1)\eta}{2}, \quad N'_4=\xi(\eta+1)/2 \tag{2}$$
$$N'_5=-\frac{\xi(\eta-1)}{2}$$

The element displacements u and v can be interpolated as follows:

$$u=\sum_{i=1}^{5} N_i u_i, \qquad v=\sum_{i=1}^{5} N_i v_i \tag{3}$$

where the shape functions N_i are

$$N_1=\frac{\eta(\eta-1)}{2}\times P(\xi), \quad N_2=-(\eta-1)(\eta+1)\times P(\xi)$$
$$N_3=\frac{\eta(\eta+1)}{2}\times P(\xi) \tag{4}$$

and the propagation function $P(\xi)$ is defined as

$$P(\xi)=\exp(-\alpha_L\xi)\times\exp(-ik_L\xi) \tag{5}$$

where α_L denotes the displacement amplitude decay factor and k_L is the wave number, both expressed in the local coordinates. In equation (5), the term $\exp(-\alpha_L\xi)$ is introduced to represent the amplitude attenuation due to wave dispersion and the term $\exp(-ik_L\xi)$ the phase decay due to wave propagation in the local coordinates. The two local parameters α_L and k_L can be related to the parameters α and k in the global coordinates as: $\alpha_L = \alpha L$ and $kL = k_L$. The latter are known to be available in practice, if it is realized that $k = \omega/v$, where ω and v denote, respectively, the frequency and velocity of the traveling waves.

According to the previous studies by Yang et al. (1996) and Yang and Hung (1997), for the case of an elastic half-space subjected to a line load on the free surface, the amplitude decay factor α should be selected as $\alpha = 1/2R$ for modeling the regions where the body waves are dominant, with R denoting the distance between the source and the boundary of the far field. As for the mesh size requirements of finite /infinite element, R is allowed to vary from $0.5\lambda_s$ to $5\lambda_s$, where λ_s is the length of the shear wave. Since the Rayleigh waves do not decay on the free surface under the same loading condition, it is suggested that $\alpha = 0$ be used for regions near the free surface.

3 EQUATION OF MOTION IN FREQUENCY DOMAIN

By the procedure of assembly for finite elements and the assumption of harmonic loading, one can derive the equation of motion for the finite/infinite element assemblage under a particular exciting frequency ω:

$$\left(-\omega^2[M]+i\omega[C]+[K]\right)\{\Delta\}=\{F\} \qquad (6)$$

where $[M]$, $[C]$, and $[K]$ respectively represent the mass, damping, and stiffness matrices of the two-dimensional model, $\{\Delta\}$ the amplitudes of nodal displacements, and $\{F\}$ the applied loads acting on the viaduct.

Because of the infinite range in the integrals for the mass and stiffness matrices brought by the term $P(\xi)$ through the shape functions, conventional integration schemes cannot be directly applied. Instead, the Newton-Cotes integration scheme with infinite domains devised by Bettess and Zienkiewicz (1977) will be employed to evaluate the integrals of the mass and stiffness matrices. Readers who are interested in derivation of the finite/infinite element method described herein should refer to Yang et al. (1996) and Yang and Hung (1997) for further details, which will not be recapitulated herein.

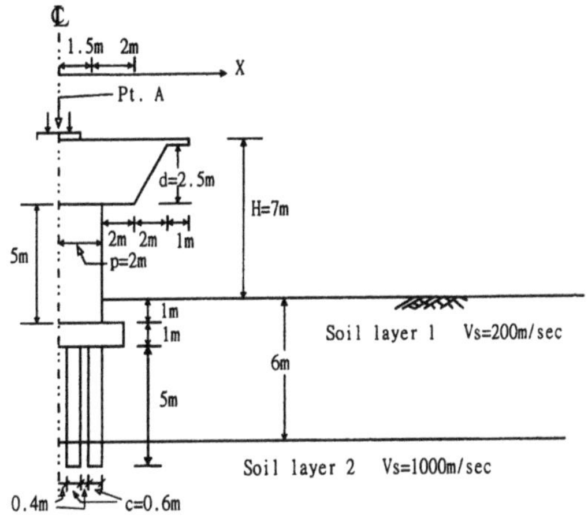

Fig. 3 Model of soil-viaduct-pile interaction system

Table 1 Basic properties of viaduct-pile-soil interaction system

	Elastic modulus E (MPa)	Poisson ratio ν	Density ρ (kg/m^3)	Damping ratio β
Sleeper	34,000	0.17	2,400	0.02
Bridge	21,000	0.2	2,400	0.02
Pier	7,000	0.2	1,900	0.02
Foundation	9,000	0.2	1,900	0.02
Pile	5,000	0.17	1,900	0.05
Soil Layer 1	198.72	0.38	1,800	0.05
Soil Layer 2	6,912	0.44	2,400	0.05

4 NUMERICAL EXAMPLES

In the present study, the finite/infinite element method described will be employed to investigate the influence of the viaduct structure and piles upon the ground vibration caused by a unit harmonic loading at the center point of the viaduct in representation of the excitation effect of the traveling train. To evaluate the influence of pile foundations on transmission of the train-induced vibrations to the nearby soils, the piles are assumed to be sitting on the underlying rocks.

Figure 3 shows the geometry characteristics of the viaduct-pile-soil interaction system. Two layers of soil are considered in this example. The shear wave velocity of soil layer-1 is assumed to be 200 m/s, as typical of soft soils, and that of soil layer-2 is 1,000 m/s, as typical of bedrock. The basic properties of the two-dimensional profile considered have been listed in Table 1. To meet the condition of plain strain, the geometric and material properties, such as the elastic modulus and density, of the piles and foundations are reduced in this example to account for their original discrete distribution along the direction normal to the profile.

4.1 *Effect of viaduct-pile system*

To predict the influence of the viaduct-pile system on the ground vibrations due to trains moving over the viaduct, some numerical analyses will be carried out herein. For the purpose of comparison, two cases are considered. One is the case with no viaduct, and the other is the case with a viaduct, the dimensions and material properties of which are given in Fig. 3 and Table 1.

The transfer functions of the system response at point A and other three points of measurement selected along the soil surface at $x = 15$ m, 24 m, and 35 m have been plotted in Fig. 4(a). As can be seen, since the stiffness of the viaduct-pile system is higher than that of the soils, the existence of the viaduct-pile system tends to increase the resonant frequency of the soils, while reducing the response amplitude of the soil surface for some frequencies.

In Fig. 4(b), the response amplitudes of the soils for various frequencies have been plotted for a ground distance of $x = 40$ m. In general, the response contributed by higher frequencies tends to decay more rapidly as the distance increases, but not for frequency $f = 15$ Hz. Moreover, for the case with viaduct, the response amplitude is generally smaller than that for the case with no viaduct. It should be noted that for the present example, the frequency 5 Hz is below the cut-off frequency, for which the transmissibility is very small.

4.2 *Influence of piles on the ground vibration*

To investigate the influence of piles on the dynamic responses of the viaduct and soils caused by the moving loads traveling over the viaduct, a study will be carried out for the piles with a diameter of 0.8 m, 0.6 m, 0.4 m and 0 m (i.e., no piles). The transfer functions of the vertical displacement response at point A and other three points of measurement selected along the soil surface at $x =$ 15 m, 24 m, and 35 m have been plotted in Fig. 5(a). A significant phenomenon that can be observed from this figure is the shifting of the resonant frequency (for the case with no piles) to a much higher value due to the existence of pile foundations, which increase the transmissibility of the system by connecting the two solid bodies, bridge and underlying rock. Further, the larger the pile foundation, the greater the shift is, accompanied by a smaller amplitude. The vibration amplitude of the viaduct, as represented by point A, reaches the highest at a resonant frequency of 9 Hz when there are no piles. Great decay in the response amplitude can also be observed for the case with no piles.

In Fig. 5(b), the displacement amplitude has been plotted for various frequencies along the soil surface from $x = 0$ m to 40 m. It is observed that for some frequencies, the displacements at the distant place are amplified due to the wave transmission through the pile to the bedrock. The vibration shapes are generally similar, in terms of wavelengths, for most frequencies for piles of different sizes. However, different response amplitudes can be observed for piles of different sizes.

5 CONCLUDING REMARKS

In this paper, the finite/infinite element approach is employed to investigate the vibration of a viaduct-pile-soil interaction system induced by trains moving over the viaduct. The following conclusions can be drawn from the exemplary studies concerning the existence of the viaduct and piles, which are also strictly valid only for the conditions assumed in the analysis:

(1) Compared with the case with no viaduct, the numerical studies indicate that the existence of the viaduct-pile system might lead to increase the resonant frequency of the soils, while reducing the response amplitude of the soil surface for some frequencies.
(2) For the case with viaduct, the response amplitude is generally smaller than that for the case with no viaduct.
(3) The existence of pile foundations can effectively reduce the response amplitude of the viaduct itself when subjected to the high speed trains.
(4) Large response may be carried over to the distant place due to better transmission of vibration as the two wave-transmitted bodies, namely, viaduct and underlying rocks, are connected by the piles. This is especially true for waves with frequency $f = 15$ Hz, which is close to the dominated frequency of soils considered herein.

ACKNOWLEDGEMENT

The research results reported herein were sponsored in part by the National Science Council of through Grant Nos. NSC 89-2211-E-002-002, NSC 90-2211-E-002-057, and NSC 91-2211-E-032-018. Such financial aids are gratefully acknowledged.

REFERENCES

1. Beskos, D. E. (1987). "Boundary element

methods in dynamic analysis," *Appl. Mech. Rev.*, **40**, 1-23.
2. Beskos, D. E. (1997). "Boundary element methods in dynamic analysis: Part II (1986-1996)," *Appl. Mech. Rev.*, **50**(3), 149-197.
3. Hung, H. H. (2000). "Ground vibration induced by high speed trains and vibration isolation countermeasures," Ph.D. thesis, Department of Civil Engineering, National Taiwan University, Taipei, Taiwan, R.O.C.
4. Bettess, P., and Zienkiewicz, O. C. (1977). "Diffraction and refraction of surface waves using finite and infinite elements," *Intl. J Num. Meth. Engrg.*, **11**, 1271-1290.
5. Gutowski, T. G., and Dym, C. L. (1976). "Propagation of ground vibration: A review," *J. Sound Vibr.*, ASCE, **49**(2), 179-193.
6. Yang, Y. B., Kuo, S. R., and Hung, H. H. (1996). "Frequency-independent infinite element for analyzing semi-infinite problems," *Intl. J Num. Meth. Engrg.*, **39**, 3553-3569.
7. Yang, Y. B. and Hung, H. H. (1997). "A parametric study of wave barriers for reduction of train-induced vibration," *Intl. J Num. Meth. Engrg.*, **40**, 3729-3747.

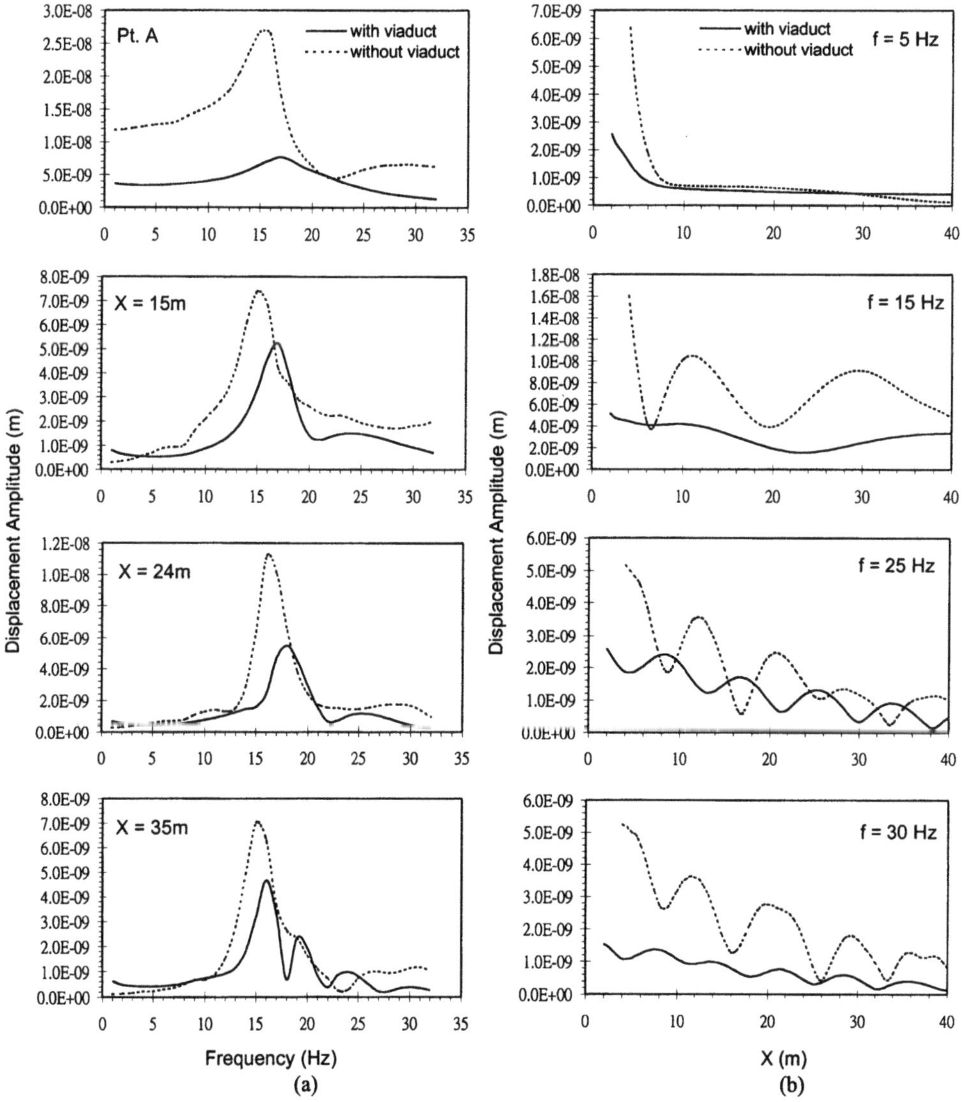

Fig. 4 Effect of the pile-viaduct system on ground vibration

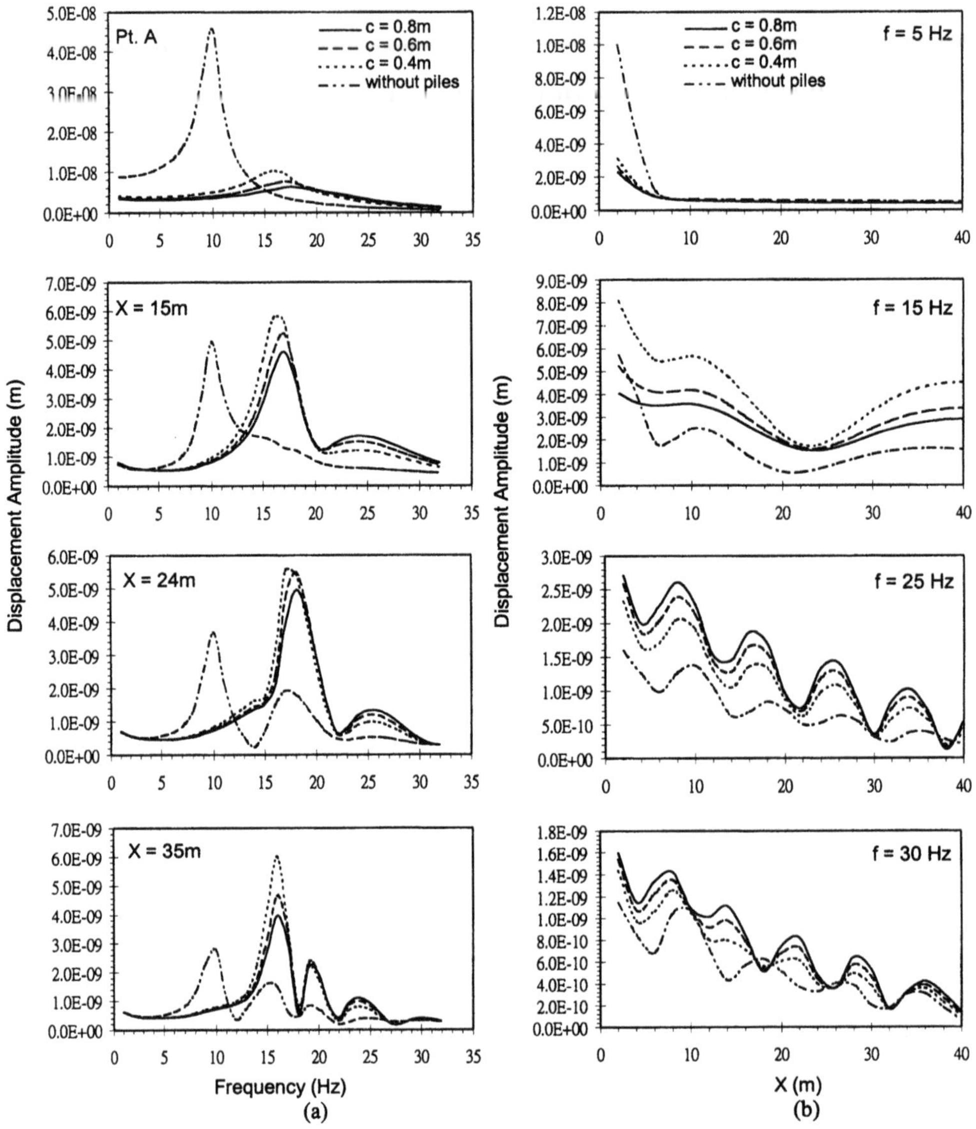

Fig. 5 Influence of pile foundations on the ground vibration

Vibration reduction

Wave propagation – Moving load – Vibration reduction, Chouw & Schmid (eds.)
 ISBN 90 5809 559 2

Reduction of response and first excursion probability for random excitation using nonlinear characteristics

S. Aoki
Tokyo Metropolitan College of Technology, Tokyo, Japan

ABSTRACT: Structures subjected to dynamic load have nonlinear characteristics in nature. Dynamic loads are sometimes random. In such a case, reliability of important structures is evaluated in probabilistic manner. First excursion failure is one of the most important failure modes of the structure. In this paper, methods for obtaining the response and the first excursion probability of the structure with nonlinear characteristic subjected to random dynamic load are presented. For simplicity, single-degree-of-freedom system is used as analytical model. Two-degree-of-freedom system is also used in order to examine coupling effect of two systems. Hysteresis loop characteristic caused by plastic deformation is considered. The mean square value of the response and the first excursion probability are obtained. It is concluded that the response and the first excursion probability are reduced by considering the hysteresis loop characteristic caused by plastic deformation.

1 INTRODUCTION

Structures subjected to dynamic load have nonlinear characteristics in nature. For example, joints and supports have friction and impact characteristics (Dupont and Bapana 1994)(Lau and Zhang 1992). When structures are subjected to excess load, stress is beyond yield point (Clough and Penzien 1993). Furthermore, in order to prevent from failure of the structures subjected to dynamic load, many types of dampers for reduction of response are used. Nonlinear characteristics are applied to some dampers in order to absorb vibration energy. For example, elasto plastic damper, friction damper and impact damper are practically used (Soong and Dargush. 1997).

Dynamic loads are sometimes random (Soong and Grigoriu 1992). For example, earthquake, wind and wave loads are random vibration. Road and rail have random surface and cars and trains are subjected to random excitations. In such a case, reliability of important structures is evaluated in probabilistic manner (Lin 1967). Some failure modes are observed in failure of actual structures. First excursion failure is one of the most important failure modes of the structure (Solnes 1997).

In this paper, methods for obtaining the response and the first excursion probability of the structure with nonlinear characteristic subjected to random dynamic load are presented. For simplicity, single-degree-of-freedom system is used as analytical model. Two-degree-of-freedom system is also used in order to examine coupling effect of two systems. Hysteresis loop characteristic caused by plastic deformation is considered. The proposed method is based on the equivalent linearization method. The mean square value of the response and the first excursion probability are obtained for various values of the damping ratio and the natural period and are compared with those for linear elastic structures.

2 ANALYRICAL METHOD FOR STATIONARY RANDOM EXCITATION

This section presents analytical methods for obtaining the mean square value of the response and the first excursion probability of the structure subjected to stationary random dynamic load.

2.1 *Single-degree-of-freedom system*

Analytical model and equivalent linearization method for obtaining the mean square value of the response and the first excursion probability of single-degree-of-freedom system are shown.

2.1.1 *Analytical model*

For simplicity, single-degree-of-freedom system as shown in Fig.1 is used as an analytical model of the structure. For nonlinear characteristic, hysteresis loop characteristic caused by plastic deformation is considered. In this paper, bilinear hysteresis loop characteristic as shown in Fig.2 is used. Equation of motion with respect to relative displacement z(=x-y) is expressed as :

$$\ddot{z} + 2\zeta\omega_n\dot{z} + f = -\ddot{y} \tag{1}$$

where ζ $(c/2\sqrt{mk})$ is the damping ratio and ω_n $(\sqrt{k/m})$ is the natural circular frequency in elastic range. f is nonlinear restoring force. Equation(1) is equivalently linearized as:

$$\ddot{z} + 2\zeta_{eq}\omega_{eq}\dot{z} + \omega_{eq}{}^2 z = -\ddot{y} \tag{2}$$

where ζ_{eq} is the equivalent damping ratio and ω_{eq} is the equivalent natural circular frequency. As input acceleration $\ddot{y}$, stationary white noise is used.

2.1.2 Equivalent damping ratio

The equivalent damping ratio is obtained as follows. When the system is subjected to a harmonic excitation with the amplitude Z, dissipated energy by hysteresis loop during one cycle is area of hysteresis loop (Fig.3) and given as:

$$E_N{}' = 4\omega_n{}^2(1-\beta)Z_e(Z-Z_e) \tag{3}$$

where Z_e is yield displacement. It is assumed that the response is narrow band random process and the probability density function of Z is given by the Rayleigh distribution as follows (Crandall and Mark 1963).

$$p(Z) = \frac{Z}{\sigma_z{}^2}\exp\left(-\frac{Z}{2\sigma_z{}^2}\right) \tag{4}$$

where σ_z^2 is the mean square value of z. The expected value of E_N is obtained as:

$$E_N = \int_{Z_e}^{\infty} E_N{}' p(Z)dZ$$
$$= 4\omega_n{}^2(1-\beta)Z_e{}^2\frac{1}{y_0}\frac{\sqrt{\pi}}{2}\mathrm{erfc}(y_0) \tag{5}$$

where

$$y_0 = \frac{Z_e}{\sqrt{2}\sigma_z},\ \mathrm{erfc}(y_0) = 1 - \frac{2}{\sqrt{\pi}}\int_0^{y_0}\exp(-y^2)dy \tag{6}$$

Dissipated energy by the damper with equivalent damping coefficient C_{eq} during one cycle is

$$E_T{}' = \pi C_{eq}\omega_n Z^2 \tag{7}$$

The expected value of E_T is obtained as:

$$E_T = \int_0^{\infty} E_T{}' p(Z)dZ = 2\pi\sigma_z^2 C_{eq}\omega_n \tag{8}$$

E_N is equal to E_T. Then,

$$C_{eq} = \frac{2\omega_n{}^2(1-\beta)y_0\mathrm{erfc}(y_0)}{\sqrt{\pi}\omega_n} \tag{9}$$

ζ_{eq} is obtained from the following equation.

$$\zeta_{eq} = \frac{2\zeta\omega_n + C_{eq}}{2\omega_{eq}} \tag{10}$$

2.1.3 Equivalent natural circular frequency

It is assumed that the equivalent natural circular frequency of the system subjected to harmonic excitation is approximated as gradient of the diagonal line of hysteresis loop. Then,

$$\left(\omega_{eq}{}^2\right)' = \begin{cases} \omega_n{}^2 & : Z \le Z_e \\ \dfrac{\omega_n{}^2 Z_e + \beta\omega_n^2(Z-Z_e)}{Z} & : Z > Z_e \end{cases} \tag{11}$$

The expected value of $(\omega_{eq}^2)'$ is obtained as:

$$\omega_{eq}{}^2 = \int_0^{Z_e}\left(\omega_{eq}{}^2\right)' p(Z)dZ + \int_{Z_e}^{\infty}\left(\omega_{eq}{}^2\right)' p(Z)dZ$$
$$= \omega_n{}^2 - (1-\beta)\omega_n{}^2\exp\left(-y_0{}^2\right)$$
$$+ (1-\beta)\omega_n{}^2\sqrt{\pi}y_0\mathrm{erfc}(y_0) \tag{12}$$

2.1.4 Mean square value and first excursion probability

The mean square value of the response and the first excursion probability for the system subjected to stationary white noise are obtained for various values of the damping ratio and the natural period and are compared with those for linear elastic structures.

The mean square value of the relative displacement response is given as:

$$\sigma_z{}^2 = \frac{\pi}{2\zeta_{eq}\omega_{eq}{}^3}S_0 \tag{13}$$

where S_0 is power spectral density of stationary white noise. The mean square value of the relative velocity is given as:

$$\sigma_{\dot{z}}{}^2 = \frac{\pi}{2\zeta_{eq}\omega_{eq}}S_0 \tag{14}$$

It is assumed that failure occurs when the absolute value of the response first crosses the tolerance level B_D. The first excursion probability is given as (Crandall and Mark 1963):

$$P_f(t) = 1 - \exp(-2\nu t) \tag{15}$$

where

$$\nu = \frac{1}{2\pi}\omega_{eq}\exp\left(-\frac{B_D{}^2}{2\sigma_z{}^2}\right) \tag{16}$$

2.2 Two-degree-of-freedom system

As an analytical model, two-degree-of-freedom system as shown in Fig.3 is also used in order to examine coupling effect of two structures. It is assumed that the upper system has nonlinear characteristic. The equations of motion with respect to relative displacement $z_s(x_s-x_p)$ and $z_p(x_p-y)$ are expressed as:

$$\left.\begin{aligned}&\ddot{z}_s + 2\zeta_s\omega_s(1+\gamma)\dot{z}_s + (1+\gamma)f \\ &\qquad - 2\zeta_p\omega_p\dot{z}_p - \omega_p^{\,2}z_p = 0 \\ &\ddot{z}_p + 2\zeta_p\omega_p\dot{z}_p + \omega_p^{\,2}z_p - 2\zeta_s\omega_s\gamma\dot{z}_s - \gamma f = \ddot{y}\end{aligned}\right\} \quad (17)$$

where ζ_s ($c_s/2\sqrt{m_s k_s}$) and ζ_p ($c_p/2\sqrt{m_p k_p}$) are the damping ratio of the upper system and that of the lower system, respectively. ω_s ($\sqrt{k_m/m_m}$) and ω_p ($\sqrt{k_p/m_p}$) are the natural circular frequency of the upper system and that of the lower system, respectively in elastic range. γ (m_s/m_p) is mass ratio of the upper system to the lower system. Nonlinear restoring force is equivalently linearized by the same method as single-degree-of-freedom system.

The mean square response of relative displacement z_s is obtained as:

$$\sigma_{z_s}^{\,2}(t) = S_0\int_{-\infty}^{\infty}\left|H_s(\omega)\right|^2 d\omega \quad (18)$$

where

$$H_s(\omega) = \frac{R_n(\omega)}{R_d(\omega)} \quad (19)$$

$$R_n(\omega) = -\left(2\zeta_p\omega_p\omega i + \omega_p^{\,2}\right) \quad (20)$$

$$\begin{aligned}R_d(\omega) = \omega^4 &- \left\{2\zeta_p\omega_p + C_{eq}(1+\gamma)\right\}\omega^3 i \\ &- \left\{\omega_p^{\,2} + \omega_{eq}^{\,2}(1+\gamma) + 2\zeta_p\omega_p C_{eq}\right\}\omega^2 \\ &+ \left(2\zeta_p\omega_p\omega_{eq}^{\,2} + C_{eq}\omega_p^{\,2}\right)\omega i + \omega_p^{\,2}\omega_s^{\,2}\end{aligned} \quad (21)$$

Integral in Eq.(18) is obtained as:

$$I_4 = \frac{\pi}{\lambda_4}\frac{\begin{vmatrix}\xi_3 & \xi_2 & \xi_1 & \xi_0 \\ -\lambda_4 & \lambda_2 & -\lambda_0 & 0 \\ 0 & -\lambda_3 & \lambda_1 & 0 \\ 0 & \lambda_4 & -\lambda_2 & \lambda_0\end{vmatrix}}{\begin{vmatrix}\lambda_3 & -\lambda_1 & 0 & 0 \\ -\lambda_4 & \lambda_2 & -\lambda_0 & 0 \\ 0 & -\lambda_3 & \lambda_1 & 0 \\ 0 & \lambda_4 & -\lambda_2 & \lambda_0\end{vmatrix}} \quad (22)$$

where

$$\lambda_4 = 1,\ \lambda_3 = 2\varsigma_p\omega_p + C_{eq}(1+\gamma)$$

$$\lambda_2 = \omega_p^{\,2} + \omega_{eq}^{\,2}(1+\gamma) + 2\zeta_p\omega_p C_{eq}$$

$$\lambda_1 = 2\zeta_p\omega_p\omega_s^{\,2} + C_{eq}\omega_p^{\,2},\ \lambda_0 = \omega_p^{\,2}\omega_{eq}^{\,2}$$

$$\xi_3 = 0,\ \xi_2 = 0,\ \xi_1 = \left(2\varsigma_p\omega_p\right)^2,\ \xi_0 = \omega_p^{\,4} \quad (23)$$

3 ANALYRICAL METHOD FOR NON-STATIONARY RANDOM EXCITATION

This section presents methods for obtaining the mean square value of the response and the first excursion probability of the structure subjected to non-stationary random dynamic load.

3.1 Single-degree-of-freedom system

For nonstationaty excitation, it is assumed that the equivalent damping ratio ζ_{eq} and the equivalent natural circular frequency ω_{eq} are approximately given by Eq.(10) and Eq.(12). For input acceleration $\ddot{y}$, non-stationary white noise as defined in the following equation is used.

$$\ddot{y}(t) = I(t)s_y(t) \quad (24)$$

where $s_y(t)$ is stationary white noise and $I(t)$ is envelop function representing nonstationary amplitude characteristic. $I(t)$ is shown in Fig.4. The mean square value of the response is obtained by the moment equations as follows:

$$\begin{aligned}&\frac{d\sigma_z^{\,2}}{dt} = 2\kappa_{z\dot{z}} \\ &\frac{d\sigma_{\dot{z}}^{\,2}}{dt} = -2C_{eq}\sigma_{\dot{z}}^{\,2} - 2\omega_{eq}^{\,2}\kappa_{z\dot{z}} + 2\pi S_0\{I(t)\}^2 \\ &\frac{d\kappa_{z\dot{z}}}{dt} = \sigma_{\dot{z}}^{\,2} - C_{eq}\kappa_{z\dot{z}} - \omega_{eq}^{\,2}\sigma_z\end{aligned} \quad (25)$$

The first excursion probability is obtained by the following equation.

$$P_f(t) = 1 - \exp\left\{-2\int_0^t \nu(t)dt\right\} \quad (26)$$

It is assumed that instants at which the response crosses the tolerance level are statistically independent. $\nu(t)$ is given as follows:

$$\nu(t) = \frac{1}{2\pi}\frac{\sqrt{D}}{\sigma_z^{\,2}}\left[\exp\left\{-\frac{B_D^{\,2}}{2\sigma_z^{\,2}}\left(1+\frac{\kappa_{z\dot{z}}}{D}\right)\right\}\right.$$
$$\left. + B_D\kappa_{z\dot{z}}\sqrt{\frac{\pi}{2D\sigma_z^{\,2}}}\exp\left(-\frac{B_D^{\,2}}{2\sigma_z^{\,2}}\right)\{1+\mathrm{erf}(C)\}\right] \tag{27}$$

where

$$C = \frac{\kappa_{z\dot{z}}B_D}{\sqrt{2D\sigma_z^{\,2}}}, \quad D = \sigma_z^{\,2}\sigma_{\dot{z}}^{\,2} - \kappa_{z\dot{z}}^{\,2}$$
$$\mathrm{erf}(u) = \frac{2}{\sqrt{\pi}}\int_0^u \exp\left(-y^2\right)dy \tag{28}$$

3.2 Two-degree-of-freedom system

The equations of motion are written as follows:

$$\dot{\mathbf{z}} = \mathbf{G}\mathbf{z} + \mathbf{f} \tag{29}$$

where

$$\mathbf{z}^T = \{z_p \quad z_s \quad \dot{z}_p \quad \dot{z}_s\} \tag{30}$$

$$\mathbf{G} = \begin{bmatrix} 0 & 0 & 1 & 0 \\ 0 & 0 & 0 & 1 \\ -\omega_p^{\,2} & \omega_{eq}^{\,2}\gamma & -2\zeta_p\omega_p & C_{eq}\gamma \\ \omega_p^{\,2} & -\omega_{eq}^{\,2}(1+\gamma) & 2\zeta_p\omega_p & -C_{eq}(1+\gamma) \end{bmatrix} \tag{31}$$

$$\mathbf{f}^T = \{0 \quad 0 \quad -\ddot{y} \quad 0\} \tag{32}$$

The second moments of the response are expressed as (Roberts and Spanos 1990):

$$\dot{\mathbf{V}} = \mathbf{G}\mathbf{V}^T + \mathbf{V}\mathbf{G}^T + \mathbf{D} \tag{33}$$

where

$$\mathbf{V} = \begin{bmatrix} \sigma_{z_p}^{\,2} & \kappa_{z_pz_s} & \kappa_{z_p\dot{z}_s} & \kappa_{z_p\dot{z}_s} \\ \kappa_{z_pz_s} & \sigma_{z_s}^{\,2} & \kappa_{z_s\dot{z}_p} & \kappa_{z_s\dot{z}_s} \\ \kappa_{z_p\dot{z}_s} & \kappa_{z_s\dot{z}_p} & \sigma_{\dot{z}_p}^{\,2} & \kappa_{\dot{z}_p\dot{z}_s} \\ \kappa_{z_p\dot{z}_s} & \kappa_{z_s\dot{z}_s} & \kappa_{\dot{z}_p\dot{z}_s} & \sigma_{\dot{z}_s}^{\,2} \end{bmatrix} \tag{34}$$

$$\mathbf{D} = \begin{bmatrix} 0 & 0 & 0 & 0 \\ 0 & 0 & 0 & 0 \\ 0 & 0 & 2\pi S_0\{I(t)\}^2 & 0 \\ 0 & 0 & 0 & 0 \end{bmatrix} \tag{35}$$

Using Eqs.(31)-(35), moment equations are obtained as follows:

$$\frac{d\sigma_{z_p}^{\,2}}{dt} = 2\kappa_{z_p\dot{z}_p}$$

$$\frac{d\kappa_{z_pz_s}}{dt} = \kappa_{z_s\dot{z}_p} + \kappa_{z_p\dot{z}_s}$$

$$\frac{d\kappa_{z_p\dot{z}_p}}{dt} = \sigma_{\dot{z}_p}^{\,2} - \omega_p^{\,2}\sigma_{z_p} + \omega_{eq}^{\,2}\gamma\ \kappa_{z_pz_s}$$
$$-2\zeta_p\omega_p\kappa_{z_p\dot{z}_p} + C_{eq}\gamma\ \kappa_{z_p\dot{z}_s}$$

$$\frac{d\kappa_{z_p\dot{z}_s}}{dt} = \kappa_{\dot{z}_p\dot{z}_s} + \omega_p^{\,2}\sigma_{z_p}^{\,2} - \omega_{eq}^{\,2}(1+\gamma)\kappa_{z_pz_s}$$
$$+2\zeta_p\omega_p\kappa_{z_p\dot{z}_p} - C_{eq}(1+\gamma)\kappa_{z_p\dot{z}_s}$$

$$\frac{d\sigma_{z_s}^{\,2}}{dt} = 2\kappa_{z_s\dot{z}_s}$$

$$\frac{d\kappa_{z_s\dot{z}_p}}{dt} = \kappa_{\dot{z}_p\dot{z}_s} - \omega_p^{\,2}\kappa_{z_pz_s} + \omega_{eq}^{\,2}\gamma\ \sigma_{z_s}^{\,2}$$
$$-2\zeta_p\omega_p\kappa_{z_s\dot{z}_p} + C_{eq}\gamma\ \kappa_{z_s\dot{z}_s}$$

$$\frac{d\kappa_{z_s\dot{z}_s}}{dt} = \sigma_{\dot{z}_s}^{\,2} + \omega_p^{\,2}\kappa_{z_pz_s} - \omega_{eq}^{\,2}(1+\gamma)\sigma_{z_s}^{\,2}$$
$$+2\zeta_p\omega_p\kappa_{z_s\dot{z}_p} - C_{eq}(1+\gamma)\kappa_{z_s\dot{z}_s}$$

$$\frac{d\sigma_{\dot{z}_p}^{\,2}}{dt} = 2\left(-\omega_p^{\,2}\kappa_{z_p\dot{z}_p} + \omega_{eq}^{\,2}\gamma\ \kappa_{z_s\dot{z}_p} - 2\zeta_p\omega_p\sigma_{\dot{z}_p}^{\,2}\right.$$
$$\left.+C_{eq}\gamma\ \kappa_{z_p\dot{z}_s}\right) + 2\pi S_0\{I(t)\}^2$$

$$\frac{d\kappa_{\dot{z}_p\dot{z}_s}}{dt} = -\omega_p^{\,2}\kappa_{z_p\dot{z}_s} + \omega_{eq}^{\,2}\gamma\ \kappa_{z_s\dot{z}_s} - 2\zeta_p\omega_p\kappa_{\dot{z}_p\dot{z}_s}$$
$$+C_{eq}\gamma\ \sigma_{z_s}^{\,2} + \omega_p^{\,2}\kappa_{z_p\dot{z}_p} - \omega_{eq}^{\,2}(1+\gamma)\kappa_{z_s\dot{z}_p}$$
$$+2\zeta_p\omega_p\sigma_{\dot{z}_p} - C_{eq}(1+\gamma)\kappa_{\dot{z}_p\dot{z}_s}$$

$$\frac{d\sigma_{\dot{z}_s}^{\,2}}{dt} = 2\left\{\omega_p^{\,2}\kappa_{z_p\dot{z}_s} - \omega_{eq}^{\,2}(1+\gamma)\kappa_{z_s\dot{z}_s} + 2\zeta_p\omega_p\kappa_{\dot{z}_p\dot{z}_s}\right.$$
$$\left.-C_{eq}\sigma_{\dot{z}_s}^{\,2}\right\} \tag{36}$$

4 NUMERICAL EXAMPLES

Numerical examples of the mean square value of the response and the first excursion probability are shown.

4.1 Response for stationary excitation

4.1.1 Single-degree-of-freedom system

The mean square value of the relative displacement z(=x-y), relative velocity, the equivalent damping ratio ζ_{eq} and the equivalent natural period T_{eq} $(2\pi/\omega_{eq})$ of single-degree-of freedom system subjected to stationary white noise excitation are obtained from Eqs.(10), (12), (13) and (14) by iteration method. The first excursion probability is obtained from Eqs.(15) and (16). Yield displacement Ze is determined using the standard deviation of the relative displacement of linear elastic structure $\sigma_{z\ell}$ as follows.

$$Z_e = \alpha\, \sigma_{z\ell} \tag{37}$$

The tolerance level B_D is determined using $\sigma_{z\ell}$ as follows.

$$B_D = \delta_t \sigma_{z\ell} \tag{38}$$

Table 1 shows the mean square value of relative displacement and relative velocity for some values of yield displacement. ζ and T_n $(2\pi/\omega_n)$ are the damping ratio and the natural period in elastic range. The mean square value of the response of linear elastic structure is shown below the table. The mean square value for the structure with nonlinear characteristic is smaller than that for linear elastic characteristic. The mean square value of the response decreases with the decrease of α and β. That is, the mean square value decreases with the decrease of yield displacement and ratio of post yield stiffness and pre yield stiffness. The equivalent damping ratio increases with the decrease of α and β. The equivalent natural period becomes longer with the decrease of α and β. However, effect of α and β on T_{eq} is very small.

Table 2 and Table 3 show the mean square values for some values of the damping ratio and the natural period, respectively. From Table 2, the equivalent natural period is almost constant and difference between ζ_{eq} and ζ increase with the increase of ζ. From Table 3, the equivalent damping ratio is constant independent of the natural period.

Figure 5 and 6 show the first excursion probability for β=0 and β=0.5, respectively. The first excursion probability of the structure with nonlinear characteristic is smaller than that of linear elastic structure. The first excursion probability decreases with the decrease of α and β.

4.1.2 Two-degree-of-freedom system

The mean square values, the equivalent damping ratio and the equivalent natural period are obtained from Eqs.(10), (12) and (18) by iteration method. Table 4 and Table 5 show the mean square values of the response of the upper and the lower system for γ=0 and γ=0.1, respectively. For γ=0, coupling effect is neglected. The mean square value of the upper system with nonlinear characteristic is smaller than those of linear elastic system for γ=0. The mean square values of both systems with nonlinear characteristic are smaller than those of linear elastic systems for γ=0.1. The equivalent damping ratio increases with the decrease of α and β. The equivalent natural period becomes longer with the decrease of α and β.

4.2 Response for nonstationary excitation

For nonstationary excitation, Z_e is determined using the maximum standard deviation of the relative displacement of linear elastic system $\sigma_{z\ell\,max}$ as follows.

$$Z_e = \alpha\, \sigma_{z\ell\,max} \tag{39}$$

The tolerance level B_D is determined using $\sigma_{z\ell\,max}$ as follows.

$$B_D = \delta_t \sigma_{z\ell\,max} \tag{40}$$

The first excursion probability is function of time. In this paper, the first excursion probability at the end of excitation is obtained.

The maximum mean square values, the maximum equivalent damping ratio and the longest equivalent natural period are obtained. Table 6 shows the results for single-degree-of-freedom system. The maximum mean square values for the structure with nonlinear characteristic are smaller than those for linear elastic structure. The maximum mean square values decrease with the decrease of α and β. The maximum equivalent damping ratio increases with the decrease of α and β. The longest equivalent natural period becomes longer with the decrease of α and β.

Table 7 and 8 shows the first excursion probability for some values of the natural period and the damping ratio, respectively. The first excursion probability is independent of the natural period. It is also independent of the damping ratio for relatively small value of the damping ratio.

Table 9 shows results for two-degree-of-freedom system. The same results as single-degree-of-freedom system are obtained.

Comparing with simulation method (Aoki and Suzuki 1980), the mean square values obtained from the proposed method agrees well with those form simulation method. The proposed method gives conservative values of the first excursion probability. However, qualitative characteristics are obtained.

5 CONCLUSIONS

The methods for obtaining the mean square value of the response and the first excursion probability of the structure with nonlinear characteristic caused by plastic deformation subjected to random dynamic load are presented. Random dynamic load is modeled as either stationary or nonstationary white noise. As model of the structure, single-degree-of-freedom system and two-degree-of-freedom system are used. It is concluded that the mean square value of the response and the first excursion probability are reduced owing to hysteresis loop characteristic. The mean square value of the response and the first excursion probability decrease with the decrease of yield displacement and ratio of post yield stiffness to pre yield stiffness. The first excursion probability is independent of the relatively low value of the damping ratio and the natural period when the tolerance level is normalized by the standard deviation of the response of linear elastic structure.

REFERENCES

Aoki, S. and Suzuki, K. 1980. First Excursion Probability Estimation of Mechanical Appendage System Subjected to Nonstationary Earthquake Excitations. Proceedings of 4th International Conference on Structural Safety and Reliability. I:291-210.

Clough, R.W. and Penzien, J. 1993. *Dynamics of Structures-Second Edtion*: 711-730. Shigapore: McGraw-Hill.

Crandall, S.H. and Mark, W.D. 1963. *Random Vibration in Mechanical Systems*: 106-110. New York: Academic Press

Dupont and Bapana,D. 1994. Stability of Sliding Frictional Surfaces with Varying Normal Force. *Transactions of American Society of Mechanical Engineers Journal of Vibration and Acoustics*. 116-2: 237-242.

Lau, S.L. and Zhang, W.-S. 1992. Nonlinear Vibrations of Piecewise-Linear Systems by Incremental Harmonic Balance Method. *Transactions of American Society of Mechanical Engineers Journal of Applied Mechanics*. 59-1: 153-160.

Lin, Y.K. 1967. *Probabilistic Theory of Structural Dynamics*: 1-5. Malabar: Krieger.

Roberts, J.B. and Spanos, P.D. 1990. Random Vibration and Statistical Linearization: 113-115. Chichester: John Wiley & sons.

Solnes, J. 1997. *Stochastic Processes and Random Vibrations*: 193-207. Chichester: John Wiley & sons.

Soong, T.T. and Dargush, G.F. 1997. *Passive Energy Dissipation Systems in Structural Engineering*: 5-125. Chichester: John Wiley & sons.

Soong, T.T. and Grigoriu, M. 1992. *Random Vibration of Mechanical and Structural Systems*: 1-3. Englewood Cliffs: Prentice-Hall.

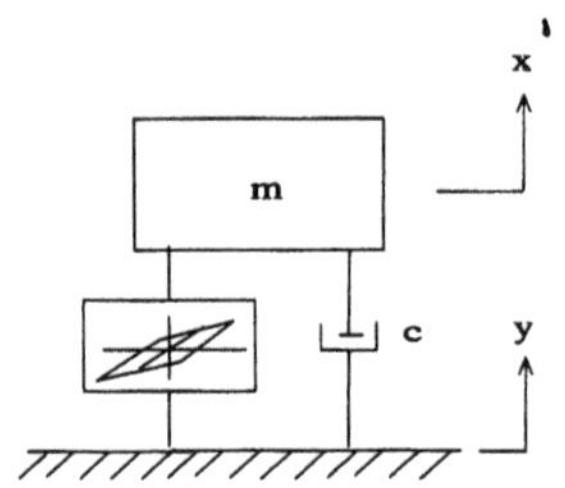

Figure 1 Single-degree-of-freedom model

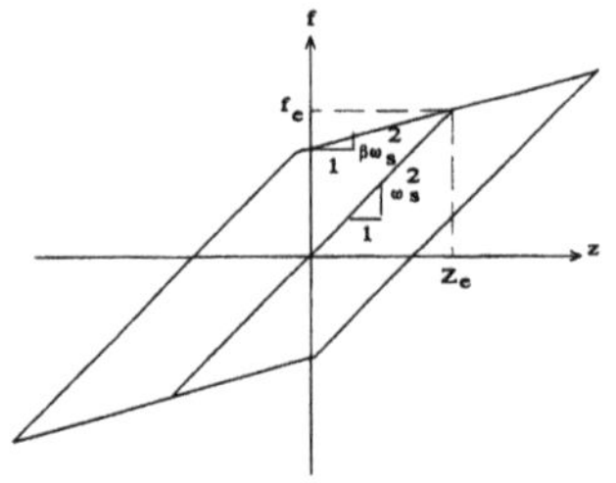

Figure 2 Bilinear hysteresis loop characteristic

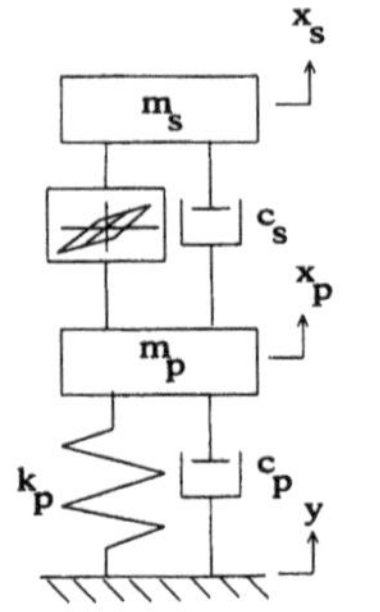

Figure 3 Two-degree-of-freedom model

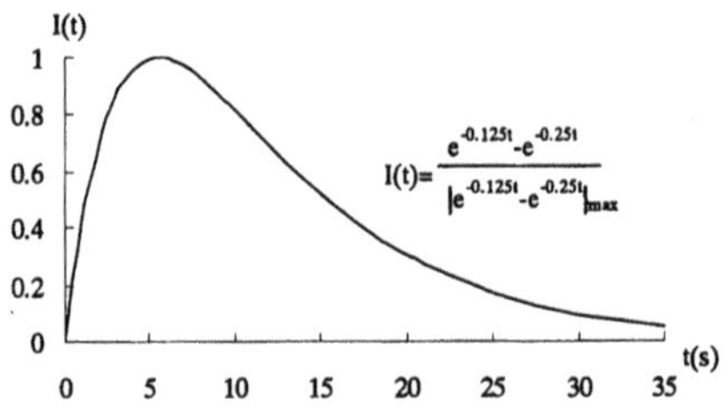

Figure 4 Envelop function for nonstationary excitation

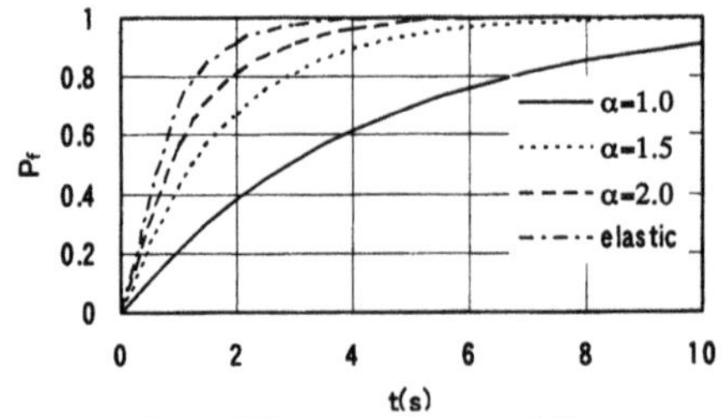

Figure 5 First excursion probability
(α=1.0, β=0, ζ=0.01, T_n=1.0s)

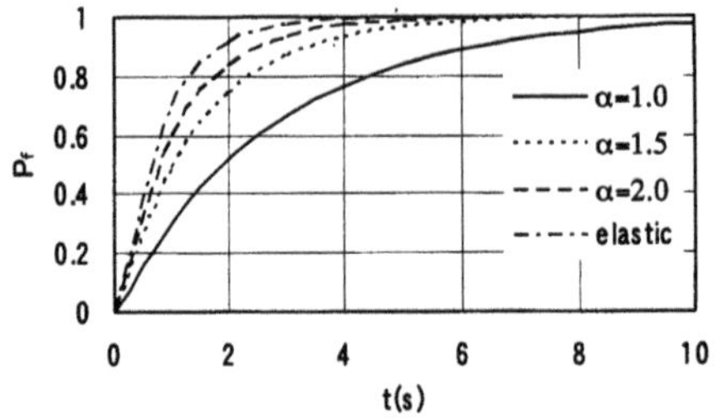

Figure 6 First excursion probability
(α=1.0, β=0.5, ζ=0.01, T_n=1.0s)

Table 1 Mean square value of stationary response, equivalent damping ratio and equivalent natural period (ζ=0.01, T_n=1.0s)

β	α	$\sigma_z^2/S_0(s^5)$	$\sigma_{\dot{z}}^2/S_0(s^3)$	ζ_{eq}	T_{eq}(s)
	1.0	1.50x10^{-1}	5.81	4.34x10^{-2}	1.01
0	1.5	2.47x10^{-1}	9.70	2.59x10^{-2}	1.00
	2.0	3.54x10^{-1}	1.39x10	1.80x10^{-2}	1.00
	1.0	1.87x10^{-1}	7.27	3.47x10^{-2}	1.01
0.5	1.5	2.91x10^{-1}	1.14x10	2.19x10^{-2}	1.00
	2.0	3.99x10^{-1}	1.57x10	1.59x10^{-2}	1.00

For elastic system σ_z^2/S_0 =6.33x10^{-1} and $\sigma_{\dot{z}}^2/S_0$ =2.50x10.

Table 2 Mean square value of stationary response, equivalent damping ratio and equivalent natural period (α=1, T_n=1.0s)

β	ζ	$\sigma_z^2/S_0(s^5)$	$\sigma_{\dot{z}}^2/S_0(s^3)$	$\sigma_z^2/S_0(s^5)$ (elastic)	$\sigma_{\dot{z}}^2/S_0(s^3)$ (elastic)	ζ_{eq}	T_{eq}(s)
	0.01	1.50x10^{-1}	5.81	6.33x10^{-1}	2.50x10	4.34x10^{-2}	1.01
0	0.02	9.41x10^{-2}	3.59	3.17x10^{-1}	1.25x10	7.07x10^{-2}	1.02
	0.05	5.30x10^{-2}	1.95	1.27x10^{-1}	5.00	1.32x10^{-1}	1.03
	0.01	1.87x10^{-1}	7.27	6.33x10^{-1}	2.50x10	3.47x10^{-2}	1.01
0.5	0.02	1.20x10^{-1}	4.61	3.17x10^{-1}	1.25x10	5.50x10^{-2}	1.01
	0.05	6.79x10^{-2}	2.55	1.27x10^{-1}	5.00	1.01x10^{-1}	1.03

Table 3 Mean square value of stationary response, equivalent damping ratio and equivalent natural period (α=1, ζ=0.01)

β	T_n(s)	$\sigma_z^2/S_0(s^5)$	$\sigma_{\dot{z}}^2/S_0(s^3)$	$\sigma_z^2/S_0(s^5)$ (elastic)	$\sigma_{\dot{z}}^2/S_0(s^3)$ (elastic)	ζ_{eq}	T_{eq}(s)
	0.5	1.87x10^{-2}	2.91	7.92x10^{-2}	1.25x10	4.34x10^{-2}	0.50
0	0.8	7.68x10^{-2}	4.65	3.24x10^{-1}	2.00.x10	4.34x10^{-2}	0.80
	1.0	1.50x10^{-1}	5.81	6.33x10^{-1}	2.50x10	4.34x10^{-2}	1.01
	1.5	5.06x10^{-1}	8.72	2.14	3.75x10	4.34x10^{-2}	1.51
	0.5	2.34x10^{-2}	3.64	7.92x10^{-2}	1.25x10	3.47x10^{-2}	0.50
0.5	0.8	9.59x10^{-2}	5.82	3.24x10^{-1}	2.00.x10	3.47x10^{-2}	0.80
	1.0	1.87x10^{-1}	7.27	6.33x10^{-1}	2.50x10	3.47x10^{-2}	1.01
	1.5	6.32x10^{-1}	1.09x10	2.14	3.75x10	3.47x10^{-2}	1.51

Table 4 Mean square value of stationary response, equivalent damping ratio and equivalent natural period (γ=0, ζ_s=0.01, T_s=1.0s, ζ_p=0.01, T_p=1.0s)

β	α	$\sigma_{z_s}^2/S_0(s^5)$	$\sigma_{\dot{z}_s}^2/S_0(s^3)$	$\sigma_{z_p}^2/S_0(s^5)$	$\sigma_{\dot{z}_p}^2/S_0(s^5)$	ζ_{eq}	T_{eq}(s)
	1.0	1.11x10	4.30x10^{2}	1.27x10^{-1}	5.00	3.49x10^{-2}	1.01
0	1.5	1.94x10	7.61x10^{2}	1.27x10^{-1}	5.00	2.29x10^{-2}	1.00
	2.0	2.86x10	1.12x10^{3}	1.27x10^{-1}	5.00	1.69x10^{-2}	1.00
	1.0	1.38x10	5.39x10^{2}	1.27x10^{-1}	5.00	2.97x10^{-2}	1.01
0.5	1.5	2.29x10	8.98x10^{2}	1.27x10^{-1}	5.00	2.01x10^{-2}	1.00
	2.0	3.24x10	1.27x10^{3}	1.27x10^{-1}	5.00	1.52x10^{-2}	1.00

For elastic system $\sigma_{z_s}^2/S_0$ =5.34x10, $\sigma_{\dot{z}_s}^2/S_0$ =2.10x10^{3}, $\sigma_{z_p}^2/S_0$ =1.27x10^{-1}, $\sigma_{\dot{z}_p}^2/S_0$ =5.00.

Table 5 Mean square value of stationary response, equivalent damping ratio and equivalent natural period (γ=0.1, ζ_s=0.01, T_s=1.0s, ζ_p=0.01, T_p=1.0s)

β	α	$\sigma_{z_s}^2/S_0(s^5)$	$\sigma_{\dot{z}_s}^2/S_0(s^3)$	$\sigma_{z_p}^2/S_0(s^5)$	$\sigma_{\dot{z}_p}^2/S_0(s^5)$	ζ_{eq}	$T_{eq}(s)$
	1.0	5.10×10^{-1}	1.66x10	7.37×10^{-2}	2.46	9.48×10^{-2}	1.04
0	1.5	7.06×10^{-1}	2.38x10	9.32×10^{-2}	3.04	5.24×10^{-2}	1.01
	2.0	8.90×10^{-1}	3.06x10	1.13×10^{-1}	3.66	3.02×10^{-2}	1.00
	1.0	6.73×10^{-1}	2.23x10	8.51×10^{-2}	2.80	6.24×10^{-2}	1.03
0.5	1.5	8.07×10^{-1}	2.83x10	1.05×10^{-1}	3.41	3.72×10^{-2}	1.01
	2.0	9.79×10^{-1}	3.40x10	1.22×10^{-1}	3.97	2.26×10^{-2}	1.00

For elastic system $\sigma_{z_s}^2/S_0$=1.17, $\sigma_{\dot{z}_s}^2/S_0$=4.14x10, $\sigma_{z_p}^2/S_0=1.43\times10^{-1}$, $\sigma_{\dot{z}_p}^2/S_0$=4.67.

Table 6 Maximum mean square value of nonstationary response, maximum equivalent damping ratio and longest equivalent natural period (ζ=0.01, T_n=1.0s)

β	α	$\sigma_z^2/S_0(s^5)$	$\sigma_{\dot{z}}^2/S_0(s^3)$	ζ_{eq}	$T_{eq}(s)$
	1.0	1.07×10^{-1}	4.09	6.14×10^{-2}	1.02
0	1.5	1.70×10^{-1}	6.62	3.62×10^{-2}	1.00
	2.0	2.36×10^{-1}	9.25	2.37×10^{-2}	1.00
	1.0	1.38×10^{-1}	5.30	4.58×10^{-2}	1.01
0.5	1.5	2.02×10^{-1}	7.89	2.86×10^{-2}	1.01
	2.0	2.64×10^{-1}	1.04x10	1.94×10^{-2}	1.00

For elastic system $\sigma_z^2/S_0=3.57\times10^{-1}$ and $\sigma_{\dot{z}}^2/S_0$=1.41x10.

Table 7 First excursion probability of nonstationary response (β=0, α=1, ζ=0.01)

δ_t	$T_n(s)$			
	0.5	0.8	1.0	1.5
1.0	9.58×10^{-1}	9.01×10^{-1}	8.67×10^{-1}	8.01×10^{-1}
1.5	1.88×10^{-1}	1.63×10^{-1}	1.52×10^{-1}	1.47×10^{-1}
2.0	5.90×10^{-3}	6.34×10^{-3}	6.82×10^{-3}	8.15×10^{-3}

Table 8 First excursion probability of nonstationary response (β=0, α=1, T_n =1.0s)

δ_t	ζ		
	0.01	0.02	0.05
1.0	8.67×10^{-1}	9.10×10^{-1}	9.60×10^{-1}
1.5	1.52×10^{-1}	2.33×10^{-1}	4.21×10^{-1}
2.0	6.82×10^{-3}	1.55×10^{-2}	5.55×10^{-2}

Table 9 Maximum mean square value of nonstationary response, maximum equivalent damping ratio and longest equivalent natural period (γ=0.1, ζ_s=0.01, T_s=1.0s, ζ_p=0.01, T_p=1.0s)

β	α	$\sigma_{z_s}^2/S_0(s^5)$	$\sigma_{\dot{z}_s}^2/S_0(s^3)$	$\sigma_{z_p}^2/S_0(s^5)$	$\sigma_{\dot{z}_p}^2/S_0(s^5)$	ζ_{eq}	$T_{eq}(s)$
	1.0	4.69×10^{-1}	1.52x10	7.00×10^{-2}	2.36	1.06×10^{-1}	1.05
0	1.5	6.30×10^{-1}	2.13x10	8.35×10^{-2}	2.75	5.90×10^{-2}	1.02
	2.0	7.75×10^{-1}	2.70x10	9.66×10^{-2}	3.17	3.34×10^{-2}	1.01
	1.0	6.17×10^{-1}	2.05x10	7.72×10^{-2}	2.56	6.68×10^{-2}	1.03
0.5	1.5	7.34×10^{-1}	2.52x10	9.11×10^{-2}	2.99	4.03×10^{-2}	1.01
	2.0	8.42×10^{-1}	2.96x10	1.02×10^{-1}	3.36	2.41×10^{-2}	1.00

For elastic system $\sigma_{z_s}^2/S_0=9.60\times10^{-1}$, $\sigma_{\dot{z}_s}^2/S_0$=3.44x10, $\sigma_{z_p}^2/S_0=1.14\times10^{-1}$, $\sigma_{\dot{z}_p}^2/S_0$=3.76.

Wave propagation – Moving load – Vibration reduction, Chouw & Schmid (eds.)
© 2003 Swets & Zeitlinger, Lisse, ISBN 90 5809 559 2

Impediment of wave propagation from a moving source via the subsoil into the building

N. Chouw
Okayama University, Okayama, Japan

G. Pflanz
BMW, Munich, Germany

ABSTRACT: This paper describes a numerical study of the ability of measures to reduce the influence of ground motions on structures. The ground motions are caused by a moving load in the neighbourhood of the structure. The considered measures are a trench, and a wave-impeding barrier (wib) in the soil. An isolation of the structures using steel springs at their base is also considered. The study covers the effect of a soil layer over bedrock on the ground motions, and the effect of the ground motion characteristics on the response of the structure. The investigation indicates that a consideration of the vertical ground motions alone is not enough for a proper determination of the response of the structure. Trench and wib can be used to reduce the ground excitation of the structure. Springs at the base of the structures can reduce induced vibrations from the ground into the structure, especially in the vertical direction.

1 INTRODUCTION

In the study the response of structures to man-made ground vibrations is considered. The source of the ground vibrations is a moving load in the neighbourhood of the structures. The reduction of the structural response can be achieved by modifying the characteristics of the source, by altering the load path, by changing the wave propagation in the soil, and by partially interrupting the wave propagation into the structure or by providing the structure with more damping (Mualla et al. 2002).

We can modify the characteristics of the source by limiting its magnitude, and speed. In case of a train this is, however, difficult, because most of the passengers prefer to arrive at their destination as fast as possible, and the train company wants to transport as many passenger as possible at the same time, for example, by using the double-deck cars. More realistic is altering the characteristics of the load path, by using rigid tracks or by adding reduction measures at the tracks (Yoshioka 2002). In case of road traffic similar developments can be expected. The disturbance in residential areas can be reduced by limiting the speed, and the time period when trucks may use the road.

It is also possible to install a wave barrier to modify the propagation of waves in soil so that certain areas will experience less ground vibrations. One possible wave barrier is a trench along the track at a certain location. The advantage is that a large area beyond the trench can be affected. The disadvantage is the high cost for maintaining the open trench, and that the reflected waves could amplify the track vibrations. To reduce the maintenance cost the use of air cushions as a wave barrier is feasible (Massarsch 2002). Another possible wave barrier is the wave-impeding barrier (wib). It is based on the cut-off frequency of a soil layer over bedrock (Chouw, Le and Schmid 1991a and b). The wib can be installed below the load path or below the considered structures.

We can partially interrupt the wave propagation from the ground into the structure by installing base-isolation systems at the base of the structures. Probably the most common used base-isolation systems are rubber bearings (Naeim & Kelly 1999). The main characteristic of the steel-reinforced rubber bearings is the very large vertical stiffness, compared to the horizontal stiffness. The thin steel plates inside the rubber prevent the lateral bulging of the rubber. They stabilize the bearing against the large vertical load of the structure, and at the same time do not hinder the rubber to shear freely. The reduction of the lateral stiffness at the isolation systems shifts the fundamental frequency of the structure to a lower frequency range, and usually away from the dominant frequencies of the ground vibrations. Consequently, the structure will then experience less excitation. This idea of isolation is effective, as far as the horizontal ground motions are concerned. The advantage of the rubber bearings to carry the heavy load of the structure is at the same time a disadvantage, when the vertical ground motions are considered, since the large stiffness in the vertical direction will allow the rubber bearings to transmit the ground motions into the structure.

In this work the reduction effect of bedrock, a trench, and a wave-impeding barrier (wib) on the ground vibrations is considered. In order to overcome the disadvantage of rubber bearings, steel springs as base isolation of the structures are chosen.

Table 1. Alteration in the load distribution with load speed

Load speed [km/h]	Distribution length [m]	Maximum magnitude P_i [MN/m]
200	1.388	1.1310
300	2.083	0.7540
400	2.778	0.5655
500	3.472	0.4524

2 GROUND VIBRATIONS AND STRUCTURAL RESPONSES

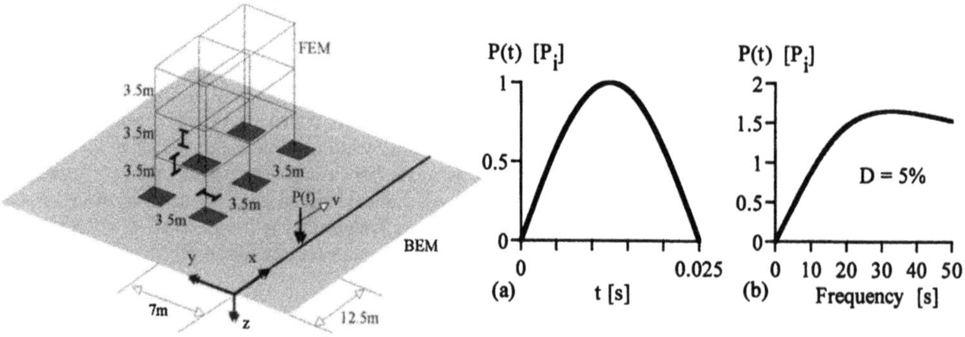

Figure 1. Moving load and the structure.
(a) Time history, and (b) response spectrum of the moving load P(t) with a damping ratio D of 5%.

The considered structures are one-storey and three-storey structure modeled as a frame structure. Figure 1 shows the considered three-storey structure and the characteristics of the load P(t). The mass of the structural members per unit length is 74kg/m. The double T-profile of the steel member is indicated in the figure. The $\hat{y}$ - and $\hat{z}$ -axis is the local axis perpendicular and parallel to the flange, respectively. The flexural stiffness in case of bending about the local $\hat{y}$ - and $\hat{z}$ -axis is $EI_{\hat{y}}$ = 4921.2kNm^2 and $EI_{\hat{z}}$ =34438.9kNm^2, respectively. The torsional stiffness factor for the element cross section (polar moment of inertia) Ip is 7.4 10^{-7}m^4. EA is 1991619kN. The modulus of elasticity E is 21 10^{10}Pa, and the Poisson ratio ν is 0.3.

It is assumed that the soil is a homogenous half-space, and a soil layer of the thickness of 3m over bedrock. The soil density is 2000m/s^3. The shear wave velocity c_s and the compressive wave velocity c_p are 120m/s and 240m/s, respectively. It is assumed that the structure and the soil have no material damping. The structure will therefore experience only radiation damping caused by the wave propagation from the foundations of the structure.

The load moves along a path 7m away from the structure. The considered speeds are 200km/h, 300km/h, 400km/h and 500km/h. In order to cover most of the frequencies of the traffic loads a load with a wide range of dominant frequency content above 20Hz is chosen. The time history of the load is displayed in Figure 1(a). To enable a comparison of the results due to different load speeds the spatial distribution along the load path and the magnitude of the load are modified. The magnitude is chosen in such a way that all boundary elements along the load path experience the same load of 1MN. With a larger speed the load has longer spatial distribution along the load path with a smaller magnitude. Table 1 indicates the dependency of the load characteristics on the speed.

The structures are modeled by beam elements with uniformly distributed mass and stiffness, and the soil by a boundary element method. It is assumed that the foundations are rigid. The ground excitation $a_g(t)$ of the structural foundations is produced by the moving load P(t). After having transformed the ground excitation $a_g(t)$ into the Laplace domain

$$\tilde{a}_g(s) = \int_0^s a_g(t)\, e^{-st}\, dt\,, \tag{1}$$

where $s = \sigma + i\omega$ is the Laplace transform parameter and $i = \sqrt{-1}$, we can determine the response of the structure in the Laplace domain. The tilde indicates the quantity in the Laplace domain.

The time history of the response $u(t)$ is obtained by transforming the results from the Laplace to the time domain.

$$u(t) = \frac{1}{2\pi i} \int_{\sigma-i\omega}^{\sigma+i\omega} \tilde{u}(s)\, e^{st}\, ds \qquad (2)$$

For an accurate Laplace transformation the function in the time domain $f(t)$ must fulfill $|f(t)| \le A e^{Bt}$, where A and B are real constants, and B is smaller than the real part of the Laplace transform parameter σ. In this study a time step of 0.001s, and σ of 3.9063 are used. Details about accurate numerical Laplace transform is given by Hillmer (1987).

For the calculation in the Laplace domain the dynamic stiffness $\tilde{K}^b$ of the structure is obtained by solving the equation of motion in the Laplace domain. Equations 3, 4, and 5 are the equation of motion in the Laplace domain for the axial, transversal and torsional vibration of a structural member, respectively.

$$\frac{m\, s^2}{EA} \tilde{u}_{\hat{x}} - \tilde{u}_{\hat{x},\hat{x}\hat{x}} = 0 \qquad (3)$$

$$\frac{m\, s^2}{EI_{\hat{z}}} \tilde{u}_{\hat{y}} + \tilde{u}_{\hat{y},\hat{x}\hat{x}\hat{x}\hat{x}} = 0, \text{ and } \frac{m\, s^2}{EI_{\hat{y}}} \tilde{u}_{\hat{z}} + \tilde{u}_{\hat{z},\hat{x}\hat{x}\hat{x}\hat{x}} = 0 \qquad (4a, b)$$

$$\frac{m\, s^2}{GI_p} \tilde{\varphi}_{\hat{x}} - \tilde{\varphi}_{\hat{x},\hat{x}\hat{x}} = 0 \qquad (5)$$

m is the mass per unit length of the structural member. Dots and commas indicate time and space derivatives, respectively. The $\hat{x}$-axis is the local axis along the structural members. E and A are the modulus of elasticity and cross-section area, respectively. GI_p is the torsional stiffness. An interpretation of a dynamic stiffness in the Laplace domain is given in (Schmid et al. 2002). To couple the two subsystems *structure and soil* we divide the degree-of-freedom (DOF) of the structure into the contact-degree-of-freedom (CDOF) at the interface between the foundations of the structure and the soil (indicated by the subscript c), and the DOF of the rest of the structure (denoted by the subscript b). The dynamic stiffness $\tilde{K}^s$ of the soil is determined by using the full-space fundamental solution. After having transformed this dynamic stiffness $\tilde{K}^s$ from the soil DOF into the CDOF we obtain the transformed dynamic stiffness $\tilde{K}^s_{cc}$. The equation of motion of the whole system in the Laplace domain is then

$$\begin{bmatrix} \tilde{K}^b_{bb} & \tilde{K}^b_{bc} \\ \tilde{K}^b_{cb} & \tilde{K}^b_{cc} + \tilde{K}^s_{cc} \end{bmatrix} \begin{bmatrix} \tilde{u}^b_b \\ \tilde{u}^b_c \end{bmatrix} = \begin{bmatrix} \tilde{P}^b_b \\ \tilde{P}^b_c \end{bmatrix} \qquad (6)$$

The superscripts b and s indicate the quantity of the structure and soil, respectively. Since only ground excitation $\tilde{P}^b_c$ due to incoming waves is considered, there is no excitation $\tilde{P}^b_b$ of the structure.

The ground excitation $a_g(t)$ of the structure due to the moving load P(t) is determined by using the boundary element method in the time domain. We first transform the wave equation for homogeneous, isotropic, linear-elastic soil

$$(c_p^2 - c_s^2)\, u_{k,ki} + c_s^2\, u_{i,kk} - \ddot{u}_i + f_i = 0, \quad i,k = 1,2,3 \qquad (7)$$

into the boundary integral equation. c_p and c_s are the compressive and shear wave velocity in the soil, respectively. For the transformation we need the fundamental solution of the full space for a Dirac load δ in space and time

$$u^{\bullet}_{ik}(x,t,\xi,\tau) = \frac{t-\tau}{4\pi\rho r^2} \left\{ \frac{3r_{,i}r_{,k} - \delta_{ik}}{r} \left[H(t-\tau-\frac{r}{c_p}) - H(t-\tau-\frac{r}{c_s}) \right] + \frac{r_{,i}r_{,k}}{c_p} \delta(t-\tau-\frac{r}{c_p}) + \frac{\delta_{ik} - r_{,i}r_{,k}}{c_s} \delta(t-\tau-\frac{r}{c_s}) \right\} \qquad (8a)$$

, and

$$t^{\bullet}_{ik}(x,t,\xi,\tau) = \frac{1}{4\pi r^2} \{ (a^{\bullet}_{ik}) \frac{t-\tau}{r^2} \left[H(t-\tau-\frac{\tau}{c_p}) - H(t-\tau-\frac{r}{c_s}) \right] + (b^{\bullet}_{ik})\delta(t-\tau-\frac{r}{c_p}) + (c^{\bullet}_{ik})\delta(t-\tau-\frac{r}{c_s}) + r(d^{\bullet}_{ik})\dot{\delta}(t-\tau-\frac{r}{c_p}) + r(e^{\bullet}_{ik})\dot{\delta}(t-\tau-\frac{r}{c_s}) \} \qquad (8b)$$

, where

$$a^{\bullet}_{ik} = 6c_s^2 (r_{,k} n_i + r_{,i} n_k + \delta_{ik} r_{,n} - 5 r_{,i} r_{,k} r_{,n}),$$

$$b^{\bullet}_{ik} = (2\frac{c_s^2}{c_p^2} - 1) r_{,k} n_i - \frac{c_s^2}{c_p^2} (12 r_{,i} r_{,k} r_{,n} - 2 r_{,k} n_i - 2 r_{,i} n_k - 2\delta_{ik} r_{,n}),$$

$$c^{\bullet}_{ik} = 12 r_{,i} r_{,k} r_{,n} - 2 r_{,k} n_i - 3 r_{,i} n_k - 3\delta_{ik} r_{,n},$$

$$d^{\bullet}_{ik} = \frac{1}{c_p} (2\frac{c_s^2}{c_p^2} - 1) r_{,k} n_i - 2\frac{c_s^2}{c_p^2} r_{,i} r_{,k} r_{,n},$$

$$e^{\bullet}_{ik} = \frac{1}{c_s} (2 r_{,i} r_{,k} r_{,n} - r_{,i} n_k - \delta_{ik} r_{,n})$$

The distance between the observation location x and the Dirac load location ξ is $r = \|x - \xi\|$. In the expression the normal vector is indicated by n, and the Heaviside's function by H( ), and the time when the Dirac load acts by τ. By assuming that the body forces can be neglected, and that we have homogeneous initial conditions, we can obtain the boundary integral equation with the help of the Reciprocal Theorem and the Somigliana identity (Dominguez, 1993)

$$c_{ik}\, u_k(\xi,t) = \int_{\Gamma_x}\int_0^t u^{*}_{ik}(x.t.\xi,\tau)\, t_k(x,\tau)\, d\tau\, d\Gamma_x - \int_{\Gamma_x}\int_0^t t^{*}_{ik}(x.t.\xi,\tau)\, u_k(x,\tau)\, d\tau\, d\Gamma_x \tag{9}$$

At the boundary either the displacement or the traction is known, and if $\xi \in \Gamma$, than c_{ik} for the half-space boundary is $0.5\,\delta_{ik}$. In order to obtain the unknown boundary values we solve Equation 9 numerically by assuming the dependency of the tractions and displacements with the time and space. Since the integration of the both terms in the right hand side of Equation 9 is similar, in the following only the integration of the first is described briefly. The integration in time of

$$\int_0^t u^{*}_{ik}(x.t.\xi,\tau)\, t_k(x,\tau)\, d\tau \tag{10a}$$

can be obtained by assuming the function $\alpha_m(\tau)$

$$t_k(x,\tau) = \sum_{n=1}^{N} \alpha_n(\tau)\, t_k^n(x), \tag{10b}$$

where the total observation period is $N \cdot \Delta t$. We then obtain

$$\sum_{n=1}^{N} \int_{n-1}^{n} u^{*}_{ik}(x.t.\xi,\tau)\, \alpha_n(\tau)\, d\tau\; t_k^n(x) = \sum_{n=1}^{N} U_{ik}^{N,n}(x.\xi)\; t_k^n(x) \tag{10c}$$

, and the integration in space can be performed

$$\int_{\Gamma} \sum_{n=1}^{N} U_{ik}^{N-n+1}(x.\xi)\; t_k^n(x)\; d\Gamma \tag{11a}$$

by assuming the shape function $\beta_m(x)$

$$t_k^n(x) = \sum_{m=1}^{M} \beta_m(x)\, t_k^{m,n}, \tag{11b}$$

where M is the total number of the nodes, and L the total number of the elements along the boundary, we then obtain the discrete form of the first term of Equation 9

$$\int_{\Gamma}\int_0^t u^{*}_{ik}(x.t.\xi,\tau)\, t_k(x,\tau)\, d\tau\, d\Gamma = \sum_{l=1}^{L}\sum_{n=1}^{N}\sum_{m=1}^{M} \int_{\Gamma} (U_{ik}^{N-n+1}(x.\xi)\, \beta_m(x)\, t_k^{m,n})\, d\Gamma(x) \tag{12}$$

After the integration of the discrete form of Equation 9 the following algebraic form is available

$$U^1 t^N = T^1 u^N + \sum_{n=2}^{N} T^n u^{N-n+1} - U^n t^{N-m+1}, \tag{13}$$

where U and T are the coefficient matrices of the considered system, and u and t are the displacement and traction vectors, respectively. With Equation 13 the ground excitation $a_g(t)$ of the structure due to the moving load P(t) can be determined. Details are given by Pflanz (2001) and Pflanz et al. (2002).

2.1 *Ground vibrations*

Figures 2(a), (b) and (c) show the peak ground acceleration (PGA) in the x-, y- and z-direction, respectively, as a function of the location along the dotted line 18m from the load entrance, and perpendicular to the load path.

The PGA in the x-direction parallel to the load path decreases with increasing distance from the load path; at the load path it increases with the load speed. The highest value occurs at the load path when the load speed is close to the Rayleigh wave speed in the soil. In the considered soil this speed is 402km/h. When the load speed exceeds the wave speed the PGA decreases again. However, the PGA close to the load path is the largest.

The PGA in the y-direction (perpendicular to the load path) first increases, and than decreases with increasing distance from the load path. At the load path the PGA should have the value zero due to the symmetrical wave propagation. The calculated non-zero values at the load path are caused by chosen unsymmetrical model. When the load speed exceeds the Rayleigh wave speed in soil the PGA along the dotted line is also the largest compared to the PGA due to other load speeds. The decrease of the PGA with the distance from the load path occurs slower than the PGA in the x- and vertical direction.

In the vertical direction the PGA at the load path changes strongly with the load speed. In the surrounding area (along the dotted line) the PGA has similar tendency as the PGA in the x-direction.

The PGA is only one of the characteristics of the ground motions, which can be important for the response of the structures.

2.2 *Structural responses*

Figure 3 shows the vertical response of the one-storey structure at the top of the middle column to the simultaneous ground excitations. As we know from previous investigation (Chouw 1994), even if the load remains stationary, due to the different relative distance to the source the foundations will experience different ground excitations. In case of a moving load all structural foundations have different relative distance to the moving load, which is time dependent. The foundation excitation will therefore have an additional effect from the load speed.
Corresponding to the increasing ground motions with the load speed the structure shows the strongest response at the load speed of 500km/h.

In case of man-made induced vibrations it is commonly believed that the most significant ground motions are the vertical ground vibrations, and the horizontal ground vibrations are not important, because they are normally smaller than the vertical vibrations. Even though a structure is always excited by the vertical and horizontal ground motions simultaneously, often only the vertical ground motion is considered in the analysis of the structural response. Since we assume in this study that the structure and soil behave linearly, the response of the structure to the ground motions can be considered as a composition of the effect of the horizontal ground motions, and the influence of the vertical ground excitation.

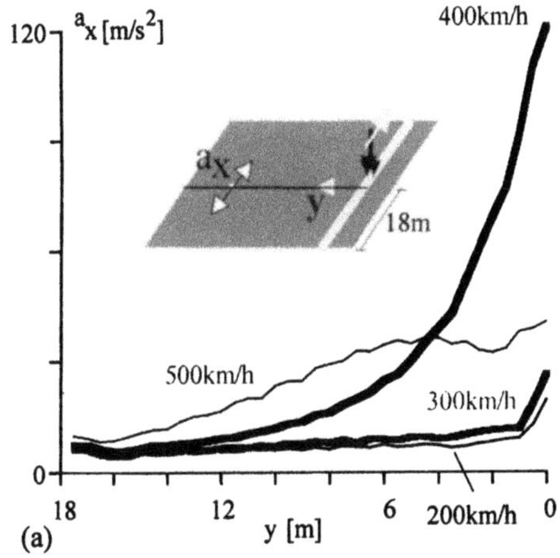

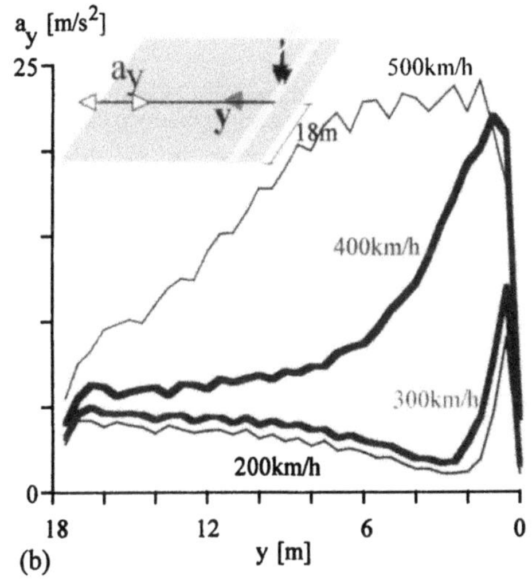

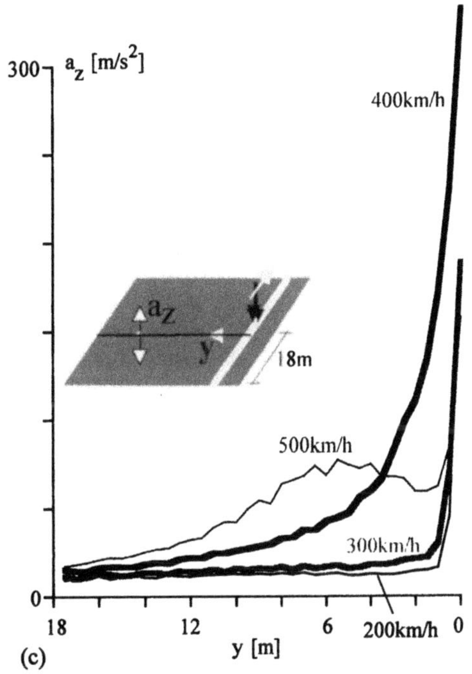

Figure 2. Development of the peak ground acceleration PGA with the distance from the load path.
(a) PGA in the x-direction, (b) PGA in the y-direction, and (c) PGA in the vertical direction.

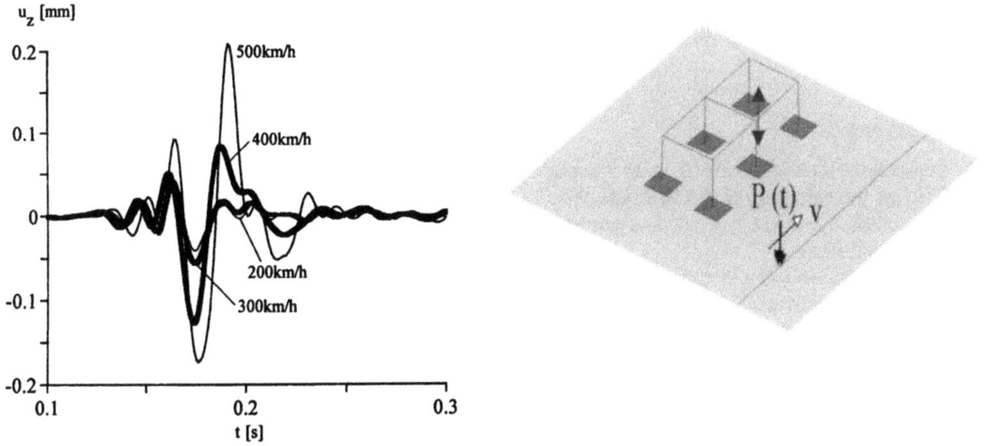

Figure 3. Development of structural response with the load speed.

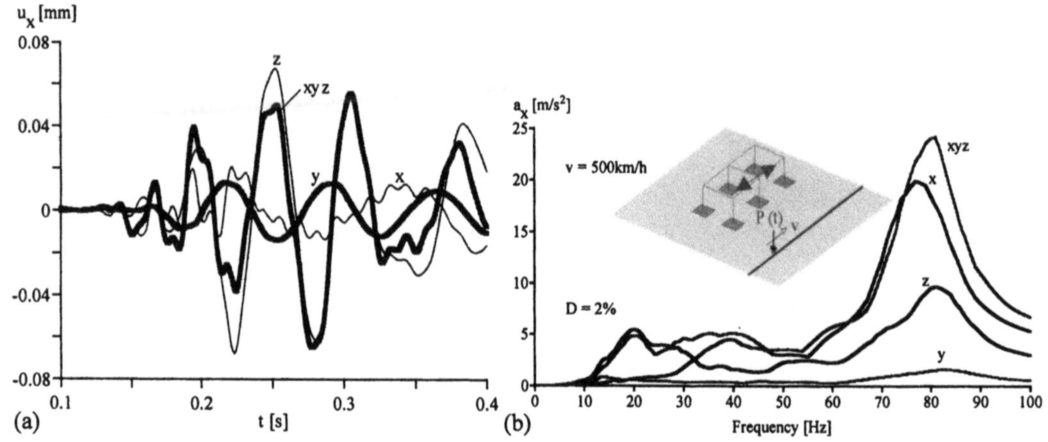

Figure 4. Influence of the direction of the ground excitation on the structural responses.
(a) Displacement response, and (b) acceleration responses.

In order to investigate the effect of each component of the ground excitation, we consider the ground excitation in the x-, y- and z-direction separately, and we compare these results with that due the simultaneous ground excitation. Figure 4 shows the response at the upper end of the middle column in the x-direction due to the x-, y-, z-ground motion alone, and due to the simultaneous ground motions. The displacement response in Figure 4(a) shows that the vertical ground motions caused indeed the strongest response, even stronger than the response due to the simultaneous ground motions. The response spectra in Figure 4(b), which can be interpreted as the maximum response of a single-degree-of-freedom (SDOF) system with the damping ratio D attached at the considered location to the induced vibrations, show however, that the strongest response occurs when the simultaneous ground motion is considered. The result also shows that in the higher frequency range the stronger response in the x-direction is not caused by the vertical ground motions, but as expected by the ground vibrations in the x-direction.

2.3 *Effect of the bedrock*

Figure 5 shows the effect of bedrock at the depth location of 3m on the response of the structure. The structure is excited by the x-, y- and z-ground-motions simultaneously.

In Figures 5(a), (b) and (c) the response at the upper end of the middle column to the ground motions due to the moving load with a speed of 300km/h is displayed. In Figures (e), (f) and (g) the response is due to the moving load with a speed of 500km/h. In case of the load speed of 300km/h (below the Rayleigh-wave speed) the bedrock causes a decrease of the structural response in all considered frequencies and response directions except around the layer's natural frequency $f_l = c_s/4H = 20$Hz.

The strongest decrease occurs in the higher frequency range around 80Hz. In case of the load speed of 500km/h (above the Rayleigh-wave speed) the response in the x- and y-direction is similar to the response in case of the load speed below the Rayleigh-wave speed. Response amplification occurs around the layer frequency f_l, and strong reduction at the frequency around 80Hz. In contrast, the bedrock caused amplification in almost all considered frequencies in the vertical direction. The result indicates that bedrock is beneficial to the structure when the load speed is below the wave speed in the soil. However, in order to enable a general conclusion further investigation is needed.

3 REDUCTION OF VIBRATION

The considered wave barrier trench is 0.5m wide, and has a depth of 4.5m. The distance of the centre of the trench to the load path is 2m. The wave-impeding barrier (wib) is built 1.5m below the load path. The shear wave velocity in the wib material is 250m/s. The wib has a thickness of 0.8m and a width of 3m.

Figures 6(a), (b) and (c) show the horizontal response of the structure in the x-direction at the upper end of the middle column to the simultaneous ground motions without measure, with trench, and with wib as reduction measure, respectively. Without a reduction measure the response increases with the load speed. The trench reduces the strong response in the frequency range around 80Hz drastically. In the frequency range around 20Hz the effectiveness of the trench decreases with the load speed. However, the trench still causes a strong

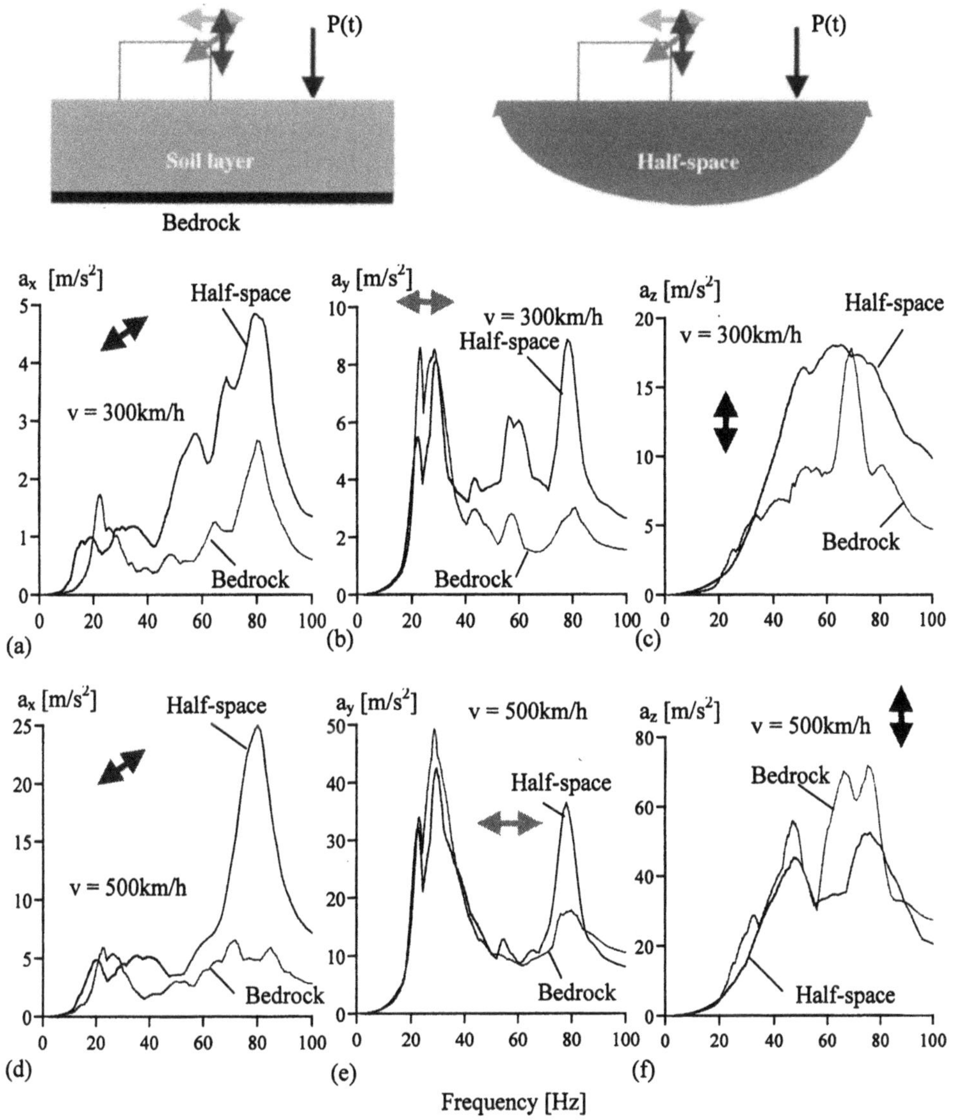

Figure 5. Influence of the bedrock on the structural responses. Load speed (a)-(c) v = 300km/h, and (d)-(f) v = 500km/h.

decrease of the structural response. As indicated in the previous publication (Chouw & Pflanz 2000) the trench can cause amplification of the ground motion in the direction perpendicular to the load path. It causes, however, a strong decrease of the ground vibration in the direction parallel to the load path, and in the vertical direction. Since the structure is strongest excitated in the x-direction parallel to the load path, and the amplified ground motion in the y-direction has no effect on the structural response in the x-direction, the response of the structure is mainly determined by the strongly reduced ground motion in the x-direction. The response therefore becomes very small.

Employing the wib the structural response can also be reduced. However, the reduction is not as pronounced as the reduction due to the trench. As we have seen from the vertical response of the ground surface with bedrock, the bedrock will lose its effect when the load speed is larger than the Rayleigh wave speed in the soil. This is one possible reason for the response in the frequency range around 80Hz still being pronounced, since the vertical ground motions have an influence on the structural response in the x-direction.

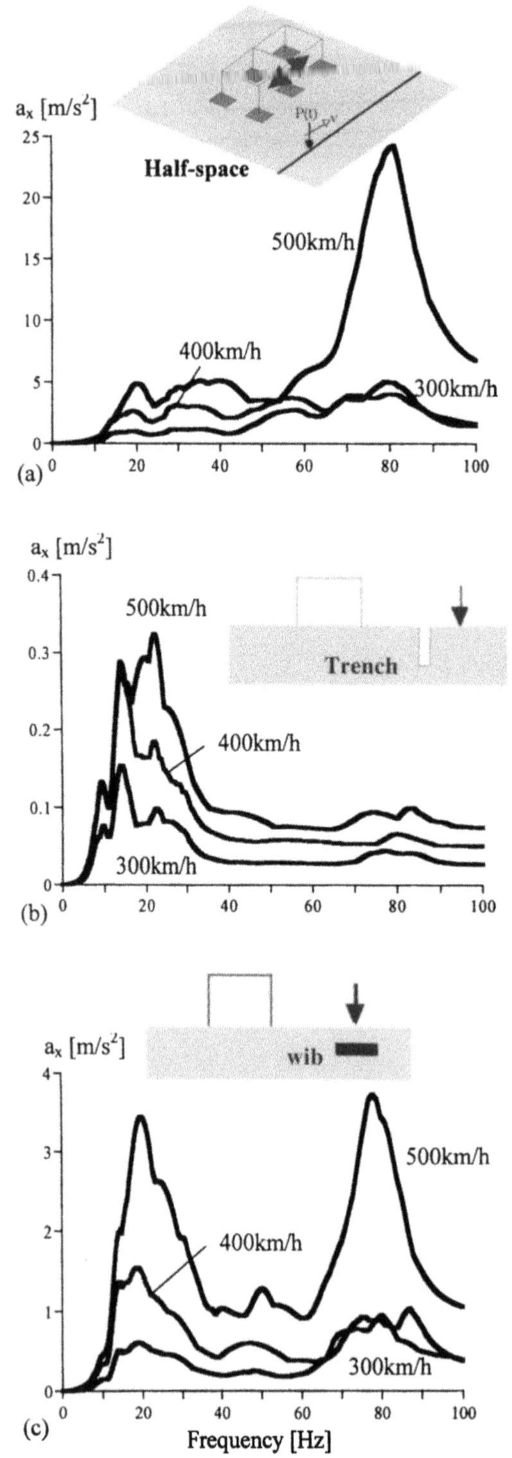

Figure 6. Structural response (a) without any measure, (b) with a trench, and (c) with a wave-impeding barrier.

The considered base-isolation system consists of steel springs with a total stiffness of 1128.533kN/m and 3389.4667kN/m for the one- and three-storey structure, respectively. Figures 7(a), (b) and (c) show the effect of the steel springs as base-isolation system on the maximum structural acceleration response in the y-, x- and z-direction, respectively. The response is displayed as a function of the load speed and the height location of the three-storey structure. The result clearly shows that the base-isolation system causes a significant reduction in all considered responses. The horizontal responses in Figures 7(a) and (b) show the influence of the activated rigid body motion in the almost identical maximum acceleration at the first, second, and third floor.

The strongest reduction is achieved in the vertical response (Figure 7(c)). Since all columns experience almost the same ground excitation in the vertical direction, the maximum acceleration at the base, and at all other floors are almost the same.

Figure 8 shows the response of the one-storey structure at the upper end of the left column in the x-direction parallel to the load path due to the simultaneous ground motions. The maximum reduction occurs at the load speed of 300km/h.

The response spectra in Figure 9 show different reduction characteristics. At the load speed of 500km/h the base-isolation system have the strongest reduction effect at the frequency around 80Hz. In the frequency range around 30Hz, however, the base-isolation system caused amplification. This conclusion cannot be seen in the time history response.

The result shows that a reduction of the maximum response of the time history cannot be used as an indicator of the effectiveness of a reduction measure.

4 CONCLUSIONS

The study of the reduction effects of a trench, wave impeding barrier (wib), and a base-isolation system on structural vibrations due to a moving load in the neighbourhood reveals:

A consideration of the vertical ground motions alone may underestimate the excitation of the simultaneous vertical and horizontal ground motions.

Existing bedrock can have a reduction effect on the ground surface vibration, if the speed of the moving load is slower than the velocity of the propagating Rayleigh waves on the ground surface.

Trench as reduction measure is more effective than wib.

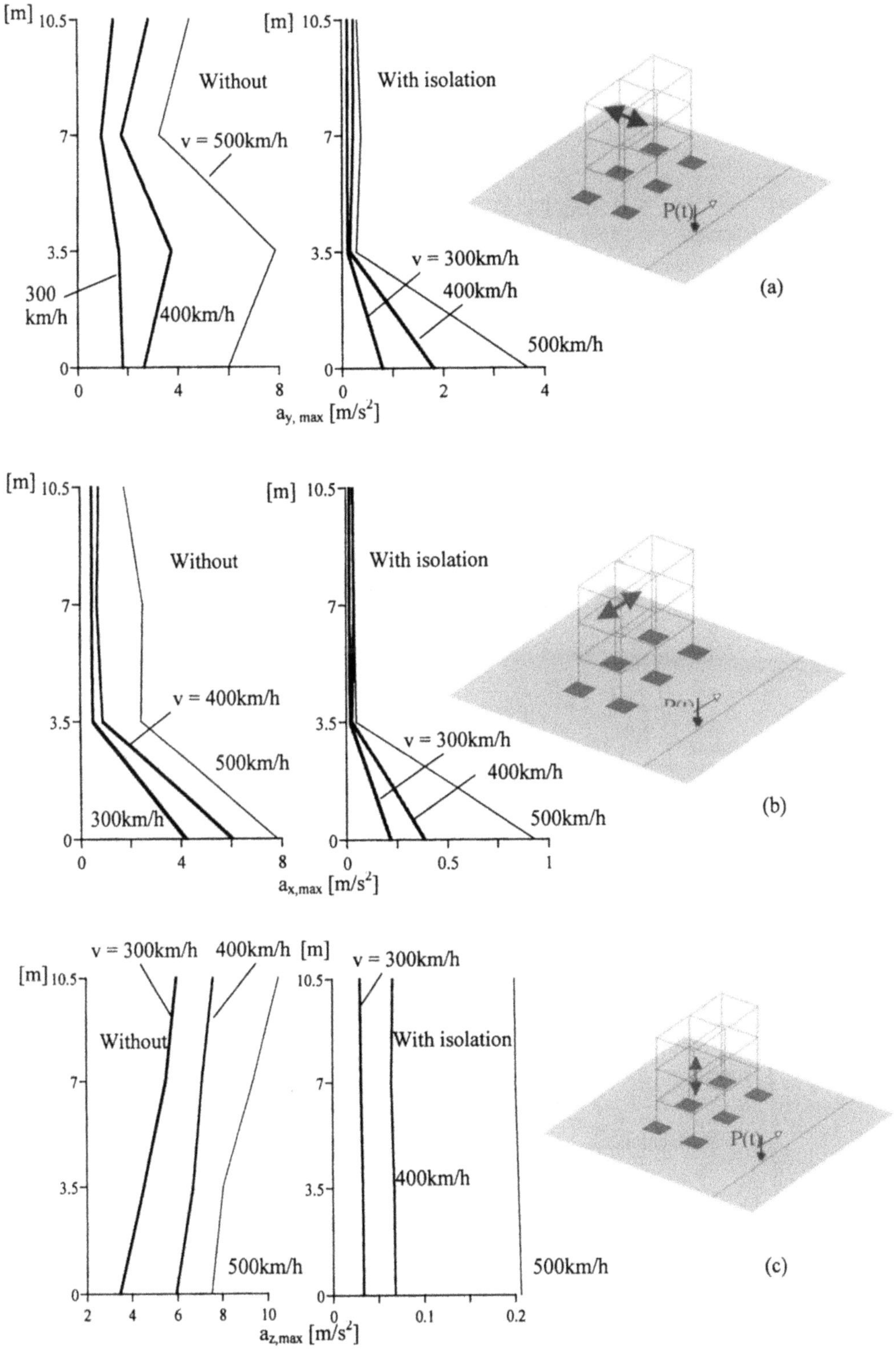

Figure 7. Reduction of ground vibration effect due to base-isolation systems. Maximum induced vibration in (a) the y-direction, (b) the x-direction, and (c) in the vertical direction as a function of the height location of the structure.

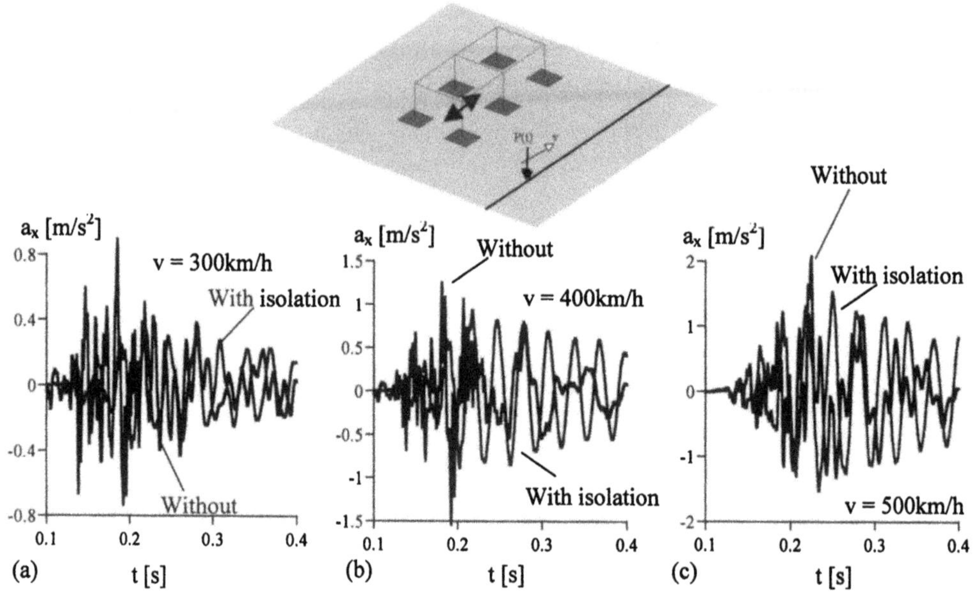

Figure 8. Effect of the base isolation systems on the structural response in the x-direction. The load speed is (a) v = 300km/h, (b) v = 400km/h, and (c) v = 500km/h.

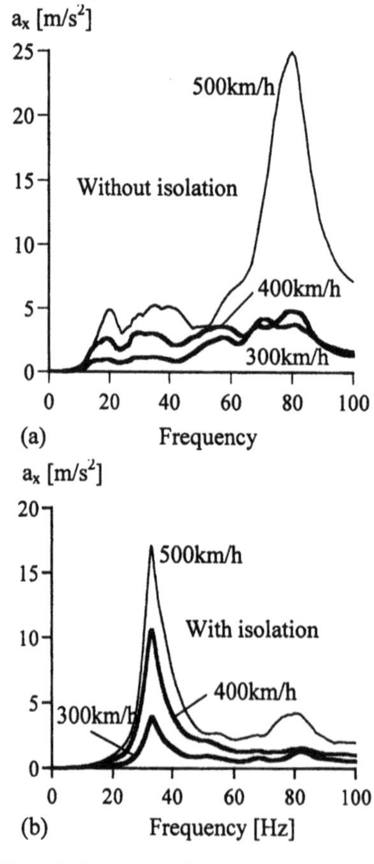

Figure 9(a) and (b). Induced vibrations (a) without and (b) with base isolation systems.

Steel springs at the base of the structure can reduce the induced vibrations in the structure drastically, especially in the vertical direction. The reduction of the maximum structural acceleration response should not be used as the effectiveness criterion for the reduction measure, because the lowering of the natural frequency of the structure due to the base-isolation systems can cause an amplification of induced vibrations in the lower frequency range. It is better to use the response spectrum values as criterion for the effectiveness of the reduction measure.

ACKNOWLEDGEMENTS

This research work was supported by Wesco Scientific Promotion Foundation, Japan. The first author gratefully acknowledges the support.

The authors would like to thank the reviewer for the constructive comments.

REFERENCES

Chouw, N., Le, R., Schmid, G. 1991a. An approach to reduce foundation vibrations and soil waves using dynamic transmitting behaviour of a soil layer. *Bauingenieur 66*: 215-221 (in German).

Chouw, N., Le, R., Schmid, G. 1991b. Propagation of vibration in a soil layer over bedrock, *Engineering Analysis with Boundary Elements 8*: 125-131.

Chouw, N. 1994. Wave propagation from the source via the subsoil into the building, *Proceedings of the 1st int. workshop WAVE'94, Bochum, Germany*, pp. 33-46. Rotterdam: Balkema.

Chouw, N., Pflanz, G. 2000. Reduction of structural vibrations due to moving load. *Proceedings of the 2nd int. workshop WAVE2000, Bochum, Germany*. Pp. 251-268. Rotterdam: Balkema.

Dominguez, J., 1993. *Boundary elements in dynamics*, Computational Mechanics Publications, Southampton and Boston.

Hillmer, P. 1987. *Berechnung von Stabtragwerken mit lokalen Nichtlinearitäten unter Verwendung der Laplace-Transformation*, Technisch-wissenschaftliche Mitteilungen, Nr. 87-1, Ruhr-Universität Bochum, 168pp.

Naeim, F., Kelly, J. M. 1999. *Design of seismic isolated structures*. John Wiley & Sons.

Massarsch, R. 2002. Ursachen von Bodenerschütterungen und deren Begrenzung. *Flesbau* 20 (5): 12 pp.

Mualla, I., Nielsen, O., Chouw, N., Belev, B., Liao W. I., Loh, C. H., Agrawal, A. 2002. Enhanced response through supplementary friction damper devices, *Proceedings of the third international workshop on wave propagation, moving load and vibration reduction, Okayama University, Japan*, Balkema.

Pflanz, G. 2001. Numerische Untersuchung der elastischen Wellenausbreitung infolge bewegter Lasten mittels der Randelementmethode im Zeitbereich, *Technisch-wissenschaftliche Mitteilungen, Institut für Konstruktiven Ingenieurbau, Ruhr-Universität Bochum, Fortschritt-Berichte VDI, Reihe 18, Nr. 265*, VDI Verlag GmbH, Düsseldorf, 149 pp.

Pflanz, G., Hashimoto, K., Chouw, N. 2002. Reduction of structural vibrations induced by a moving load. *Journal of Applied Mechanics* 5: 555-563.

Schmid, G., Pflanz, G., Garcia, J. 2002. Dynamic soil-structure interaction: methods and numerical application. In: Computational structural dynamics, Eds.: Talaganov, K. and Schmid, G., Swets & Zeitlinger B.V., Lisse, Netherlands: 19-31.

Yoshioka, O. 2002. Reduction measures in track for Shinkansen-induced ground vibrations, *Proceedings of the third international workshop on wave propagation, moving load and vibration reduction, Okayama University, Japan*, Balkema, 10pp.

Wave propagation – Moving load – Vibration reduction, Chouw & Schmid (eds.)
© 2003 Swets & Zeitlinger, Lisse, ISBN 90 5809 559 2

Efficient numerical approach for the analysis of vibrations due to moving load

S. Hirose
Department of Mechanical and Environmental Informatics, Tokyo Institute of Technology, Tokyo, Japan

ABSTRACT: In this paper, the boundary element methods are developed for the numerical analysis of the moving load problems with 2.5-D and 3-D configurations. Firstly, the boundary element method is presented for the 2.5-D problem with 2-D geometry subjected to the moving load. Numerical results are shown to discuss the effect of wave barriers on the surface displacements. Secondly, the 3-D BEM analysis is carried out using the 2.5-D solutions as an incident wave field. Numerical examples for the wave scattering problem by a finite trench are demonstrated to show computational efficiency of the 3-D BEM in conjunction with the 2.5-D analysis.

1. INTRODUCTION

In general, it is anticipated that as a train speed increases, the railway-induced vibration level becomes higher. With increase of train speed, therefore, more attention has recently been paid to the reduction of ground vibrations. Quantitative information on ground vibration obtained by analytical and numerical methods has been required for the evaluation of effective vibration reduction methods.

The simplest analytical model for a dynamic problem due to a train is a moving point load on an elastic half space. Even for such a simplest model, however, the solution can not be obtained in a closed form; for example, see Eringen & Suhubi(1975). For solving more complex problems, therefore, it is necessary to use numerical methods like a finite element method (FEM) and a boundary element method (BEM). However, a numerical method requires huge capacity and time in computation when directly applied to a complex 3-D problem including train-track system and surrounding soil, e.g., see Chouw & Pflanz (2002) and von Estorff et al.(2002).

A 3-D dynamic response of a model where the material parameters vary only in 2-D geometry is demanded in many engineering problems. Such a configuration having a 3-D wave field in a 2-D medium is sometimes called a 2.5-D problem. In the field of earthquake engineering, 2.5-D modelling has been implemented to obtain 3-D behaviors of seismic waves in a 2-D configuration of canyon and alluvial valley, using various approaches such as asymptotic ray theory (e.g. Bleistein 1986), the finite-difference method (e.g. Okamoto 1993), the FEM (e.g. Hwang & Lysmer 1981), the indirect boundary method (Luco et al. 1990), the pseudospectral method (Furumura & Takenaka 1996) and the BEM (Khair et al. 1989 and Zhang & Chopra 1991). In nondestructive ultrasonics, Li et al. (1992) have developed a 2.5-D boundary element modelling to investigate the reflection and transmission of obliquely incident surface waves by an edge of quarter space. In the analysis of acoustic sound field, a 2.5-D problem has been solved to determine the 3-D sound pressure field around a noise barrier of constant cross-section (Duhamel 1996).

The moving load problem dealing with a train-track-soil system with 2-D geometry is also a typical 2.5-D problem. Several researchers have solved the moving load problem by using a FEM. Since the FEM can deal with the finite region only, however, the FEM has to be combined with other numerical techniques like the thin layer method (Hanazato et al. 1991) and the infinite element approach (Yang & Hung 2001) to take into account the radiation conditions in a half space.

In this paper, the direct BEM is presented for the 2.5-D problem of a half space subjected to a moving load. It is remarked that the BEM can treat the half space problem without any modification, since the radiation condition in far field can be automatically taken into account in the BEM. Then the efficient 3-D analysis in conjunction with the 2.5-D analysis is developed to improve the computational efficiency for the dynamic problem due to a moving load passing by a trench with finite length.

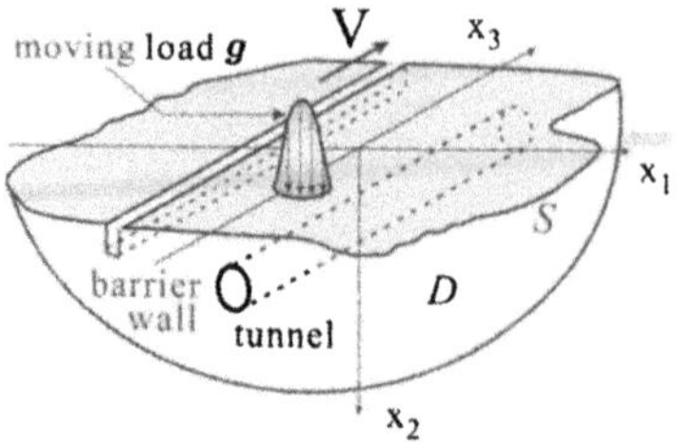

Figure 1. A 2.5-D model of a half space subjected to a moving load.

2. 2.5-D BEM ANALYSIS

2.1 *Problem*

As shown in Figure 1, we consider a dynamic problem in a homogeneous isotropic elastic solid D subjected to a moving load $\boldsymbol{g}$, which travels at a constant velocity V without changing the spatial distribution. It is assumed that the geometry has a 2-D configuration with a uniform cross-section along the propagation direction of the moving load, say x_3.

In a homogeneous isotropic elastic solid, the displacement field $u_i(\boldsymbol{x},t)$ at the point $\boldsymbol{x}$ in D at time t satisfies the Navier-Cauchy's equations:

$$L_{ik}(\boldsymbol{\partial}_x)u_k(\boldsymbol{x},t) = \rho\frac{\partial^2}{\partial t^2}u_i(\boldsymbol{x},t), \quad \boldsymbol{x} \in D \tag{1}$$

where ρ is the mass density and $L_{ik}(\boldsymbol{\partial}_x)$ is the differential operator defined by $L_{ik}(\boldsymbol{\partial}_x) = C_{ijkl}\partial^2/\partial x_j \partial x_l$. Here C_{ijkl} are the elastic moduli given by $C_{ijkl} = \lambda\delta_{ij}\delta_{kl} + \mu(\delta_{ik}\delta_{jl} + \delta_{il}\delta_{jk})$, where λ and μ are Lamé constants. In general, the boundary conditions are given by

$$\begin{aligned} u_i(\boldsymbol{x},t) &= h_i(\boldsymbol{x},t), \quad \boldsymbol{x} \in S_u \\ t_i(\boldsymbol{x},t) &\equiv \overset{n}{T}_{ik}(\boldsymbol{\partial}_x)u_k(\boldsymbol{x},t) \\ &= g_i(\boldsymbol{x},t), \quad \boldsymbol{x} \in S_t = S \backslash S_u \end{aligned} \tag{2}$$

where h_i and g_i are given boundary values, $\overset{n}{T}_{ik}(\boldsymbol{\partial}_x)$ are the traction operators defined by $\overset{n}{T}_{ik}(\boldsymbol{\partial}_x) = C_{ijkl}n_j(\boldsymbol{x})\partial/\partial x_l$ and $n_j(\boldsymbol{x})$ is a unit vector normal to the surface S.

Now the Fourier transforms with respect to x_3 and t are defined as follows:

$$\bar{f}(x_1,x_2,\xi_3,t) = \int_{-\infty}^{\infty} f(x_1,x_2,x_3,t)e^{-i\xi_3 x_3}dx_3 \tag{3}$$

$$\hat{f}(x_1,x_2,x_3,\omega) = \int_{-\infty}^{\infty} f(x_1,x_2,x_3,t)e^{i\omega t}dt. \tag{4}$$

Applying the Fourier transforms defined above to Equations 1 and 2 yields the following governing equations in a two dimensional space $\boldsymbol{X} = (x_1, x_2)$:

$$\begin{aligned} &L_{ik}(\partial_{x_1}, \partial_{x_2}, i\xi_3)\hat{\bar{u}}_k(\boldsymbol{X},\xi_3,\omega) \\ &= -\rho\omega^2\hat{\bar{u}}_i(\boldsymbol{X},\xi_3,\omega), \quad \boldsymbol{X} \in \bar{D} \end{aligned} \tag{5}$$

subjected to the boundary conditions

$$\begin{aligned} \hat{\bar{u}}_i(\boldsymbol{X},\xi_3,\omega) &= \hat{\bar{h}}_i(\boldsymbol{X},\xi_3,\omega), \quad \boldsymbol{X} \in \partial S_u \\ \hat{\bar{t}}_i(\boldsymbol{X},\xi_3,\omega) &= \hat{\bar{g}}_i(\boldsymbol{X},\xi_3,\omega), \\ &\boldsymbol{X} \in \partial S_t = \partial S \backslash \partial S_u \end{aligned} \tag{6}$$

where $\bar{D}$ is the two dimensional domain bounded by the boundaries ∂S_u and ∂S_t in the x_1-x_2 plane.

If the two dimensional solutions $\hat{\bar{u}}_i$ are obtained by solving the boundary value problems defined by Equations 5 and 6, the solutions u_i at the point $\boldsymbol{x}$ and time t can be calculated by using the inverse Fourier transform as follows:

$$u_i(\boldsymbol{x},t) = \frac{1}{(2\pi)^2}\int_{-\infty}^{\infty}\int_{-\infty}^{\infty}\hat{\bar{u}}_i(\boldsymbol{X},\xi_3,\omega)e^{i(\xi_3 x_3 - \omega t)}d\omega d\xi_3. \tag{7}$$

For this calculation, however, we have to obtain the two dimensional solutions $\hat{\bar{u}}_i$ for all values of ξ_3 and ω, which may be time consuming numerically.

As mentioned before, the moving load is assumed to have a constant velocity V and unchanging spatial distribution along the x_3-direction. Thus the traction $g_i(\boldsymbol{x},t)$ on S_t can be expressed by $g_i(\boldsymbol{x},t) = g_i(\boldsymbol{X}, x_3 - Vt)$. Since the geometry of the model has a uniform cross-section in the x_3 direction, the displacement u_i is also assumed to have the same form as the traction g_i, i.e.,

$$u_i(\boldsymbol{x},t) = u_i(\boldsymbol{X}, x_3 - Vt). \tag{8}$$

If the Fourier transforms defined by Equations 3 and 4 are applied to Equation 8, then we obtain

$$\hat{\bar{u}}_i(\boldsymbol{X},\xi_3,\omega) = 2\pi\hat{\bar{u}}_i'(\boldsymbol{X},\xi_3)\delta(\omega - \xi_3 V) \tag{9}$$

where $\delta(\cdot)$ is the Dirac delta function and $\hat{\bar{u}}_i'$ denotes the solution of Equation 5 with the circular frequency $\omega = \xi_3 V$. On substitution of Equation 9 into Equation 7, the displacement can be obtained by

$$u_i(\boldsymbol{x},t) = \frac{1}{2\pi}\int_{-\infty}^{\infty}\hat{\bar{u}}_i'(\boldsymbol{X},\xi_3)e^{i\xi_3(x_3 - Vt)}d\xi_3. \tag{10}$$

Consequently, three dimensional displacements u_i due to a moving load can be constructed by solving the 2-D boundary value problem with the condition $\omega = \xi_3 V$ and superposing the solutions $\hat{\bar{u}}_i'(\boldsymbol{X},\xi_3)$ by means of the single inverse Fourier transform with respect to ξ_3 as shown in Equation 10. In numerical calculations, the inverse Fourier transform is carried out in a discretized form by means of the fast Fourier transform.

2.2 *BEM formulation*

In this paper, a BEM is used to solve the 2-D boundary value problems defined by Equations 5 and 6. In the BEM formulation, the fundamental solution satisfying the equations of motion

$$L_{ij}\left(\partial_{x_1},\partial_{x_2},-i\xi_3\right)U_{jk}(\boldsymbol{X},\boldsymbol{Y})+\delta_{ik}\delta(\boldsymbol{X}-\boldsymbol{Y}) = -\rho\omega^2 U_{ik}(\boldsymbol{X},\boldsymbol{Y}) \tag{11}$$

is utilized. The explicit expression for the fundamental solution U_{ik} was obtained by Li *et al.* (1992):

$$U_{ik}(\boldsymbol{X},\boldsymbol{Y})=\frac{i}{4\mu}\left\{H_0^{(1)}(\tilde{k}_T R)\delta_{ik} + M_{ik}\left[H_0^{(1)}(\tilde{k}_T R)-H_0^{(1)}(\tilde{k}_L R)\right]\right\} \tag{12}$$

where $H_0^{(1)}(\,)$ is the Hankel function of the zeroth order of the first kind, and M_{ik} is the following differential operator:

$$M_{ik}=\frac{1}{k_T^2}\left[\delta_{\alpha k}\delta_{\beta i}\frac{\partial^2}{\partial x_\alpha \partial x_\beta} - i\xi_3(\delta_{3k}\delta_{\alpha i}+\delta_{3i}\delta_{\alpha k})\frac{\partial}{\partial x_\alpha} - \xi_3^2\delta_{3k}\delta_{3i}\right]. \tag{13}$$

In Equation 13 and hereafter, the Greek subscripts for summation convention take the values 1 and 2 only. Other symbols appeared in Equations 12 and 13 are defined by $R=[(x_1-y_1)^2+(x_2-y_2)^2]^{1/2}$, $\tilde{k}_T=(k_T^2-\xi_3^2)^{1/2}$, and $\tilde{k}_L=(k_L^2-\xi_3^2)^{1/2}$ where k_T and k_L are the transverse and longitudinal wave numbers defined by $k_T=\omega/c_T$ and $k_L=\omega/c_L$, respectively, and c_T and c_L are the velocities of the P and S waves, respectively.

Multiplying Equation 5 by $U_{ik}(\boldsymbol{X},\boldsymbol{Y})$ and Equation 11 by $\hat{\tilde{u}}_i(\boldsymbol{X},\xi_3,\omega)$, subtracting the latter from the former and integrating over the domain $\bar{D}$ and after some manipulations, we have

$$\int_{\partial S}\left\{U_{ik}(\boldsymbol{X},\boldsymbol{Y})\hat{\tilde{t}}_i(\boldsymbol{X},\xi_3,\omega) - T_{ik}(\boldsymbol{X},\boldsymbol{Y})\hat{\tilde{u}}_i(\boldsymbol{X},\xi_3,\omega)\right\}ds_X = \begin{cases}\hat{\tilde{u}}_k(\boldsymbol{Y},\xi_3,\omega) & \boldsymbol{Y}\in\bar{D}\\ \frac{1}{2}\hat{\tilde{u}}_k(\boldsymbol{Y},\xi_3,\omega) & \boldsymbol{Y}\in\partial S\\ 0 & \text{otherwise}\end{cases} \tag{14}$$

where the boundary ∂S is assumed to be smooth and $T_{ik}(\boldsymbol{X},\boldsymbol{Y})=\overset{n}{T}_{ij}\left(\partial_{x_1},\partial_{x_2},-i\xi_3\right)U_{jk}(\boldsymbol{X},\boldsymbol{Y})$. Equation 14-b is discretized into the system of equations and is solved using the boundary conditions given by Equation 6. The displacement $\hat{\tilde{u}}_i(\boldsymbol{Y},\xi_3,\omega)$ in the domain $\bar{D}$ can be obtained by substituting the solution of Equation 14-b into Equation 14-a. The 2-D solution $\hat{\tilde{u}}_i(\boldsymbol{Y},\xi_3,\omega)$ with $\omega=\xi_3 V$ is substituted into Equation 10 to reconstruct the 3-D solution.

2.3 *Numerical examples*

Numerical calculations are carried out for 2.5-D models of a homogeneous half space with or without a wave barrier wall. The geometry of the cross-section at $x_3=0$ is depicted in Figure 2. The

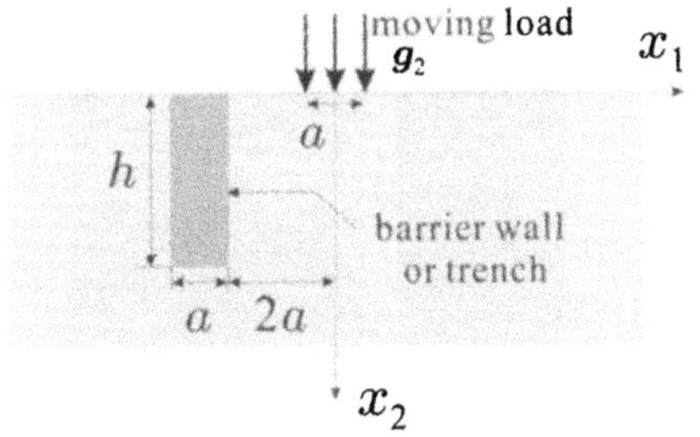

Figure 2. Geometry of a cross-section of a half space with a wave barrier wall or a trench.

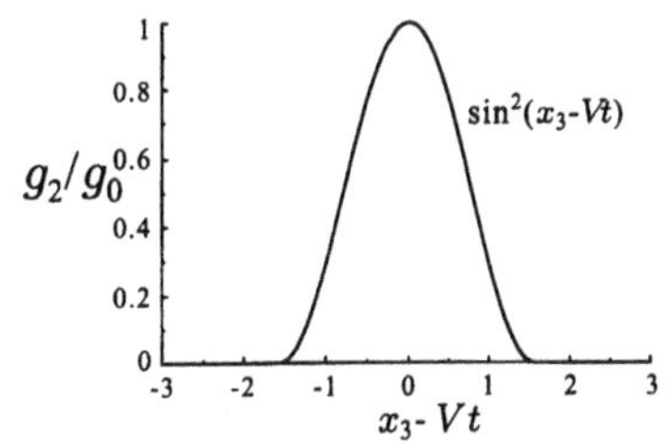

Figure 3. Distribution of a moving load along the x_3 direction.

region of $-3 < x_1/a < -2$ and $0 < x_2 < h$ is a wave barrier wall made of concrete or a trench with traction free condition, where the width a of the moving load is 1m. The material constants for the concrete wall are given by the mass density of 2400kg/m^3 and the wave speeds of 2424 m/s and 1400 m/s for P and S waves, respectively, while the half space is characterized by the mass density of 1700kg/m^3, the wave speeds of 200 m/s and 100 m/s for P and S waves, respectively. The perfect bond is assumed on the interface between the half space and the concrete wall.

The moving load is given by vertical traction forces distributed with the constant values in the x_1 direction and the sinusoidal square function along the x_3 direction as shown in Figure 3. The peak value of the traction force is given by g_0.

Figure 4 shows the amplitudes of the vertical displacements on the surface of the half space at time $t=0$, when the peak value of the moving load is located at the origin. The speed of the moving load in the x_3 direction is 300km/s, which is lower than the Rayleigh wave speed of 324km/s. Figures 4 (a), (b) and (c) show the results for the half space with a concrete wall, the half space with a traction free trench and the plain half space without any barrier, respectively. The depth h of the concrete wall and trench is $4a$. Figure 5 shows the vertical surface displacements along the x_1 axis at $x_3=0$. From Figures 4 and 5, it is seen that the concrete barrier wall can reduce effectively the vertical displacements on the surface except for

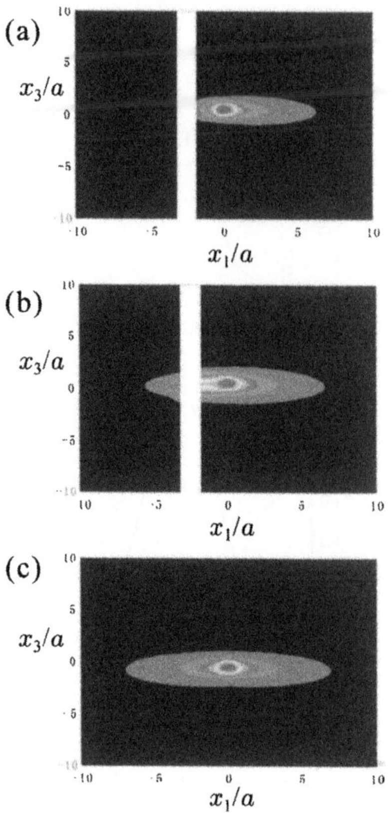

Figure 4. Vertical displacements on the surfaces of (a) the half spaces with a concrete wall, (b) the half space with a traction free trench and (c) the plain half space without any barrier. The speed of the moving load is 300km/s.

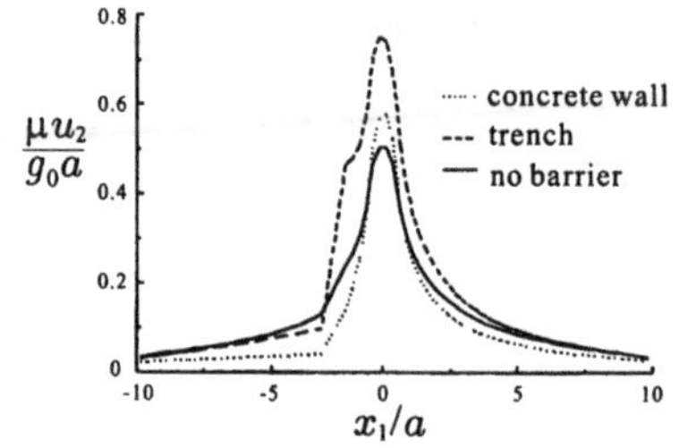

Figure 5. Vertical surface displacements along the x_1 axis of the half spaces with a concrete wall, the half space with a traction free trench and the plain half space without any barrier.

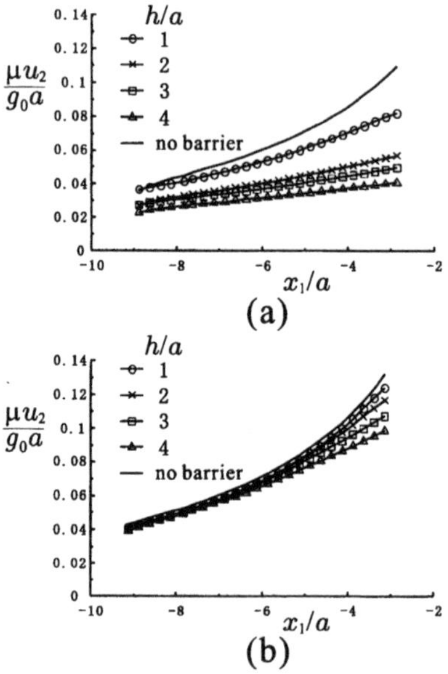

Figure 6. Peak values of vertical surface displacements for $-9 < x_1/a < -3$ of the half spaces with (a) concrete wall and (b) trench having various depths.

the vicinity of the loading zone. As for the traction free trench, on the other hand, the displacement amplitudes decrease somewhat on the surface $x_1/a < -3$, but the vertical displacements are much amplified for $x_1/a > -2$, compared with the case of the plain half space with no barrier.

Figures 6 (a) and (b) show the peak values of vertical displacements on the surface $-9 < x_1/a < -3$ of the half spaces with the concrete wall and the trench, respectively, with various depths of $h/a = 1, 2, 3$ and 4. It can be shown that as the depth of the wall and trench increases, the concrete wall reduce the vertical displacements more effectively than the trench does.

Figure 7 illustrates the same results as in Figure 4, but for the moving load with the speed of 400km/s, which is beyond the Rayleigh wave speed 324km/s of the half space. The straight wavefront due to the constructive interference of waves generated from the high speed load can be clearly seen in all cases. It is also shown that the concrete wall reduce well the displacement amplitudes on the surface of $x_1/a < -3$, whereas in the case of the trench, large amplitudes still remain in a wide area due to wave diffraction and reflection.

3. 3-D BEM ANALYSIS

3.1 *Problem*

A 2.5-D modelling is very effective to investigate the dynamic behaviors of the ground subjected to a moving load, as shown in the last section. However, the 2.5-D modelling is subject to a strong restriction that the geometry has to be a 2-D con-

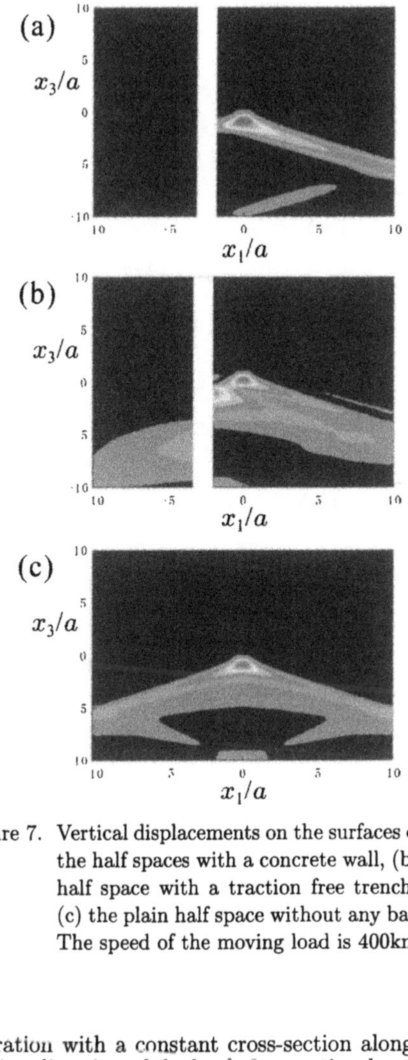

Figure 7. Vertical displacements on the surfaces of (a) the half spaces with a concrete wall, (b) the half space with a traction free trench and (c) the plain half space without any barrier. The speed of the moving load is 400km/s.

figuration with a constant cross-section along the moving direction of the load. In practice, however, the wave barrier may have a finite length. In this section, therefore, the effect of 3-D geometry is investigated by using a 3-D BEM in conjunction with the 2.5-D analysis (Hirose & Kitahara 1995).

The geometry of the problem considered here is shown in Figure 8. The wave barrier is a traction free trench with the finite length of $10a$ in the x_3 direction. The depth h of the trench is $4a$ and the speed of the moving load is 300km/s. Other parameters regarding loading conditions and material constants are the same as in the 2.5-D analysis shown in Figure 2.

If there is no wave barrier, the problem is reduced to a 2.5-D problem of a plain half space subjected to a moving load, which is already analyzed in the last section. The 3-D problem as shown in Figure 8 is then divided into the following two problems. One is the 2.5-D problem dealing with the wave radiation from a load running on the surface of a plain half space. The other is the 3-D problem which is formulated for the scattered wave field of the 2.5-D solution by a finite trench.

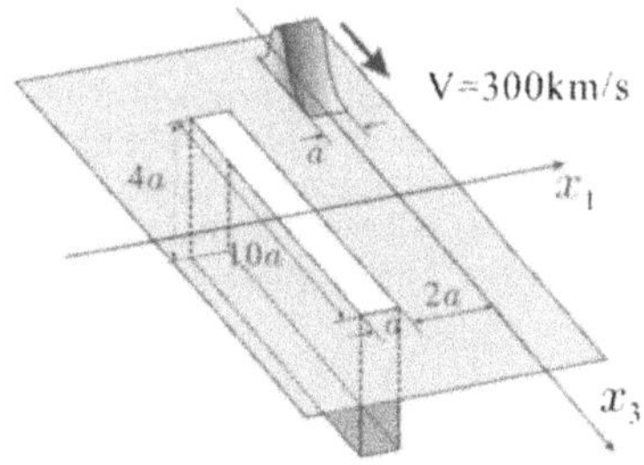

Figure 8. 3-D Geometry of a half space with a finite trench.

3.2 *BEM formulation*

The solution of the 3-D problem can be expressed as the sum of the incident field from a moving load and the wave field scattered by a finite barrier wall, i.e.,

$$\boldsymbol{u}^{3D} = \boldsymbol{u}^{in} + \boldsymbol{u}^{sc}. \tag{15}$$

If the 3-D scattering problem is formulated for the scattered field of $\boldsymbol{u}^{sc}$ and $\boldsymbol{t}^{sc}$ in the frequency domain, then we have the boundary integral equation as follows:

$$\int_S \left\{ U_{ik}^{3D}(\boldsymbol{x},\boldsymbol{y}) t_i^{sc}(\boldsymbol{x},\omega) - T_{ik}^{3D}(\boldsymbol{x},\boldsymbol{y}) u_i^{sc}(\boldsymbol{x},\omega) \right\} dS_x = \begin{cases} u_k^{sc}(\boldsymbol{y},\omega) & \boldsymbol{y}\in D \\ \frac{1}{2}u_k^{sc}(\boldsymbol{y},\omega) & \boldsymbol{y}\in S \\ 0 & \text{otherwise} \end{cases} \tag{16}$$

where U_{ik}^{3D} is the fundamental solution for a 3-D elastodynamic problem, given by

$$U_{ik}^{3D}(\boldsymbol{x},\boldsymbol{y}) = \frac{1}{4\pi\mu}\left[\frac{e^{ik_T r}}{r}\delta_{ik} + \frac{1}{k_T^2}\frac{\partial^2}{\partial x_i \partial x_k}\left\{\frac{e^{ik_T r}}{r} - \frac{e^{ik_L r}}{r}\right\}\right] \tag{17}$$

and T_{ik}^{3D} is defined by

$$T_{ik}^{3D}(\boldsymbol{x},\boldsymbol{y}) = \overset{n}{T}_{ij}(\boldsymbol{\partial}_x) U_{jk}^{3D}(\boldsymbol{x},\boldsymbol{y}). \tag{18}$$

Substituting $\boldsymbol{u}^{sc} = \boldsymbol{u}^{3D} - \boldsymbol{u}^{in}$ into Equation 16-b and taking account of the boundary condition $t_i^{sc}(\boldsymbol{x},\omega) = 0$ for $\boldsymbol{x} \in S$ yield the boundary integral equation for the 3-D wave field:

$$\frac{1}{2}u_k^{3D}(\boldsymbol{y},\omega) + \int_S T_{ik}^{3D}(\boldsymbol{x},\boldsymbol{y}) u_i^{3D}(\boldsymbol{x},\omega) dS_x = \frac{1}{2}u_k^{in}(\boldsymbol{y},\omega) + \int_S T_{ik}^{3D}(\boldsymbol{x},\boldsymbol{y}) u_i^{in}(\boldsymbol{x},\omega) dS_x. \tag{19}$$

As mentioned before, the incident wave used in the above 3-D boundary integral equation is the solution for the 2.5-D radiation problem from a load running on the surface of a plain half space. The 2.5-D solution in the frequency domain is obtained from Equation 10 as follows. Changing the integral variable ξ_3 to $\omega = V\xi_3$ in Equation 10, we have

$$u_i(\boldsymbol{x},t) = \frac{1}{2\pi}\int_{-\infty}^{\infty} V^{-1}\hat{\tilde{u}}_i'(\boldsymbol{X},\omega/V)e^{i\omega x_3/V} \times e^{-i\omega t}d\omega. \quad (20)$$

From Equation 20, it can be found that

$$u_i^{in}(\boldsymbol{x},\omega) = V^{-1}\hat{\tilde{u}}_i'(\boldsymbol{X},\omega/V)e^{i\omega x_3/V}. \quad (21)$$

Thus, the boundary integral equation (19) for the 3-D problem can be solved in conjunction with the 2.5-D analysis.

3.3 *Numerical examples*

Numerical calculations for the 3-D problem as shown in Figure 8 are carried out for several frequencies. Figures 9 to 11 show the vertical displacement amplitudes on the surface around the finite trench at the frequencies of $c_T\omega/a = 0.05, 0.5$ and 1, respectively, where c_T is the velocity of the S wave in a half space. In each figure, figure (a) shows the results for the 3-D problem with the finite trench and figure (b) demonstrates the results for the 2.5-D problem of a plain half space for comparison. As seen in Figures 9 to 11, the scattering effect by the finite trench can be found only in the vicinity of the trench. Actually the displacement distributions along the x_1 axis at $x_3 = \pm 7.5$ in Figures (a) are almost the same as those in Figures (b). It is, therefore, noted that the 3-D BEM in conjunction with the 2.5-D analysis requires the numerical modelling only in vicinity of the trench, which leads to an efficient numerical calculation.

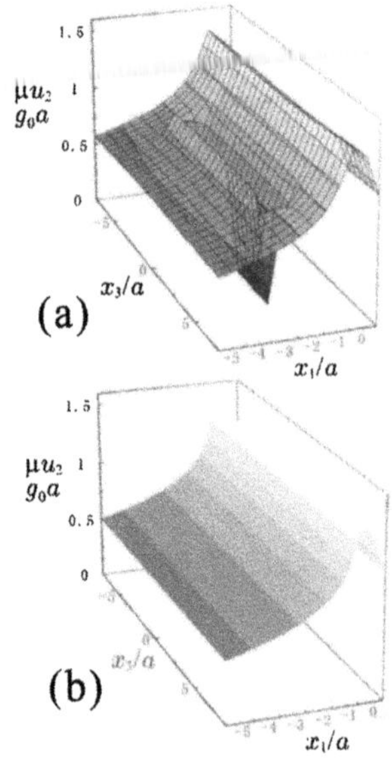

Figure 9. Vertical displacement amplitudes on the surface around the finite trench at the frequency $c_T\omega/a = 0.05$. (a): results for the 3-D problem with the finite trench and (b): results for the 2.5-D problem of a plain half space.

4. CONCLUSION

In this paper, the BEM was developed for the 2.5-D problem with 2-D geometry subjected to the moving load. Numerical results were shown to discuss the effect of a concrete barrier wall and a trench on the amplitudes of vertical surface displacements. Then the 3-D BEM analysis was carried out using the 2.5-D solutions as an incident wave field. Numerical examples for the 3-D scattering problem by a finite trench were demonstrated to show the efficiency of the 3-D BEM in conjunction with the 2.5-D analysis.

REFERENCES

Bleistein, N. 1986. Two-and-one-half dimensional inplane wave propagation. *Geophys. Prospect.* 34: 686–703.

Chouw, N. & Pflantz, G. 2000. Reduction of structural vibrations due to moving load. In N. Chouw & G. Schmid (eds), *WAVE 2000; Proc. intern. workshop wave 2000*, Bochum/Germany, 13-15 December 2000: 251–268. Rotterdam: Balkema.

Duhamel, D. 1996. Efficient calculation of the three-dimensional sound pressure field around a noise barrier. *J. Sound and Vibration* 197 (5): 547–571.

Eringen, A.C. & Suhubi, E.S. 1975. *Elastodynamics Volume II Linear Theory.* New York: Academic Press.

Furumura, T. & Takenaka, H. 1996. 2.5-D modelling of elastic waves using the pseudospectral method. *Geophys. J. Int.* 124: 820–832.

Hanazato, T. et al. 1991. Three-dimensional analysis of traffic-induced ground vibrations. *J. Geotech. Eng. ASCE* 117: 1133–1151.

Hirose, S. & Kitahara, M. 1995. Scattering from an interface defect between fiber and matrix, In D. O. Thompson and D. E. Chimenti (eds), *Review of Progress in Quantitative Nondestructive Evaluation* 14 : 99-106. New York: Plenum Press.

Hwang, R. N. and Lysmer, J. 1981. Response of buried structures to traveling waves. *J. Geotech. Eng. ASCE* 107: 183–200.

Khair, K. R. et al. 1989. Amplification of obliquely incident seismic waves by cylindrical alluvial valleys of arbitrary cross-sectional shape: Part I. Incident P and SV waves. *Bull. Seism. Soc. Am.* 79 (3):

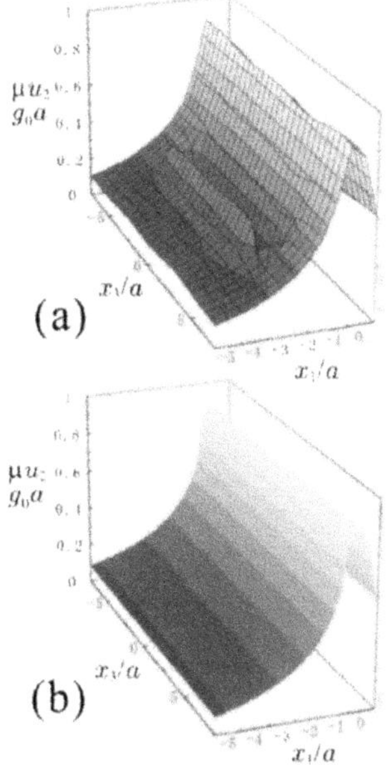

Figure 10. Vertical displacement amplitudes on the surface around the finite trench at the frequency $c_T\omega/a = 0.5$. (a): results for the 3-D problem with the finite trench and (b): results for the 2.5-D problem of a plain half space.

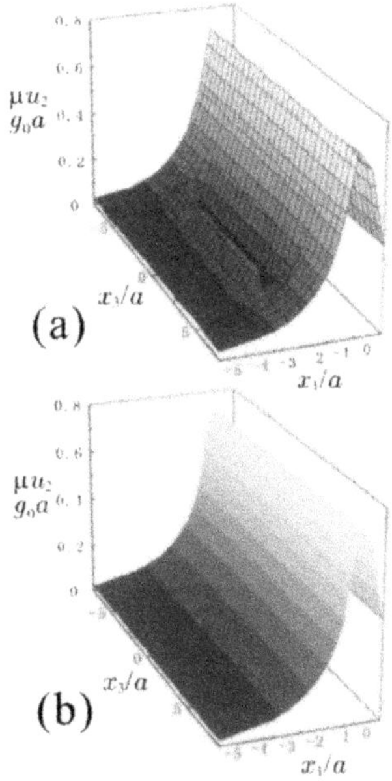

Figure 11. Vertical displacement amplitudes on the surface around the finite trench at the frequency $c_T\omega/a = 1$. (a): results for the 3-D problem with the finite trench and (b): results for the 2.5-D problem of a plain half space.

610–630.

Li, Z.L. et al. 1992. Reflection and transmission of obliquely incident surface waves by an edge of a quarter space: Theory and experiment. *J. Appl. Mech.* 59: 349–355.

Luco, J.E., Wong, H.L. & De Barros, F.C.P. 1990. Three-dimensional response of a cylindrical canyon in a layered half-space. *Earthq. Eng. Struct. Dyn.* 19: 799–817.

Okamoto, T. 1993. Teleseismic synthetics obtained from three-dimensional calculations in two-dimensional media. *Geophys. J. Int.* 112: 471–480.

von Estorff, O. et al. 2002. A three-dimensional FEM/BEM model for the investigation of railway tracks. In this volume.

Yang, Y.-B. & Hung, H.-H. 2001. A 2.5D finite/infinite element approach for modelling visco-elastic bodies subjected to moving loads. *Int. J. Numer. Meth. Engng* 51: 1317–1336.

Zhang, L. & Chopra, A.K. 1991. Three-dimensional analysis of spatially varying ground motions around a uniform canyon in a homogeneous half-space. *Earthq. Eng. Struct. Dyn.* 20: 911–926.

Wave propagation – Moving load – Vibration reduction, Chouw & Schmid (eds.)
© 2003 Swets & Zeitlinger, Lisse, ISBN 90 5809 559 2

Enhanced response through supplementary friction damper devices

I. Mualla & L.O. Nielsen
DampTech Ltd., Technical University of Denmark, Lyngby, Denmark
N. Chouw
Okayama University, Okayama, Japan
B. Belev
University of Architecture, Civil Engineering and Geodesy, Sofia, Bulgaria
W.I. Liao & C.H. Loh
National Centre for Research on Earthquake Engineering, Taipei, Taiwan
A. Agrawal
City University of New York, USA

ABSTRACT: The paper addresses results of an international effort in further developing a novel friction damper device for a reduction of induced vibrations in structures. A description of the device, results of small-scale experiments, and results of full-scale shaking table tests are presented. The result of the investigations shows that the effectiveness of the device is determined not only by the friction material but also by the location as well as the way of its installation. The devices have a stable energy dissipating behaviour. They are flexible in their application, since they only need limited space. The device can be installed easily and readjusted after installation. The damping capacity of the device can be increased by simply adding friction layers. The friction damper device proves to be an efficient and economical device for a reduction of dynamic response of structures.

1 INTRODUCTION

Excessive vibrations of a structure can occur due to accumulation of energy induced by a source in the structure like production activities in a factory. The source can also be outside like a heavy truck or high-speed train that travels in a densely populated area. Even if the vibration is not strong, long-term vibration pollution can severely affect human health.

There are many possibilities to reduce the vibration of structures. As an example we consider the Millennium Bridge. The bridge is well known for its original and elegant "blade of light" design (Fig. 1). When it opened on June 10, 2000, the bridge experienced excessive movement due to accumulating loading caused by pedestrians. Their synchronizing footsteps strongly amplified the initially small sway motion of the bridge. In order to control the level of the vibration, and to have a desired degree of comfort for the pedestrians, it is possible to stiffen the bridge by adding structural members. This will increase the stiffness of the bridge, and move its natural frequencies from the range that will be excited by the pedestrians. Also it is possible to increase the damping of the bridge by installing supplementary damper devices to enhance the energy absorbing ability of the bridge during its movement. One can reduce the load effect by limiting the number of people or by artificially modifying their walking patterns using flower corners in the path. One can also combine these three possible solutions. The first solution will significantly alter the design, since a much stiffer bridge is required. The third solution will hinder the free flow of the pedestrians. In Figure 1, and 2 the chosen second solution can be seen. Tuned mass dampers are placed below the bridge deck, and viscous dampers at certain locations as shown in Figure 2. While the tuned mass dampers control the excitation at the bridge frequencies, the viscous dampers mainly suppress the strong lateral movements.

Another possible reduction measure is the friction device. Because of the high energy dissipation potential, and the simplicity in installation and main-

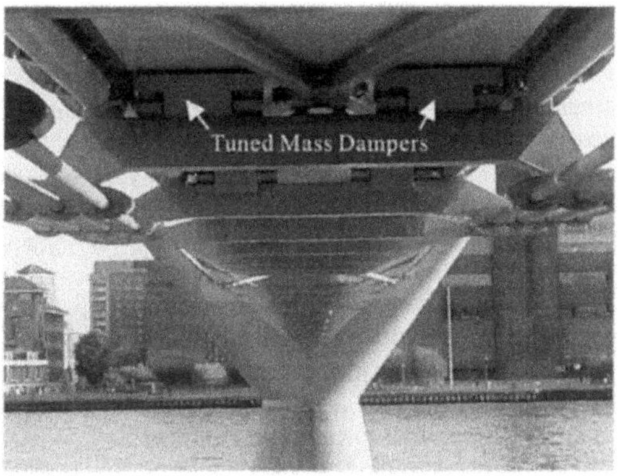

Figure 1. The London Millennium Footbridge viewed from the north.

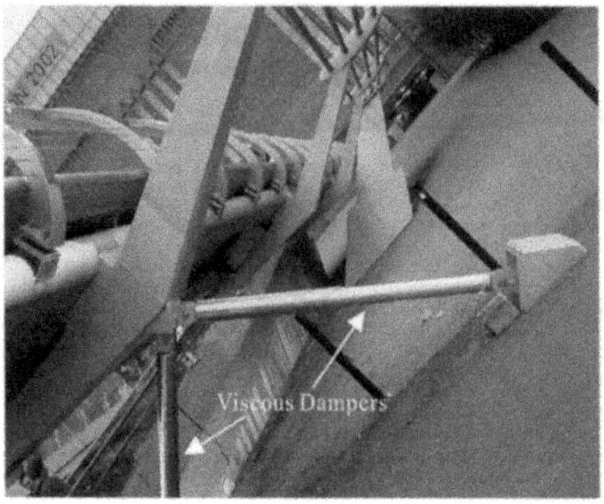

Figure 2. Viscous dampers at the south abutments of the bridge.

tenance many friction devices have been used in buildings around the world. Experimental investigations, e.g. Pall and Marsh (1982), Aiken and Kelly (1990), Fitzgerald et al. (1989) Constantiou et al. (1991), Grigorian and Popov (1993) or Nims et al. (1993) were already performed. Most of the investigations were devoted to developing the theory. Recent developments on analysis, and design of structures with friction dampers are described by e.g. Colajanni and Papia (1997), Dorka, Pradlwarter and Schueller (1998), Fu and Cherry (1999).

In this work the results of an international joint effort in further developing a novel friction damper device are presented.

2 DESCRIPTION OF THE DEVICE

The basic configuration of the damping device consists of three steel plates pre-stressed together by a steel bolt (Fig. 3(a)). Between the steel plates are two circular friction pad discs made of composite material. In order to have constant pre-stressing force several disc springs are used. Between these springs, and the two external steel plates, hardened washers are placed so that uniformly distributed pressure can be achieved. The energy absorbing potential of the device can be easily increased with almost the same space requirement by adding additional friction pads. Figure 3(b) shows a seven-plate damping device. Figure 4 shows the composition of a five-plate friction device.

Figure 5 shows a possible installation of the device in a three-storey frame structure. The centre plates are connected to the girder by a pin. In order to activate the energy absorbing mechanism of the device the horizontal plates are connected to the columns by using steel bracing members. The pre-stressing of the members prevents them from buckling. When the structure vibrates torsion between the horizontal, and the centre plates takes place. The friction at the interface between the composite material, and the steel plate resists the torsion (Fig. 4), thus a portion of the induced energy is dissipated. The resistance due to the device hinders the lateral movement of the storeys. Consequently, if no further energy is induced, the vibration of the frame structure becomes smaller and smaller with each relative rotation between the steel plates. With the pre-stressing force of the bolt and the arrangement of the damper devices in the structure we can control the degree of the resistance of the relative rotation at the devices. The simple mechanism allows an easy handling, and installation.

(a)

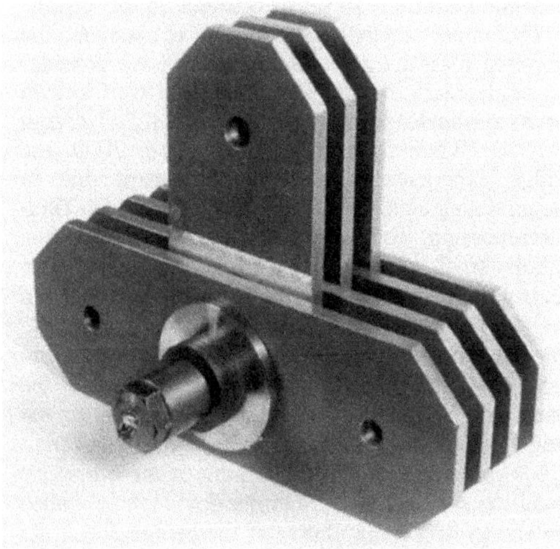

(b)

Figure 3(a) and (b). Possible configurations of the damping device.

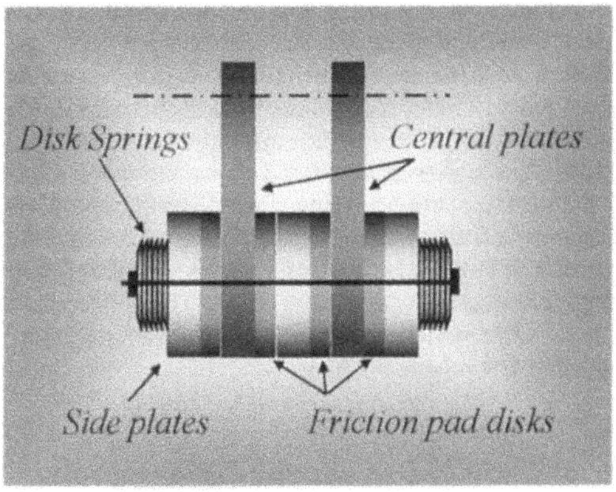

Figure 4. Composition of the damping device.

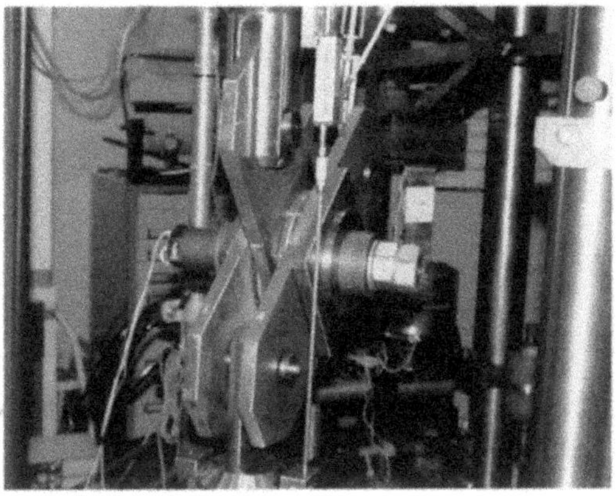

Figure 6. Laboratory test

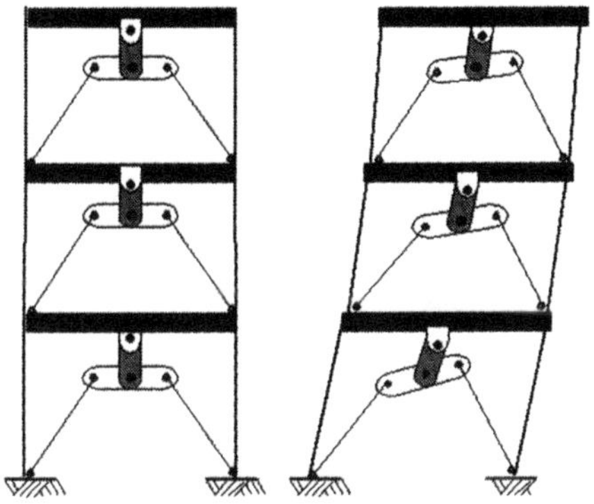

Figure 5. Activation of the energy absorbing mechanism of the device.

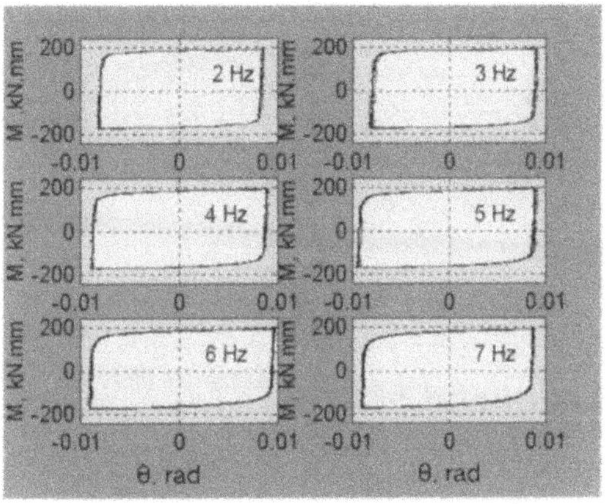

Figure 7. Hysteresis curves of the damper devices.

3 LABORATORY TESTS

Figure 6 displays a laboratory cyclic test of a damper device with a basic configuration at the Technical University of Denmark in Lyngby. Various parameters were considered, like the displacement amplitude, the frequency of the excitation, the bolt clamping force, and the number of loading cycles. In the cyclic test the effectiveness of several frictional materials was also investigated. The best material capable of sustaining up to 400 cycles without any property degradation was chosen for further extensive investigations where the damper devices were installed in a 1/3-scale portal frame model. Complete details of the experimental investigations are given by Mualla (2000).

The hysteresis curves in Figure 7 indicate that within the considered frequency range between 2 Hz to 7 Hz the amount of dissipated energy per cycle was almost frequency independent. The damping energy per cycle was proportional to the excitation amplitude.

The laboratory tests were accompanied by extensive numerical investigations. Since the hysteresis behaviour of the damper devices is determined almost only by the excitation amplitude, the Coulomb law can be used in the numerical model. The summary of the numerical investigations is given by Mualla and Belev (2001).

In order to refine the damping behaviour of the device additional viscous material can be placed between the steel plates as indicated in Figure 8. The advantage of this composition is that for lower excitation the viscous damper is activated, and for strong excitation the friction damper will be active. The device is therefore effective in a broader range of excitation. Figure 9 shows the development of the hysteresis loops due to different excitation amplitudes.

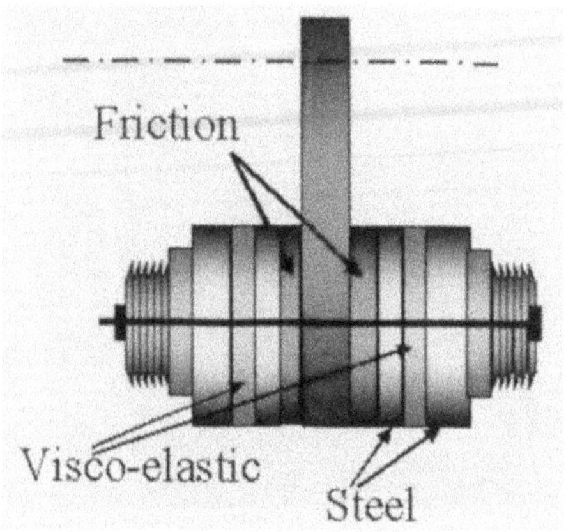

Figure 8. Friction and viscous damper device

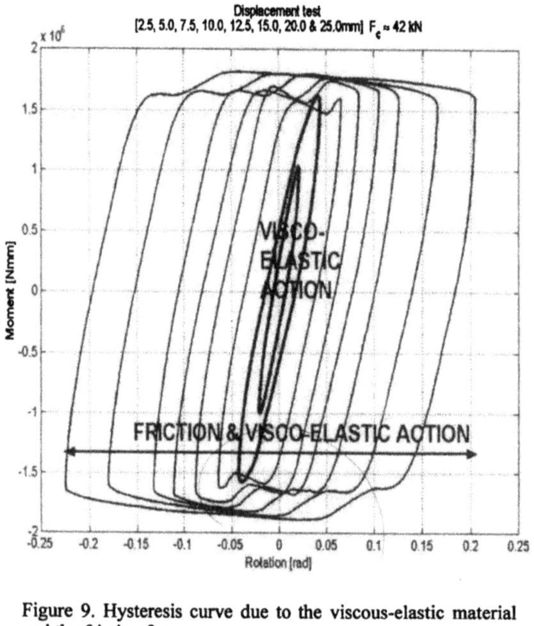

Figure 9. Hysteresis curve due to the viscous-elastic material and the friction force

4 FULL-SCALE SHAKING TABLE TESTS

In order to verify the effectiveness of the damper devices in a full-scale structure, shaking table tests were performed during the first half of the year 2001 in the national centre for research on earthquake engineering in Taipei, Taiwan. The three-storey steel frame structure had a storey height of 3m and a bay width in the direction of shaking of 4.5m. Figure 10 displays the set up of the full-scale investigation. Two damper devices in each storey were installed in plane of the shaking direction. The brace members consist of 20mm diameter round steel bars pin-connected to the damper plates and the frame joints. The column and girder cross sections are H200 x 200 x 8 x 12 and I200 x 150 x 6 x 9 in mm, respectively. The columns are fixed at the base, and the joints of the columns, and girders are fully welded. The columns will bend about the weak axis of the cross section during the shaking, and the frequency of the natural vibration in the shaking direction is 1.0684Hz. To simulate the roof and floor weights concrete blocks were used. The total mass of the frame structure including the auxiliary base perimeter frame was 38.3t.

Figure 10. Elevation of the shaking table test.

Displacement, and acceleration were measured by displacement transducers, and accelerometers attached to the base, and each floor level. Strain gauges were also mounted at the same locations on the columns, bracing bars, and girders. Two transducers were mounted on each damper device to measure its relative rotations. The total number of all channels used was 82.

The size of the shaking table is 5m x 5m. The facility is designed to simulate a ground excitation of up to 3g. Altogether 14 cases of ground excitations with the maximum ground acceleration ranging from 0.05g to 0.3g were considered (Table 1).

Several combinations of different sliding resistances of the devices (Fig. 11) along the height of the frame structure were investigated. For each combination several shaking intensities are considered (Table 2). M_f is the rotational friction resistance of the device. During the tests none of the friction pads or other device components were replaced.

Table 1. Considered excitation and intensities

Case	Ground excitation	Intensity [g]
1	El Centro (US), 1940	0.05
2	Imperial Valley	0.10
3		0.15
4		0.20
5		0.25
6		0.30
7	Kobe (Japan), 1995	0.10
8	Takatori	0.05
9		0.125
10		0.15
11	Chi-Chi (Taiwan),	0.05
12	1999	0.10
13	TCU052	0.15
14		0.20

Table 2. Considered device strength M_f and excitations

M_f (kNm)			Excitation intensity (g)		
1.60	1.90	--	El Centro		
			0.05	0.20	0.30
1.06	1.65	--	Kobe		
			all intensities		
1.34	2.50	2.6	Chi-Chi		
			0.15		

Table 3. Effectiveness of the friction damper

Storey no.	1	2	3
Storey drift [mm] without damper	80.4	79.2	50.1
Storey drift [mm] with damper	17.4	19.0	14.3
Reduction [%]	78.4	76.1	71.1

The consecutive tests with increasing peak ground acceleration were performed without readjusting the bolt clamping forces. Table 3 shows the maximum response displacement at each storey with, and without damper devices. The considered El-Centro ground excitation had the peak ground acceleration PGA of 0.3g. The shaking table tests showed that a reduction of the structural vibration up about 80% could be achieved.

Figure 12 shows the time histories of the roof displacement in case of a shaking intensity of 0.2g.

Figure 11. Close view of the damper device at the middle of the girder with measurement instrumentation

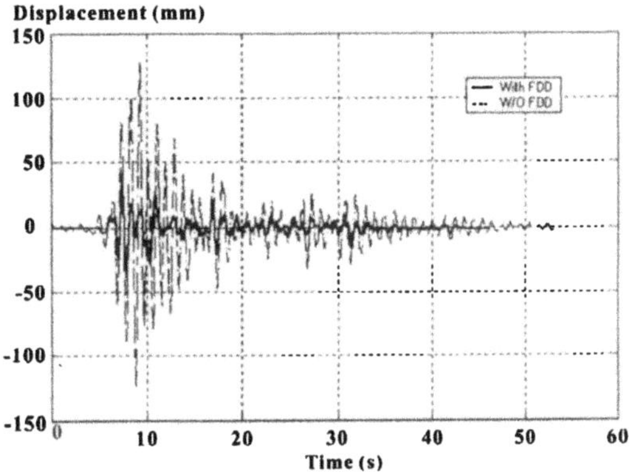

Figure 12. Reduction of the roof displacement due to the El-Centro NS-ground excitation with PGA of 0.2g.

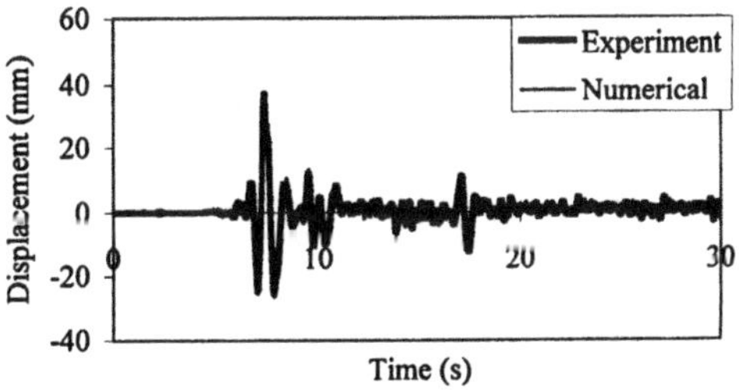

Figure 13. Roof displacement histories in case El-Centro with PGA of 0.2g.

The results of the large-scale experiment are used to verify the finite element model. The braces are modeled by tension-only links with initial pre-stressing force. Based on the approach proposed by Schneider and Amidi (1998) for the column panel zone modeling, bilinear rotational springs were introduced to account for the shear deformation of the column flanges at the connections between the beam and column when the columns bend about the weak axis.

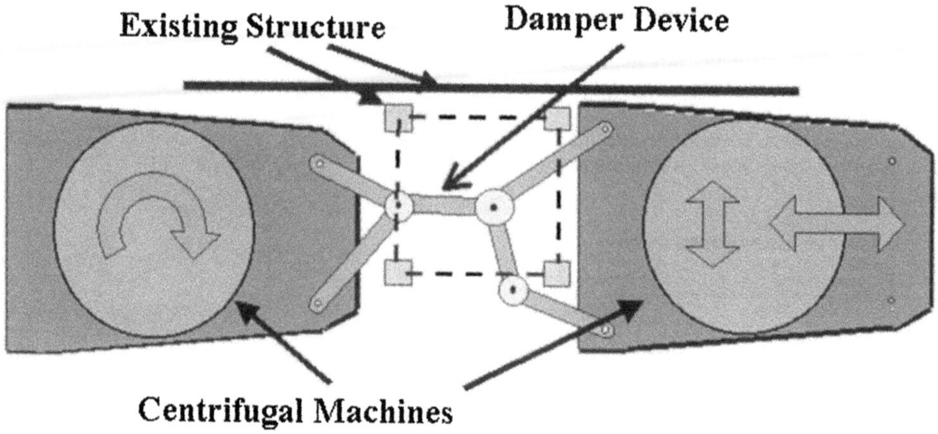

Figure 14. Vibration control of centrifugal machines.

The modal damping ratios for the first and second vibration modes were set to 1.5% and 0.5%, respectively. Figure 13 shows that with the current numerical model the experimental response can be predicted.

5 APPLICATIONS

Figure 14 shows the control of two centrifugal machines in a multi-storey structure by using the damper devices in reality. The two machines produce resonance-like vibrations during their start and stop. At those moments all floors of the building experience strong perceptible vibrations. Since the adjacent machines vibrate differently, the relative vibration between them is used to activate the friction forces in the devices. The solution shows that the damper devices are flexible in real applications.

Figure 15(a) and (b) show the damper devices in reality. Three- and five-plate damper devices are used. Figure 16(a) and (b) show another application. The power plant experiences disturbing vibrations during its operation. With the damper devices they were drastically reduced.

6 CONCLUSION

A novel friction damper device is introduced. The device consists of steel plates pre-stressed together by a steel bolt. Between the steel plates friction pad, and viscous material can be inserted. The effectiveness of the device is confirmed not only in laboratory tests, and full-scale shaking table tests; it is also confirmed in on-site applications.

(a)

(b)

Figure 15(a) and (b). Friction damper devices for a simultaneous control of two centrifugal machines.

The effectiveness of the device is strongly determined by the bolt clamping force, and the arrangement of the devices in the structures. Since the damper devices are simple, they can be applied in various configurations.

The device will be further developed, so that it controls vibrations most effectively in various applications.

(a)

(b)

Figure 16(a) and (b). Application of damper devices in a power plant in Denmark.

ACKNOWLEDGMENTS

The authors thank the reviewer for the constructive comments.

REFERENCES

Aiken, I, Kelly, S. 1990, Earthquake simulator testing and analytical studies of two energy absorbing systems for multi-storey structures, *Report No. UCB/EERC-90/03*, Berkeley.

Colajanni, P., Papia, M. 1997. Hysteretic characterization of friction-damped braced structures, *Canadian Journal of Civil Engineering*, 16: 753-766.

Constantimou, M. C., Reinhorn, A. M., Mokha, A. S., Watson, R. 1991. Displacement control device for base isolated bridges, *Earthquake Spectra*, EERI, 7(2): 179-200.

Dorka, U., Pradlwarter, H. J., Schueller, U. 1998. Reliability of MDOF-system with hysteretic devices, Engineering Structures, 20(8): 685-691.

Fitzgerald, T. F., Anagnos, T., Goodson, M., Zsutty, T. 1989. Slotted bolted connections in a seismic design of concentrically braced connections, *Earthquake Spectra*, EERI, 5(2): 383-391.

Fu, Y., Cherry, S. 1999. Simplified seismic code design procedure for friction-damped steel frames, *Canadian Journal of Civil Engineering*, 26: 55-71.

Grigorian, C. E., Popov, E. P. 1993. Slotted bolted connection energy dissipaters, *Earthquake Spectra*, EERI, 9(3): 491-504.

Mualla, I. H. 2000. *Experimental & computational evaluation of a new friction damper device.* Ph.D. Thesis, Dept. of Structural Engineering and Materials, Technical University of Denmark.

Mualla, I. H. & Belev, B. 2002. Performance of steel frames with a new friction damper device under earthquake excitation. *Journal of Engineering Structures*, 24: 365-371.

Nims, D. K., Richter, P. J., Bachman, R. E. 1993. The use of the energy dissipation restraint for seismic hazard mitigation, *Earthquake Spectra*, EERI, 9(3): 467-487.

Pall, A. S., Marsh, C. 1982. Response of friction damped braced frames. *Journal of Structural Engineering*, ASCE, 108: 1313-1323.

Schneider, S. & Amidi, A. 1998. Seismic behaviour of steel frames with deformable panel zones. *Journal of Structural Engineering*, 124(1): 35-42.

Wave propagation – Moving load – Vibration reduction, Chouw & Schmid (eds.)
© 2003 Swets & Zeitlinger, Lisse, ISBN 90 5809 559 2

Nonstationary robust vibration control for moving wire

M. Otsuki & K. Yoshida
Department of System Design Engineering, Keio University, Yokohama, Japan

ABSTRACT: This paper presents a vibration control method for a wire changing its length such as elevator cables and crane wires. We propose synthesis methods of vibration controller, one is nonstationary optimal control method considering a time-varying dynamics and the other is a robust vibration control method to design a compensator with the robustness for a variation of tension, a transition of frequency of resonance and a spillover of high order modes. The performances of controllers are demonstrated by numerical calculation for the case that the system is subjected to an acceleration disturbance and has the parameter variation due to the change of tension of wire. We compare the robust controllers and nonstationary optimal one with a gain scheduling controller based on LQG theory. Consequently, the robust controllers show the advantage for the robustness on the variation of parameter.

1 INTRODUCTION

For a moving wire, we conceive two systems as followings. One is an axially moving wire system without variation of its parameters, such as cable tramways, band saws and power transmission chains. And the other is a system of wire changing its parameters as tension and length, such as crane wires, elevator cables and tethers of satellite. Both systems of wire generate their transverse and longitudinal vibration due to its behavior and disturbance. Then, the investigations of vibration control for the formers are obtained. For one thing, the control method to suppress the vibration of wire at its boundary is presented based on an adaptive control and computed-torque method (Fung, F.R. et al. 2002). Meanwhile, with respect to the latter, Takagi and Nishimura in 1998 proposed a gain-scheduled control method based on LMI for a tower crane considering variation of length of crane-rope. However, few savants investigate for the transverse vibration control of wire as the elevator cable caused by the resonance with a building which envelops the moving wire system. Actually, there was some accidents as the breakdown of elevator cage and rope on World Trade Center in New York in 1977. After that, on some high-rise buildings, the operation of elevator has been controlled depending on the building-sway to prevent this rope-sway problem. Therefore, the active control method for solving this problem is demanded to increase efficiency of elevator operation.

For the elevator cable, the stark difference exists at the point that the wire desires to control itself directly by comparison with the others. Moreover, the elevator cable and crane string are a time-varying system due to changing its length and natural frequency with time. Since the sensor and actuator for the control are located on both ends of wire alone, the wire is controlled at its boundary position alone. And also, the controller considering with its time-varying characteristics is effective in reducing vibration of wire. Some researchers have proposed a study about control theory for a time-varying system. We have been proposed the optimal control method for reducing rope-sway of elevator on skyscraper as a solution of vibration problem of time-varying system (Otsuki et al. 2001, 2002). O'Brien and Iglesias present a synthesis method of controller designed by using time-varying weighting in a frequency domain in 1998. Furthermore, for the nonstationary optimal controller, Hara & Yoshida in 1994 implemented the simultaneous synthesis method of controller to determine the positioning and reduce the vibration of control object, which includes a time-varying uncertainty.

The main objective of this paper is suppressed the transverse vibration of wire subjected to the variation of tension and natural frequency, and a model error due to ignored high order modes. Especially, we consider about the boundary control, the construction constrain due to locating sensors, a synthesis of controller considering the system noise, the observation one and the time-varying characteristics of control object, and a time lag of actuator for the control. We examine the performance of controllers through the numerical calculation in the frequency and time domain.

2 CONTROL OBJECT

In this study, we consider a following system: a structure contains two mass points and a wire as shown in Figure 1. This system changes its parameter with time, since the length of wire takes a different value all the time. We suppose that the structure, the wire and the mass points have a horizontal deflection, and in addition the lower mass point moves in up-and-down direction. Furthermore, a longitudinal vibration is ignored in this research, because we focus on the transverse vibration of wire caused by the resonance between the structure and wire. And a control input forces the upper mass point to displace for reducing the vibration of wire.

First, by using *FEM* method, we construct the equation of motion of wire. A wave equation of moving wire is expressed as follows.

$$\rho A\left(\frac{\partial}{\partial t}+v\frac{\partial}{\partial s}\right)^2 r(s,t) - \frac{\partial}{\partial s}T(s)\frac{\partial r(s,t)}{\partial s} + c(s)\left(\frac{\partial}{\partial t}+v\frac{\partial}{\partial s}\right)r(s,t) = 0 \tag{1}$$

where s is the coordinate along the wire, $r(s,t)$ the deflection of wire in the transverse direction, t the arbitrary time, ρA the line density of wire, $T(s)$ the tension of wire depending on the up-and-down position, v the velocity of variation of length of wire and $c(s)$ the damping coefficient per unit length of wire.

The model of wire is transformed into a discrete one shown in Figure 1. From the previous wave equation, the motion of equation of an element for the discrete model is given by the following equation with *FEM* method.

$$\frac{\rho A l(t)}{6}\begin{bmatrix} 2 & 1 \\ 1 & 2 \end{bmatrix}\begin{bmatrix} \ddot{r}_i \\ \ddot{r}_{i+1} \end{bmatrix} + \frac{1}{6}\begin{bmatrix} 2\beta-d_c & \beta+d_c \\ \beta-d_c & 2\beta+d_c \end{bmatrix}\begin{bmatrix} \dot{r}_i \\ \dot{r}_{i+1} \end{bmatrix} + \frac{1}{l(t)}\begin{bmatrix} \chi-d_k & -\chi+d_k \\ -\chi-d_k & \chi+d_k \end{bmatrix}\begin{bmatrix} r_i \\ r_{i+1} \end{bmatrix} = 0 \tag{2}$$

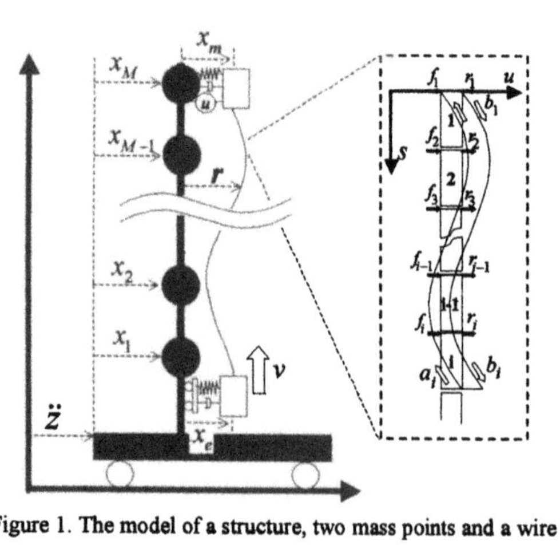

Figure 1. The model of a structure, two mass points and a wire expressed by a discrete model

where $\beta = c(s)l(t)$, $d_c = \rho A v$, $\chi = T(s) - d_c v$, $d_k = \frac{1}{2}\beta v$, $l(t) = (l(0)+vt)/N$, N is the discrete number of wire, it is assumed that the wire is not interacted directly with an external force.

Second, it is supposed that the structure is approximated by the M mass points, dampers and springs. Each mass point is modeled so as to have a single degree of freedom in the transverse direction. Consequently, we derive the combined motion of equation of the structure, the wire and the two mass points by their superposition as follows.

$$\boldsymbol{M}(t)\ddot{\boldsymbol{x}}_d(t) + \boldsymbol{C}(t)\dot{\boldsymbol{x}}_d(t) + \boldsymbol{K}(t)\boldsymbol{x}_d(t) = \boldsymbol{F}(t) \tag{3}$$

$$\boldsymbol{x}_d(t) = \begin{bmatrix} x_{b1} & x_{b2} & \cdots & x_{bM} & x_m & r_2 & r_3 & \cdots & r_N & x_e \end{bmatrix}^T \tag{4}$$

where x_{bi} is the relative displacement of i-th mass point between the structure and a basement, x_e and x_m are the relative displacements of two mass points from the structure, which are attached to the structure and coincide with r_1 and r_{N+1}, respectively. In this paper, we omit the details about the matrices of this motion of equation.

In addition, we have the above equation of motion been non-dimensional to generalize the control object. For the non-dimensional model, we adopt the maximum transverse displacement A_m of top story of structure as the central value of length, and also the propagation time $\tau = L_{max}\sqrt{\rho A/T}$ of traveling wave on the wire as central value of time. Here, L_{max} is the maximum length of wire. By using these parameter, the non-dimensional model is designed by the followings.

$$\boldsymbol{M}^*(t^*)\ddot{\boldsymbol{x}}_d^*(t^*) + \tau\boldsymbol{C}^*(t^*)\dot{\boldsymbol{x}}_d^*(t^*) + \tau^2\boldsymbol{K}^*(t^*)\boldsymbol{x}_d^*(t^*) = \boldsymbol{F}^*(t) \tag{5}$$

$$\ddot{\boldsymbol{x}}_d^*(t^*) = \tau^2\ddot{\boldsymbol{x}}_d(t)/A_m,\ \dot{\boldsymbol{x}}_d^*(t^*) = \tau\dot{\boldsymbol{x}}_d(t)/A_m,\ \boldsymbol{x}_d^*(t^*) = \boldsymbol{x}_d(t)/A_m$$

where * indicates non-dimensional value. And the real control input is described by the equation:

$$u = \frac{A_m}{\tau^2}\left(m_m + \frac{\rho A l(t)}{3}\right)u^* \tag{6}$$

where m_m is the weight of upper mass point. From this sentence onward, we abbreviate the asterisk * for the non-dimensional value. Based on the non-dimensional motion of equation of control object, we derive the state equation expressed by the following equation.

$$\dot{\boldsymbol{x}}(t) = \boldsymbol{A}(t)\boldsymbol{x}(t) + \boldsymbol{B}(t)u(t) + \boldsymbol{D}(t)\boldsymbol{z}(t) \tag{7}$$

$$\boldsymbol{x}(t) = \begin{bmatrix} \boldsymbol{x}_d & \dot{\boldsymbol{x}}_d \end{bmatrix}^T \tag{8}$$

$$\boldsymbol{z} = \begin{bmatrix} \ddot{z} & \ddot{x}_{b1} & \cdots & \ddot{x}_{b2} & \ddot{z}_e \end{bmatrix}^T \tag{9}$$

where $\ddot{z}$ is the absolute acceleration of basement, $\ddot{x}_{bi}$ the relative acceleration of structure depending on the up-and-down position of each mass point of wire from the basement, and $\ddot{z}_e$ the relative acceleration of lower mass point attached to the bottom of wire, which depends on the up-and-down position.

Finally, for the parameter of control object, we deal with the control object whose non-dimensional parameter $\alpha = A_m / L_{max} = 0.5\times10^{-3}$. The fundamental natural frequency of structure is about 0.9, the damping ratio 0.01 about all modes, the height 240 meters and M equals to five. The time-varying wire is attached to two mass points at its top and bottom, its line density is 1.0 kg/m and its length changes from 20 to 200 meters, that is, its first natural frequency varies from 0.5 to 4.0. Consequently, the resonance between the structure and wire is caused in the middle of period for the up-and-down movement. And also, one of mass points is constrained at 20 meters below the top of structure. We show the phenomena of resonance between the wire and structure during descending motion of lower mass point in Figure 2.

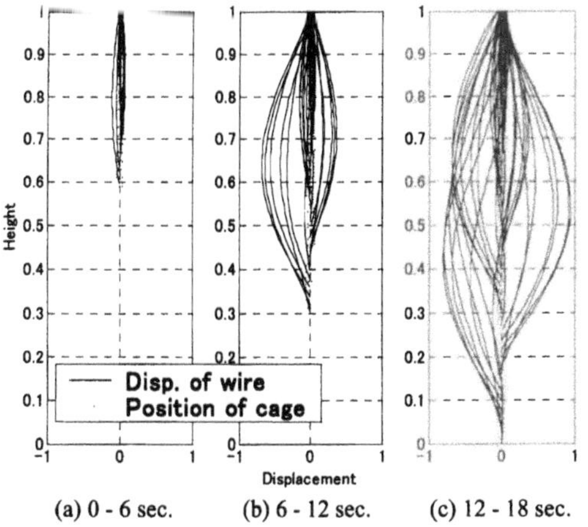

(a) 0 - 6 sec. (b) 6 - 12 sec. (c) 12 - 18 sec.

Figure 2. The phenomena of resonance between the wire and structure during descending motion of lower mass point.

3 CONTROL METHOD

3.1 *Nonstationary optimal control method*

First, let consider the control method for the time-varying system expressed by Equation.(7). If a time-invariant criterion function is adopted to design an optimal controller for reducing vibration of time-varying objects, the control input might not be adequate for this performance. That is, the vibration control requires the time-varying input according to the state of this object. Hence, a criterion function with the time-varying weightings is utilized to reduce the vibration of time-varying objects. We can express this criterion function in the form:

$$J = \boldsymbol{x}_z^T(t_f)\boldsymbol{S}\boldsymbol{x}_z(t_f) + \int_0^{t_f}\left[\boldsymbol{x}_z^T(t)\boldsymbol{Q}(t)\boldsymbol{x}_z(t) + u^2(t)/r_n(t)\right]dt \quad (10)$$

where t_f is a determinate end time of control period, $\boldsymbol{Q}(t)$ the time-varying weighting matrix on the state values during control period, $r_n(t)$ the time-varying weighting matrix on the control input, $\boldsymbol{S}$ the weighting matrix on the state values at the end of control period. By minimizing this criterion function, the control input $u(t)$ is uniquely determined by the optimal control method and is given by $u(t) = -\boldsymbol{f}(t)\boldsymbol{x}(t)$, where $\boldsymbol{f}(t)$ is a time-varying feedback gain vector and $\boldsymbol{f}(t) = \boldsymbol{B}^T(t)\boldsymbol{P}(t)/r_n(t)$. $\boldsymbol{P}(t)$ is derived by solving the following time-varying Riccati differential equation:

$$\begin{aligned} -\boldsymbol{P}(t) = {} & \boldsymbol{P}(t)\boldsymbol{A}(t) + \boldsymbol{A}^T(t)\boldsymbol{P}(t) \\ & -\boldsymbol{P}(t)\boldsymbol{B}(t)R^{-1}(t)\boldsymbol{B}^T(t)\boldsymbol{P}(t) + \boldsymbol{Q}(t) \end{aligned} \quad (11)$$

$$\boldsymbol{P}(t_f) = \boldsymbol{S} \quad (12)$$

In this method, the vibration control is carried out with this time-varying feedback gain, which is a priori derived from Equations (11) and (12) by using Runge-Kutta method in the inverse time direction. And the weighting matrix $\boldsymbol{Q}(t)$ generally depends on the displacements and velocities of control objects. Furthermore, we implement an output feedback control by using Kalman filter as an estimator of state values.

3.2 *Gain-scheduled control based on LQG theory*

Second, we use the gain-scheduled controller for the vibration reduction as the comparison method of performance. Gain scheduled controller is designed based on the fixed model of control object at the arbitrary time by using LQG theory. At all the time, this controller is designed with respect to each constant interval of time, and then, the controller designed at the arbitrary time t is derived as follows.

$$f_i(t) = \theta(t) f_i(t_2) + (1-\theta(t)) f_i(t_1) \quad (13)$$

$$\theta(t) = (t - t_1)/\Delta t \quad (14)$$

where $t_1 \le t \le t_2 = t_1 + \Delta t$, Δt is the constant interval of time and f_i is a vector element of feedback gain.

Between the nonstationary optimal control method and gain-scheduled one, there is the difference that the nonstationary optimal controller at the arbitrary time is designed with considering the temporal response of control object. Moreover, for the gain-scheduled control method, it is disadvantage to spend considerable time on designing its controller. However, the controller is adaptive for the case when the length of wire changes randomly.

3.3 *Robust control method*

Finally, we meditate on the controller having the robustness for scaled structured and unstructured uncertainties. The scaled structured uncertainties indicate the variation of parameter such as the tension of wire, a system disturbance and an observation noise in the time-domain. Meanwhile, the unstructured uncertainties are the influence of model error, the spillover due to the ignored high order modes and the nonlinear characteristics in the frequency domain. Hence, for designing the robust controller, we construct the augmented system, which is based on the nominal system and includes the previous uncertainties. In this paper, we adopt the time-invariant system fixed in the middle of control period and the time-varying system described by Equation (7) as the nominal system. The augmented system is given by the following equation in the state space description.

$$\dot{x}_z(t) = \begin{bmatrix} A_n + \Delta A_n & 0 \\ 0 & A_r + \Delta A_r \end{bmatrix} x_z + \begin{bmatrix} B_n + \Delta B_n \\ B_r + \Delta B_r \end{bmatrix} u + \begin{bmatrix} D_n \\ 0 \end{bmatrix} \ddot{z} \quad (15)$$

$$y(t) = \begin{bmatrix} C + \Delta C & 0 \\ 0 & C_r + \Delta C_r \end{bmatrix} x_z + \begin{bmatrix} 0 \\ D_r \end{bmatrix} u + \begin{bmatrix} N \\ 0 \end{bmatrix} w_n \quad (16)$$

$$x_z = \begin{bmatrix} x & x_r \end{bmatrix}^T$$

where Δ shows the scaled structured uncertainty in the time domain, x_r is the state value of high-pass filter that shows the shape of unstructured uncertainty, A_n, B_n, C_n and D_n are the matrices of nominal model, A_r, B_r, C_r and D_r the matrices of high-pass filters to indicate the unstructured uncertainty, N the observation disturbance matrix and w_n the observation noise.

We estimate the quantity of the scaled structured uncertainty with considering the time-varying variation by the followings.

$$\Delta A = A_{max} - A,\ \Delta B = B_{max} - B,\ \Delta C = C_{max} - C \quad (17)$$

In this equation, the subscript 'max' means the matrix including the maximum values in all the time and all the symbol express the matrix. Then, the structured uncertainties are substituted to the followings.

$$\Delta A = I_A \delta_A I W_A,\ \Delta B = I_B \delta_B I W_B,\ \Delta C = I_C \delta_C I W_C \quad (18)$$

where I_i is the identity matrix, $\delta_i I$ is a repeated scalar block and W_i the weighting matrix expressing the quantitative scale. Meanwhile, we adopt the 2nd order high-pass filter as the notation of the unstructured uncertainty and also it is designed as it envelops the model error due to ignored high order modes shown in Figure 3.

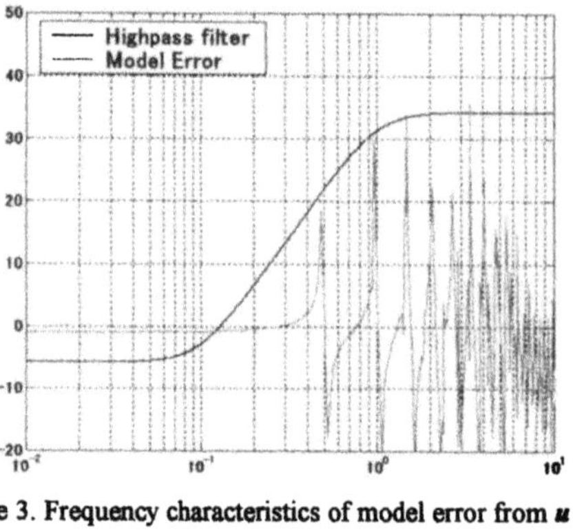

Figure 3. Frequency characteristics of model error from u to y and shaping filter enveloping them at the end of control period

In this argument, we consider two robust control methods as the followings. One is a stationary robust control method, which is designed by using time-invariant system in the arbitrary time as nominal model. For this case, the scaled structured uncertainty shows the variation due to passing time in addition to a parameter variation. The other is a nonstationary robust controller, which is designed by using time-varying system as nominal model. For this controller, the scaled structured uncertainty indicates the parameter variation alone and the high-pass filters enveloping the model error with considering a shift due to time express the unstructured uncertainty. The order of stationary robust controller is getting smaller than nonstationory one. However, for the case that the variation of parameter due to time-varying characteristics, the stationary robust controller is getting conservative and wrong for reduction of vibration of control object. We expect the good performance for reducing the vibration of wire from the nonstationary robust controller. But then, since the variation of controller in the time direction is smooth and simple, we implement to reduce the dimension of controller with the polynominal interpolation.

For designing the stationary robust controller, we show the differential game type criterion function in the time domain with the worst disturbance as the following equation (Doyle et al. 1989):

$$J = \int_0^{\infty} \left[x_z^T(t) Q(t) x_z(t) + u^2(t)/r_n - \gamma^2 w_s(t) w_s(t) \right] dt \quad (19)$$

where r_n is the weighting on the control input, γ the time-invariant weighting on the worst disturbance $w_s(t)$ into the system. At the same time, for designing the nonstationary robust controller, we show the criterion function as the following equation.

$$J = x_z^T(t_f) S x_z(t_f) + \int_0^{t_f} \left[x_z^T(t) Q(t) x_z(t) + u^2(t)/r_n(t) - \gamma^2(t) w_s(t) w_s(t) \right] dt \quad (20)$$

where $\gamma(t)$ is the time-varying weighting on the worst disturbance.

Then, we express the uncertainties with the open-loop expression by using virtual disturbance input w_a, w_b, w_c, w_g, the real disturbance $\ddot{z}$, w_n and the weighting performance variables z_Q, z_r, z_a, z_b, z_c, z_g. The open-looped augmented system is shown in Figure 4. The control problem of this augmented system is H-infinity one and a generalized plant to derive the controller is described by the state space description as follows.

$$\dot{x}_z(t) = A_z(t) x_z(t) + B_1(t) w(t) + B_2(t) u \quad (21)$$
$$z_h(t) = C_1(t) x_z(t) + D_{12}(t) u \quad (22)$$
$$y(t) = C_2(t) x_z(t) + D_{21}(t) w(t) \quad (23)$$

$$w = \left[\ddot{z}\ w_o\ w_a\ w_b\ w_c\ w_g \right]^T$$

$$z_h = \left[z_g\ z_r\ z_a\ z_b\ z_c\ z_g \right]^T$$

$$A_z = \begin{bmatrix} A_n & 0 \\ 0 & A_r \end{bmatrix},\ B_1 = \left[\begin{bmatrix} D_n \\ 0 \end{bmatrix} 0\ I\ I\ 0\ 0 \right],\ B_2 = \begin{bmatrix} B_n \\ B_r \end{bmatrix}$$

$$C_1 = \begin{bmatrix} Q^{1/2} & 0 \\ 0 & 0 \\ W_a & 0 \\ 0 & 0 \\ W_c & 0 \\ 0 & C_r \end{bmatrix},\ D_{12} = \begin{bmatrix} 0 \\ R^{1/2} \\ 0 \\ W_b \\ 0 \\ D_r \end{bmatrix},\ C_2 = \left[C_n\ 0 \right]$$

$$D_{21} = \left[0\ N\ 0\ 0\ I\ I \right]$$

where $Q = Q^{\frac{1}{2}T} Q^{\frac{1}{2}}$, $R^{\frac{1}{2}} = \sqrt{r_n}$ and N is given by the power spectrum density of observation noise.

In Equations (21), (22) and (23), the optimal control input is obtained by the following controller (Sanpei et al. 1990):

$$\dot{\hat{x}}(t) = \hat{A}(t)\hat{x}(t) + \hat{B}(t)y(t) \tag{24}$$

$$u(t) = \hat{C}(t)\hat{x}(t) \tag{25}$$

$$\begin{aligned} \hat{A}(t) = {} & A_z(t) + \\ & \left\{ B_1(t)B_1^T(t)/\gamma^2(t) - B_2(t)\left[D_{12}^T(t)D_{12}(t) \right]^{-1} B_2^T(t) \right\} X(t) \\ & - \left[I - Y(t)X(t)/\gamma^2(t) \right]^{-1} Y(t)C_2^T(t)\left(D_{21}(t)D_{21}^T(t) \right)^{-1} C_2(t) \end{aligned}$$

$$\hat{B}(t) = -\left[I - Y(t)X(t)/\gamma^2(t) \right]^{-1} Y(t)C_2^T(t)$$

$$\hat{C}(t) = -\left[D_{12}^T(t)D_{12}(t) \right]^{-1} B_2^T(t)X(t)$$

where $X(t)$ and $Y(t)$ is derived by solving the following time-varying differential Riccati equation including the term of worst disturbance with Runge-

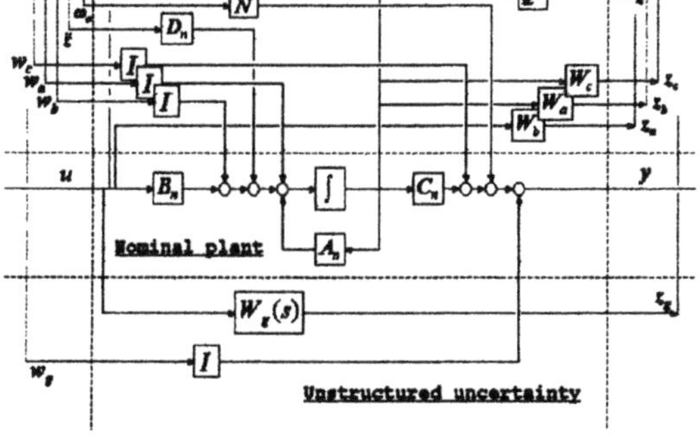

Figure 4. Schematic diagram of structure of generalized plant

Kutta method.

$$\begin{aligned} -\dot{X}(t) = {} & X(t)A_z + A_z^T X(t) + C_1^T C_1 \\ & + X(t)\left\{ B_1 B_1^T/\gamma^2(t) - B_2\left[D_{12}^T D_{12} \right]^{-1} B_2^T \right\} X(t) \end{aligned} \tag{26}$$

$$X(t_f) = S \tag{27}$$

$$\begin{aligned} \dot{Y}(t) = {} & Y(t)A_z^T + A_z Y(t) + B_1 B_1^T \\ & + Y(t)\left\{ C_1 C_1^T/\gamma^2(t) - C_2^T\left[D_{21}D_{21}^T \right]^{-1} C_2 \right\} Y(t) \end{aligned} \tag{28}$$

$$Y(t_0) = M_0 = 0 \tag{29}$$

where Equation (27) is the final condition, Equation (29) the initial condition. Therefore, Equation (26) is solved by using Runge-Kutta method in the inverse direction of time and then Equation (28) is performed in the forward direction of time. Here, if $\gamma(t)$ is getting much bigger than $Q(t)$, the above Riccati equation approaches to Equation (11). Therefore, we implement the transitions of the nonstationary robust controller for reducing the vibration and having the robustness for the uncertainties by changing the weighting with time.

4 EXAMINATION OF NUMERICAL CALCULATIONS

4.1 *Conditions of numerical calculation*

In this chapter, we examine the reduction performance of controller, which is derived from the proposed method through the numerical calculations in the frequency and time domain. This calculation is performed for two cases. One is the verification of performance in the frequency domain and the other is the examination of reduction performance of controller for the case that the structure is subjected to the basement acceleration in the time domain.

Each case is verified for the same condition where the lower mass point is going down alone in the cases that the value of wire tension is normal or 5 times as much as normal one. And the mass point descends at the speed of 10 meters per a second constantly. Hence, the final time of control is set to be 18 seconds, which is the time that it takes for the wire to change up to 200 from 20 meters. The wire in the augmented system for designing the controller is modeled on the case $N=10$. However, we verify the performance of controller with the model of wire, which is modeled on the case $N=40$. As a result, we validate the evidence of robustness for the model error using the difference of dimension of model. Moreover, the outputs from the control object are the relative displacements of all the stories of structure and each two discrete mass points of wire at both ends, that is, the four discrete mass points of wire at both ends is measured. Finally, it is assumed that the measurement information includes the observation noise and the actuator for the control has the nature of the time-lag, which is 10 m seconds and is the first order delay.

4.2 Designing controllers

A nonstationary optimal controller (Nonst.) is designed in order to reduce the vibration of wire all the time and have the robustness for the variation of tension and model errors in the start of control period. Hence, we let the weighting $r_n(t)$ change with time depending on the characteristics of wire. For the gain-scheduled controller (Gain.), we carry out the design similarly to the previous one. For the nonstationary robust controller (Nonst. Robust), the synthesis method of weightings $Q(t)$ is roughly the same as the nonstationary one. However, the weighting $\gamma(t)$ on the worst disturbance is designed by depending on the variation of uncertainty due to time. In the case that the influence of uncertainty is small, we let the weighting $g(t)$ be big, that is, the good performance of controller is derived. Conversely, when the weighting $g(t)$ is small, the controller gets conservative for the worst disturbance. Furthermore, we design the stationary robust controller estimating the scaled structured uncertainties are adequate for reducing the vibration of wire and considering the worst disturbance by g.

Consequently, we get the controller having the robustness. In addition, we design all the controller as the root mean square value of each control input makes nearly equal in the case that the structure is subjected to sine wave with frequency component 0.9.

4.3 Consideration about numerical calculation

Figures 5,6 and 7 show the frequency response of control object from $\ddot{z}$ to the center mass point of wire with controls and non-control on 0, 9 and 18 seconds, respectively. And we indicate the results in the time domain during descending motion of wire alone for the case that the wire has the parameter variation and neither it does. First, Figure 8 shows the time history of relative displacement of wire at the center. Second Figures 9 and 10 show the time histories of root mean square and maximum value of displacement of wire. Third, Figures 11 expresses the time histories of control inputs derived from each controller. Furthermore, Figures 12, 13, 14 and 15 indicate the time histories of all the mass points of wire through all the time for the cases that the wire includes the variation or the no variation of tension. Finally, in Table 1 and 2, the results through all the time are shown with respect to the root mean square value and maximum value of wire or control input.

In Figures 5,6 and 7, all the controllers show the good performances in the frequency domain on the arbitrary time. Especially, using the nonstationary and gain-scheduled controllers reduces the peak of first mode adequately. As a result, there is no accurate difference among the performances of all the controllers except for the result on 9 seconds. At that time, the nonstationary optimal controller and gain-scheduled one are derived the suppression of first mode of control object. However, the second mode of control object is not suppressed by using any controller, because this mode is the natural mode of vibration of structure, and the upper mass point is pretty lighter than the top story of structure. In Figures 8,9 and 10, the resonance between the wire and the structure is certified in the latter period and last half of control period, respectively. For the control input, the nonstationary optimal and gain-scheduled controller generate much the same

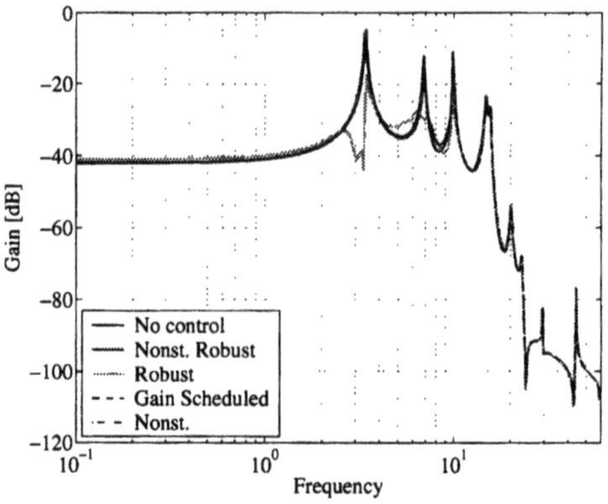

Figure 5. Frequency response of control object from w to the center mass point of wire on the initial state

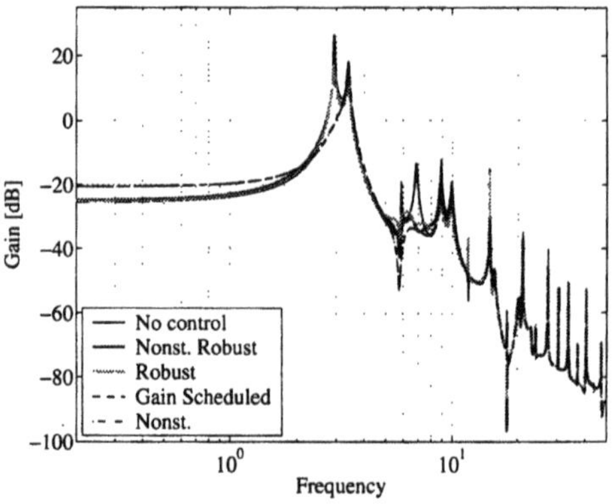

Figure 6. Frequency response of control object from w to the center mass point of wire on 9 seconds

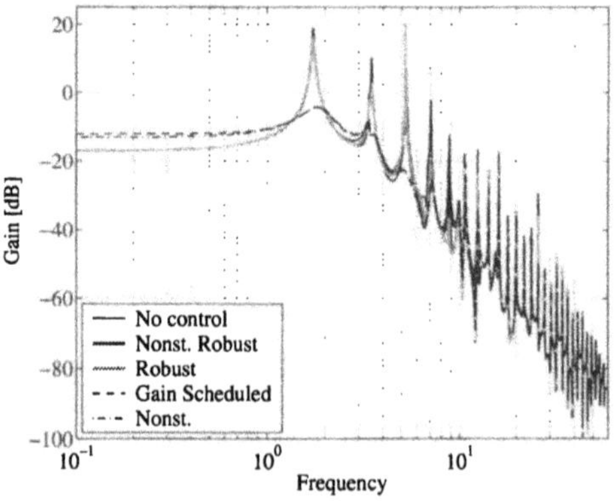

Figure 7. Frequency response of control object from w to the center mass point of wire on 18 seconds

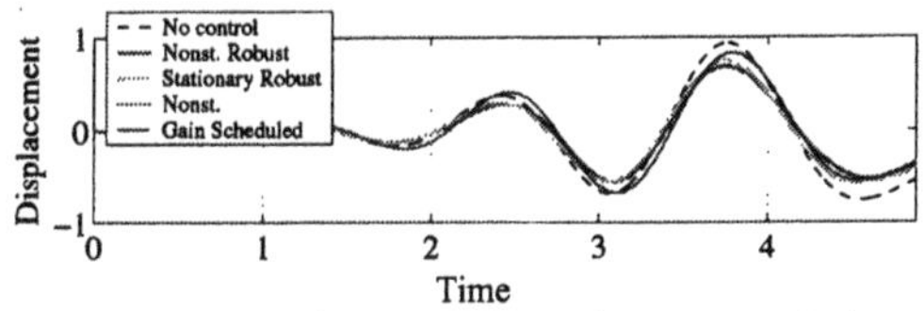

Figure 8. The time history of relative displacement of wire at the center during descending motion

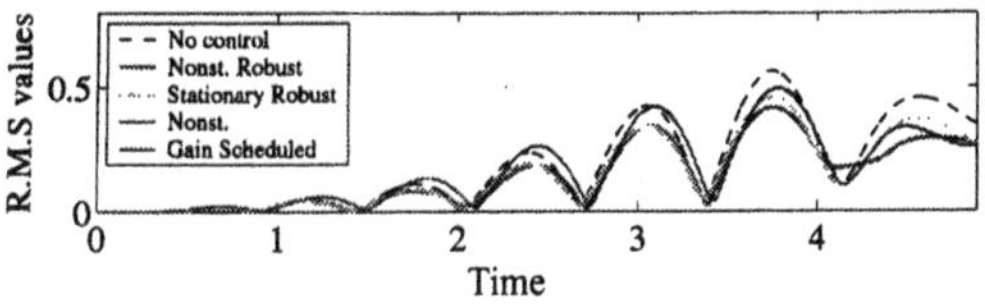

Figure 9. The time history of root mean square value of relative displacement of wire during descending motion

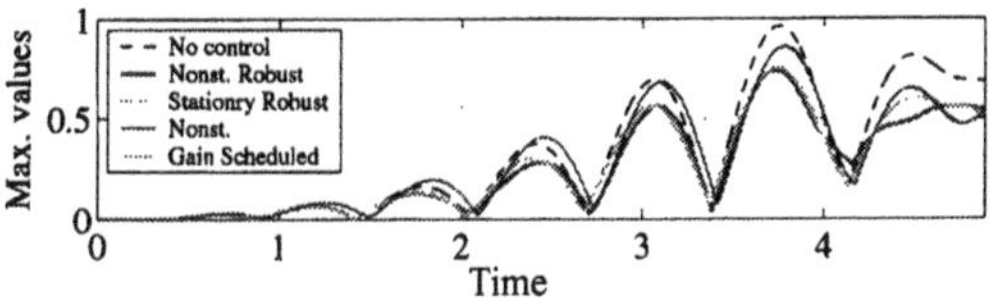

Figure 10. The time history of maximum value of relative displacement of wire during descending motion

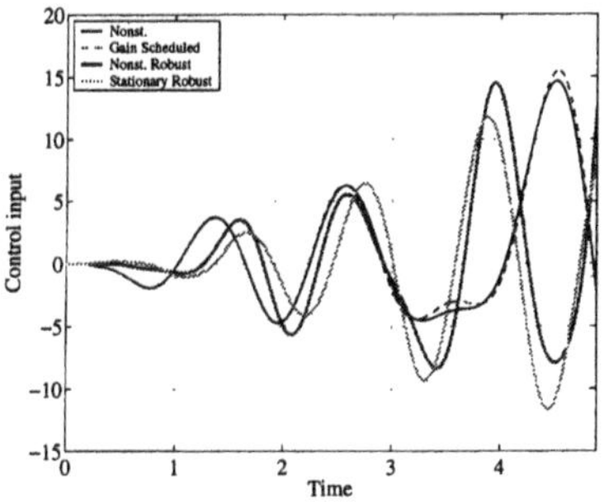

Figure 11. The time history of control input for reducing the vibration of wire during descending motion

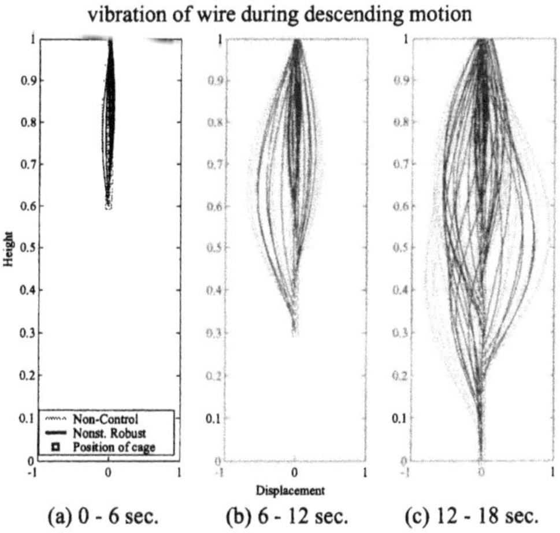

(a) 0 - 6 sec. (b) 6 - 12 sec. (c) 12 - 18 sec.

Figure 12. The time history of all the discrete mass point of wire without variation of tension

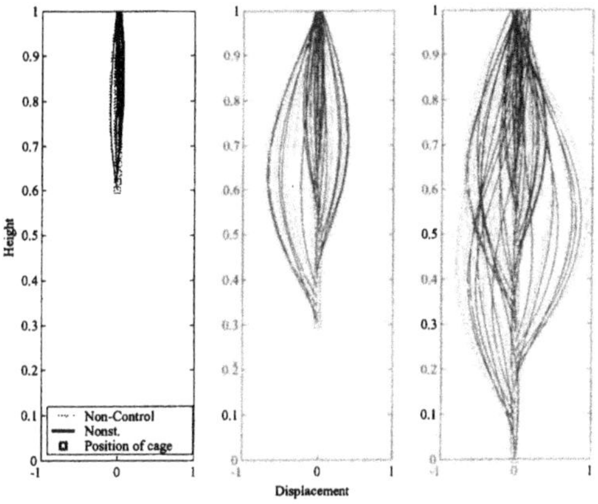

(a) 0 - 6 sec. (b) 6 - 12 sec. (c) 12 - 18 sec.

Figure 13. The time history of all the discrete mass point of wire without variation of tension

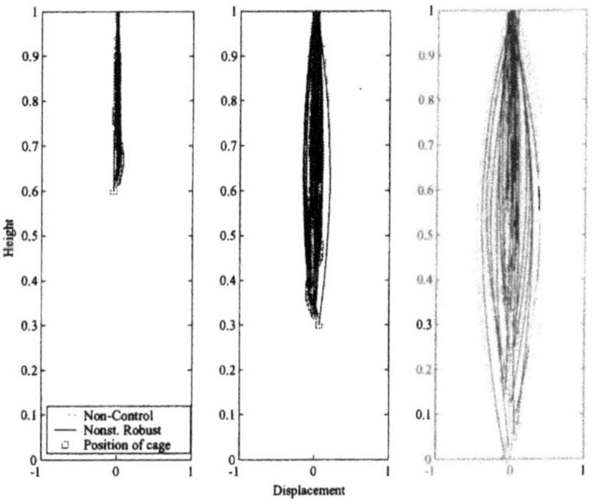

(a) 0 - 6 sec. (b) 6 - 12 sec. (c) 12 - 18 sec.

Figure 14. The time history of all the discrete mass point of wire with variation of tension

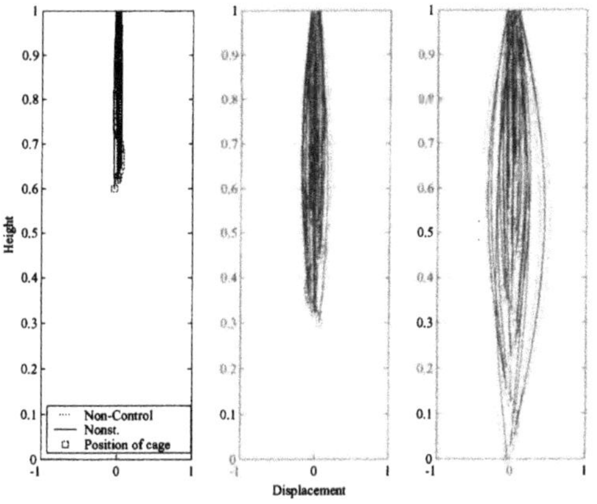

(a) 0 - 6 sec. (b) 6 - 12 sec. (c) 12 - 18 sec.

Figure 15. The time history of all the discrete mass point of wire with variation of tension

Table 1. The maximum value and the root mean square value of the relative displacement of wire and the control input without variation of tension

	Non-Control	Nonst.	Gain-Scheduled	Stationary Robust	Nonst. Robust
R.M.S r	0.182	0.162	0.164	0.148	0.138
Max. r	0.962	0.856	0.865	0.756	0.743
R.M.S u	–	3.61	3.63	3.60	3.61
Max. u	–	14.7	15.5	11.8	14.5

Table 2. The maximum value and the root mean square value of the relative displacement of wire and the control input with variation of tension

	Non-Control	Nonst.	Gain-Scheduled	Stationary Robust	Nonst. Robust
R.M.S r	0.0858	0.0852	0.0857	0.0540	0.0720
Max. r	0.518	0.477	0.483	0.330	0.381
R.M.S u	–	3.87	3.79	3.75	4.04
Max. u	–	12.4	12.2	12.7	20.9

power found in Figure 11. In Figures 12 and 13, the nonstationary robust controller clearly indicates the better performance for the reduction of wire vibration. In particular, the nonstationary optimal controller cannot derive the better performance than the robust one as shown in Figures 12, 13, 14, 15, Table 1 and 2. Consequently, it is difficult for the nonstationary optimal controller to show the better performance in the case that the uncertainties with high frequency are involved with an actual system and the parameters of control object vary with time.

5 CONCLUSIONS

This study presented a synthesis method of nonstationary vibration controller using a time- varying game type criterion function. We concluded that the nonstationary robust control method is superior to the nonstationary optimal one in the performance for reducing the vibration and having the ordinary robustness. And the proposed method enables the time-varying weightings to be designed on the arbitrary worst disturbance, the variation of parameter, the state values and the control input through all the time. Although the usefulness of the proposed control method was verified in the numerical calculation, it might be necessary to make the method more applicable to a practical use.

REFERENCES

Fung, F.R., Wu, W.J. & Lu, Y.P. 2002. Adaptive boundary control of an axially moving string system, ASME J. Vibration and Acoustics, Vol. 124, 435-440

Takagi, K. & Nishimura, H. 1998. Gain-scheduled control of a tower crane considering varying load-rope length, JSME series-C, Vol.64, No.626, 113-120

Otsuki, M., Nakata, R., Yoshida, K., Nagata, K., Fujimoto, S. & Munakata, T. 2001. Vibration control of ropesway of elevator for high-rise building, Proc. ASME DETC2001

Otsuki, M. & Yoshida, K. 2002. Experimental study on vibration control for rope-sway of elevator of high-rise building, Proc. 2002 American Control Conf., 238-243

O'Brien R. T. & Iglesias P. A. 1998 Robust controller design for time-varying systems, Proc. IEEE Conf. Decis.Cont., 37th-4, 3813-3818

Hara, S. & Yoshida, K. 1994. Simultaneous optimization of posi tioning and vibration controls using time-varying criterion function, J. Robotics and Mechatronics, Vol.6, No.4, 278-284

Doyle, J.C., Glover, K., Khargonekar, P.P. & Francis, B.A. 1989. State-space solutions to standard H2 and H-infinity control problems, IEEE Trans. Auto. Cont., Vol.34, No.8, 831-847

Sampei, M., Mita, T. & Nakamichi, M. 1990. An algebraic approach to H-infinity output feedback control problems, Sys. & Cont. Letters, Vol. 14, 13-24

Wave propagation – Moving load – Vibration reduction, Chouw & Schmid (eds.)
© 2003 Swets & Zeitlinger, Lisse, ISBN 90 5809 559 2

Reduction measures in track for Shinkansen-induced ground vibrations

Osamu Yoshioka
Central Japan Railway Company, Japan

ABSTRACT: Shinkansen-induced wayside vibrations must be reduced from the viewpoint of environmental preservation. This paper presents four investigations on vibration reduction measures in railway track: (1) field measurements on vibration reduction effectiveness of ballast-mats, rubber-coated sleepers and more-resilient pads, (2) explanation of the effectiveness by using a simple dynamic model, (3) physical implication on variation of the effectiveness on data acquisition sites, and (4) a reduction measure by stiffening track-bending rigidity. The measurements demonstrate not only the average effectiveness of each measure but also that some common features exist in the effectiveness of track-support softening measures. The common properties are successfully explained by our simple dynamic model, which suggests that moving loads and bending properties of track are important factors to be taken into account in the model. Our data on the effectiveness are largely scattered depending on data acquisition sites. Implication of the data scattering is examined and an interpretation is induced that the effectiveness depends on track-support stiffness before inserting the measure. In the interpretation, if track-bed is soft, track-support softening measures are ineffective. For the soft bed, we propose a new measure by stiffening track-bending rigidity and show its effectiveness on the basis of model computations and relevant field data.

1 INTRODUCTION

Passenger transportation of 100 km order in our country is mainly performed by the Japanese high-speed train, Shinkansen. It was first opened in 1964 from Tokyo, the biggest city in our country, to Osaka, the second big one, which is called Tokaido Shinkansen. Since then new routes have been successively constructed and today total length of the routes exceeds two thousands kilometers. Various renewals, such as speedup projects, introduction of newly designed cars and so on, have been also carried out enthusiastically. However, in some cases, Shinkansen train passage generates undesirable excessive vibrations in the wayside. Therefore, it is necessary to develop practical measures, so that the excessive vibrations can be reduced.

We present four investigations related to measures in railway track for reducing the vibrations: field measurements on reduction effectiveness of ballast-mats, rubber-coated sleepers and more-resilient pads (chapter 2), explanation of the effectiveness by using a simple dynamic model (chapter 3), physical interpretation on variation of the effectiveness on data acquisition sites (chapter 4), and a reduction measure by stiffening track-bending rigidity (chapter 5). A part of these investigations is briefly introduced in Yoshioka (2000) of the WAVE2000 proceedings, but the details are presented in the present paper.

Vibration amplitudes are described in terms of vibration levels, which are widely used in our country to evaluate not only the train-induced vibrations but also environmental vibrations in general. We abbreviate the vibration level to VL in the present paper. Confer Yoshioka (2000) for a definition of the vibration level. Vibrations are also discussed only in the vertical component, since the component is often used in official regulations of the vibrations in our country (also see Yoshioka (2000)).

2 FIELD MEASUREMENTS ON VIBRATION REDUCTION EFFECTIVE NESS OF TRACK-SUPPORT SOFTENING MEASURES

2.1 *Procedures of the field measurements*

We summarize, in this chapter, vibration reduction effectiveness of ballast-mats, rubber-coated sleepers and more-resilient pads, on the basis of field measurements.

Generally speaking, when effectiveness of a measure is examined by field tests, it is often planned to compare vibrations acquired at a

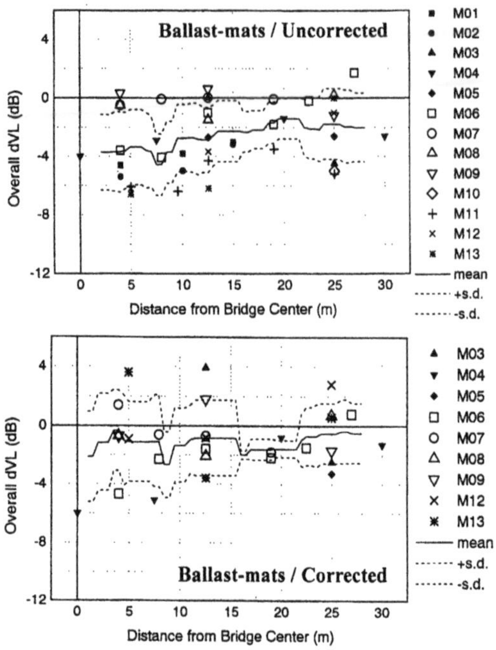

Fig.1 Measured effectiveness in overall VLs of ballast-mats.
vertical comp. / on ground surface / rigid-frame bridge section / train speed ~ 200 km/h.

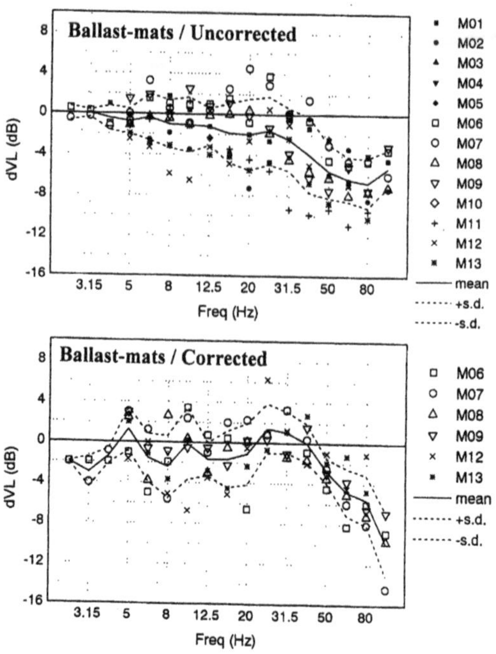

Fig.2 Measured effectiveness in bandpass VLs of ballast-mats.
the same data with the figure 1 / averaged over data with different distances.

measure-inserted site with ones at another site where the measure is not inserted. This method is easy in planning the examination work, but it is unreliable because of difficulty in specifying the origin of difference in vibration levels measured at the two sites. Instead, works for our measurements are conducted as follows: (1) vibration measurements before insertion of the measure, (2) insertion work of the measure, and (3) vibration measurements after the insertion. The after-insertion measurements (3) are carried out, at the same points in the same site, by the same measuring instruments, and by the same method, with those for the before-insertion measurements (1).

The Shinkansen line has double tracks and the measure is usually inserted into only either side of the tracks in an insertion work. Vibration level meters specified by the Japanese Industrial Standard are set at several points on wayside ground surface on the measure-inserted side (say, 0, 12.5, 25 meters away from the railway structure center) and sometimes at an additional point on bridge slab. Vibrations are measured for both trains passing through the measure-inserted and the opposite sides. Thus, we have four types of vibration level data depending on measurement time (before/after insertion) and train running side (inserted/opposite side): VL_{bi}, VL_{ai}, VL_{bo}, VL_{ao}, where the first subscript b or a denotes "measured before or after the measure insertion", and the second one i or o does "for trains running through the inserted or the opposite sides", respectively.

It is natural that VL_{ao} should be equal to VL_{bo}, since the measure is not inserted into the opposite side. However, in our field data, the two are sometimes not equal to each other. Unfortunately, we do not understand the reason at present. Thus, we shall use the following quantities in combination to express the effectiveness in vibration levels:

uncorrected effectiveness: $dVL_{uncor} = VL_{ai}-VL_{bi}$,

corrected effectiveness: $dVL_{cor} = dVL_{uncor}-(VL_{ao}-VL_{bo})$,

mean effectiveness: $dVL = (dVL_{cor}+dVL_{uncor})/2$.

2.2 *Effectiveness of ballast-mats*

Here, we show the effectiveness of ballast-mats in vibration levels.

The ballast-mat (A45 type) is a mat made of hard rubber with a thickness of 25 mm and is laid under railway ballasts. Its spring constant is about 4.5 MN/m, for a standard specimen of 100 mm × 100 mm in area and 25 mm in thickness, in the compression perpendicular to the area.

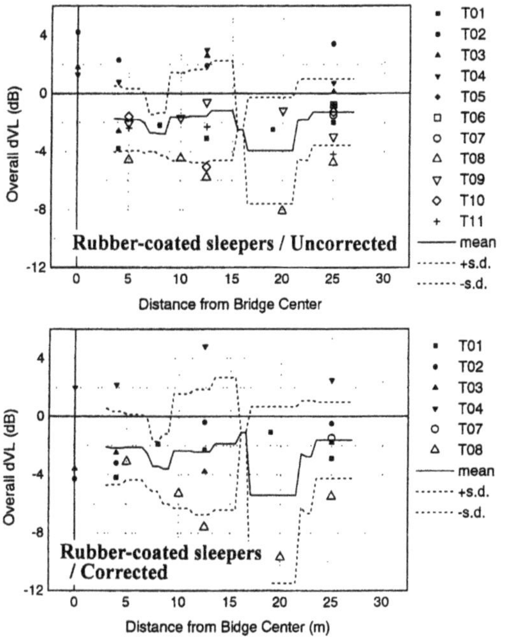

Fig.3 Measured effectiveness in overall VLs of rubber-coated sleepers.
vertical comp. / on ground surface / rigid-frame bridge section / train speeds ~ 200 km/h.

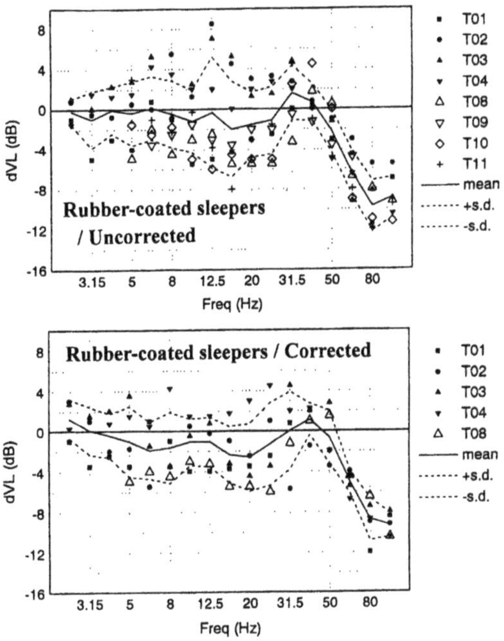

Fig.4 Measured effectiveness in bandpass VLs of rubber-coated sleepers.
the same data with the figure 3 / averaged over data with different distances.

Figure 1 shows reduction effectiveness of the ballast-mats in overall vertical vibration levels, which are measured for trains with speed of about 200 km/h, on wayside ground surface at several sites in rigid-frame bridge sections of the Shinkansen. The upper and the lower figures represent the uncorrected and the corrected effectiveness, respectively. Data are plotted by averaging vibration levels measured for several trains in the before- and the after insertion steps each and then by converting the averaged levels into the effectiveness. Data in the lower figure are fewer than ones in the upper, since the data correction is not possible for a part of data. Each symbol from M01 to M13 corresponds to a measurement site. The solid and the dotted lines denote the mean and the standard deviation range, respectively. They are drawn by averaging data in the range of ±3 meters at each distance, so their fine zigzags should not be paid attention.

The figure shows that data are largely scattered depending on measurement sites, but they indicate that the average effectiveness is 2 or 3 dB in overall vibration levels. The tendency is also found that as the distances become large the effectiveness diminishes.

The corrected effectiveness is smaller in absolute values than the uncorrected one. It implies that, by the measure insertion, vibrations are reduced to some extent even for trains passing through the opposite side where the measure is not inserted. The cause is not understood at present.

Figure 2 shows reduction effectiveness of the ballast-mats in the third-octave band vertical vibration levels. The data are identical to ones used in the figure 1. Similarly, the uncorrected and the corrected data are shown in the two figures each. One plotted point represents a mean value over data with various distances from railway structure center (0~30 meters). Such data handling may be verified since the data are decomposed into frequency components.

The data are scattered largely depending on the measurement sites, but they also show an average dependence of the effectiveness on frequencies. Note that the average effectiveness at the predominant frequencies (16~20 Hz) is roughly equal to the overall effectiveness. This demonstrates reliability of our bandpass data.

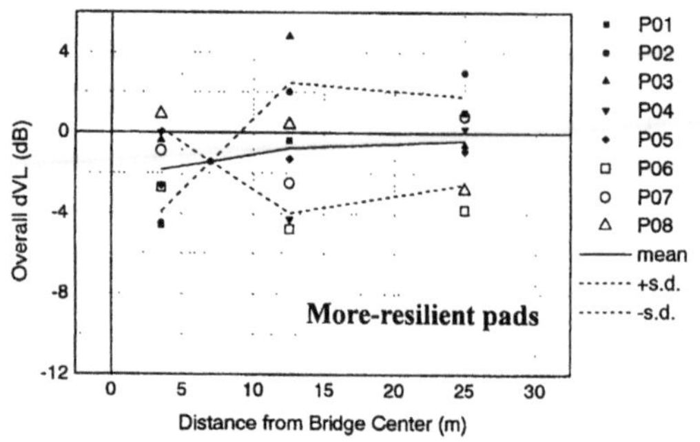

Fig.5 Measured effectiveness in overall VLs of more-resilient pads.
vertical comp. / on ground surface / rigid-frame and girder bridge section / train speeds ~ more than 200 km/h.

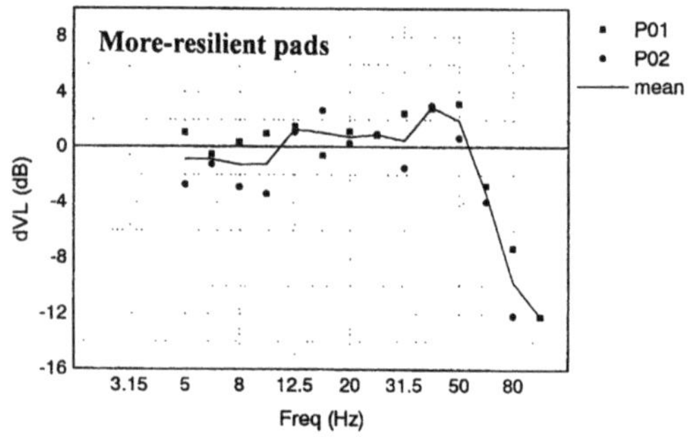

Fig.6 Measured effectiveness in bandpass VLs of more-resilient pads.
the same data with the figure 5 / averaged over data with different distances.

2.3 *Effectiveness of rubber-coated sleepers*

Next, we present the effectiveness of rubber-coated sleepers in vibration levels.

The rubber-coated sleeper is the one which hard rubber is coated with a thickness of 15 mm on the bottom surface of sleeper. The spring constant of the rubber is about 9 MN/m for the above-mentioned standard specimen.

Figure 3 shows reduction effectiveness of the rubber-coated sleepers in overall vertical vibration levels. The data are acquired, similarly to the case of ballast-mats, for trains with speed of about 200 km/h, on wayside ground surface in several rigid-frame bridge sections and a girder bridge section of the Shinkansen. Each symbol from T01 to T11 again corresponds to an investigated site. The figure is also drawn in the way similar to the figure 1, so the fine zigzags of the solid and the dotted lines should not be paid attention again.

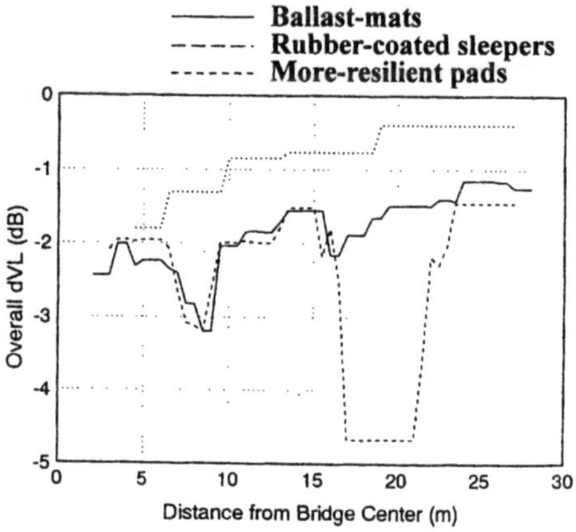

Fig.7 Averaged effectiveness in overall VLs of the three measures.
vertical comp. / on ground surface / rigid-frame and girder bridge sections / train speeds ~ more than 200 km/h.

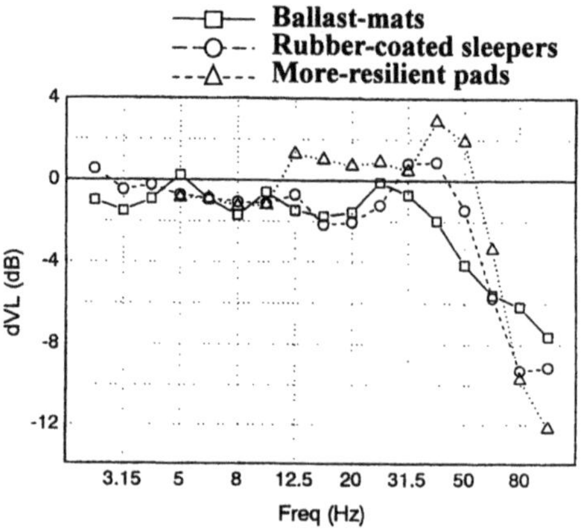

Fig.8 Averaged effectiveness in bandpass VLs of the three measures.
the same data with the figure 7 / averaged over data with different distances.

Features presented in the figure 3 are very similar to ones in the figure 1 in the view of the data scattering, the average effectiveness and the dependence of the effectiveness on distances.

Figure 4 shows reduction effectiveness of the rubber-coated sleepers in the third-octave band vertical vibration levels. The figure is also drawn in the way similar to the figure 2.

The figure 4 shows the features similar to those in the figure 2. Data are scattered largely depending on the measurement sites, but they demonstrate an average dependence of the effectiveness on frequencies.

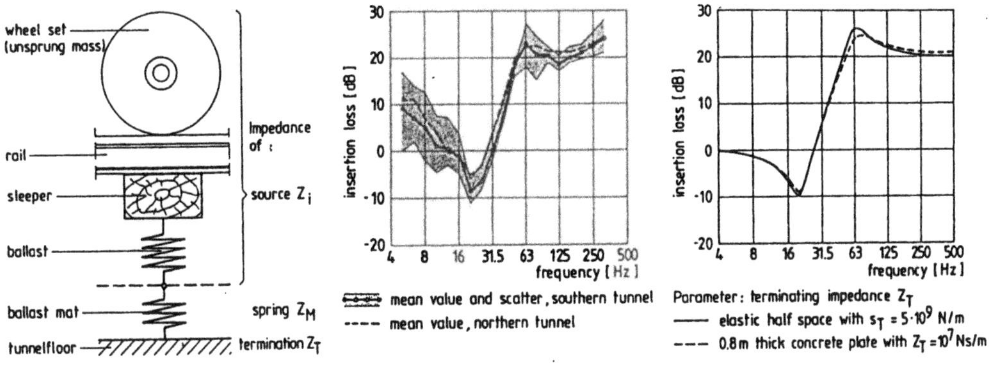

(a) Model (b) Measured (c) Computed

Fig.9. Insertion loss model (Wettschureck, 1985).
The insertion loss is defined as level *before* insertion minus one *after* insertion, which is opposite to our convention.

2.4 *Effectiveness of more-resilient pads*

Here, we refer to effectiveness of more-resilient pads in vibration levels. The more-resilient pad is made of softer rubber than one used in conventional pads: 1/2~1/3 times in the spring constant by indoor tests.

Figure 5 shows reduction effectiveness of the more-resilient pads in overall vertical vibration levels. As for this measure, our data are at present incomplete and the data correction is impossible except for two data of P01 and P02. Thus, only one figure is shown instead of two figures of the uncorrected and the corrected. In this figure, the two correctable data are plotted in terms of the mean effectiveness and other data are in the uncorrected one. Various conditions for vibration measurements are similar to ones for the above two measures, except that all data are acquired in slab track sections.

Similarly to the above-mentioned cases, data are largely scattered, but in an average view they show a certain degree of overall effectiveness, which seems smaller than that of the above two measures.

Figure 6 is the third-octave band reduction effectiveness of the more-resilient pads in vibration levels. Only the two correctable data P01 and P02 are shown in terms of the mean effectiveness. The more-resilient pads appear ineffective in the present figure, but the data are too poor and should be further collected.

2.5 *Comparison among three measures*

Here, we summarize the average effectiveness of the above-mentioned three measures.

Figure 7 illustrates comparison of the average effectiveness of the three measures in overall vertical vibration levels. It can be said that they show mutually similar effectiveness, though the more-resilient pad is somewhat less effective than the other two types.

Figure 8 compares the average third-octave band spectra of effectiveness for the three measures. They show a similar spectral shape, though the more-resilient pad is less effective than the rest. In other words, it can be said that the track-support softening measures in general have a common dependence of the effectiveness on frequencies.

Data on the effectiveness of more-resilient pads are poor at present, especially in spectra. Further acquisition of field data is required for high reliability of our discussions.

3 EXPLANATION OF REDUCTION EFFECTIVENESS USING A SIMPLE DYNAMIC MODEL

3.1 *Insertion loss model*

The track-support softening measures have a common dependence of the effectiveness on frequencies, as it is mentioned in the section 2.5. In this chapter, we explain the common dependence by using a dynamic model.

The insertion loss model is well known as the one for explaining the vibration reduction effectiveness of ballast-mats [Wettschureck (1985), Hayakawa (1991)]. The structure of the model is very simple, as it is illustrated in figure 9(a). The car and the track are simply approximated by a mass-spring system with single degree-of-freedom, neglecting the moving load, and the insertion of ballast-mat is represented by an additional spring. It is insisted that the model gives successful explanations on the

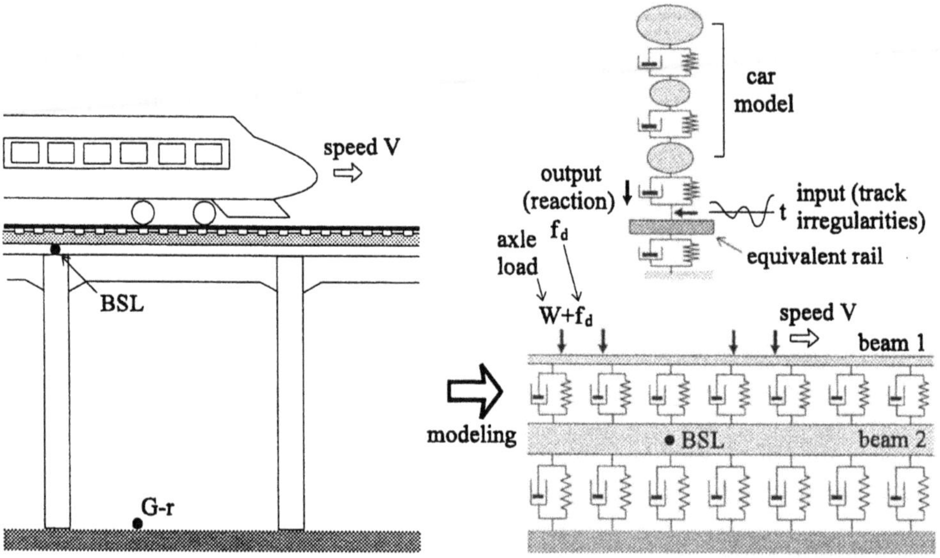

Fig.10 Schematic illustration of our model.

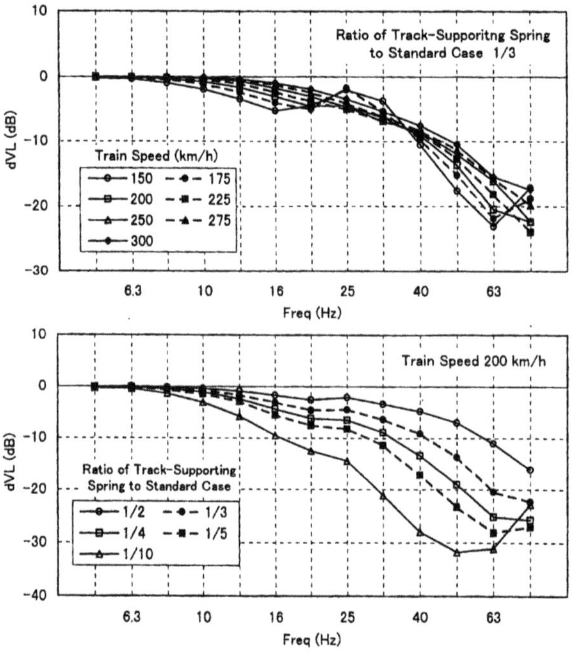

Fig.11 Effectiveness in bandpass VLs due to decrease of track-supporting spring constant computed by our model

effectiveness in spite of its simplicity. Figures 9(b) and 9(c) are such an example introduced by Wettschureck (1985). Figure 9(b) shows the effectiveness of ballast-mats measured in their insertion into tracks of two parallel subway tunnels. Note that the sign of his insertion loss in the vertical axis of the figure is taken oppositely to that of our effectiveness. Figure 9(c) represents the loss values computed by his model; two values are indicated by changing the model parameters slightly. Comparison of the figures 9(b) and 9(c) shows that the computed values agree very well with the measured.

This model predicts that at the frequencies between 10 and 20 Hz the vibrations rather increase by the measure insertion, when the model parameters are selected in the reasonable range. This results in increase of overall vibration levels, since components at these frequencies are predominant in our Shinkansen vibrations. However, our field data on Shinkansen vibrations suggest that ballast-mats are effective not only in bandpass vibration levels at the frequencies but also in overall levels, as shown in the section 2.2. How should we understand this?

3.2 *Explanation by using our dynamic model*

We try to formulate a simple dynamic model, which is illustrated schematically in figure 10 and is described briefly in Yoshioka (2000). The moving load and the track bending property are

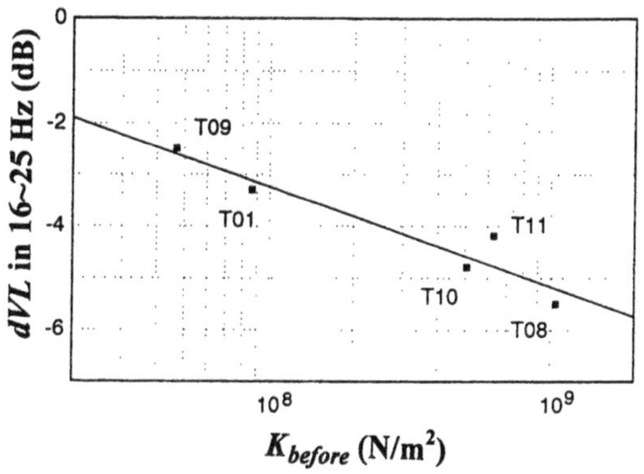

Fig.12. Relationship between the effectiveness at predominant frequencies and inversely estimated rail-supporting spring constant before insertion K_{before}.

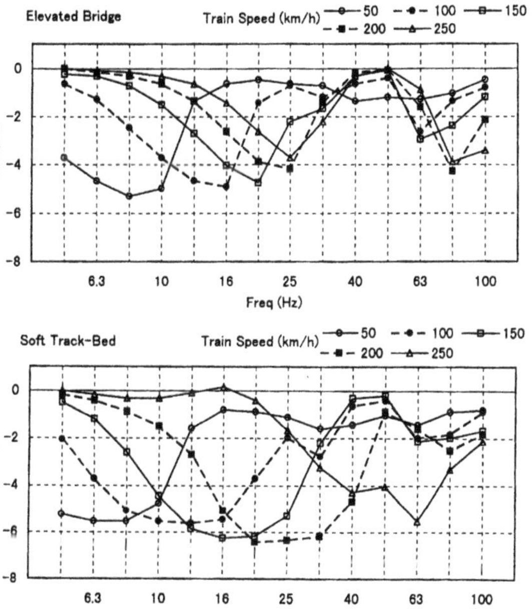

Fig.13. Computed effectiveness in bandpass VL due to stiffening bending rigidity of track
The upper figure shows the elevated bridge case, and the lower the soft track-bed case.

taken into account in our model, though they are neglected in the insertion loss model.

Our excitation is decomposed into a load for each axle and its periodic effect, and the former is assumed to be the sum of a static axle load and a dynamic force reacting from car vibrations due to track irregularities to track, where the irregularities are taken as a spatially-stationary random process. The track and the bridge are modeled by a system composed of two beams and their visco-elastic supports, as shown in the figure 10. The direct output from our model is the spectrum in vibration levels of bridge slab vibrations. Vibrations on surface ground are obtained by adding an empirical transfer function between the slab vibrations and the ground surface ones to the slab vibrations.

Simplicity of our model enables us to write down the vibration response in an analytical form, and thus to solve the model inversely from the response to model parameters (not analytically but numerically). In fact, some model parameters, such as equivalent constants, are determined by the inversion computation from vibrations measured on the bridge slab. The inversion computation is carried out by using a nonlinear least square method formulated by Tarantola and Valette (1982).

Figure 11 illustrates the results computed by our model for effectiveness of track-support softening measures. The upper figure shows the case with a constant ratio of track-supporting spring after insertion to one before insertion (1/3) and with various train speeds (150~300 km/h), and the lower displays the opposite case with various spring ratios (1/2~1/10) and with a constant train speed (200 km/h).

They agree approximately with our Shinkansen data mentioned in the chapter 2, even for low frequencies between 10 and 25 Hz. Therefore, it can be concluded that, at low frequencies, the moving load and the track bending property are important for explaining the vibration reduction effectiveness of track-support softening measures.

4 IMPLICATION OF VARIATION OF FIELD DATA ON VIBRATION REDUCTION EFFECTIVENESS

Our field data on the effectiveness of track-support softening measures show large variation depending on data acquisition sites, as it is mentioned in the chapter 2. Here, we interpret physical meanings of the variation.

It is conjectured that the track-support softening measure indicates larger effectiveness as the track support stiffness decreases more visibly in the measure insertion. This conjecture can be expressed in other words as follows. Denoting the track support stiffness before and after the insertion by K_{before} and K_{after}, respectively, the effectiveness of the considered measure depends on $K_{ratio} = K_{after}/K_{before}$ and it

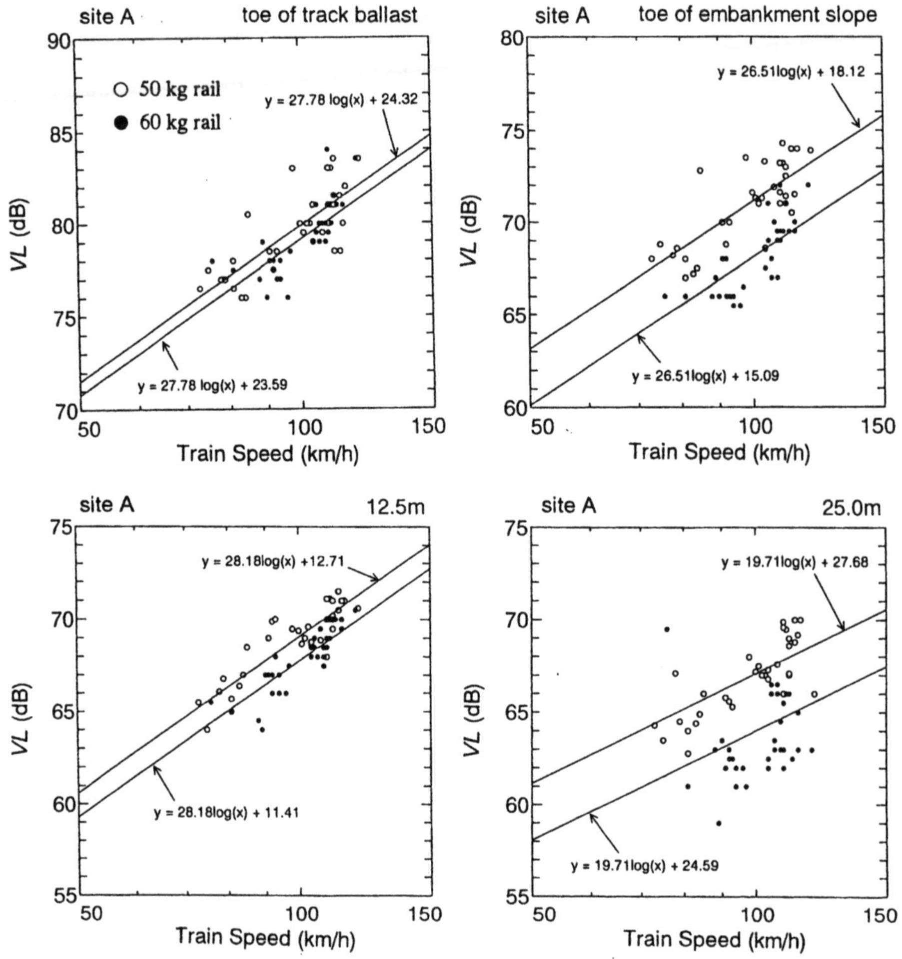

Fig.14. Overall vertical VLs measured before and after replacement from 50 kg/m rails to 60 kg/m ones (in an embankment section of a conventional line)

becomes large as K_{ratio} diminishes. Moreover, K_{after} is nearly constant independently of measurement sites in our present data, since maybe it is almost determined by the stiffness of rubber coated on the sleeper. If so, the effectiveness directly depends on K_{before} and it becomes large as K_{before} increases. How is our conjecture verified?

In order to verify it, we examine the relationship between the effectiveness and the K_{before} value by using data on the rubber-coated sleepers.

In the examination, the effectiveness is taken as a mean value in the range from 16 to 25 Hz of the measured effectiveness spectrum. Though the K_{before} value has never been measured directly in our vibration measurements, it can be estimated inversely through our model, if vibration data measured on bridge slab before insertion are available. That is, when the K_{before} value and other parameters are known, the forward computation of our model leads to the vibration spectrum on bridge slab before insertion; in contrast, when the spectrum is given by measurements and so on, the backward computation determines the K_{before} value if other parameters are fixed. We have determined the K_{before} value by such an inversion method.

Figure 12 illustrate the relationship between the effectiveness and the K_{before} value. Data at five sites are plotted in the figure, since only at the sites bridge slab vibrations have been measured before the insertion. Note that the unit of K_{before} is N/m^2, not N/m^3, because K_{before} is an equivalent parameter in our present model. The figure shows that, if K_{before}

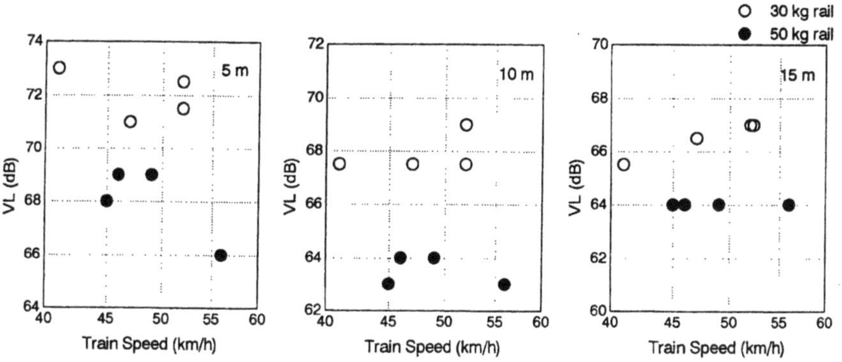

Fig.15. Overall vertical VLs measured before and after replacement from 30 kg/m rails to 50 kg/m ones (in an embankment section of a local conventional line)

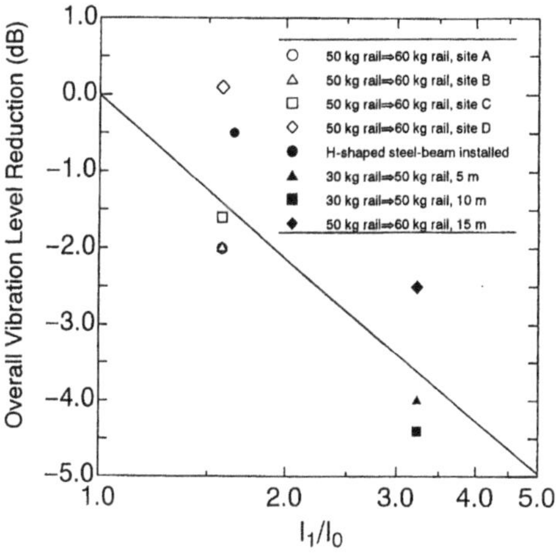

Fig.16. Empirical dependence of overall VL reduction on track-bending rigidity

becomes large, rubber-coated sleepers become more effective. This verifies our above-mentioned conjecture.

In this interpretation, the large variation of data on the effectiveness results from the large scattering of K_{before} values and rubber-coated sleepers are more effective at the site with large K_{before}.

The K_{before} value can be evaluated before the measure insertion, by the inversion of our model, if bridge slab vibrations are measured. Then, the regression line in the figure 12 can be utilized to predict the effectiveness of rubber-coated sleepers at the stage to plan the insertion work.

5 REDUCTION MEASURE BY STIFFENING TRACK-BENDING RIGIDITY

The discussion in the preceding chapter suggests that the track-support softening measures, mentioned in the chapter 2, are ineffective for cases with soft roadbed such as soil roadbed. For such cases, alternative measures in track are to be developed.

In our model, the track is expressed in terms of rail-supporting spring constant *K*, rail mass *M*, and rail-bending rigidity *EI*. Change in *K* has been already applied to the track-support softening measures. However, change in *M* or *EI* has never been examined yet and is worth being further studied.

At first, by using our model, we have examined dependence of vibrations on *M* or *EI*. The change in *M* produces too small variation in vibration levels to be displayed in graphic representation except for extremely large change of *M*. In contrast, the change in *EI* can have visible reduction effectiveness. This is shown in the figure 13, where twice increase of *EI* is assumed in all results. The upper figure shows the case in a section of typical elevated bridges and the lower figure displays the case with softer track-bed such as soil bed. Strictly speaking, the latter is unreasonable to be treated by our model, since our model simulates an elevated bridge section as shown in the figure 10. Our case of 'soft track-bed' is simply specified by the conditions that the bending rigidity of the beam 2 representing the bridge slab (*EI*) and the constant of the spring supporting the beam (*K*) have 0.1 times the values for a standard bridge (EI_{st} and K_{st}) each: $EI=0.1EI_{st}$ and $K=0.1K_{st}$. Thus, it does not represent the case of soft bed literally, but maybe it is reasonable enough to examine dependence of the effectiveness on stiffness of the track-bed roughly.

The *EI* change in rails is effective even if we double the *EI* value, though complex variations are generated depending on frequencies. The figure also shows that the *EI* increase is more effective in softer roadbed. This tendency is opposite to that in the track-support softening and so is useful as supplement of the softening.

We also have examined field data related to the *EI* increase. The relevant data are found in the case of conventional railways. Figure 14 is a comparison of vibrations (in overall vertical vibration levels) measured before and after rail replacement from 50 to 60 kg/m rails in a low embankment section of a conventional line. We can see reduction of a few decibels.

Figure 15 is another example in rail replacement from 30 to 50 kg/m rails in a local conventional line. This shows larger effectiveness than the preceding case in the figure 14. It may be the reason why the *EI* increase in this case is larger than that in the preceding one.

Figure 16 summarizes the present field data on vibration reduction due to the *EI* increase. The horizontal axis is expressed in the ratio of I (cross-sectional area moment of inertia), instead of *EI*, because our data are almost acquired in rail replacement and E (Young's modulus of elasticity) does not change in the work. The regression line is given by $dVL = -3.1 \ln(I_1/I_0)$, where I_0 and I_1 denote the I values before and after the work, respectively. This diagram enables us to roughly predict effectiveness due to the *EI* increase. It is desired that related field data are acquired further to guarantee high reliability of the tendency.

Since the rail replacement is difficult in the Shinkansen, instead of it, the measure is studied in which the track is reinforced by beams to increase the bending rigidity of track.

6 CONCLUDING REMARKS

(1) Vibration reduction effectiveness of track-support softening measures is examined and the following results are obtained.

(a) Overall and spectral effectiveness of ballast-mats, rubber-coated sleepers, and more-resilient pads are summarized on the basis of field measurements. The spectral effectiveness shows a common property in the frequency dependence.

(b) The common property can be approximately explained by our dynamic model. This also shows that the moving load and the track-bending property are important in modeling to explain the effectiveness at low frequencies.

(c) Field data on the effectiveness are largely scattered depending on data acquisition sites. This is interpreted to reflect the state of track support stiffness before the measure insertion.

(2) If track-bed is soft, track-support softening measures are ineffective. For such cases, the measure of stiffening track-bending rigidity may be useful.

ACKNOWLEDGEMENT

The author greatly thanks the scientific committee and the organizing committee of the international workshop WAVE2002, as well as the editors, Assoc. Prof. Nawawi Chouw in Okayama University, Japan, Prof. Sohichi Hirose in Tokyo Institute of Technology, Japan, Prof. Guenther Schmid in Ruhr University Bochum, Germany, and Prof. Takeo Taniguchi in Okayama University, Japan, for giving the author the opportunity to make a presentation.

REFERENCES

Hayakawa, K. 1991. Study on reduction measures for traffic-induced ground vibration. Doctoral thesis Osaka Univ. (in Japanese).

Tarantola, A. and Vallette, B. 1982. Generalized nonlinear inverse problems solved using the least square criterion. *Rev. Geophys. Space Phys.* 20: 219-232.

Wettschureck, R. 1985. Ballast mats in tunnels - analytical model and measurements. *Proc. Inter-Noise 85*: 721-724.

Yoshioka, O. 2000. Basic characteristics of Shinkansen-induced ground vibration and its reduction measures. *Proc. WAVE 2000*: 219-237.

Track-soil analysis

Wave propagation – Moving load – Vibration reduction, Chouw & Schmid (eds.)
© 2003 Swets & Zeitlinger, Lisse, ISBN 90 5809 559 2

Three-dimensional analysis of subway track vibrations due to running wheels

K.Abe & D.Satou
Niigata University, Niigata, Japan

T.Suzuki
Railway Technical Research Institute, Tokyo, Japan

M.Furuta
Tokyo Metropolitan Subway Const. Co., Tokyo, Japan

ABSTRACT: A three-dimensional wheel / railway dynamic interaction analysis method is developed for subway track. The sub-domain consisting of the ballast, tunnel and ground is modeled by a three-dimensional wave field approach. The solution is expressed by the discrete wave number method, and then the problem is reduced to a quasi two-dimensional problem corresponding to each discrete wave number. The developed method is applied to a subway track, and the numerical results are compared with the in situ measurements. It is found that the three-dimensional model can reproduce the dynamic response of subway track.

1 INTRODUCTION

The ground vibration induced by a running train may affect the environment in urban area. Therefore the railway track vibration is an important problem from the viewpoints of the design and maintenance of railway track. Nowadays, several remedies such as sleeper pad and ballast mat are attempted to reduce the vibration level. However the prediction of these effects is difficult. In this context, the development of a numerical method which can simulate the track vibration has practical significance.

In order to understand the dynamic characteristics of railway track, a numerical model in which the rail is modeled with running wheels and discrete sleepers has been employed by many researchers (Grassie et al. 1982, Knothe & Grassie 1993, Kalker 1996). However, since the main objective of these studies is to investigate the dynamic responses of wheels, rail and sleepers at high frequency, the sleeper support is modeled by a mass-spring system or a rigid foundation.

On one hand the three-dimensional ground model is used by researchers who are interested in the ground vibration due to running trains. Mohamadi & Karabalis (1995) have discussed the interaction between the railway track and the half space under a harmonic excitation using a three-dimensional BE-FE coupling method. Takemiya et al. (2001) have investigated the wave propagation induced by Shinkansen trains using the FEM described by the discrete wave number method. They have modeled the railway track consisting of the rail, sleepers and ballast as an elastic beam. Knothe & Wu (1998) have constructed a numerical method for infinitely long track modeled by a rail, rail pads, sleepers and three-dimensional soil layers. In these studies, since the discussion is focused on the ground vibration or the track-ground interaction, the external force is modeled by a prescribed moving load or a harmonic excitation.

The authors have developed a wheel-subway track interaction analysis method (Abe et al. 1999). The objective of our study is to estimate the effect of vibration reduction such as sleeper pad and ballast mat. To achieve this, we must simulate the transmission of vibration through the rail, sleeper, pads, ballast and ground. Therefore, it is necessary to take into consideration the running wheels, railway track and ground in the modeling. In Abe et al. (1999) the wheels are modeled by running masses, the rail is expressed by a Bernoulli-Euler beam supported by sleepers, and the ballast and ground are modeled by a two-dimensional wave field.

In order to validate the developed method, this method has been applied to a subway track. Furthermore, identification methods have been developed for the rail-surface profile and the material properties which make it possible to reproduce the measured data (Abe et al. 2000, Suzuki et al. 2000). Although these methods enable us to deal with the dynamic response of the subway track, the in situ measurements cannot be reproduced. It is thought that the cause of this is not the identification methods but the modeling of the ground. In the case of subway track, it is necessary to simulate the waves propagating in the tunnel and the surrounding ground. For that purpose, the tunnel and ground should be modeled in 3-D.

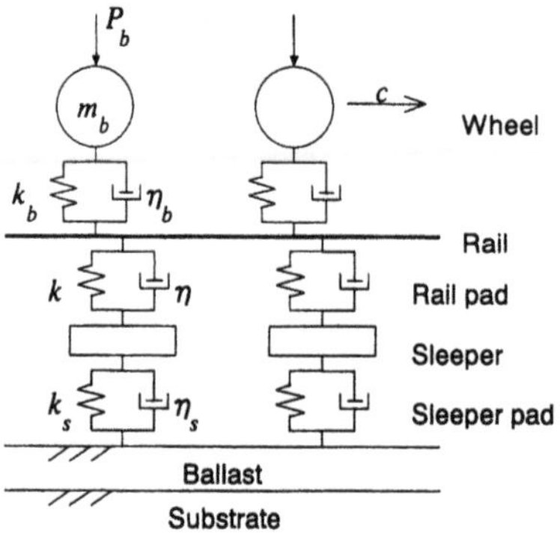

Figure 1. Modeling of wheels and track.

In this paper, the substructure consisting of the ballast, tunnel and ground is modeled by an unbounded three-dimensional wave field approach, and coupled with the wheel-railway track subsystem. The three-dimensional substructure is described by the discrete wave number method (Bouchon & Aki 1977), and then the three-dimensional problem is reduced to a quasi two-dimensional one in the wave number space. As a numerical method the FE-BE coupling method is used. The developed method is applied to an observation site of a subway. Numerical results are compared with the in situ measurements and the two-dimensional analysis, and the importance of three-dimensional modeling is discussed.

2 MODELING OF SUBWAY TRACK-GROUND INTERACTION

The subway track is modeled as depicted in Figure 1. The dynamic response of each member such as wheel, rail and sleeper is formulated in time domain by using a convolution concerning the time history of load as in Abe et al. (1999). In this section, the numerical modeling for subway-track dynamic interaction is described briefly.

2.1 *Modeling of running wheels*

The wheels running on the rail with speed c are modeled as mass points. The motion of the wheel is expressed in the form of convolution as

$$u_{bi}^{M} = \sum_{m}^{M} \overline{G}_{b}^{M-m+1}(m_b g + P_b - P_i^m),$$
$$\overline{G}_b^m = (m - \frac{1}{2})\frac{\Delta t^2}{m_b}, \quad (1)$$

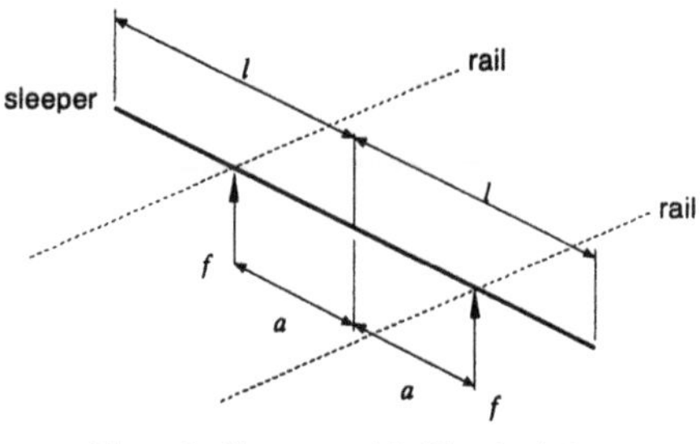

Figure 2. Sleeper modeled by elastic beam.

where, u_{bi}^M is the vertical displacement of the i th wheel (i =1,2) at the M th time step, $\overline{G}_b^m$ is the discretized Green's function defined as the motion of mass due to a unit force acting during a time-step interval. m_b is the mass of wheel set, g is the acceleration due to gravity, P_b is the weight of car body, P_i^m is the contact force between the i th wheel and the rail at the m th step, and Δt is the time increment.

2.2 *Modeling of rail*

The rail is modeled as an infinitely long Bernoulli-Euler beam. The deflection of the rail is described by a time domain integral representation.

The deflection beneath the i th wheel $\overline{u}_i^M$ is given by

$$\overline{u}_i^M = \sum_{j}^{2}\sum_{m}^{M} A_{ij}^{M-m+1}P_j^m - \sum_{j}^{N}\sum_{m}^{M} B_{ij}^{M,M-m+1}F_j^m, \quad (2)$$

where F_j^M is the contact force between the rail and the j th sleeper acting at the m th step. Note that N sleepers are considered in the analysis.

The deflection of the rail above the i th sleeper u_i^M is given by

$$u_i^M = \sum_{j}^{2}\sum_{m}^{M} C_{ij}^{M,M-m+1}P_j^m - \sum_{j}^{N}\sum_{m}^{M} D_{ij}^{M-m+1}F_j^m, \quad (3)$$

In Equations (2) and (3), A_{ij}^m, $B_{ij}^{M,m}$, $C_{ij}^{M,m}$ and D_{ij}^m are the coefficients defined by

$$\begin{aligned}
A_{ij}^m &= \int_{t_{m-1}}^{t_m} u^*(x_j - x_i - c\tau, \tau)d\tau,\\
B_{ij}^{M,m} &= \int_{t_{m-1}}^{t_m} u^*(a_j - x_i - ct_M, \tau)d\tau,\\
C_{ij}^{M,m} &= \int_{t_{m-1}}^{t_m} u^*(x_j - a_i + c(t_M - \tau), \tau)d\tau,\\
D_{ij}^m &= \int_{t_{m-1}}^{t_m} u^*(a_j - a_i, \tau)d\tau,
\end{aligned} \quad (4)$$

where $t_m = m\Delta t$, a_i is the position of the i th sleeper and u^* is the time domain Green's function of the

Bernoulli-Euler beam (Abe & Furuta 1997, Graff 1975).

2.3 *Modeling of sleepers*

The sleeper is represented by an elastic beam as illustrated in Figure 2. The contact force acting on the sleeper is modeled as a concentrated force. The displacement of the i th sleeper on the railseat at the M th step u_{si}^M is expressed as

$$u_{si}^M = \sum_m^M \overline{G}_s^{M-m+1}(F_i^m - F_{si}^m), \tag{5}$$

where F_{si}^m is the contact force between the i th sleeper and ballast. $\overline{G}_s^m$ is the numerical Green's function corresponding to the displacement of sleeper in a free space due to a couple of unit forces acting on the both railseats during the first time-step interval. The sleeper is discretized by a number of beam elements. The numerical Green's function is calculated in advance.

2.4 *Modeling of tunnel and ground*

The dynamic behavior of ballast, tunnel and ground are modeled by a three-dimensional wave field. The numerical Green's function, which corresponds to the vertical displacement on the ballast surface beneath the i th sleeper at the m th time step due to a unit force acting on the contact area of the j th sleeper during the first time step, is defined as $\overline{G}_{ij}^m$ and obtained through the three-dimensional analysis in advance. Once the numerical Green's function $\overline{G}_{ij}^m$ is obtained, the displacement on the ballast surface can be expressed as

$$u_{0i}^M = \sum_j^N \sum_m^M \overline{G}_{ij}^{M-m+1} F_{sj}^m, \tag{6}$$

The formulation of the three-dimensional analysis will be described in the next section.

2.5 *Contact forces*

The contact element inserted between the wheel and rail is modeled by a Voigt unit as shown in Figure 1. In this study, the contact force P_i^M is derived by a time integration method developed by Abe (1997) as

$$P_i^M = (\frac{\eta_b}{\Delta t} + \frac{k_b}{2})(u_{bi}^M - \overline{u}_i^M) - (\frac{\eta_b}{\Delta t} - \frac{k_b}{2})(u_{bi}^{M-1} - \overline{u}_i^{M-1}), \tag{7}$$

where k_b and η_b are the spring and damping constants, respectively.

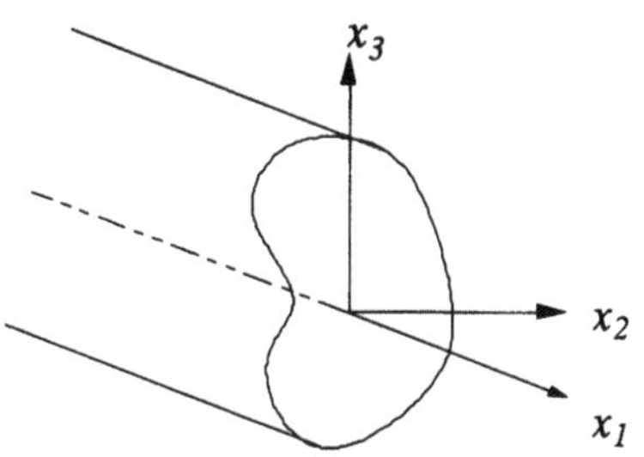

Figure 3. Modeling of tunnel.

The contact forces between the rail and sleeper and the sleeper and ballast are acting on the rail pad and the sleeper pad, respectively. These pads are represented by the Voigt model. Therefore the contact forces can be obtained as for P_i^M . However, since these pads have a nonlinearity in the displacement-force relation, the formulation is carried out with considerarion of nonlinearity. The contact forces are given by

$$\begin{aligned} F_i^M &= (\frac{\eta}{\Delta t} + \frac{k^l}{2})\hat{u}_i^M - (\frac{\eta}{\Delta t} - \frac{k^l}{2})\hat{u}_i^{M-1} + \overline{F}_i^M, \\ F_{si}^M &= (\frac{\eta_s}{\Delta t} + \frac{k_s^l}{2})\hat{u}_{si}^M - (\frac{\eta_s}{\Delta t} - \frac{k_s^l}{2})\hat{u}_{si}^{M-1} + \overline{F}_{si}^M, \end{aligned} \tag{8}$$

where $\hat{u}_i^M = u_i^M - u_{si}^M$, $\hat{u}_{si}^M = u_{si}^M - u_{0i}^M$. k^l and k_s^l are the spring constants, and η and η_s are the damping constants of pads. $\overline{F}_i^M$ and $\overline{F}_{si}^M$ are the terms concerning the nonlinearity, and defined as functions of displacement.

2.6 *Interaction analysis*

In the analysis, the displacement of wheel u_{bi}^M , the deflections of rail $\overline{u}_i^M$ and u_i^M , the displacement of sleeper u_{si}^M , the displacement at the ballast surface u_{0i}^M , and the contact forces P_i^M , F_i^M and F_{si}^M are unknowns. The equations for the interaction analysis are constructed by coupling Equations (1), (2), (3), (5), (6), (7) and (8). Since Equation (8) has nonlinearity, the solution is obtained with iterative method.

3 ANALYSIS METHOD FOR THREE-DIMENSIONAL WAVE FIELD

The numerical Green's function of Equation (6) is obtained by a three-dimensional analysis for the substructure consisting of ballast, tunnel and ground. In

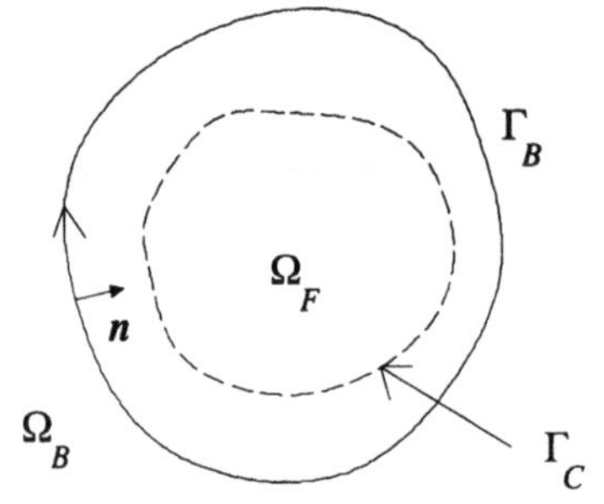

Figure 4. Real and auxiliary boundaries, and finite element and boundary element domains.

the analysis, it is assumed that the tunnel has a straight and infinitely long structure as depicted in Figure 3, and the Fourier transform is achieved with respect to the tunnel axis x_1 and to the time axis _. Thus the problem is reduced to a quasi two-dimensional one in the wave number-frequency space.

In this study, the ballast, tunnel and a part of ground surrounding the tunnel are discretized by finite elements and the rest of the unbounded ground is represented by boundary elements. Hence the FE-BE coupling method is used for the solution.

3.1 *Boundary element equation*

In order to avoid the singularity occurring in an integral equation, the indirect formulation is used with distributing a source density on an auxiliary boundary Γ_C located inside the domain bounded by the coupling interface Γ_B as shown in Figure 4.

The displacement in the unbounded external domain Ω_B is described by the single-layer α_j as

$$\bar{u}_i(\mathbf{x}) = \int_{\Gamma_C} \bar{u}^*_{ij}(\mathbf{x},\mathbf{y})\alpha_j(\mathbf{y})d\Gamma \qquad (\mathbf{x}\in\Omega_B, \mathbf{y}\in\Gamma_C), \tag{9}$$

where i, j = 1,2,3, and repeated indices imply the summation convention. $\bar{u}_{ij}$ is the Fourier transform of the fundamental solution in the frequency domain with respect to the tunnel axis x_1. Note that the problem has already been transformed into the frequency domain.

The traction on Γ_B is obtained from Equation (9) as

$$\bar{p}_i(\mathbf{x}) = \int_{\Gamma_C} \bar{p}^*_{ij}(\mathbf{x},\mathbf{y})\alpha_j(\mathbf{y})d\Gamma \qquad (\mathbf{x}\in\Gamma_B, \mathbf{y}\in\Gamma_C), \tag{10}$$

where

$$\bar{p}^*_{1j} = \mu(ik\bar{u}^*_{\alpha j}n_\alpha + \frac{\partial \bar{u}^*_{1j}}{\partial n}),$$
$$\bar{p}^*_{\alpha j} = \lambda(ik\bar{u}^*_{1j} + \bar{u}^*_{\beta j,\beta})n_\alpha + \mu(\bar{u}^*_{\beta j,\alpha}n_\beta + \frac{\partial \bar{u}^*_{\alpha j}}{\partial n}). \tag{11}$$

Here k is the wave number, λ, μ are Lame constants, α , β =2,3, i^2 =-1, and n is the outward normal at $\mathbf{x}$ on Γ_B. Note that the Fourier transform has already been achieved in x_1 direction, and thus the differentiation is carried out with respect to $\mathbf{x}(x_2, x_3)$.

In order to discretize Equations (9) and (10), α_j is approximated by

$$\mathbf{a}(\mathbf{y}) = \mathbf{N}(\mathbf{y})\mathbf{A}, \tag{12}$$

where $\mathbf{N}(\mathbf{y})$ and $\mathbf{A}$ are the matrix of interpolation functions and the vector of nodal values of α_j, respectively.

Substituting Equation (12) into Equations (9) and (10), and applying the collocation method, one obtains the following equation,

$$\bar{\mathbf{u}} = \mathbf{HA}, \quad \bar{\mathbf{p}} = \mathbf{GA}, \tag{13}$$

where $\bar{\mathbf{u}}$ and $\bar{\mathbf{p}}$ are vectors concerning $\bar{\mathbf{u}}(\mathbf{x}_i)$ and $\bar{\mathbf{p}}(\mathbf{x}_i)$, and $\mathbf{H}$ and $\mathbf{G}$ are matrices defined by

$$\mathbf{H} = \int_{\Gamma_C} \bar{\mathbf{u}}^*(\mathbf{x}_i,\mathbf{y})\mathbf{N}(\mathbf{y})d\Gamma_y,$$
$$\mathbf{G} = \int_{\Gamma_C} \bar{\mathbf{p}}^*(\mathbf{x}_i,\mathbf{y})\mathbf{N}(\mathbf{y})d\Gamma_y, \tag{14}$$

Here $\mathbf{x}_i$ is a collocation point on Γ_B.

Eliminating $\mathbf{A}$ in Equation (13), we obtain the boundary element equation

$$\mathbf{GH}^{-1}\bar{\mathbf{u}} = \bar{\mathbf{p}} \tag{15}$$

In order to combine the BE equation with the FE equation, Equation (15) is transformed into a finite element type

$$\mathbf{K}_B\bar{\mathbf{U}}_B = \bar{\mathbf{F}}_B, \quad \mathbf{K}_B = \mathbf{SGH}^{-1}, \quad \bar{\mathbf{F}}_B = \mathbf{S}\bar{\mathbf{p}}, \tag{16}$$

where $\bar{\mathbf{U}}_B$ stands for a displacement vector associated with the boundary element solution, and $\mathbf{S}$ is the matrix transforming a traction vector into a nodal force vector.

3.2 *Finite element equation*

In the wave number-frequency domain the finite element equation is given by

$$\bar{\mathbf{K}}\,\bar{\mathbf{U}} = \bar{\mathbf{F}}, \quad \bar{\mathbf{K}} = \left[\mathbf{K} + i\omega\mathbf{C} - \omega^2\mathbf{M}\right]$$
$$\mathbf{K} = \int_{\Omega_F} \mathbf{N}^T\mathbf{B}^*\mathbf{DBN}d\Omega_F, \tag{17}$$
$$\mathbf{M} = \int_{\Omega_F} \rho\mathbf{N}^T\mathbf{N}d\Omega_F, \quad \mathbf{C} = \int_{\Omega_F} \xi\mathbf{N}^T\mathbf{N}d\Omega_F,$$

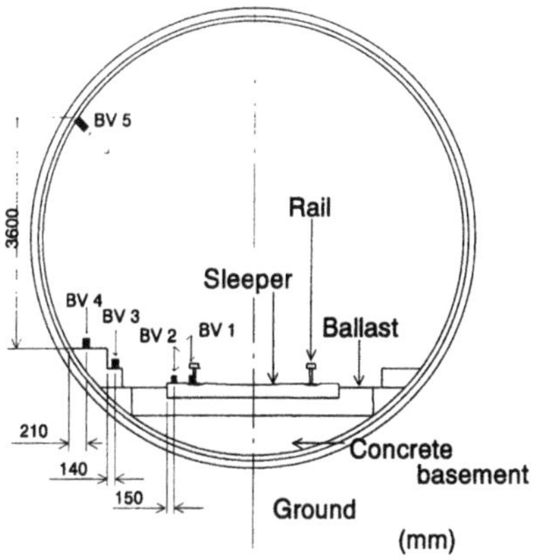

Figure 5. Section of tunnel at the observation site. Measurements at BV1 – BV5.

where **K**, **M** and **C** are stiffness, mass and damping matrices. **D** and **B** are matrices associated with the stress-strain and the strain-displacement relations, and $(\ )^*$ stands for the conjugate of a transpose matrix. ρ is the mass density and ξ is the damping parameter.

Dividing **U** and **F** into two parts $\mathbf{U}_1$, $\mathbf{U}_2$ and $\mathbf{F}_1$, $\mathbf{F}_2$, in which $(\ )_1$ and $(\ )_2$ denote subvectors corresponding to nodes on Γ_B and to the remaining ones, respectively, we obtain the finite element equation

$$\begin{bmatrix} \overline{\mathbf{K}}_{11} & \overline{\mathbf{K}}_{12} \\ \overline{\mathbf{K}}_{21} & \overline{\mathbf{K}}_{22} \end{bmatrix} \begin{Bmatrix} \overline{\mathbf{U}}_1 \\ \overline{\mathbf{U}}_2 \end{Bmatrix} = \begin{Bmatrix} \overline{\mathbf{F}}_1 \\ \overline{\mathbf{F}}_2 \end{Bmatrix}. \tag{18}$$

3.3 *Coupling of FE and BE equations*

The compatibility and equilibrium conditions are imposed on the nodal displacement and nodal force, respectively, that is,

$$\overline{\mathbf{U}}_1 = \overline{\mathbf{U}}_B, \quad \overline{\mathbf{F}}_1 + \overline{\mathbf{F}}_B = \mathbf{0}. \tag{19}$$

Equations (16), (18) and (19) yield the combined equation

$$\begin{bmatrix} \overline{\mathbf{K}}_{11} + \mathbf{K}_B & \overline{\mathbf{K}}_{12} \\ \overline{\mathbf{K}}_{21} & \overline{\mathbf{K}}_{22} \end{bmatrix} \begin{Bmatrix} \overline{\mathbf{U}}_1 \\ \overline{\mathbf{U}}_2 \end{Bmatrix} = \begin{Bmatrix} \mathbf{0} \\ \overline{\mathbf{F}}_2 \end{Bmatrix}. \tag{20}$$

Equation (20) is to be solved for each discrete point in the wave number-frequency space. Applying the Fast Inverse Fourier Transform(IFFT) to these solutions in both wave number and frequency directions, we obtain the three-dimensional numerical Green's function in time domain.

4 NUMERICAL ANALYSIS

4.1 *Outline of the analysis*

The developed method is applied to the dynamic analysis of a subway track. The numerical solution is compared with the in situ measurements and the two-dimensional analysis. The tunnel section is shown in Figure 5. The ballast mat with thickness of 25 mm is inserted between the ballast and concrete basement. The rail pad inserted between the rail and sleeper has a nominal spring constant of 110 MN/m. The nominal spring constant of sleeper pad is 9 MN/m. Acceleration was measured at 5 points (BV1-BV5) as shown in Figure 5.

4.2 *Setting of material properties*

The material properties associated with the wheel, rail, sleeper and pads, and the ballast, tunnel and ground are tabulated in Tables 1 and 2, respectively. The damping constants of rail pad and sleeper pad are set as shown in Table 1 based on an identification achieved for two-dimensional model. Since the material constants concerning the ballast have not been measured, these values are determined through numerical experiments. The ground is modeled as an unbounded homogeneous linear elastic soil. The material properties of the tunnel and ground are set based on the wave speed shown in Furuta & Nagasima (1994). The damping constants of the ballast and tunnel are set to 8% and 3%, respectively. Since the ballast mat is very thin with 2.5cm thick, only the vertical deformation will be essential. Therefore, the ballast mat is modeled as a membrane with Poisson's ratio 0.

It is difficult to determine the material constants which can reproduce the measured data properly in two-dimensional analysis. Hence the shear modulii of ballast and concrete basement and the damping constants of pads used in the two-dimensional analysis are determined with the identification method developed by Suzuki et al. (2000) so that the numerical result may reproduce the measured acceleration. The material constants of ballast and concrete

Table 1. Parameters of wheel, rail, sleeper and pads.

Weight of car body		(kN)	36.75
Mass of wheelset		(kg)	350
Spring const.		(MN/m)	2000
Damping coeff.		(kN · s/m)	0
Flexural rigidity of rail		(MN · m^2)	4
Mass of rail		(kg/m)	50
Rail pad	spring const.	(MN/m)	110
	damping coeff.	(kN · s/m)	650
Sleeper pad	spring const.	(MN/m)	9
	damping coeff.	(kN · s/m)	100
Flexural rigidity of sleeper		(MN · m^2)	1
Mass of sleeper		(kg)	205

Table 2. Parameters of sleeper support.

Ballast	Density	(kg/m^3)	1570
	Shear modulus	(MN/m^2)	21
	Poisson's ratio		1/3
	Damping const.	(%)	8
Ballast mat	Density	(kg/m^3)	1227
	Shear modulus	(MN/m^2)	3.06
	Poisson's ratio		0
Tunnel	Density	(kg/m^3)	2300
	Shear modulus	(MN/m^2)	1435
	Poisson's ratio		0.26
	Damping const.	(%)	3
Ground	Density	(kg/m^3)	1420
	Shear modulus	(MN/m^2)	88.75
	Poisson's ratio		0.45
	Damping const.	(%)	0

Table 3. Parameters of ballast and concrete base in 2-D.

Ballast	Density	(kg/m^3)	1667
	Shear modulus	(MN/m^2)	103.0
	Poisson's ratio		0.3
Concrete Base	Density	(kg/m^3)	2300
	Shear modulus	(GN/m^2)	19.6
	Poisson's ratio		0.2

basement in two-dimensional analysis are shown in Table 3.

4.3 *Wheels, rail and sleepers*

The analytical conditions concerning the wheels, rail and sleepers are shown in Figure 6. A number of 25 sleepers with 58 cm spacing are considered in the analysis. The speed of wheels is set to 20 m/sec based on the observation condition. The numerical analyses are carried out with the time increment Δt =0.001 sec. In order to reproduce the observed acceleration, the rail-surface irregularity which represents the whole imperfection in track is identified using the method proposed by Abe et al. (2000).

4.4 *Calculation of numerical Green's function*

The contact force F_{sj}^m acts on the sleeper pad. The present subway track has sleeper pads with dimension of 24 cm × 20 cm. Hence the contact force is distributed over that area. In the calculation of the numerical Green's function $\overline{G}_{ij}^m$, the contact force is distributed uniformly. Since $\overline{G}_{ij}^m$ is given by the displacement at the i th sleeper pad, this value is obtained by averaging the displacement at each node located on the contact area of the i th sleeper pad.

The three-dimensional elastodynamic analysis by discrete wave number method is reduced to a quasi two-dimensional analysis at each discrete point in the wave number-frequency space. The discretiza-

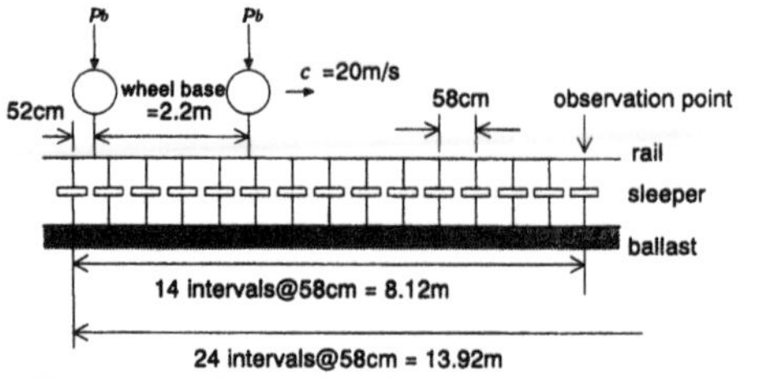

Figure 6. Analytical conditions of wheels, rail and sleepers.

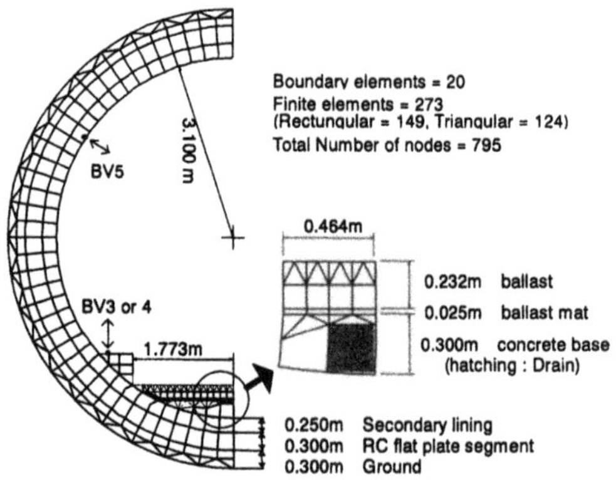

Figure 7. Descretization with finite elements and boundary elements.

tion used in this study for the quasi two-dimensional problem is shown in Figure 7. Due to symmetry, a half of the section is discretized. Ballast, ballast mat, concrete basement, tunnel and a part of ground surrounding the tunnel are discretized by finite elements. The rest of the unbounded ground is represented by the boundary element method. The discretization is achieved by quadratic elements. It is known that subway tracks have a peak in the acceleration response at the range of 0-100 Hz. Therefore the mesh size is determined so that waves propagating in the ground with frequencies lower than 100 Hz can be represented adequately.

In the present analysis, a numerical periodicity is imposed on the time and tunnel axis direction, respectively. Hence these periods must be long enough so that the existence of periodicity becomes negligible. Periods of L =14.848 m and T =0.512 sec are employed with 512 and 1024 divisions, respectively.

4.5 *Numerical results*

The Fourier spectra of acceleration at BV1-BV3 are shown in Figures 8-10. Since in the two-dimensional analysis the substrate is modeled by a half plane extending along the rail, the geometrical configuration of the tunnel cannot be taken into account. Hence, in two-dimensional case, the acceleration at BV3 is

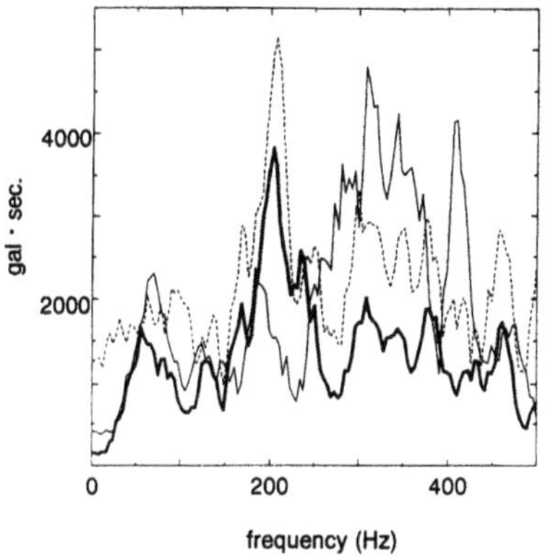

Figure 8. Fourier spectra of acceleration at BV1. Numerical results with 3-D and 2-D models and measured data are represented with thick, thin and broken curves, respectively.

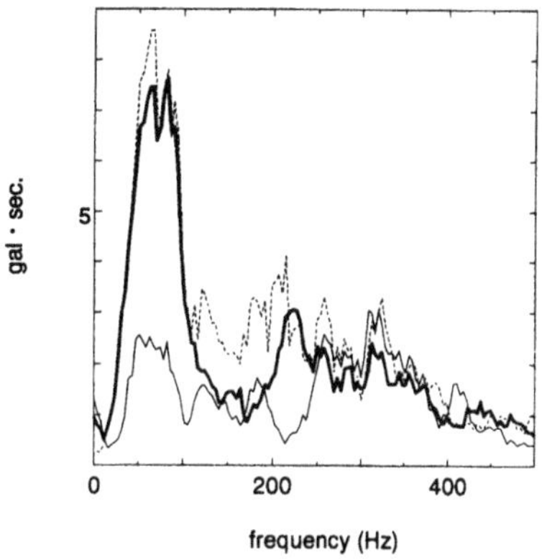

Figure 10. Fourier spectra of acceleration at BV3. Numerical results with 3-D and 2-D models and measured data are represented with thick, thin and broken curves, respectively.

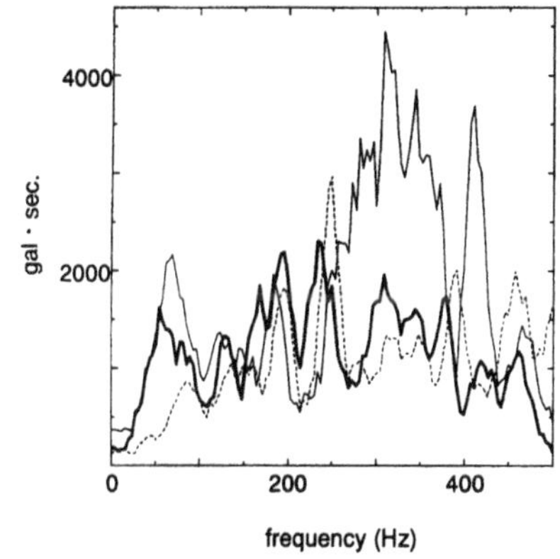

Figure 9. Fourier spectra of acceleration at BV2. Numerical results with 3-D and 2-D models and measured data are represented with thick, thin and broken curves, respectively.

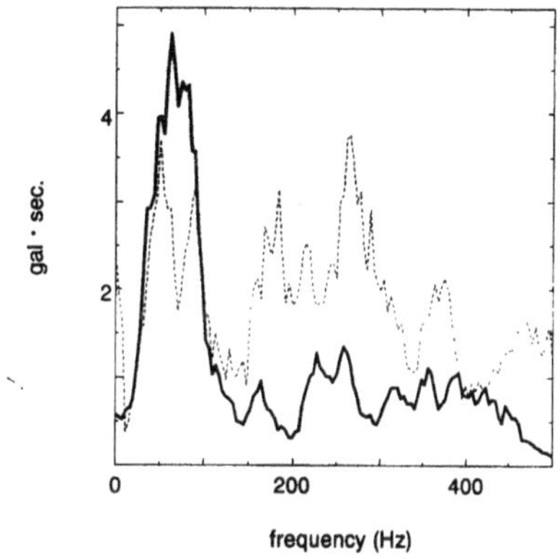

Figure 11. Fourier spectra of acceleration at BV5. Numerical results with 3-D model and measured data are represented with thick and broken curves, respectively.

represented by the one at the concrete base beneath the ballast. In Figures 8 and 9, the two-dimensional model has higher vibration level than the observation and the three-dimensional model at frequencies higher than 250 Hz for BV1 and BV2. The acceleration at BV3 also cannot be reproduced by the two-dimensional model. Especially, the dominant response observed in the measurement at the range of 0-100 Hz is not reproduced. On the other hand, the three-dimensional model has good agreement with the data observed at each point in the tunnel section. From Figure 10, it is found that the peak of the Fourier spectrum distribution on frequencies lower than 100 Hz at BV3 can also be reproduced by the three-dimensional model as well as the response of rail and sleeper.

The acceleration at BV5 is shown in Figure 11 with observed data. Since the two-dimensional model does not have any nodes corresponding to the tunnel, it is impossible to compare the numerical results obtained by the two-dimensional analysis with the measurements. The figure shows that the developed method can reproduce the dynamic response of the tunnel in the range of 0-100 Hz. As mentioned above, waves with frequencies higher than 100Hz cannot be represented by the present mesh resolution. However, since waves with frequencies of higher than 100 Hz are not important for the estimate of ground vibration, the present mesh is enough for this purpose.

5 CONCLUSION

A subway track-tunnel-ground interaction analysis method was constructed in 3-D. The acceleration response obtained by the present method was compared with the measurements and the two-dimensional model. The two-dimensional model cannot reproduce the dynamic response of the tunnel. On the contrary, the three-dimensional model can predict the vibration level of subway track and tunnel. The developed method will be available to estimate the vibration reduction due to the change of subway track structure.

REFERENCES

Abe, K. 1997. Application of time integration scheme based on integral equation to elastodynamic FE/BE coupling analysis. *Proc. of Japan Nat. Sympo. Bound. Elem. Meth.* 14 : 93-98 (in Japanese).

Abe, K. & Furuta, M. 1997. Time domain integral representation for track vibration analysis. *J. Struct. Eng.* 43A : 365-372 (in Japanese).

Abe, K., Suda M. & Furuta, M. 1999. A time domain numerical method for wheel / track interaction analysis. *J. Struct. Eng.* 44A : 271-280 (in Japanese).

Abe, K., Suzuki, T. & Furuta, M. 2000. Identification of rail-surface irregularity for railway track vibration analysis. *J. Appl. Mech. JSCE* 3 : 107-114 (in Japanese).

Bouchon, M. & Aki, K. 1977. Discrete wave-number representation of seismic source wave fields. *Bull. Seism. Soc. Am.* 67(2) : 259-277.

Furuta, M. & Nagasima, F. 1994. Dynamic response analysis of subway-induces vibration on shield tunnel-ground systems, and calculation examples. *Proc. of Tunnel Eng. JSCE* 4 : 93-100 (in Japanese).

Graff, K.F. 1975. *Wave motion in elastic solids*. Oxford Univ. Press.

Grassie, S.L., Gregory, R.W., Harrison, D. & Johnson, K.L. 1982. The dynamic response of railway track to high frequency vertical excitation. *J. Mech. Eng. Sci.* 24(2) : 77-90.

Kalker, L.L. 1996. Discretely supported rails subjected to transient loads, *Vehicle Sys. Dyn.* 25 : 71-88.

Knothe, K.L. & Grassie, S.L. 1993. Modelling of railway track and vehicle / track interaction at high frequency. *Vehicle Sys. Dyn.* 22 : 209-262.

Mohammadi, M. & Karabalis, D.L. 1995. Dynamic 3-D soil-railway track interaction by BEM-FEM. *Earthq. Eng. Struct. Dyn.* 24 : 1177-1193.

Suzuki, T., Abe, K. & Furuta, M. 2000. Identification of parameters for railway track vibration analysis with neural network. *Proc. of BTEC* 10 : 55-60 (in Japanese).

Takemiya, H., Satonaka, S. & Xie, W-P. 2001. Train track-ground dynamics due to high speed moving source and ground vibration transmission. *Proc. of JSCE* 682 : 299-309 (in Japanese).

Wave propagation – Moving load – Vibration reduction, Chouw & Schmid (eds.)
© 2003 Swets & Zeitlinger, Lisse, ISBN 90 5809 559 2

A three-dimensional FEM/BEM model for the investigation of railway tracks

O. von Estorff & M. Firuziaan
Technical University Hamburg-Harburg, 21073 Hamburg, Germany

K. Friedrich, G. Pflanz & Günther Schmid
Ruhr-University Bochum, 44780 Bochum, Germany

ABSTRACT: The simulation of the interaction between a railway bed and a train calls for the development of a reliable computational model. Whereas models for the train, developed with conventional methods, exist and are applied successfully in industry, only recently suitable models have been initiated to describe, in detail, the dynamic behavior of the bedding, the foundation, and the undisturbed soil. In the present paper the usage of the boundary element method and a combined finite element/boundary element procedure is suggested for the numerical analysis of bedding-foundation-soil interaction problems. Different material laws, e.g., an elastoplastic and a damage model, are implemented in the finite element procedure in order to investigate the influence of different nonlinear models on the dynamic interaction. Different time dependent load cases, including moving loads, are considered as well. The analysis is carried out in the frequency and time domain where nonlinear problems are investigate in the time domain only. The accuracy and applicability of the model are shown by several examples, namely a benchmark problem and a number of railway track configurations.

1 INTRODUCTION

Today's increasing ground vibration and stress levels on railroad track, substructure and subsoil due to the velocity of modern high speed trains call for the development of optimized track systems. Features of the track which are subjected to moving or non-moving loads have to be analyzed with respect to the dynamic behavior. Additionally, economically suitable systems with high endurance and low maintenance cost need to be built. The simulation of a passing train with numerical models is a suitable means in order to display the dynamic behavior next to extensive measurements, since in computational simulations different track-geometries, material parameters and the layered soil can be varied easily. Thus the influence of the most diverse parameters on the dynamic behavior of the track can be studied and it is possible to optimize the railroad track and the substructure without the necessity to build extensive test tracks.

Suitable mathematical formulations for the various subsystems were chosen in order to display the dynamic features of the individual components in the model. Subsystems such as the infinitely extended soil and those parts of the track for which the wave radiation is essential, are often modelled with boundary elements while parts of the superstructure such as rails, sleepers and elastic pads are modelled with finite elements. At the nodal points of the common surface, the subsystems are coupled by means of the substructure technique (see Fig. 1).

In recent years a large number of publications in this discipline emphasizes the importance of proper simulation tools. Thus a numerical analysis of the wave propagation caused by loads moving with sub- and supercritical velocity using a time-domain approach is presented in Pflanz (2001). Other investigations focussing on railway dynamics and moving loads can be found in Petyt & Jones (1999), Lefeuve-Mesgouez (1999) and Mohammadi & Karabalis (1995). In order to meet the requirements of high speed trains, an elastically supported track system with ladder sleepers is developed in Japan (Read et al. 1999). Discrete and continuous models for the railway track dynamics subjected to high frequency vertical excitation can be found in Grassie et al. (1982). Extended analytical and numerical examinations on the dynamic interaction of rails, pads, sleepers and subsoil are given in Knothe (2001) and Ripke (1995). Moreover, in some cases special emphasis is placed on nonlinear effects, e.g. as discussed in von Estorff & Firuziaan (2000a, 2000b).

The objective of this paper is to introduce a rather complete FEM/BEM model where different numerical formulations are integrated in one program which satisfies the aspects mentioned above. It facilitates to carry out the analysis under harmonic and transient loads as well as under moving load conditions. Although the numerical method is also suitable for horizontal dynamics, the investigations done in this paper are focussing on the more important vertical dynamics of track-subsoil interaction.

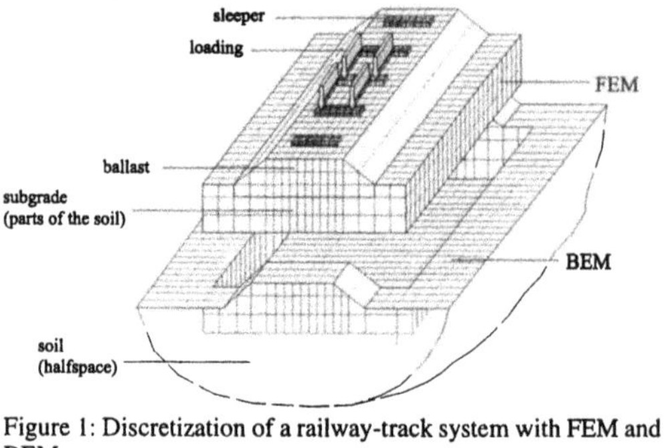

Figure 1: Discretization of a railway-track system with FEM and BEM.

2 ANALYSIS PROCEDURE

In this chapter the basic aspects of the numerical analysis procedures which are implemented in the developed program are presented.

2.1 *Boundary Element Method (BEM)*

The boundary element method is based on the fundamental solution of the differential equation of motion for a homogeneous, isotropic, linear-elastic continuum (Dominguez 1993). This equation, also known as Lamé-Navier equation, can be derived as follows: Starting from equilibrium considerations, one obtains

$$\sigma_{ij,j} + f_i = \varrho \ddot{u}_i \,, \tag{1}$$

where σ_{ij} , f_i and u_i are the components of stress, volume force and displacement, respectively, and $\varrho\, \ddot{u}_i$ is the force due to inertia. The commas and dots indicate space and time derivatives, respectively. The material law for an elastic material is given as

$$\sigma_{ij} = \lambda \varepsilon_{kk} \delta_{ij} + 2\mu \varepsilon_{ij} \,, \tag{2}$$

where λ and μ are the Lamé constants and δ_{ij} is the Kronecker delta. The kinematic compatibility is expressed as

$$\varepsilon_{ij} = \frac{1}{2}(u_{i,j} + u_{j,i}) \,. \tag{3}$$

Substituting (2) and (3) into (1) leads to the well known wave equation:

$$\varrho\left[\left(c_p^2 - c_s^2\right) u_{i,ik} + c_s^2 u_{k,ii} - \ddot{u}_k\right] = -f_k \,, \tag{4}$$

where $c_p = \sqrt{\frac{\lambda+2\mu}{\varrho}}$ and $c_s = \sqrt{\frac{\mu}{\varrho}}$ are the pressure and shear wave velocities, respectively. In order to obtain the wave equation in the frequency domain, a Fourier Transform needs to be applied which leads to

$$\varrho[(c_p^2 - c_s^2)\tilde{u}_{i,ik} + c_s^2 \tilde{u}_{k,ii} + \omega^2 \tilde{u}_k] = -\tilde{f}_k \,. \tag{5}$$

In this equation $\tilde{u}_k$ and $\tilde{f}_k$ can be interpreted as the complex displacement and force amplitudes, respectively, of a harmonic motion with circular frequency ω.

When body forces are assumed to be zero and homogeneous initial conditions apply, the boundary integral equation can be obtained by the help of Betti's law; this yields:

$$c_{ik}\tilde{u}_i^\star = \int_\Gamma \tilde{u}_{ik}^\star\, \tilde{s}_i d\Gamma - \int_\Gamma \tilde{s}_{ik}^\star\, \tilde{u}_i d\Gamma \,, \tag{6}$$

where $\Gamma = \Gamma_1 \cup \Gamma_2$ is the boundary of the considered elastic domain Ω. $\tilde{s}_{ik}^\star(\boldsymbol{x}, \xi, \omega)$ and $\tilde{u}_{ik}^\star(\boldsymbol{x}, \xi, \omega)$ are the fundamental solutions for the traction and displacement components, respectively, in direction i at point $\boldsymbol{x}$ due to a unit load in direction k at point ξ. For smooth boundaries the elements of the boundary factor c_{ik} are equal to δ_{ik} if $\xi \in \Omega$ and $0.5\,\delta_{ik}$ if $\xi \in \Gamma$. On the boundaries Γ_1 and Γ_2 the displacement boundary conditions $\bar{\tilde{u}}_i(\boldsymbol{x}, \omega)$ and the traction boundary condition $\bar{\tilde{s}}_i(\boldsymbol{x}, \omega)$ are prescribed, respectively. Using a complex young's modulus $E = E(1 + i\omega\eta)$ with viscous damping coefficient η, the frequency domain formulation allows to incorporate also damping.

The discretization of the boundary with elements leads to an algebraic form of the integral equation. In matrix notation this can be written as:

$$\tilde{\boldsymbol{U}}\, \tilde{\boldsymbol{s}} = \tilde{\boldsymbol{T}}\, \tilde{\boldsymbol{u}} \,, \tag{7}$$

where $\tilde{\boldsymbol{u}}$ and $\tilde{\boldsymbol{s}}$ are the complex frequency dependent displacements and tractions of all nodal points at the boundary, and $\tilde{\boldsymbol{U}}$ and $\tilde{\boldsymbol{T}}$ are the influence matrices. The time dependent displacements and tractions can be obtained by an Inverse Fourier Transform as done in Sect. 3.3.

For the numerical solution of equation 4, a discretization of the body surface with boundary elements as well as a discretization of the observation time by equal time increments Δt is needed. Collocation at each boundary node and at all time steps leads to a system of equations (von Estorff & Prabucki 1990)

$$\begin{aligned} \boldsymbol{U}^{(1)} \boldsymbol{s}^{(m)} &= \boldsymbol{T}^{(1)} \boldsymbol{u}^{(m)} + \sum_{k=1}^{m-1} \Big[\boldsymbol{T}^{(m-k+1)} \boldsymbol{u}^{(k)} \\ &\quad - \boldsymbol{U}^{(m-k+1)} \boldsymbol{s}^{(k)} \Big] \end{aligned} \tag{8}$$

which is valid for each time step m. $\boldsymbol{U}^{(i)}$ and $\boldsymbol{T}^{(i)}$ are influence matrices, which contain integral terms, evaluated over each boundary element and over the time step i. The vectors $\boldsymbol{u}^{(k)}$ and $\boldsymbol{s}^{(k)}$ contain the according displacements and tractions, respectively.

To obtain consistency between the FE and BE formulations, the boundary tractions $\boldsymbol{s}$ have to be transformed by means of a matrix $\boldsymbol{A}$ such that

$$\boldsymbol{P} = \boldsymbol{A}\boldsymbol{s} \tag{9}$$

where the vector $\boldsymbol{P}$ contains resultant nodal forces. For each boundary element Γ_e the elements m_{ij} of the transformation matrix $\boldsymbol{A}$ can be calculated using the shape functions N_i (see, e.g., von Estorff & Prabucki 1990, Becker 1992)

$$A_{ij} = \int_{\Gamma_e} N_i N_j d\Gamma. \tag{10}$$

By means of equation 9, a global nodal force vector can be formulated from equation 8 which relates the BE nodal forces to the displacements at the time step m and a sum of the influence of all previous time steps ($k = 1, 2, ..., m-1$):

$$\begin{aligned} \boldsymbol{P}^{(m)} &= \boldsymbol{A}\left[\boldsymbol{U}^{(1)}\right]^{-1}\boldsymbol{T}^{(1)}\boldsymbol{u}^{(m)} + \\ &\quad \boldsymbol{A}\left[\boldsymbol{U}^{(1)}\right]^{-1}\sum_{k=1}^{m-1}\Big[\boldsymbol{T}^{(m-k+1)}\boldsymbol{u}^{(k)} - \\ &\quad \boldsymbol{U}^{(m-k+1)}\boldsymbol{A}^{-1}\boldsymbol{F}^{k}\Big] \\ &= \boldsymbol{K}_B\boldsymbol{u}^{(m)} + \sum_{k=1}^{m-1}\boldsymbol{H}_k^{(m)}. \end{aligned} \tag{11}$$

2.2 *Finite Element Method (FEM)*

Using the Principle of Virtual Displacements, the equation of motion of the finite element method in the frequency domain can be obtained as

$$-\omega^2\boldsymbol{M}\tilde{\boldsymbol{u}} + \boldsymbol{K}\tilde{\boldsymbol{u}} = \tilde{\boldsymbol{P}}, \tag{12}$$

where

$$\begin{aligned} \boldsymbol{M} &= \int_\Omega \varrho \boldsymbol{N}^{\mathrm{T}}\boldsymbol{N}d\Omega, \\ \boldsymbol{K} &= \int_\Omega \boldsymbol{B}^{\mathrm{T}}\boldsymbol{D}\boldsymbol{B}d\Omega \end{aligned} \tag{13}$$

are the mass and stiffness matrices, respectively, and

$$\tilde{\boldsymbol{P}} = \int_\Gamma \boldsymbol{N}^{\mathrm{T}}\tilde{\boldsymbol{s}}d\Gamma + \int_\Omega \boldsymbol{N}^{\mathrm{T}}\tilde{\boldsymbol{f}}d\Omega \tag{14}$$

are the nodal forces. The matrix $\boldsymbol{B}$ contains the relation between strains and displacements, $\boldsymbol{D}$ describes the connection between stress and strain and $\boldsymbol{N}$ contains the shape functions. Equation 12 can be written as

$$\tilde{\boldsymbol{P}} = \tilde{\boldsymbol{K}}^{\mathrm{FEM}}\tilde{\boldsymbol{u}}, \tag{15}$$

with

$$\tilde{\boldsymbol{K}}^{\mathrm{FEM}} = \boldsymbol{K} - \omega^2\boldsymbol{M}. \tag{16}$$

In the case of beam elements which are used in this paper, the mass and stiffness matrix are obtained by an analytical integration.

In order to be able to take nonlinear effects into account, the transient solution scheme is needed. The basic problem in a nonlinear analysis is to find the state of equilibrium of a body Ω subjected to an external time dependent load. It can be solved by assuming that the solution for the discrete time t_i is known and that the solution for the discrete time $t_{i+1} = t_i + \Delta t$ is required, where Δt is a suitably chosen time increment. For each time step, the equilibrium conditions of a nonlinear finite element system representing the body can be expressed as

$$\boldsymbol{R} - \boldsymbol{P} = 0 \tag{17}$$

where the vector $\boldsymbol{P}$ contains the sum of all externally applied nodal forces. In the vector $\boldsymbol{R}$, on the other hand, the nodal forces, which correspond to the element stresses and the dynamic part of the equilibrium due to inertia effects in the current configuration, can be found.
Using a linearized virtual work formulation, an effective tangent stiffness matrix $\boldsymbol{K}_{eff}$ can be derived (von Estorff & Firuziaan 2000a) which corresponds to the geometric and material conditions at time t and also contains the mass matrix. Since this stiffness matrix depends on the deformation of the system, in each time step an iteration has to be performed employing, e.g., a Newton scheme as given in Figure 2.

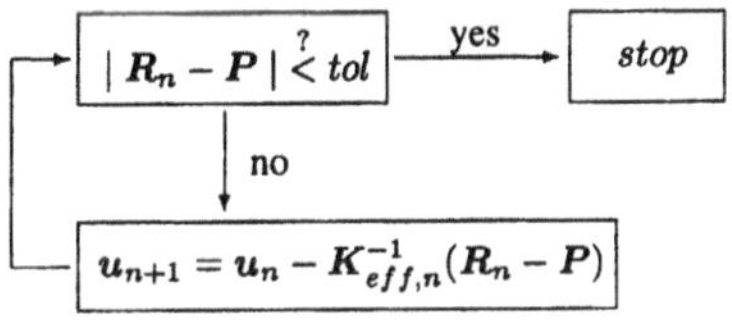

n: number of the current step in the Newton-Raphson iteration
$\boldsymbol{R}_n$: vector containing the sum of the nodal forces corresponding to the element stresses at step n and the dynamic part of the equiblirium
$\boldsymbol{P}$: global load vector
$\boldsymbol{u}_n$: vector of the nodal displacements at step n
$\boldsymbol{K}_{eff,n}$: effective stiffness matrix at step n
tol: tolerance value defining the end of the iteration loop

Figure 2: Solution scheme for uncoupled nonlinear FEM calculations.

2.3 *Coupling of BEM and FEM*

2.3.1 *Frequency domain*

A coupling of finite elements and boundary elements allows the examination of structures consisting of domains with different dynamic behavior. Additionally it enables to benefit from the characteristics of each particular formulation. In the frequency domain those parts of the structure whose dynamic behavior essentially depends on the wave radiation like subsoil, ballast or rigid track are modelled with boundary elements, while sleepers, rails and elastic pads are modelled with finite elements (see Fig.3). The sleepers and rails are discretized with Timoshenko beam elements containing shear deflection. The structure of the corresponding element mass and stiffness matrix which can be obtained from (13) is given in Przemieniecki (1968). The elastic pads are discretized with

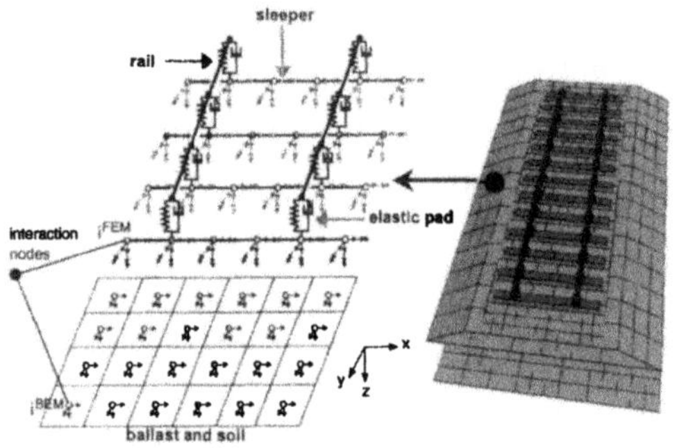

Figure 3: Model of the track-soil system

linear viscous spring-damper elements. The degrees of freedom of each finite element node are:

$$\tilde{\boldsymbol{u}}_i^{\mathrm{FEM}} = \begin{bmatrix} \tilde{u}_x \\ \tilde{u}_y \\ \tilde{u}_z \\ \tilde{\varphi}_x \\ \tilde{\varphi}_y \\ \tilde{\varphi}_z \end{bmatrix}_i^{\mathrm{FEM}} . \tag{18}$$

The track and the soil are modelled with constant boundary elements with three degrees of freedom at each node:

$$\tilde{\boldsymbol{u}}_i^{\mathrm{BEM}} = \begin{bmatrix} \tilde{u}_x \\ \tilde{u}_y \\ \tilde{u}_z \end{bmatrix}_i^{\mathrm{BEM}} . \tag{19}$$

The coupling conditions for the displacements at the contact points between the finite element region and the boundary element region are:

$$\begin{aligned} \tilde{u}_{x_i}^{\mathrm{BEM}} &= \tilde{u}_{x_i}^{\mathrm{FEM}} , \\ \tilde{u}_{y_i}^{\mathrm{BEM}} &= \tilde{u}_{y_i}^{\mathrm{FEM}} , \\ \tilde{u}_{z_i}^{\mathrm{BEM}} &= \tilde{u}_{z_i}^{\mathrm{FEM}} . \end{aligned} \tag{20}$$

In order to obtain a stiffness matrix of the boundary element region, equation 7 has to be transformed to relate nodal displacements with nodal forces

$$\tilde{\boldsymbol{K}}^{\mathrm{BEM}} \tilde{\boldsymbol{u}} = \tilde{\boldsymbol{P}} \tag{21}$$

and

$$\tilde{\boldsymbol{P}} = \tilde{\boldsymbol{A}} \tilde{\boldsymbol{s}} . \tag{22}$$

The dynamic stiffness matrix $\tilde{\boldsymbol{K}}^{\mathrm{FEM}}$ of the rails, the sleepers and the pads, and the dynamic stiffness matrix $\tilde{\boldsymbol{K}}^{\mathrm{BEM}}$ of the boundary element part have to be assembled at the interaction nodes of the surface. To do this, the dimension of the boundary element matrix has to be expanded to the degrees of freedom of the finite element matrix inserting zeros in the boundary element matrix at the corresponding rotational degrees of freedom. This method implies that no bending and twisting moments will be transmitted through the soil at the common interface, but in the case of the relatively stiff sleepers related to the stiffness of the underlying soil, this is an adequate approximation, which gives good results. The equation of the coupled system can be written in expanded form as:

$$\begin{bmatrix} \tilde{\boldsymbol{K}}_{FF}^{FEM} & \tilde{\boldsymbol{K}}_{FC}^{FEM} & \boldsymbol{0} \\ \tilde{\boldsymbol{K}}_{CF}^{FEM} & \tilde{\boldsymbol{K}}_{CC}^{FEM} + \tilde{\boldsymbol{K}}_{CC}^{BEM} & \tilde{\boldsymbol{K}}_{CB}^{BEM} \\ \boldsymbol{0} & \tilde{\boldsymbol{K}}_{BC}^{BEM} & \tilde{\boldsymbol{K}}_{BB}^{BEM} \end{bmatrix} \begin{bmatrix} \tilde{\boldsymbol{u}}_F \\ \tilde{\boldsymbol{u}}_C \\ \tilde{\boldsymbol{u}}_B \end{bmatrix} = \begin{bmatrix} \tilde{\boldsymbol{P}}_F \\ \tilde{\boldsymbol{P}}_C \\ \tilde{\boldsymbol{P}}_B \end{bmatrix} , \tag{23}$$

where the subscripts F and B mark the finite element and the boundary element area and the subscript C indicates the coupled part of the structure as sketched in Figure 3. General descriptions on the coupling of finite elements and boundary elements are presented in Becker (1992). Similar coupling algorithms especially for the track-subsoil interaction can be found in Rücker (1981), Bode et al. (2000) and Mohammadi & Karabalis (1995). A numerical solution for structure-soil interaction problems based on Green's functions is described in Bode (2000).

2.3.2 *Time domain*

In the time domain examples, finite elements are used to discretize the sleepers, the ballast and parts of the underlaying soil, while boundary elements are used to model an infinite halfspace. The coupling is performed by satisfying the continuity conditions along the FE/BE interfaces, namely the displacements of the FE and the BE nodes have to be equal along common interfaces (see equation 20) and, in addition, equilibrium has to be ensured.

At the interfaces, the nodal forces resulting from the BE equation 11 can be treated as additional loads in each iteration. Consequently, the coupling algorithm can be obtained by modifying the FE algorithm given in Figure 2. This yields the new procedure as depicted in Figure 4. Since nonlinearities may occur only in the FE-subdomains, the geometrical linearity is checked along the interfaces by observing the

$$\left| R_n - P + K_B u_n + \sum_{k=1}^{m-1} H_k^{(m)} \right| \overset{?}{<} tol$$

no   yes

$$stop$$

$$u_{n+1} = u_n - (K_{eff,n} + K_B)^{-1}\left(R_n - P + K_B u_n + \sum_{k=1}^{m-1} H_k^{(m)}\right)$$

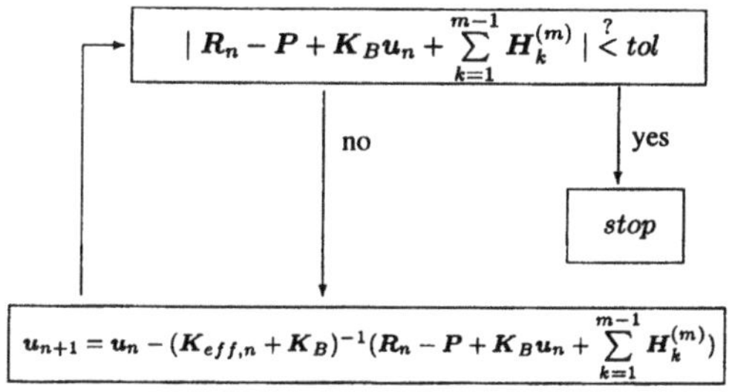

m: number of the current step
H: vector containing the influence of all previous time steps
K_B: effective stiffness matrix (BEM)

Figure 4: Solution scheme for the coupled nonlinear FEM/BEM calculations.

strains at the interface nodes.

Assembling the FE- and BE submatrices, one obtains the coupled system of equations,

$$\underbrace{\begin{bmatrix} K_{eff,n} & \\ & K_B \end{bmatrix}}_{K_{eff}^{(m)}} \underbrace{\begin{bmatrix} u_F^{(m)} \\ u_{BF}^{(m)} \\ u_B^{(m)} \end{bmatrix}}_{u^{(m)}} = \underbrace{\begin{bmatrix} P_F^{(m)} \\ \hline 0 \end{bmatrix} + \begin{bmatrix} 0 \\ \hline P_B^{(m)} - \sum_{k=1}^{m-1} H_k^{(m)} \end{bmatrix}}_{R^{(m)}} \qquad (24)$$

where the global system matrix $K_{eff}^{(m)}$ is, due to the BE part, not symmetric and not positive definite anymore. A special solution strategy, using block equation solvers and parallel computation algorithms, has been developed to handle these systems in a rather efficient way.

3 NUMERICAL RESULTS IN THE FREQUENCY DOMAIN

In order to obtain information about the dynamic behavior of the railroad track and the subsoil, ballasted tracks and rigid tracks on layered soil are analyzed. The examination is performed with tracks subjected to a fixed harmonic load as well as to a moving load to obtain information about the influence of the train velocity.

3.1 Model size

As the area of discretization plays an important role in the boundary element method when the domain extends to infinity and a full-space fundamental solution is used, a first investigation is done on the influence of the length l of the discretized model, where the length of the discretization is equal to the length of the rigid track. In this example a harmonic unit load is directly applied on the concrete layer of a rigid track on the half-space and the magnitude of the vertical compliances at the middle of the plate are calculated. The results shown in Figure 5 indicate that the length of the model of a rigid track has to be at least 11.0 m

in order to avoid large discretization errors. The amplitudes of the vertical compliances at the center of the concrete layer increase with a bigger length of the modelled plate. Convergence of the solution can be observed by increasing the model length from 5 m to 11 m.

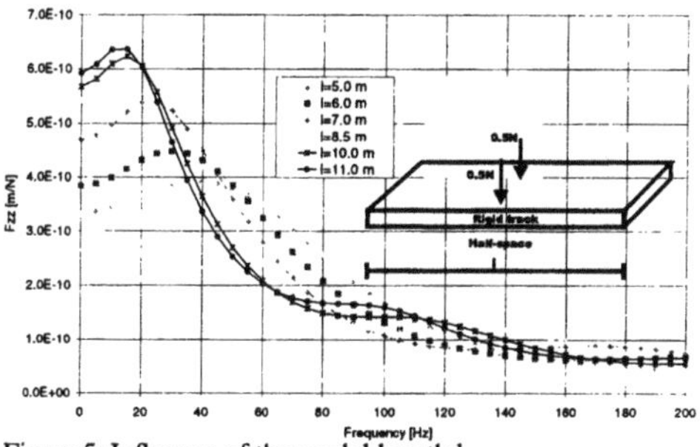

Figure 5: Influence of the model length l

3.2 Influence of different track geometries

Regarding the results of Sect. 3.1 by using a model length of 11.0 m, the vertical displacements of different rigid track systems on the half-space subjected to a harmonic load of $P = 0.5N$ at two rail support points as shown in Figure 6 are investigated. The material properties, width b and hight h of concrete layer (CL), hydraulically consolidated layer (HCL), frost protection layer (FP) and the half-space are listed in Table 1. Figure 7 shows the vertical displacements of the ana-

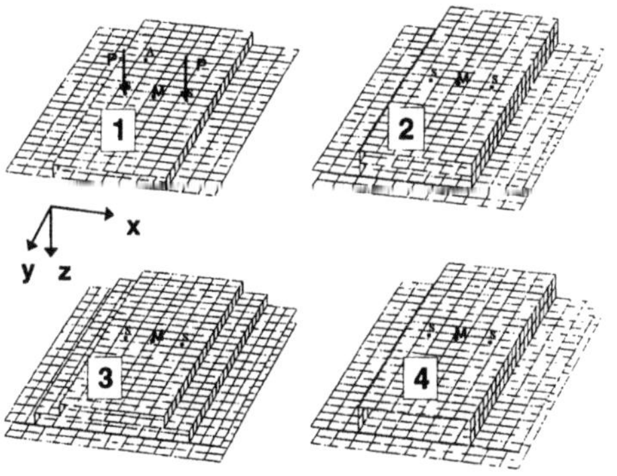

Figure 6: Geometry of the rigid track systems 1 to 4

lyzed rigid track systems in the middle of the concrete layer (point M). It can be seen that the displacements mostly depend on the material properties and thickness of the concrete layer.

Table 1: Material properties and geometry of the models

	$c_s\,[m/s]$	$\varrho\,[kg/m^3]$	$\nu\,[-]$	$b\,[m]$	$h\,[m]$
System 1					
CL	2236	2500	0.2	2.7	0.3
Half-space	200	2000	0.33		
System 2					
CL	2407	2445	0.2	2.7	0.3
HCL	1903	2300	0.2	2.7	0.3
FPL	217	1938	0.32		0.4
Half-space	233	1837	0.33		
System 3					
CL	2407	2445	0.2	2.7	0.3
HCL	1903	2300	0.2	4.0	0.3
FPL	217	1938	0.32		0.4
Half-space	233	1837	0.33		
System 4					
CL	2236	2500	0.2	2.7	0.6
FPL	217	1938	0.32		0.4
Half-space	233	1837	0.33		

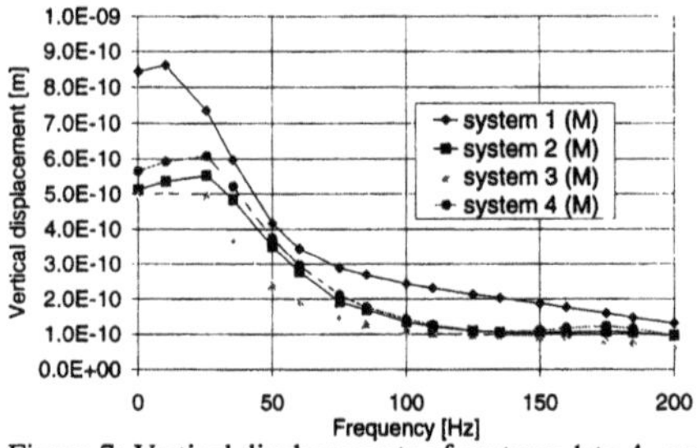

Figure 7: Vertical displacements of systems 1 to 4, evaluated in point M

The amplitude characteristic shows that the main differences between the four systems become apparent in the frequency range between 0 Hz and 25 Hz. For frequencies higher than 25 Hz the displacements decrease monotonously with increasing frequency. A comparison of systems 2 and 3 which differ in the width of the hydraulically consolidated layer, shows that the static displacements are almost identical while the bigger width of the hydraulically consolidated layer of system 3 leads to a reduction of the displacements in the frequency range from 10 Hz up to 100 Hz. An investigation of the vertical displacements of system 1 evaluated in the middle of the concrete layer (point M), at the rail support points S and at a point A, which has a distance of 1.9 m from point S in y-direction, is shown in Figure 8. Owing to the high stiffness of the cross section of the concrete layer, only a small difference between the displacements in point M and point S can be observed. Similar results of an investigation of a rigid track are presented in Savidis et al. (1999), where an approach based on the coupling of the thin layer method with a finite element method is used. Measured results can be found in Rücker (1999).

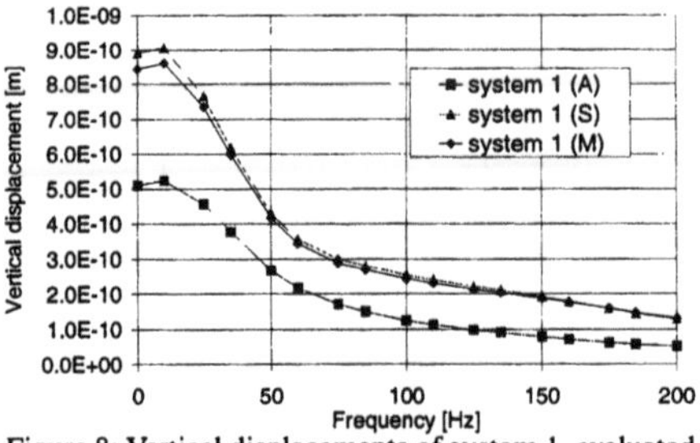

Figure 8: Vertical displacements of system 1, evaluated in points S, M and A

In addition to the results obtained by a load directly applied on the concrete layer, a further investigation is carried out on the influence of the railpads when the load acts on the rail. The displacements of the rail for a rigid track and a ballasted track on the top of an elastic halfspace as well as on the top of a rigid halfspace are given in Figure 9 and Figure 10. The material properties are listed in Table 2 and Table 3. The results are compared with the displacements evaluated directly at the surface of the track at the rail support points below the load.

Table 2: Parameters of rigid track and ballasted track

Rigid track	
width CL and HCL [m]	3.0
height CL and HCL [m]	0.3
c_s CL [m/s]	2274
ρ CL $[kg/m^3]$	2500
c_s HCL [m/s]	1533
ρ HCL $[kg/m^3]$	2200
stiffness railpad [N/m]	$2.25 \cdot 10^7$
damping railpad [Ns/m]	$2.0 \cdot 10^4$
Ballasted track	
height [m]	0.35
width (top/bottom) [m]	3.60/5.60
c_s [m/s]	300
ρ $[kg/m^3]$	1700
ν [-]	0.25
stiffness railpad [N/m]	$6.0 \cdot 10^8$
damping railpad [Ns/m]	$2.0 \cdot 10^4$

Due to the relatively small stiffness of the pads ($k = 2.25 \cdot 10^7\,N/m$) of a rigid track, the displacement at the rail is about 10 times higher than the displacement at the concrete layer of the rigid track (see Fig. 9). It can be seen, that only in the lower frequency range up to 40 Hz the displacements of the rail are determined by the properties of the subsoil.

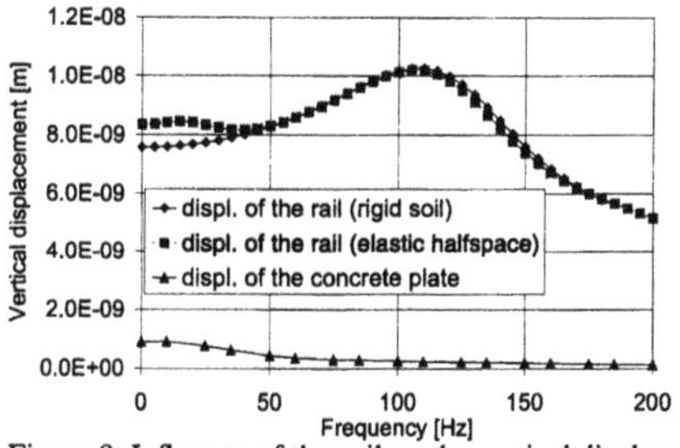

Figure 9: Influence of the soil on the vertical displacements of a rigid track

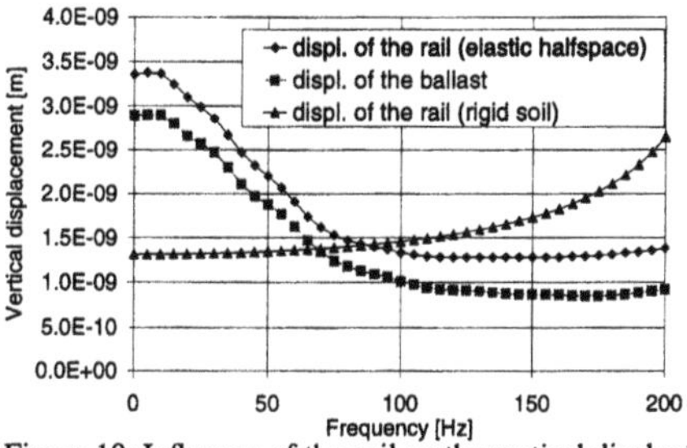

Figure 10: Influence of the soil on the vertical displacements of a ballasted track

Table 3: Parameters of the rail and the sleepers

Sleepers	
sleeper distance [m]	0.6
length [m]	2.6
width [m]	0.26
heigth [m]	0.2
mass [kg]	290
$\rho_{concrete}$ $[kg/m^3]$	2145
ν [-]	0.16
Young's modulus $[N/m^2]$	$3.0 \cdot 10^{10}$
Rail UIC 60	
rail distance [m]	1.50
$I_y [m^4]$	$3.055 \cdot 10^{-5}$
cross section area $[m^2]$	0.007686
mass $[kg/m^3]$	7850
ν_{steel} [-]	0.3
Young's modulus $[N/m^2]$	$2.1 \cdot 10^{11}$

In the case of a ballasted track, as shown in Figure 10, the displacements of the rail and the displacements of the ballast below the rail do not differ very much due to the fact that on the one side the ballasted layer is much softer than the rigid track and on the other side the stiffness of the railpad used for ballasted tracks ($k = 6.0 \cdot 10^8\, N/m$) is higher than the elastic pad of a rigid track. In this case the numerical model for the ballast and the underlying subsoil is of great importance because the frequency dependency of the vertical displacements of a ballasted track show contrary behaviour in the case of a rigid or an elastic half-space (see Fig. 10). The calculation of the track-subsoil interaction considering the effect of vibration barriers is presented in Hubert et al. (2000). A deeper investigation of the dynamic behavior of the railpads used for rigid tracks and ballasted tracks is carried out in Knothe (2001) and Müller-Boruttau & Breitsamter (2000). An extended examination of the influence of the subsoil on the vertical displacements of the rigid track as well as of a ballasted track is presented in the frame of the benchmark test which is published in Rücker et al. (2002).

3.3 *Moving load*

Another goal of this project is the determination of the dynamic behavior of the track caused by loads moving with different velocities along the track. In a first investigation a pulse-type axle load, resulting from the distribution of the load from the contact point of the wheel over the rail, the elastic pads and the sleepers (Verbic et al. 1997, Schmid & Verbic 1997) is directly applied to the surface of the concrete layer (see Fig. 11). In a second examination the load is moving along the rail supported by elastic pads and ties. Both investigations are based on the assumption that a point on the track subjected to a moving load experiences a temporal load of the same form as the spatially distributed load. The temporal load distribution P(t) used for both cases is shown in Figure 11.

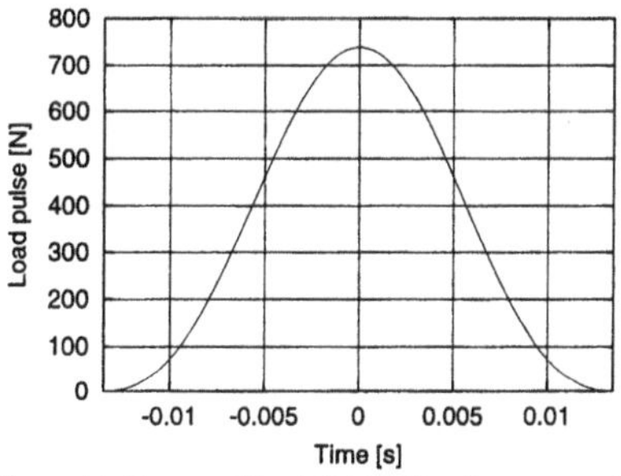

Figure 11: Temporal load distribution for a one-axle load, v=400 km/h

Since the developed programm computes the responses in the frequency domain, the pulse load $P(t)$ has to be transformed via Discrete Fast Fourier transform into frequency domain. The calculated displacements in the frequency-domain have to be transformed with the Discrete Inverse Fast Fourier Transform back into the time domain (Brigham 1995). For a functions f and its transform F, the formulas for the Discrete Fast Fourier Transform and the Inverse Fast Fourier Transform are given as:

$$F_N = \frac{T}{N} \sum_{k=0}^{N-1} f_k e^{-i2\pi \frac{nk}{N}}, \quad n = 0, 1, 2, ..., N-1 \quad (25)$$

and

$$f_k = \frac{1}{T}\sum_{n=0}^{N-1} F_n e^{i2\pi\frac{nk}{N}}, \quad n = 0, 1, 2, ..., N-1 , \tag{26}$$

where N is the number of the chosen time steps Δt and

$$T = N\,\Delta t \tag{27}$$

is the period. The frequency content $\tilde{P}_T(\omega)$ of the considered axle load is shown in Figure 12. The displacements in the frequency domain $\tilde{\boldsymbol{u}}_T(\omega)$ are obtained from (23) with the load $\tilde{P}_T(\omega)$. The implementation of the above mentioned algorithm is tested by the calculation of the displacement of two adjacent rigid ties at which one tie is subjected to a time dependent square-wave impulse of magnitude 1.0 kN whose frequency spectrum is shown in Figure 13. The chosen time step is $\Delta t = 7.22^{-4}s$ and $N = 64$ is the number of time steps. The results (RUB) are compared with a calculation performed with the time-domain program (TUHH). The results of the calculated displacements are shown in Figure 14 and indicate good agreement between the two methods.

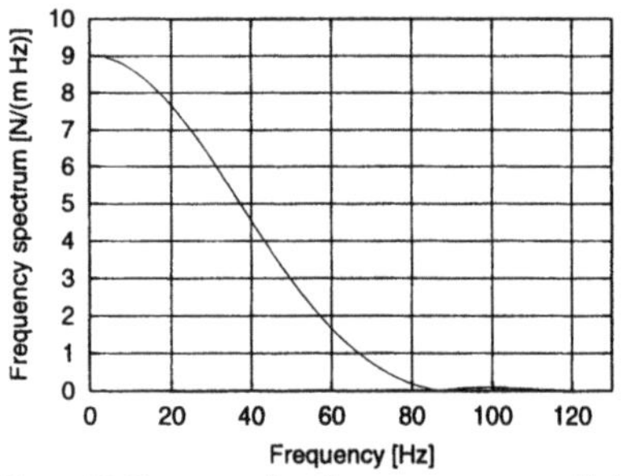

Figure 12: Corresponding frequency spectrum $P_T(\omega)$

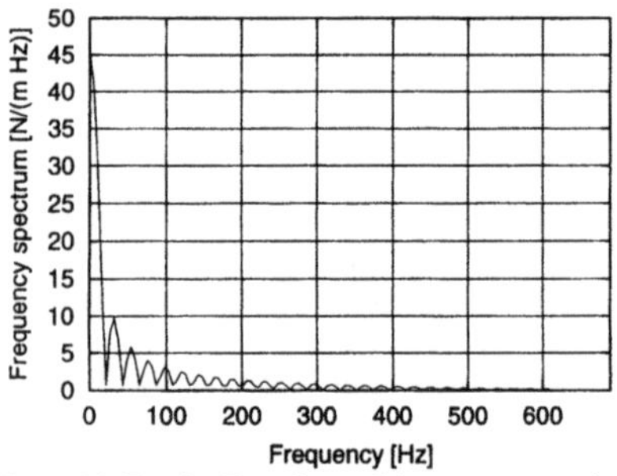

Figure 13: Fourier Transform of the square-wave impulse

Considering a moving load the time dependent response $u^k(t)$ at a selected point k on a track with length L is obtained by a superposition of the responses due to a load acting at discrete points i along the path. The time lag of the responses corresponds to the velocity of the moving load (see Fig. 15). The formula for the superposition is

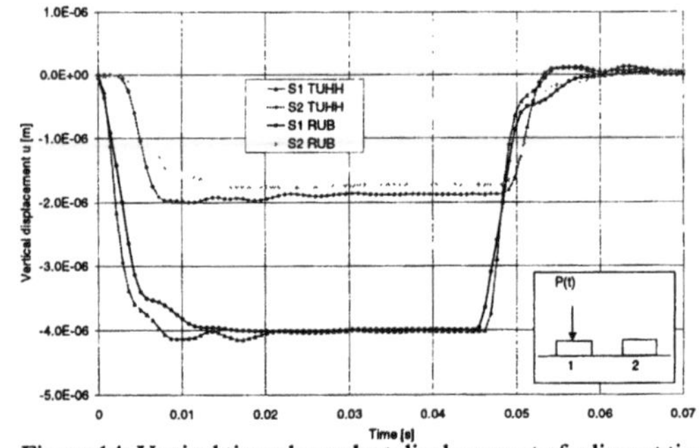

Figure 14: Vertical time-dependent displacement of adjacent ties (TUHH: time domain formulation, RUB: frequency domain formulation)

$$u^k(n\,\Delta t) = \sum_{i=1}^{n} u^i((n-i+1)\Delta t) . \tag{28}$$

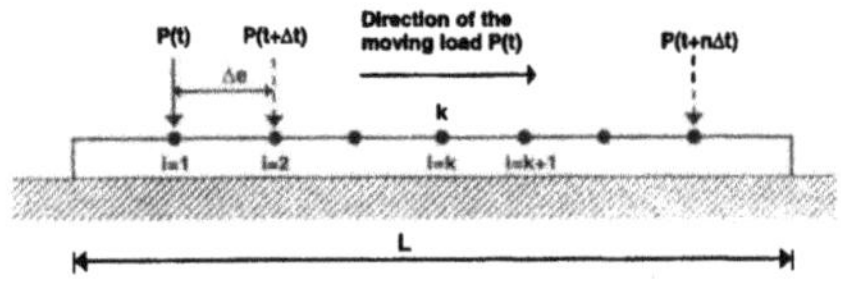

Figure 15: Numerical calculation due to a moving load

The distance Δe of two consecutive nodes depends on the velocity v of the load and the chosen time step Δt:

$$\Delta t = \frac{\Delta e}{v} . \tag{29}$$

The maximum frequency depends on the chosen time step given by

$$f_{max} = \frac{1}{2\Delta t} , \tag{30}$$

and the frequency step results from

$$\Delta f = \frac{1}{N\,\Delta t} . \tag{31}$$

In order to avoid long computing time, not every frequency in the required frequency range is computed. A well tested algorithm for a non-linear interpolation of substeps which is based on the response of a 2-degree of freedom system (Tajirian 1981) is implemented in the program. A comparison of the completely calculated frequency spectrum of a rigid track system with 41 frequencies and $\Delta f = 5\,Hz$ and an interpolated frequency spectrum based on only 15 calculated frequencies with $\Delta f = 15\,Hz$ is shown in Figure 16. The calculated frequency step of the interpolation procedure is $\Delta f = 1.6\,Hz$. Apart from a small difference in the lower frequency spectrum, the interpolated displacements show very good agreement with the results calculated with all 41 frequencies. Accordingly, the procedure is a suitable tool especially for a rigid track on the half-space whose dynamic behavior is approximately similar to a 2-degree of freedom system.

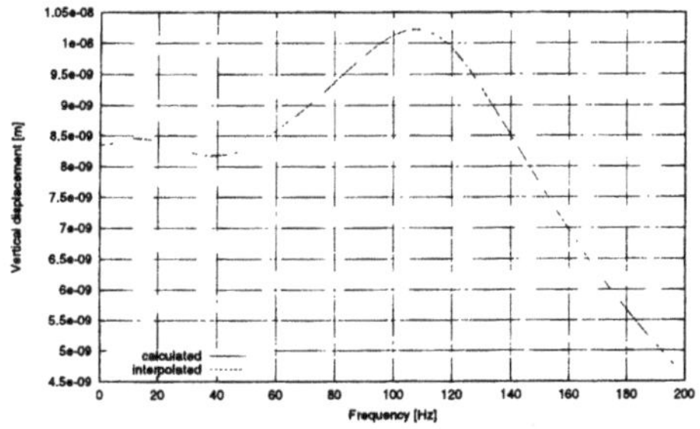

Figure 16: Comparison of calculated and interpolated frequency spectrum

The algorithm presented in this chapter is applied to a rigid track resting on a half-space, where the concrete layer, the hydraulically consolidated layer and the frost protection layer are modelled in a simplified manner as one homogeneous layer with constant width as shown in Figure 17. The material properties considered for the track and the varying soil are given in Table 4. The first calculation of the displacements

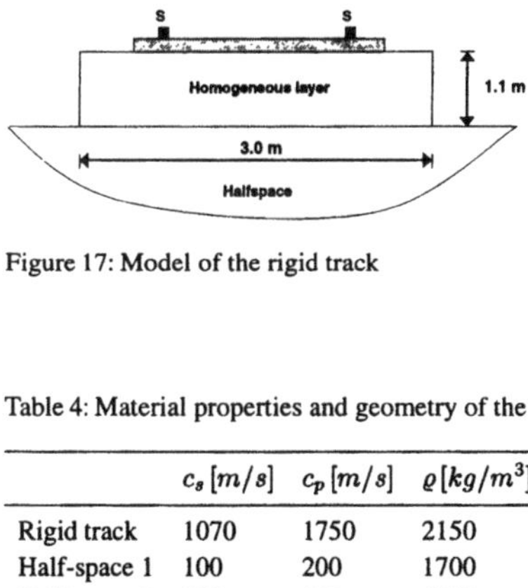

Figure 17: Model of the rigid track

Table 4: Material properties and geometry of the model

	$c_s\,[m/s]$	$c_p\,[m/s]$	$\varrho\,[kg/m^3]$
Rigid track	1070	1750	2150
Half-space 1	100	200	1700
Half-space 2	250	500	1700
Half-space 3	500	1000	1700

of the concrete layer which is performed without rails, pads and ties is shown in Figure 18, while the second calculation shown in Figure 19 obtains the displacements at the rail due to the load acting along the rail. The parameters for the rail and the pads are taken from Table 2

In these calculations the pulse load results from a load velocity of 400 km/h (Fig. 11). It should be mentioned, however, that the program also allows the consideration of other pulse-type loads resulting from a superposition of the axle-loads of a moving train (Tosecký 2001). As shown in Section 3.2, the time

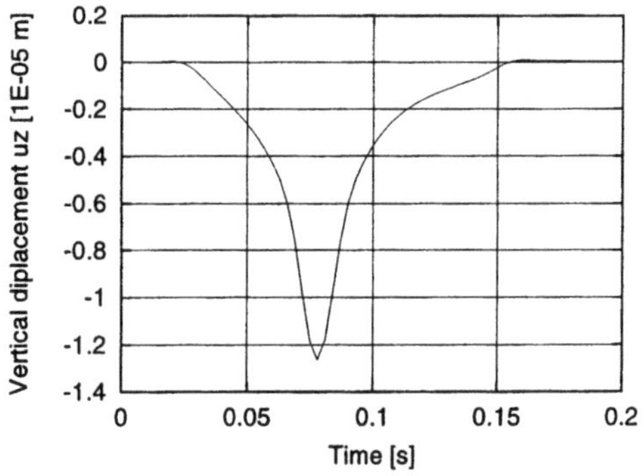

Figure 18: Vertical displacement of the concrete layer; moving load on concrete layer

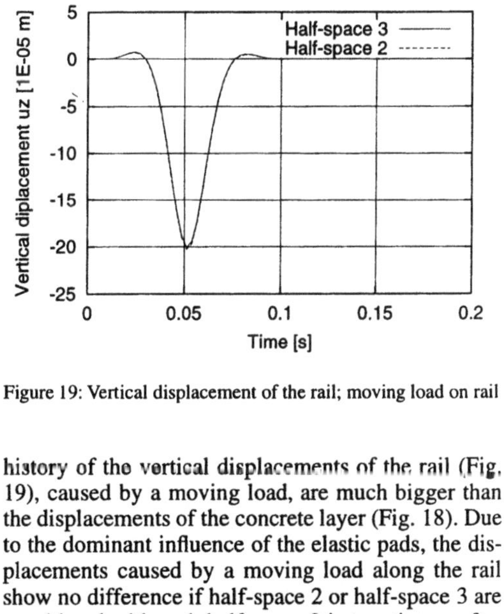

Figure 19: Vertical displacement of the rail; moving load on rail

history of the vertical displacements of the rail (Fig. 19), caused by a moving load, are much bigger than the displacements of the concrete layer (Fig. 18). Due to the dominant influence of the elastic pads, the displacements caused by a moving load along the rail show no difference if half-space 2 or half-space 3 are considered, although half-space 2 is two times softer than half-space 3 as given in Table 4.
Considering a moving load directly along the plate, the various soils effect big differences in the vertical displacements of the concrete layer as shown in Figure 20. More details on this calculation are given in Pflanz et al. (2000) and Schmid et al. (2001).
Earlier results on the investigation of the influence of moving loads on a rigid track as well as directly on a half-space are presented in (Schmid & Verbic 1997, Verbic et al. 1997, Friedrich et al. 1999, Tosecký

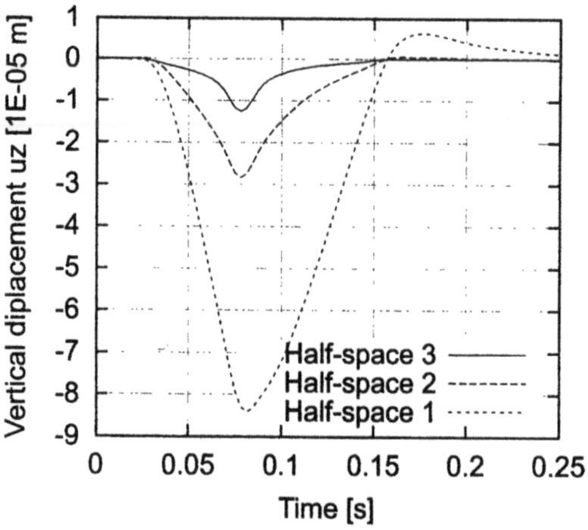

Figure 20: Influence of different soils on the vertical displacements of the rigid track due to a moving load

2001). A semi-analytical algorithm for an investigation of a rigid track subjected to a harmonic moving load is shown in Dinkel (2000).

4 NUMERICAL RESULTS IN THE TIME DOMAIN

According to the development steps of the new model, the numerical examples documented next are distinguished in two- and three-dimensional investigations.

4.1 *Two-Dimensional Systems*

The two-dimensional model provides a very helpful approach to gain a first idea of the dynamical behavior of a complicated track system. It can be used under the assumption that the loading condition and the system do not change in one direction (perpendicular to the investigated plane).

The example deals with the nonlinear behavior of a railway track on a halfspace. The system consists of three subregions: the sleepers (region I, discretized with finite elements), the ballast including the subgrade (region II, discretized with finite elements), and the undisturbed soil (region III, discretized with finite elements and with boundary elements). The complete geometry and its discretization can be found in Figure 21. In particular, also the parameters needed afterwards for the different material laws are depicted in that figure. Throughout the study, the sleepers (region I) are represented by a Neo-Hooke material, while in the case of the ballast and the subgrade (region II) an elastoplastic material with isotropic and kinematic hardening is used as proposed by von Mises (see, e.g. Chen (1994)). Using special finite elements for parts of the undisturbed soil (region III, FE model), also the investigation of local damages is possible. In general, this can be done by defining so-called "damage functions", which are able to take into account, for in-

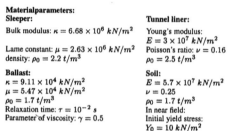

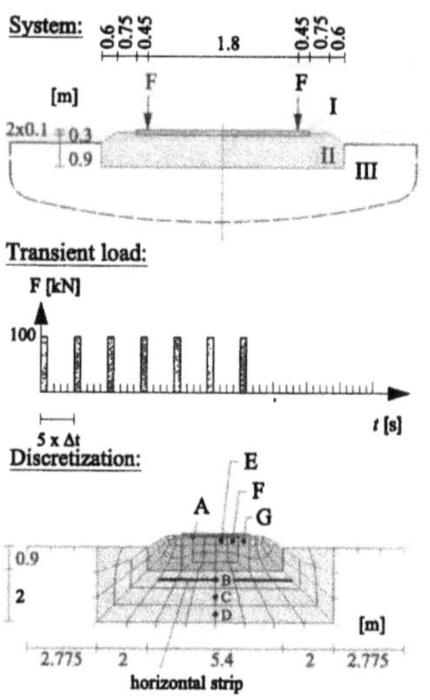

Materialparameters:

Sleeper:
Bulk modulus: $\kappa = 6.68 \times 10^6\ kN/m^2$
Lame constant: $\mu = 2.63 \times 10^6\ kN/m^2$
density: $\rho_0 = 2.2\ t/m^3$

Ballast:
$\kappa = 9.11 \times 10^4\ kN/m^2$
$\mu = 5.47 \times 10^4\ kN/m^2$
$\rho_0 = 1.7\ t/m^3$
Relaxation time: $\tau = 10^{-2}\ s$
Parameter of viscosity: $\gamma = 0.5$

Tunnel liner:
Young's modulus: $E = 3 \times 10^7\ kN/m^2$
Poisson's ratio: $\nu = 0.16$
$\rho_0 = 2.5\ t/m^3$

Soil:
$E = 5.7 \times 10^7\ kN/m^2$
$\nu = 0.25$
$\rho_0 = 1.7\ t/m^3$
In near field:
Initial yield stress:
$Y_0 = 10\ kN/m^2$

Figure 21: Railway track on a halfspace

stance, a stiffness reduction of a given material. In this example, the damage model given by Simo (1987) is used. It is based on an elastic model which is extended by a damage function depending on two additional material parameters α and β. For more details see Simo (1987) and von Estorff & Firuziaan (2000a). The remaining part of the soil (region III, BE model) is assumed to be linear elastic.

The system is subjected to two vertical loads acting in time as a sequence of impulses of $P = 100\ kN$, which each has a duration of one Δt (Fig. 21). The size of the time step Δt has been chosen such that the pressure wave in the undisturbed soil is able to cross one element of the BE discretization within one step.

Figure 22 shows the time history of the vertical displacements at point A located below the sleeper (see Fig. 21). Different combinations of the hardening parameters H_{iso} (isotropic) and H_{kin} (kinematic) are considered. The maximum displacement occurs for $H_{iso} = 0$. For this case also a positive displacement can be observed, which means that some tension occurs in the system (i.e., the model is not valid anymore). Generally the displacements reduce to a constant value as soon as the load is set to zero. To investigate the distribution of the damage inside the soil material, a horizontal strip directly underneath the ballast is considered (see Fig. 21). The material parame-

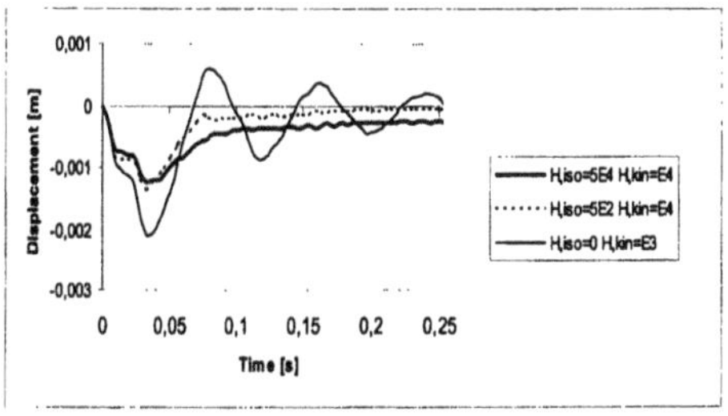

Figure 22: Time history of the vertical displacements at point A.

ters of the ballast are assumed to be $Y_0 = 10\ kN/m^2$ (initial yield stress), $H_{iso} = 5.10^5\ kN/m^2$ and $H_{iso} = 104\ kN/m^2$. In Figure 23 the distribution of the damage after 180 load pulses is given. As expected, the

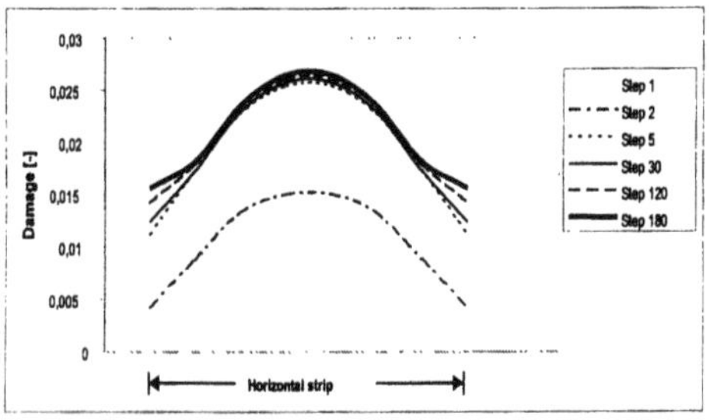

Figure 23: Distribution of the damage in a horizontal strip directly underneath the ballast at different time steps.

maximum occurs at the center of the considered strip and it may be observed that the damage of the soil is converging to a final maximum value. Looking at the time histories of the damage at the points B, C and D, which are depicted in Figure 24, the same convergence effect can be observed.

Finally, the yielding behavior of the ballast is investigated. As before, the micro stress, qH_{iso} resulting from the isotropic hardening, is considered. Figure 25 shows, at the points E, F and G (see Fig. 21), the dependence of this parameter from the number of load pulses. It can be seen that also in this case the harden-

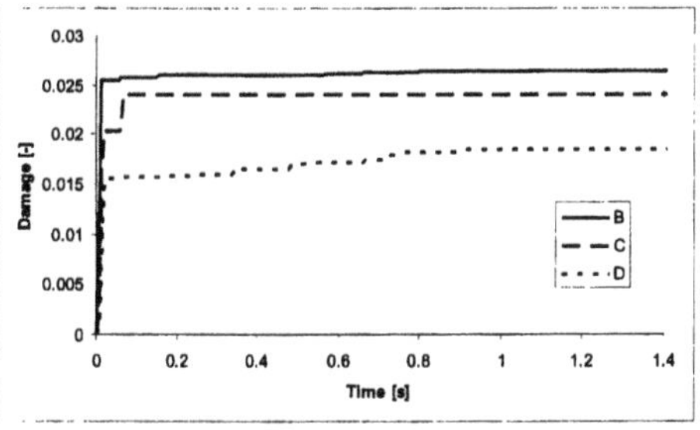

Figure 24: Time history of the damage at points B, C and D (see Fig. 21).

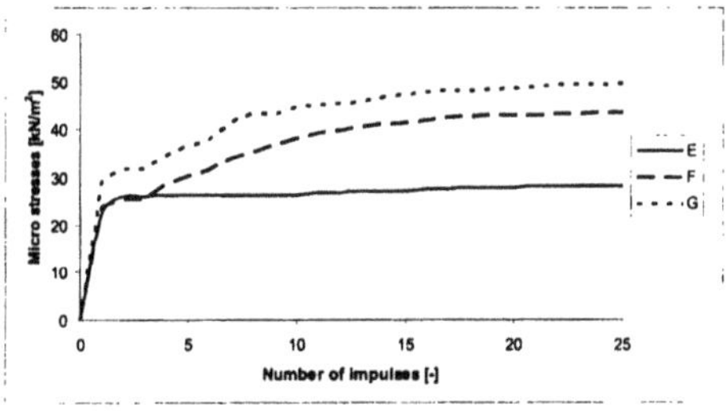

Figure 25: Micro stresses (isotropic hardening) at points E, F and G for a different number of impulses.

ing is approaching a final maximum value. Once this value is reached, additional load pulses do not lead to further changes.

4.2 *Three-dimensional systems*

In many cases, e.g. if a moving train is considered, a three-dimensional element model of the track system is needed (see section 3). Of course, due to limitations arising from the available computer power, only a small part of the complete track can be discretized. Usually, a limited number of sleepers is considered, mainly in order to take into account local three-dimensional effects. Two representative examples are investigated: a rigid sleeper on a halfspace and on a more sophisticated soil model.

An additional example has been discussed in section 3 already.

4.2.1 *Rigid sleeper on an elastic halfspace*

Consider the three-dimensional system given in Figure 26. It consists of a rigid sleeper of 20 × 26 × 260 [cm], $\rho = 2145\ kg/m^3$, which is placed on an elastic halfspace, whose material properties are $E = 1.7 \times 10^5\ kN/m^2$, $\nu = 0,25$, and $\rho = 1700\ kg/m^3$. While the sleeper is modeled with 2 × 2 × 20 quadratic finite elements, the halfspace is represented by 28 × 10 quadratic boundary elements. The system is loaded by two vertical forces located at the position of the rails. They act in time as a rectangular pulse over a time duration of 50 time steps $\Delta t = 0.00065s$.

In order to verify the transient model described above, its results (TUHH) are compared to solutions calculated by the frequency domain approach (RUB) and by the thin-layer method (TUB) described in Kausel (1994) and extended by Bode (2000) and Hirschauer (2001). Figure 26 shows the transient vertical displacement of the rigid sleeper. In the loading phase, the system reaches its static state after about 0.012 s and returns to the initally undeformed situation again after unloading. It can be observed that the results obtained with the three different models agree very well.

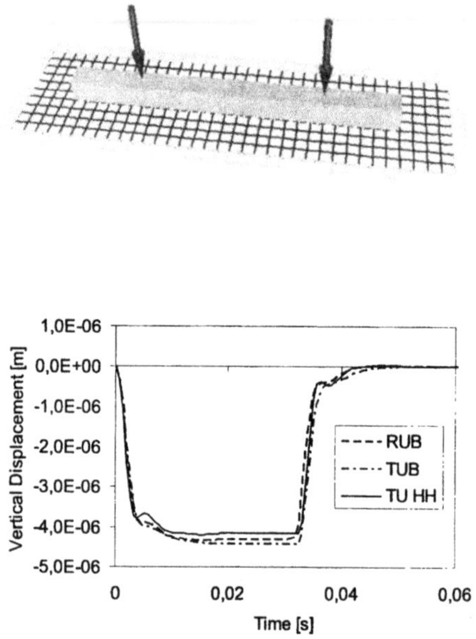

Figure 26: Discretization of the rigid sleeper on a halfspace (top) and a comparison of different computational models (bottom).

It should be pointed out that the different methodologies also have been compared to measurements. The respective results are discussed by Plenge & Lammering (2002) and shall not be repeated here.

4.2.2 *Rigid Sleeper on a nonlinear soil model.*

In order to demonstrate the applicability of this approach in the case of a more realistic soil model, a rigid sleeper of $20 \times 25 \times 250$ [cm], $\rho = 2145\ kg/m^3$, is placed on a halfspace as before. However, the soil in the vicinity of the sleeper, in a subdomain of $75 \times 75 \times 300$ [cm], is modeled using an elastic-plastic material law with a yield surface suggested by Chen (1994). The effects of a kinematic hardening are also included in this model. The soil outside the subdomain underneath the sleeper, i.e. the halfspace, is assumed to behave linearly elastic. Its material parameters are the same as in the benchmark example ($E = 1.7 \times 10^5\ kN/m^2$, $\nu = 0,25$). Within the subdomain, the elastic-plastic model necessitates additional material parameters. These are the cohesion, $c = 10.5\ kN/m^2$, the angel of internal friction, $\phi = 45$, and a hardening parameter, $H = 100\ kN/m^2$. The discretization of the FEM and BEM sub-regions are depicted in Figure 27. The sleeper is discretized with 10 quadratic finite elements, the soil material in the near field is modeled with $3 \times 3 \times 12$ quadratic finite elements, and the remaining soil is represented by 252 quadratic boundary elements. The system is loaded, as before by two vertical forces located at the position of the rails. They act in time as four rectangular pulses of a time duration of $2 \times \Delta t$ each with a period of $10 \times \Delta t$, where $\Delta t = 7.216 \times 10^{-4}\ s$. The amplitude of the load is $P = 50\ kN$.

Figure 28 (top) shows the vertical displacement of the sleeper due to the four impulses. The remaining displacement after unloading can be seen in Figure 28 (bottom). The growth of the plastic strain is given in Figure 29, where A is a point directly underneath the sleeper below the acting force. The points B and C are located on the common interface between the FE- and the BE-mesh: C vertically below A, and B below the center of the sleeper. In all points, the influence of the four pulses can be observed.

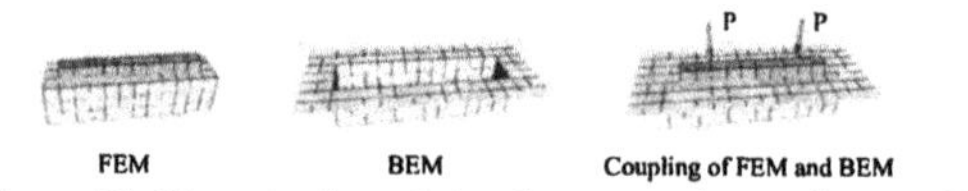

Figure 27: Discretization of the sleeper on the nonlinear soil model.

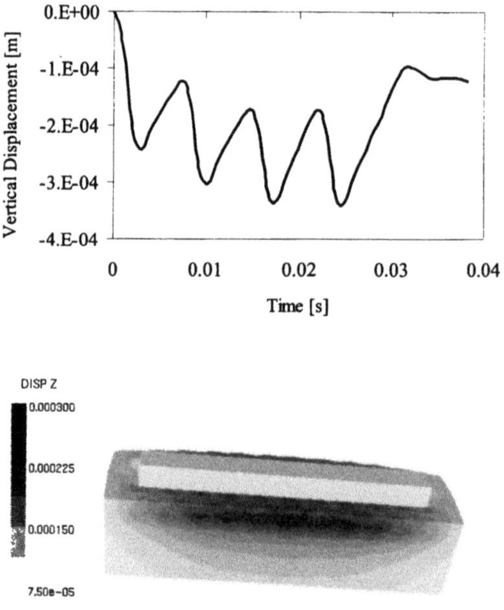

Figure 28: Transient vertical displacement of the sleeper (top) and the remaining soil deformations after unloading (bottom).

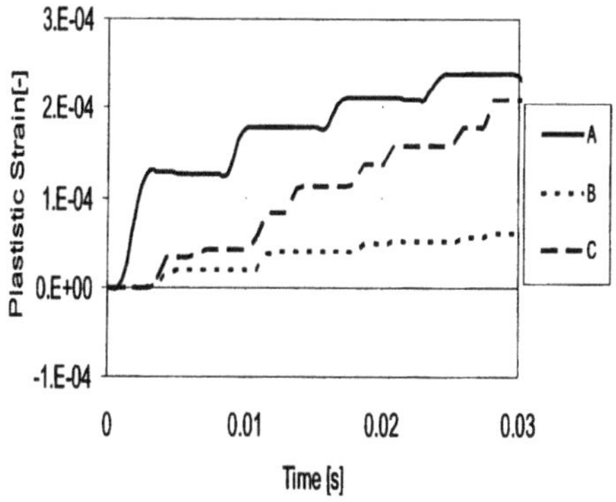

Figure 29: Transient development of the plastic strain of three representative points A, B and C.

5 SUMMARY AND CONCLUSIONS

The aim of this paper was the introduction of an analysis procedure using the finite element method and the boundary element method to satisfy the various boundary and interface conditions of track, substructure and layered soil. The coupling of both methods proved to be an appropriate technique for the investigation of the dynamic behavior of the interaction between railway track systems and the subsoil. The analysis has been performed with harmonic as well as with transient loads. Moreover, moving loads have been considered by a frequency-time-domain transformation of the system displacements due to load pulses resulting from the train passage.

The presented results in the frequency domain show the importance of a sufficient discretization of the track model. The investigation of different track systems demonstrate that a simplified model of the subsoil is adequate for the calculation of the dynamic behavior of a rigid track. Due to the stiffness of the superstructure the load is distributed along the rigid track and only small stresses are induced into the underlying soil. The most significant part of the system displacements are induced by the railpad. Relatively soft railpads cause high displacement amplitudes of the rails whereas at the concrete layer only 10 % of the rail displacement occur. Considering moving loads on a rigid track, the results show that the maximum displacements of the rails are mostly independent of the stiffness of the subsoil, whereas the dynamic behavior of the concrete layer resulting from a load moving directly along this layer is determined by the properties of the underlying soil.

The FEM/BEM formulation directly in the time domain allows the investigation of nonlinear effects. Large deformations and nonlinear material laws can be taken into account in the FE subsystems, while the energy absorption due to the expansion of the soil is taken care of by the boundary elements. Two- and three-dimensional examples demonstrated the flexibility and applicability of the transient formulation as well as the importance of nonlinear influences.

An extensive comparison of the presented results with other numerical procedures within a benchmark test has been documented by Rücker et al. (2002). The report shows the usefullness of the presented developments with respect to generality, flexibility and accuracy.

ACKNOWLEDGEMENT. The financial support by the Deutsche Forschungsgemeinschaft through the project Schm 546/9-1-3 and ES2-2 is greatfully acknowledged. The authors would like to thank the DFG.

REFERENCES

Becker, A. (1992). *The boundary element method in engineering*. Berkshire: McGraw-Hill.

Bode, C. (2000). *Numerisches Verfahren zur Berechnung von Baugrund-Bauwerk-Interaktionen im Zeitbereich mittels GREEN-scher Funktionen für den Halbraum*. Ph. D. thesis, Veröffentlichungen des Grundbauinstitutes der Technischen Universität Berlin.

Bode, C., R. Hirschauer, & S. Savidis (2000). Three-dimensional time domain analysis of moving loads on railway tracks on layered soils. In N. Chouw & G. Schmid (Eds.), *Wave 2000, Proc. of the International Workshop Wave 2000 Bochum, Germany, 13-15 December, 2000*, pp. 3–12. Rotterdam: Balkema.

Brigham, E. (1995). *FFT Schnelle Fourier-Transformation*. 6. Auflage. München: R. Oldenburg Verlag.

Chen, W. (1994). *Constitutive Equations for Engineering Materials*, Volume 2: Plasticity and Modeling. Elsevier, Amsterdam.

Dinkel, J. (2000). *Ein semi-analytisches Modell zur dynamischen Berechnung des gekoppelten Systems Fahrzeug-Fahrweg-Untergrund für das Oberbausystem Feste Fahrbahn*. Ph. D. thesis, Technische Universität München.

Dominguez, J. (1993). *Boundary elements indynamics*. Southampton, Boston: Computational Mechanics Publications.

Friedrich, K., G. Pflanz, & G. Schmid (1999). Modellierung des dynamischen Verhaltens der Festen Fahrbahn bei Zugüberfahrt. Siegmann, J. In *Feste Fahrbahn - Mechanische Modellierung, Betriebserfahrung und Akustik, 2.*

Auflage, Interdisziplinärer Forschungsverbund Bahntechnik IFV, Symposium 4.-5. Nov. 1999, Berlin.

Grassie, S., R. Gregory, D. Harrison, & K. Johnson (1982). The dynamic response of railway track to high frequency vertical excitation. *J Mech Eng Sci 24 No 2*, 77–90.

Hirschauer, R. (2001). *Kopplung von Finiten Elementen mit Rand-Elementen zur Berechnung der dynamischen Baugrund-Bauwerk-Interaktion*. Ph. D. thesis, Veröffentlichungen des Grundbauinstitutes der Technischen Universität Berlin.

Hubert, W., K. Friedrich, G. Pflanz, & G. Schmid (2000). Frequency- and time-domain BEM analysis of rigid track andhalf-space with vibration barriers under investigation ofcausality errors for concave domains. In *EUROMECH Colloquium 414Boundary Element Methods for Soil/Structure Interaction*, Catania, Italy.

Kausel, E. (1994). Thin-layer-method: Formulation in the time domain. *Int. J. for Num. Methods in Eng. 37*, 927–941.

Knothe, K. (2001). *Gleisdynamik*. Berlin: Ernst & Sohn.

Lefeuve-Mesgouez, G. (1999). *Propagation dondes dans un massifsoumis à des charges se déplacant àvitesse constante*. Ph. D. thesis, Ecole Centrale de Nantes.

Mohammadi, M. & D. Karabalis (1995). Dynamic 3-d soil-railway track interaction by BEM-FEM. *Earthquake engineering and structural dynamics 24*, 1177–1193.

Müller-Boruttau, F. & N. Breitsamter (2000). Elastische Elemente verringern die Fahrwegbeanspruchung. *ETR Eisenbahntechnische Rundschau 49, Heft 9, 2001*, 587–596.

Petyt, M. & C. Jones (1999). Modelling of groundbourne vibration from railways. In Fryba & Naprstek (Eds.), *Structural Dynamics, EURODYN 1999*, Volume I. Balkema: Rotterdam.

Pflanz, G. (2001). *Numerische Untersuchung der elastischen Wellenausbreitung infolge bewegter Lasten mittels der Randelementmethode im Zeitbereich*. Ph. D. thesis, VDI Fortschritt-Berichte Reihe 18, Nr. 265, Düsseldorf: VDI Verlag.

Pflanz, G., J. Garcia, & G. Schmid (2000). Vibrations due to loads moving with sub-critical and super-critical velocities on rigid track. In N. Chouw & G. Schmid (Eds.), *Wave 2000, Proc. of the International Workshop Wave 2000 Bochum, Germany, 13-15 December, 2000*, pp. 131–148. Rotterdam: Balkema.

Plenge, M. & R. Lammering (2002). The dynamics of railway track and subgrade with respect to deteriorated sleeper support. In K. Popp & W. Schiehlen (Eds.), *System Dynamics and Long-Term Behaviour of Railway Vehicles, Track and Subgrade*. Berlin: Springer-Verlag.

Przemieniecki, J. (1968). *Theory of matrix structural analysis*. New York: McGraw-Hill.

Read, D., N. Matsumoto, & H. Wakui (1999). Fast testing japanese-developed ladder sleeper system. *RT&S*, 16–18.

Ripke, B. (1995). *Hochfrequente Gleismodellierung und Simulation der Fahrzeug-Gleis-Dynamik unter Verwendung einer nichtlinearen Kontaktmechanik*. Ph. D. thesis, VDI Fortschritt-Berichte Reihe 12, Nr. 249, Düsseldorf: VDI Verlag.

Rücker, W. (1981). *Dynamische Wechselwirkung eines Schienen-Schwellensystems mit dem Untergrund*. Forschungsbericht 78, Bundesanstalt für Materialprüfung, Berlin.

Rücker, W. (1999). Kurzzeitdynamik und Setzungsverhalten der Festen Fahrbahn. Siegmann, J. In *Feste Fahrbahn - Mechanische Modellierung, Betriebserfahrung und Akustik, 2. Auflage, Interdisziplinärer Forschungsverbund Bahntechnik IFV, Symposium 4.-5. Nov. 1999*, Berlin.

Rücker, W., L. Auersch, & M. Baessler (2002). A Comparative Study of Results from Numerical Track-Subsoil Calculations. In K. Popp & W. Schiehlen (Eds.), *System Dynamics and Long-Term Behaviour of Railway Vehicles, Track and Subgrade*. Berlin: Springer-Verlag.

Savidis, S., S. Bergmann, C. Bode, & R. Hirschauer (1999). Dynamische Wechselwirkung zwischen der Festen Fahrbahn und dem geschichteten Untergrund. Siegmann, J. In *Feste Fahrbahn - Mechanische Modellierung, Betriebserfahrung und Akustik, 2. Auflage, Interdisziplinärer Forschungsverbund Bahntechnik IFV, Symposium 4.-5. Nov. 1999*, Berlin.

Schmid, G., G. Pflanz, K. Friedrich, J. Garcia, & W. Hubert (2001). Dynamic soil-structure interaction for moving load - methods and numerical application. In *Proc. of 2nd ICTACEM at Kharagpur, India, Dec. 27-30, 2001*. Indian Institute of Technology, Paper No. 113.

Schmid, G. & B. Verbic (1997). Modellierung der Erschütterung aus dem Schienenverkehr mit der Randelementmethode. In H. Bachmann (Ed.), *Erdbebensicherung bestehender Bauwerke und aktuelle Fragen der Baudynamik, Tagungsband D-A-CH'97, SIA, Dokumentation DO145, 1997*.

Simo, J. (1987). On a fully three-dimensional finite-strain viscoelastic damage model: Formulation and computational aspects. *Computer Methods in Applied Mechanics and Engineering 60*, 153–173.

Tajirian, F. (1981). *Impedance matrices and interpolation techniques for 3-D interaction analysis by the flexible volume method.* Ph. D. thesis, University of California. Berkeley.

Tosecký, A. (2001). *Numerische Untersuchung der Erschütterungsausbreitung infolge bewegter Lasten auf einem Feste Fahrbahn-System mittels der Methode der dünnen Schichten/Methode der flexiblen Volumen.* Diploma thesis, Ruhr-Universität Bochum, Bochum 2001.

Verbic, B., G. Schmid, & H. Köpper (1997). Anwendung der Randelementmethode zur dynamischen Berechnung der Festen Fahrbahn. *EI - Der Eisenbahningenieur, Februar 1997.*

von Estorff, O. & M. Firuziaan (2000a). Coupled BEM/FEM approach for nonlinear soil/structure interaction. *Engineering Analysis with Boundary Elements 24, 10,* 715–725.

von Estorff, O. & M. Firuziaan (2000b). FEM and BEM for nonlinear soil/structure interaction analyses. In N. Chouw & G. Schmid (Eds.), *Wave 2000, Proc. of the International Workshop Wave 2000 Bochum, Germany, 13-15 December, 2000*, pp. 357–368. Rotterdam: Balkema.

von Estorff, O. & M. Prabucki (1990). Dynamic response in the time domain by coupled boundary and finite elements. *Computational Mechanics 6*, 35–46.

Wave propagation – Moving load – Vibration reduction, Chouw & Schmid (eds.)
© 2003 Swets & Zeitlinger, Lisse, ISBN 90 5809 559 2

Instability of vibrations of a moving vehicle on an elastic structure

A.V. Metrikine
Delft University of Technology, Delft, The Netherlands

ABSTRACT: Stability of coupled vibrations is investigated of a moving vehicle and an elastic structure. It is shown that these vibrations may become unstable if the vehicle's velocity exceeds the minimum phase velocity of elastic waves in the structure. Instability occurs because of the dynamic reaction of the structure to the moving vehicle. At high velocities, this reaction is similar to that of a dashpot with a negative damping coefficient. Physically, such reaction is caused by anomalous Doppler waves that destabilize vibrations of the vehicle by transferring the energy of its translational motion into the vibrational energy. It is shown that if the track is built on a soft soil, instability of a train's bogie may occur at speeds that are easily achievable for modern high-speed trains. Effect of periodic inhomogeneity of the railway track on vehicle's stability is considered showing possibility of parametric resonance at velocities that are much lower than those that lead to instability in the homogeneous case.

1 INTRODUCTION

Modern high-speed trains can move with a speed that is comparable with or even exceeds the speed of elastic waves in a railway track. Accordingly, a high-speed train is able to reach a *critical velocity* and thereby cause a pronounced amplification of the dynamic response of the railway track, see Filippov (1961), Labra (1975), Krylov (1995), Dieterman & Metrikine (1996, 1997a, 1997b), Suiker et al. (1998), Grundmann et al. (1999), Sheng et al. (1999), Kaynia et al. (2000), Kruse & Popp (2001). The physical phenomenon lying behind this amplification is *resonance*. It is worth mentioning that the critical velocities depend not only on the physical parameters of the railway structure but also on the elastic, inertial and viscous properties of the train, Dean (1990), Vesnitskii (2001).

The present paper is devoted to a less known phenomenon which may cause dynamic amplification of vibrations of a railway structure. This phenomenon is usually referred to as *instability* and for the first time was independently described by Bogacz (1983) and Denisov et al. (1985). The instability contains in the exponential increase of vibrations of the "train - railway track" system and is caused by interaction of the train and the railway track. The energy for this increase is transferred from the kinetic energy of the translational motion of the train through the contact points between the train's wheels and the rails, see Metrikine (1994), Metrikine & Popp (1999), Metrikine & Verichev (2001). This energy transfer is accomplished during radiation of *anomalous Doppler waves* (more information on these waves and their role in physics is available in Ginzburg (1979)), which cause the railway track react to the train as if a system of dashpots with negative damping coefficients were put under the contact points.

There are two crucial differences between *resonance* and *instability*, see Metrikine & Dieterman (1997). First, instability takes place in a range of the train velocities, while resonance occurs at certain individual velocities. Second, in contrast to resonance, the amplitude of the unstable vibrations grows in time not linearly but exponentially. The latter fact provides a significant difference in the effect of damping on resonance and instability. While resonance can be totally removed by increasing the damping, the instability domains are only moved in the space of structural parameters. Evidently, the aforementioned features of instability make this phenomenon even more unfavorable in practice than resonance.

The current paper is structured in the following way. Firstly, the instability phenomenon in mechanics is defined on the hand of a simple model that consists of Euler-Bernoulli beam on Winkler's foundation and a point mass m that moves on the beam. By definition, instability occurs if the characteristic equation for vertical vibrations of the mass moving on the beam has a root with a positive real part. Considering the dynamic stiffness of the beam in the contact point, it is shown that the necessary condition of instability is that the velocity of the mass exceeds the minimum phase velocity of waves in the beam. If this necessary condition is satisfied, the beam's reaction in the contact point can become similar to that of a dashpot with a negative damping coefficient (in the frequency band $0<\omega<\omega^*$). This kind of reaction is caused by anomalous Doppler waves, which transfer the energy of translation motion of the mass along the beam into the energy of its vertical vibrations. This can be shown by using the energy-momentum variation laws for the mass-beam system, see Metrikine (1994).

Stability of a moving vehicle depends crucially on visco-elastic and inertial parameters of the vehicle itself. The effect of these parameters is demonstrated in the paper on the hand of the model that consists of a Timoshenko beam on visco-elastic foundation and a bogie that moves along the beam.

To demonstrate that the instability phenomenon can be of practical significance, a three-dimensional model for a railway track is considered. The model consists of a beam on a visco-elastic half-space and a two-mass oscillator that moves on the beam. It is shown that if the half-space, which is used to model the subsoil of a railway track is soft, then the velocities at which instability can occur are easily reachable for modern high-speed trains.

The last part of the paper deals with the effect of periodic inhomogeneity of a railway track on stability of moving vehicle. It is discussed that the inhomogeneity can cause parametric resonance at relatively low velocities of the mass. The zones of this resonance, however, are found to be narrow so that this kind of instability seems to be practically unimportant.

2 PHENOMENON OF INSTABILITY AND ITS PHYSICAL BACKGROUND

2.1 *Mathematical definition of instability*

Consider a mass that moves on an infinitely long, visco-elastically supported Timoshenko beam as depicted in Figure 1. Under the assumptions that the mass moves with a constant speed V, is subject to a constant vertical force P and remains in contact with the beam, governing equations for vertical vibrations of the system can be written as

$$
\begin{aligned}
&\rho F\frac{\partial^2 u}{\partial t^2}-\chi\mu F\frac{\partial^2 u}{\partial x^2}+\chi\mu F\frac{\partial\varphi}{\partial x}+k_f u+v_f\frac{\partial u}{\partial t}= \\
&\qquad\qquad =-\delta(x-Vt)m\frac{d^2u^0}{dt^2}, \\
&\rho I\frac{\partial^2\varphi}{\partial t^2}-IE\frac{\partial^2\varphi}{\partial x^2}+\chi\mu F\left(\varphi-\frac{\partial u}{\partial x}\right)=0, \\
&u^0(t)=u(Vt,t), \\
&\lim_{|x-Vt|\to\infty}|u(x,t)|<\infty,\quad u(x,0)=\frac{\partial u(x,0)}{\partial t}=u^0(0)=0
\end{aligned}
\tag{1}
$$

with $u(x,t)$ and $u^0(t)$ the vertical deflections of the beam and the mass m, respectively; $\varphi(x,t)$ the angle of rotation of the cross-section of the beam, E and μ the Young's modulus and the shear modulus of the beam material; ρ and I the mass density of the beam material and the moment of inertia of the beam cross-section; F the cross-sectional area of the beam, χ the Timoshenko factor; k_f and v_f the stiffness and the viscosity of the foundation per unit length and $\delta(\dots)$ the Dirac delta-function.

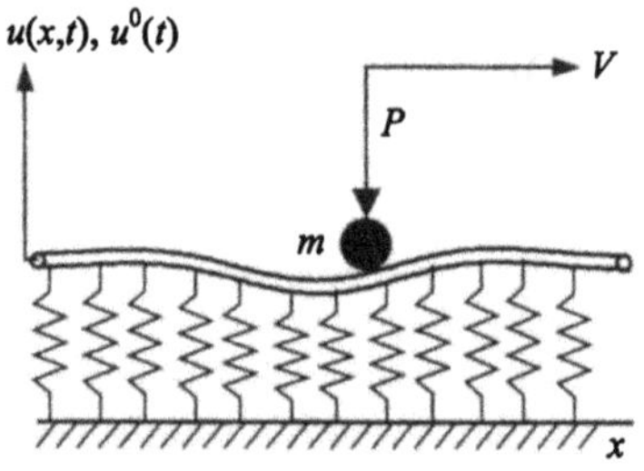

Figure 1. Moving mass on a flexibly supported beam.

Introducing in Eqs.(1) the moving reference system $\{\xi=x-Vt,\ \tau=t\}$ and applying integral Laplace and Fourier transforms, the following expressions can be found for deflections of the mass and the beam in the Laplace domain:

$$
\begin{aligned}
&\tilde{u}(\xi,s)=\frac{P}{2\pi s}\int_{-\infty}^{\infty}\frac{\exp(ik\xi)}{D(k,s)}dk\left(-1+\frac{ms^2}{ms^2+\chi^{eq}(s,V)}\right), \\
&\tilde{u}^0(s)=-\frac{P}{s\left(ms^2+\chi^{eq}(s,V)\right)}
\end{aligned}
\tag{2}
$$

with the dynamic stiffness χ^{eq} of the beam in the moving contact point given by

$$
\begin{aligned}
&\chi^{eq}(s,V)=\left(\frac{1}{2\pi}\int_{-\infty}^{\infty}\frac{dk}{D(k,s)}\right)^{-1}, \\
&D(k,s)=\frac{A(k,\omega)B(k,\omega)-(\chi\mu Fk)^2}{A(k,\omega)}, \\
&A(k,s)=\rho I(s-ikV)^2+IEk^2+\chi\mu F,
\end{aligned}
\tag{3}
$$

$$B(k,s)=\rho F(s-ikV)^2+\chi\mu Fk^2+\nu_f(s-ikV)+k_f$$

The mathematical procedure that leads to Eqs.(2) is carefully described in Metrikine & Dieterman (1997) and Metrikine & Verichev (2001).

Normally, analysing the dynamic response of the beam-mass system, researchers assume *a priori* that the steady-state regime of vibrations exists and the transient vibrations of the system caused by the initial conditions and the external force at the moment of its application decay in time. In other words, it is commonly assumed that natural vibrations of the system are *stable*. Consequently, in the expression for $u(\xi,s)$ in Eq.(2), the Laplace variable s is normally set to zero (because of the constant load) and, accordingly, the mass does not play any role in the steady-state solution of the problem. This assumption, however, is correct only in the case that the mass-beam natural vibrations are stable, which mathematically implies that the characteristic equation

$$ms^2+\chi^{eq}(s,V)=0 \qquad (4)$$

has no roots s^* with a positive real part.

Equation (4) can be easily studied for having roots with a positive real part by applying the D-decomposition method described in Neimark (1978), Denisov et al. (1985), Metrikine & Dieterman (1997) and Metrikine & Verichev (2001). Application of this method results in a curve that is depicted in Figure 2 and breaks the plane of parameters m,α (with $\alpha=V\,(\rho/\chi G)^{1/2}$ the dimensionless velocity) into two domains. The shaded domain corresponds to unstable vibrations.

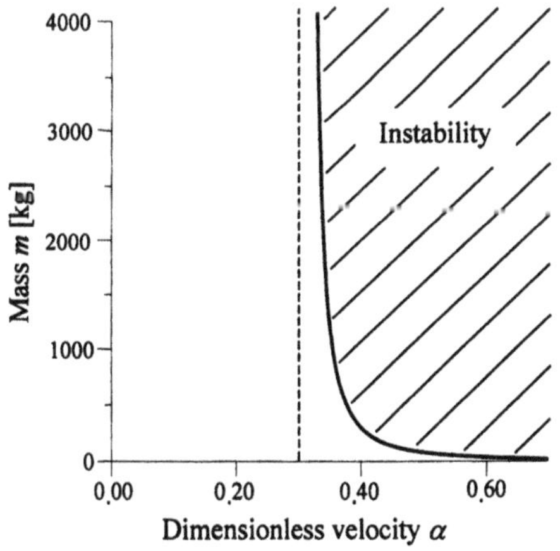

Figure 2. Instability domain for a mass on a flexibly supported Timoshenko beam.

To plot Figure 2 the following parameters of the beam and its foundation have been used:

$$\rho=7849[\text{kg}],\ F=7.687\cdot10^{-3}\left[\text{m}^2\right],$$
$$I=3.055\cdot10^{-5}\left[\text{m}^4\right],\chi=0.82,$$
$$E=2\cdot10^{11}\left[\text{N/m}^2\right],\mu=7.813\cdot10^{10}\left[\text{N/m}^2\right],$$
$$k_f=10^8\left[\text{N/m}^2\right],\ \nu_f=10^4\left[\text{Ns/m}^2\right] \qquad (5)$$

Figure 2 shows that instability occurs if the mass moves with a speed that exceeds a certain dimensionless critical velocity that is marked in the figure by the vertical dashed line. It can be shown that this velocity, as damping in the beam's foundation tends to zero, is given as

$$\alpha^*=\sqrt{\gamma-\gamma\kappa-2\kappa+2\kappa\sqrt{\gamma\kappa+1-\gamma}}\Big/(\kappa-1)$$

with

$$\gamma=c_p^{\,2}/c^2,\quad \kappa=\left(c^2F\right)/\left(\omega_0^2I\right),\quad c=\sqrt{\chi}\,c_S, \qquad (6)$$
$$c_p=\sqrt{E/\rho},\ c_S=\sqrt{\mu/\rho},\ \omega_0=\sqrt{k_f/(\rho F)}$$

and represents the minimum phase velocity of waves in the Timoshenko beam on the elastic foundation. For the parameters under consideration, $\alpha^*\approx0.31$. Increasing the damping coefficient of the foundation, one would obtain a larger magnitude of α^*.

2.2 *Physical background of instability*

To understand the reason that causes instability, it is worthwhile to study the dynamic stiffness χ^{eq} of the beam in the moving contact point as a function of frequency of vibrations of the mass and its velocity. Actually, it is sufficient to study the imaginary part of χ^{eq}, since it is this part that is responsible for attenuation or amplification of vibrations in time. If $\mathrm{Im}(\chi^{eq})$ were zero, natural vibrations of the mass on the beam would take place with a constant amplitude.

The imaginary part of χ^{eq} is plotted qualitatively in Figure 3 as a function of frequency $\omega=-is$.

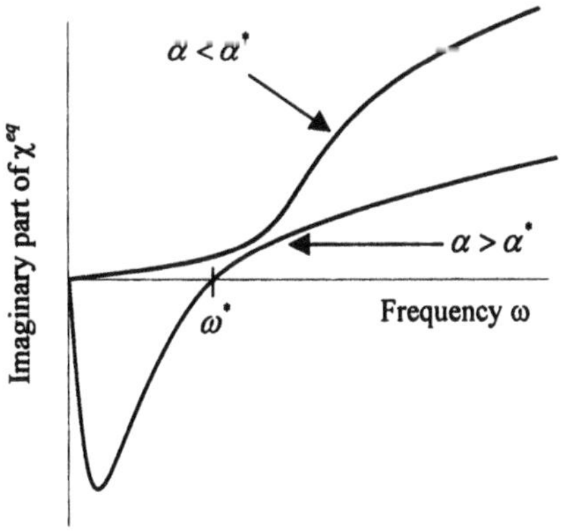

Figure 3. Imaginary part of χ^{eq} versus frequency.

The upper curve in Figure 3 corresponds with the sub-critical $\alpha < \alpha^*$ motion of the mass, whereas the lower curve is plotted for the super-critical case $\alpha > \alpha^*$. The crucial difference between these two curves is that in the super-critical case the imaginary part of the dynamic stiffness is negative in the low-frequency band $\omega < \omega^*$. Since $\omega = -is$, this implies that in this frequency band the reaction of the beam to the moving mass is similar to that of a dashpot with a negative damping coefficient. Physically this means, that the reaction of the beam is in-phase with vibrations of the mass, which leads to amplification of these vibrations. Let us underline, however, that this amplification takes place in the frequency band $\omega < \omega^*$, while at higher frequencies the beam reaction is in anti-phase with the mass vibrations and, therefore, attenuates these vibrations. Knowing only the dynamic stiffness, it is impossible to say which mechanism (amplification or attenuation) will prevail. This depends on the physical properties of the moving object. Thus, the fact that the imaginary part of the dynamic stiffness is negative in a frequency band can serve as a *necessary condition* of instability but not as its criterion. To determine parameters of the system "moving vehicle – elastic structure" which lead to instability, the characteristic equation has to be studied for having eigenvalues with a positive real part.

Figure 3 visualizes the qualitative difference between the beam reaction in the sub- and super-critical regimes of motion. It remains unclear, however, why there exists a critical frequency ω^*, which distinguishes two qualitatively different types of this reaction in the super-critical case. To understand the physical background of this frequency, waves in the beam have to be analyzed, which are radiated by the moving vehicle. As shown by Metrikine (1994) a super-critically moving vehicle can generate anomalous Doppler waves in the beam. These waves, having the phase velocity larger than the vehicle's velocity, have a very distinguished feature discovered by Ginzburg (1979) when he studied an electromagnetic system. These waves, instead of reducing the internal energy of a source that has radiated them by means of the radiation damping, increase this energy by transferring it from the energy of translational motion of the source. Thus, it is the anomalous Doppler waves that destabilize the system. Having understood this, it is easy to find out the physical meaning of the critical frequency ω^*. This is just the frequency of vibrations of the contact point, which, being exceeded prohibits radiation of anomalous Doppler waves. Having anomalous Doppler waves not radiated, the reaction of the beam remains always in anti-phase with the mass vibrations, thereby ensuring that no energy amplification is possible at this frequency band.

3 INSTABILITY OF A VEHICLE ON ELASTICALLY SUPPORTED BEAM

3.1 *Two-mass oscillator on a flexibly supported Timoshenko beam*

As mentioned in the previous section, stability of a moving vehicle on elastic structure depends not only on the physical parameters of the structure and the vehicle's velocity but also upon the physical parameters of the vehicle itself. To demonstrate this fact and to determine the most influential parameters of the vehicle, it is worthwhile to consider the model depicted in Figure 4.

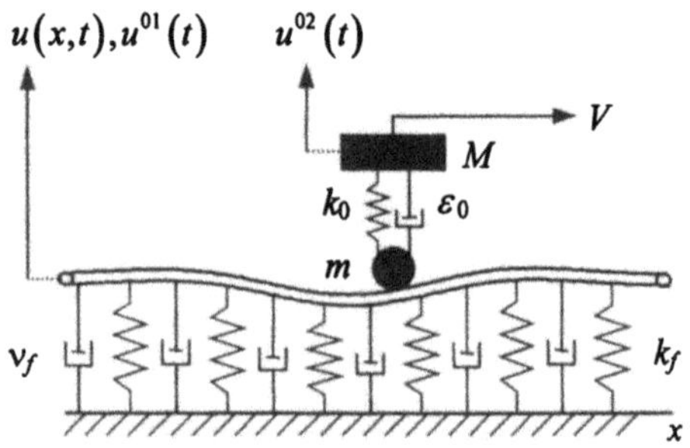

Figure 4. Moving two-mass oscillator on flexibly supported Timoshenko beam.

The model consists of a two-mass oscillator and a Timoshenko beam on a visco-elastic foundation, along which the former moves with a constant speed V. It is assumed that the lower mass m of the oscillator is permanently in contact with the beam. The upper mass M is connected to the lower mass by the spring k_0 and the dashpot ε_0. With these assumptions, the governing equations for vertical vibrations of the beam and oscillator read

$$\rho F \frac{\partial^2 u}{\partial t^2} - \chi\mu F \frac{\partial^2 u}{\partial x^2} + \chi\mu F \frac{\partial \varphi}{\partial x} + k_f u + v_f \frac{\partial u}{\partial t} = -$$
$$\delta(x - Vt)\left(m\frac{d^2 u^{01}}{dt^2} + k_0\left(u^{01} - u^{02}\right) + \varepsilon_0\left(\frac{du^{01}}{dt} - \frac{du^{02}}{dt}\right)\right),$$

$$\rho I \frac{\partial^2 \varphi}{\partial t^2} - IE\frac{\partial^2 \varphi}{\partial x^2} + \chi\mu F\left(\varphi - \frac{\partial u}{\partial x}\right) = 0,$$

$$M\frac{d^2 u^{02}}{dt^2} + k_0\left(u^{02} - u^{01}\right) + \varepsilon_0\left(\frac{du^{02}}{dt} - \frac{du^{01}}{dt}\right) = 0,$$

$$u^{01}(t) = u(Vt,t), \lim_{|x-Vt|\to\infty} u(x,t) = 0, \lim_{|x-Vt|\to\infty} \varphi(x,t) = 0,$$

$$u(x,0) = \frac{\partial u(x,0)}{\partial t} = u^{01}(0) = u^{02}(0) = 0$$

(7)

In Eqs.(7), the same notations as in Eqs.(1) are used and, additionally, u^{01} and u^{02} are the vertical deflections of the lower and the upper mass, respectively.

In accordance with the analysis presented by Metrikine & Verichev (2001), the characteristic equation for the vertical vibrations of the oscillator is given as

$$\left(ms^2 + k_0 + \varepsilon_0 s + \chi^{eq}(s,V)\right)\left(Ms^2 + k_0 + \varepsilon_0 s\right) - \left(k_0 + \varepsilon_0 s\right)^2 = 0. \tag{8}$$

with χ^{eq} defined by Eq.(3). Obviously, Eq.(8) is the characteristic equation for the two-mass oscillator, the lower mass of which is supported by an equivalent lump element with the stiffness χ^{eq}.

The curves, which break the plane "dimensionless velocity of the oscillator – stiffness of the oscillator" into the stable and unstable domains (the latter is above the curves) are shown for two magnitudes of the lower mass of the oscillator. The upper mass and the damping in the oscillator are set as

$$M = 2.10^4\,[\text{kg}],\ \varepsilon_0 = 10^3\,[\text{Ns/m}] \tag{9}$$

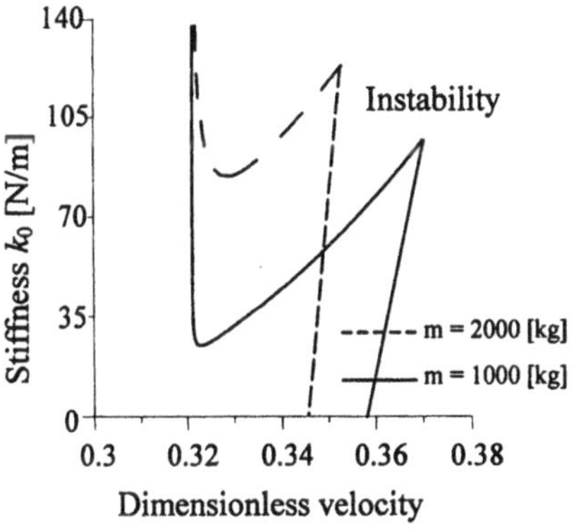

Figure 5. Separation of the plane "stiffness-velocity" into the stable and unstable domains for two-mass oscillator.

Figure 5 shows that there is an interval of the oscillator's velocities, in which vibrations of the system can be either stable or unstable ($0.32 < \alpha < 0.37$). Stabilization can be achieved by lowering the oscillator stiffness.

One can also see from Figure 5 that the increase of the lower mass has a stabilizing effect on the system (this could change if the lower mass would be chosen greater than the upper one). As to the other physical parameters of the oscillator, it can be shown that the most influential one is the viscosity of the oscillator. Even a small increase of this viscosity stabilizes the system perceptibly by shifting the instability domain in Figure 5 towards higher values of k_0, see Metrikine & Verichev (2001).

3.2 *Bogie on a flexibly supported Timoshenko beam*

Two-mass oscillator is quite a simplistic model for a moving vehicle. To enhance it, let us consider a model that is depicted in Figure 6. This model is composed of a rigid bar and two identical supports that connect the bar (all together, these elements may be referred to as a bogie) and a Timoshenko beam on visco-elastic foundation.

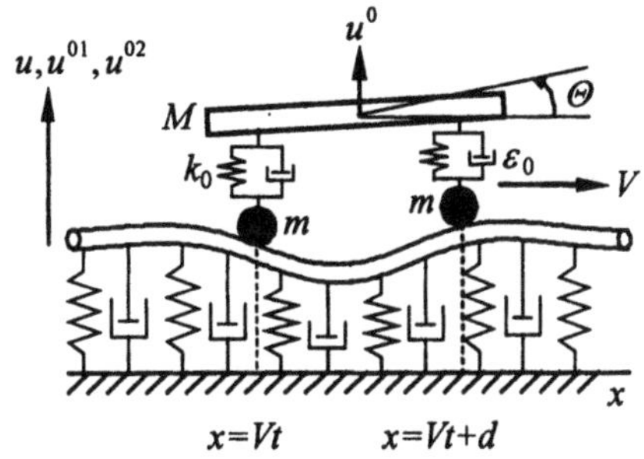

Figure 6. Moving bogie on flexibly supported Timoshenko beam.

With the same assumptions as for the oscillator (constant speed and permanent contact between the lower masses and the beam), the governing equations for the model may be written as

$$\rho F \frac{\partial^2 u}{\partial t^2} - \chi\mu F \frac{\partial^2 u}{\partial x^2} + \chi\mu F \frac{\partial \varphi}{\partial x} + k_f u + v_f \frac{\partial u}{\partial t} = -$$
$$\delta(x - Vt - d)\left(m\frac{d^2u^{01}}{dt^2} + \left(k_0 + \varepsilon_0 \frac{d}{dt}\right)\left(u^{01} - u^0 - \frac{d\Theta}{2}\right)\right)$$
$$-\delta(x - Vt)\left(m\frac{d^2u^{02}}{dt^2} + \left(k_0 + \varepsilon_0 \frac{d}{dt}\right)\left(u^{02} - u^0 + \frac{d\Theta}{2}\right)\right),$$
$$\rho I \frac{\partial^2 \varphi}{\partial t^2} - EI\frac{\partial^2 \varphi}{\partial x^2} + \chi\mu F\left(\varphi - \frac{\partial u}{\partial x}\right) = 0,$$
$$M\frac{d^2u^0}{dt^2} + \left(k_0 + \varepsilon_0 \frac{d}{dt}\right)\left(2u^0 - u^{01} - u^{02}\right) = 0,$$
$$J\frac{d^2\Theta}{dt^2} + \frac{d}{2}\left(k_0 + \varepsilon_0 \frac{d}{dt}\right)\left(d\Theta - u^{01} + u^{02}\right) = 0,$$
$$u^{01}(t) = u(Vt + d, t),\quad u^{02}(t) = u(Vt, t),$$
$$u(x,0) = \frac{\partial u(x,0)}{\partial t} = 0,$$
$$u^{01}(0) = u^{02}(0) = u^0(0) = \Theta(0) = 0,$$
$$\lim_{|x-Vt|\to\infty} u(x,t) = 0,\quad \lim_{|x-Vt|\to\infty} \varphi(x,t) = 0 \tag{10}$$

Notations in Eqs.(10) are the same as in Eqs.(7) except for k_0 and ε_0, which now are the stiffness and the damping factor of the supports of the bar.

Additionally, $u^0(t)$ is the vertical displacement of the centre of mass of the bar, $\Theta(t)$ is the angle of rotation of the bar around the centre of mass, M and J are the mass and the moment of inertia of the bar, d is the distance between the bogie's supports.

As shown by Verichev & Metrikine (2002), the characteristic equation for vibrations of the bogie as it moves on the beam is given as

$$a_{11}a_{22} - a_{12}a_{21} = 0, \tag{11}$$

$$a_{11} = 1 + Z_0 \cdot I_0 - Z_1(I_0 + I_-) + Z_2(I_- - I_0)$$
$$a_{12} = Z_0 \cdot I_- - Z_1(I_0 + I_-) - Z_2(I_- - I_0)$$
$$a_{21} = Z_0 \cdot I_+ - Z_1(I_0 + I_+) + Z_2(I_0 - I_+)$$
$$a_{22} = 1 + Z_0 \cdot I_0 - Z_1(I_0 + I_+) - Z_2(I_0 - I_+),$$

$$Z_0 = ms^2 + k_0 + \varepsilon_0 s, \quad Z_1 = \frac{(k_0 + \varepsilon_0 s)^2}{Ms^2 + 2(k_0 + \varepsilon_0 s)},$$

$$Z_2 = \frac{(k_0 + \varepsilon_0 s)^2}{2\left(Js^2/d^2 + (k_0 + \varepsilon_0 s)\right)}, \quad I_0 = \frac{1}{2\pi}\int_{-\infty}^{\infty} \frac{dk}{D(k,s)},$$

$$I_+ = \frac{1}{2\pi}\int_{-\infty}^{\infty} \frac{\exp(ikd)\,dk}{D(k,s)}, \quad I_- = \frac{1}{2\pi}\int_{-\infty}^{\infty} \frac{\exp(-ikd)\,dk}{D(k,s)}$$

where $D(k,s)$ is defined by Eq.(3).

To determine whether the characteristic equation (11) has roots with a positive real part, it is necessary to use both the D-decomposition method and the principle of the argument, see Fuchs et al. (1964). The procedure of subsequent application of these two methods is described by Verichev & Metrikine (2002).

The curves that bound the instability domain are shown in Figures 7, 8. These curves were obtained by using parameters given by Eq.(5) and the following parameters

$$M = 2.10^4\,[\text{kg}], \; J = 598000\,\left[\text{kg m/s}^2\right],$$
$$d = 15.7\,[\text{m}], \; m = 1000[\text{kg}], \; M = 20000[\text{kg}], \tag{12}$$
$$\varepsilon_0 = 86000[\text{Ns/m}],$$

one of which was varied.

In Figure 7 the boundary of the stability domain is depicted for three different magnitudes of the damping factor ε_0. This boundary divides the parameter plane into two domains. In the domain below the boundary, vibrations of the model are stable, whereas in the domain that is located above the boundary the vibrations are unstable. This figure shows that increase of the damping in the supports leads to expansion of the stability domain. This expansion mainly takes place along the $K-$ axis, which implies that a higher damping factor allows to use stiffer supports keeping the system stable. The effect of the damping in the supports on the velocity after which the model becomes unconditionally unstable ($\alpha \approx 0.37$) is minor.

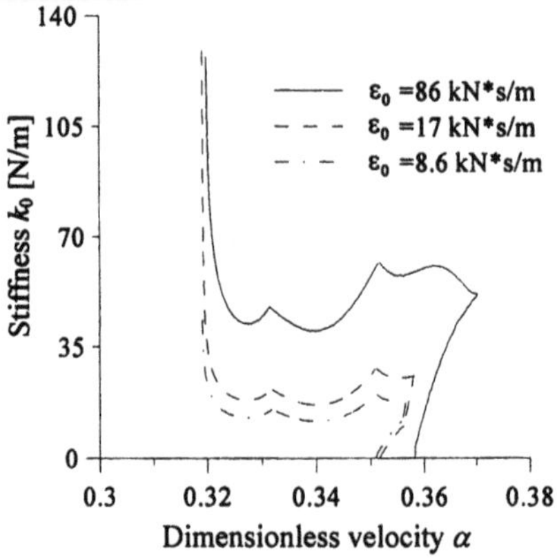

Figure 7. Effect of the damping in the bogie's supports on instability domain.

Another physical parameter that influences the stability domain significantly is the damping in the beam's foundation. The effect of this parameter is shown in Figure 8. This figure shows that as the damping in the foundation decreases, the stability domain visibly shrinks along the velocity axes. Besides that, the boundary of the stability domain for the smaller magnitude of the damping possesses a well-observed maximum at $\alpha \approx 0.335$. This maximum is concerned with reflection of waves in the beam that takes place between the bogie supports. This reflection plays a crucial role when the damping in the beam's foundation is small enough to allow standing waves to form between the supports.

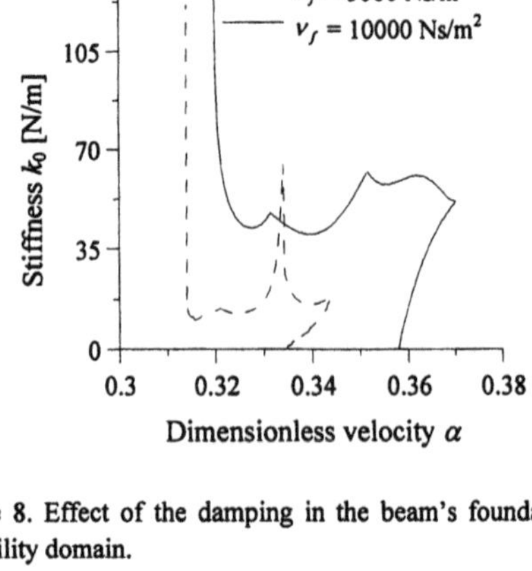

Figure 8. Effect of the damping in the beam's foundation on instability domain.

Completing this section, let us compare the instability domain for the bogie and the two-mass oscillator. This domain is depicted in Figure 9. For the comparison the lower mass of the oscillator was taken equal to the mass of the bogie's support, while the upper mass of the oscillator was considered equal to the half of the mass of the bogie bar. The other parameters are defined by Eqs.(5) and (12).

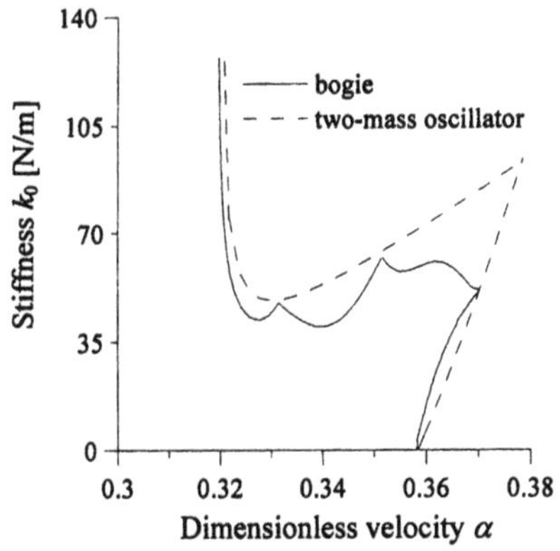

Figure 9. Boundaries of the instability domain for the bogie and two-mass oscillator.

The figure shows that the instability domains for these two models are quite similar. The domain for the bogie, however, is slightly smaller and much less monotonic. The latter is obviously concerned with the wave reflection between the bogie supports. The most important conclusion to be made from this comparison is that the simple two-mass oscillator model can be used for a rough estimation of the stability of the bogie, since their stability domains do not differ much. One should be aware, however, of possible deviations of the stability domain for the bogie from that for the oscillator, especially in the case of a small foundation's damping when the cross-influence of the bogie supports that takes place by means of waves in the beam becomes pronounced.

4 TWO-MASS OSCILLATOR ON A BEAM THAT OVERLIES A HALF-SPACE

In the previous sections it was shown that instability may occur only under the condition that a vehicle moves with a speed which exceeds the minimum phase velocity of waves in elastic structure. How big is this velocity for a railway track? To answer this question and to assess whether the instability phenomenon is of practical significance, a three-dimensional model for a railway track has to be considered.

The model under consideration is shown in Figure 10. It consists of a beam with finite width $2a$ that overlies a visco-elastic half-space (due to Kelvin-Voigt model) and a two-mass oscillator that moves along the beam with a constant speed V. The contact between the beam and the half-space is described in the same way as presented by Metrikine et al. (2001).

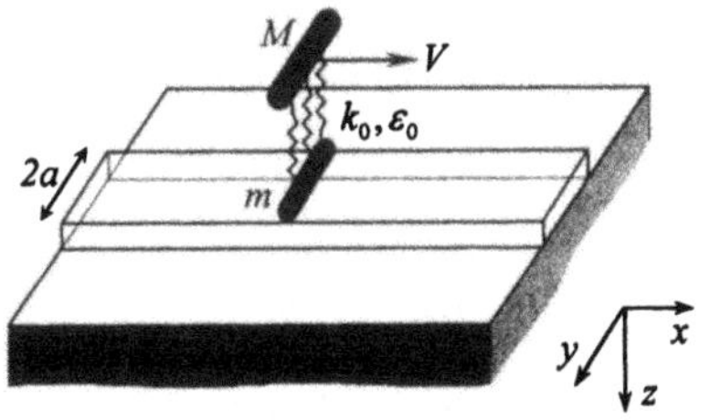

Figure 10. Two-mass oscillator on a beam that overlies a visco-elastic half-space.

Analyzing coupled vibrations of the system in the way that was presented by Metrikine and Popp (1999), the instability domain can be found that is depicted in Figure 11.

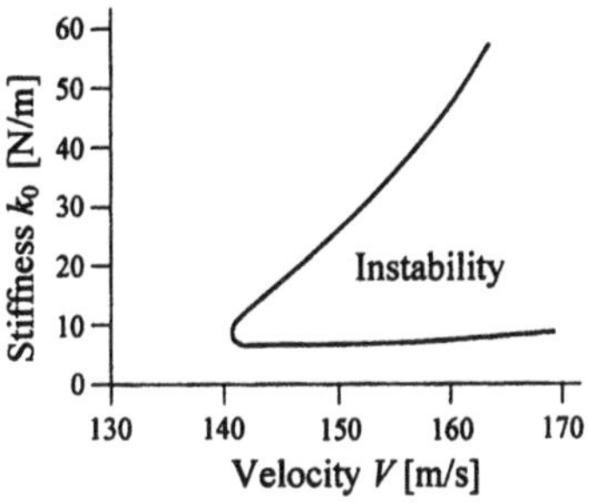

Figure 11. Instability domain for the 3D model.

To plot Figure 11, the following parameters were used (notations for the beam and the half-space are the same as in Metrikine et al. (2001), while those for the oscillator coincide with ones introduced in the previous sections of this paper)

$$\nu = 0.3,\ \rho = 1960\left[\text{kg/m}^3\right],\ \mu = 3.2\cdot 10^7\left[\text{N/m}^2\right],$$

$$\mu^*/\mu = 5\cdot 10^{-4}\left[\text{s}\right],\ 2a = 3\left[\text{m}\right],\ K = 10^8[\text{Pa}],$$

$$EI = 1.3\cdot 10^8\left[\text{Nm}^2\right],\ m_b = 7500\ [\text{kg/m}],$$

$$M = 2\cdot 10^4\left[\text{kg}\right],\ m = 2\cdot 10^3\left[\text{kg}\right],\ \varepsilon_0 = 8.6\left[\text{kN}\cdot\text{s/m}\right]$$

The main conclusion that has to be drawn from Figure 11 is as follows. Instability of vibrations of a high speed train can occur at velocities that are in the order of the shear wave velocity of waves in the subsoil (the above given parameters correspond to the shear wave velocity of 130 m/s). Thus, the softer the railway track's subsoil, the lower the train's velocity at which instability can occur. This instability however, can be avoided by a proper choice of the stiffness of the train bogie's suspension. Indeed, as Figure 11 shows, for a fixed velocity, instability takes place in a range of the stiffness' magnitudes. Outside this range, vibrations of the train are stable.

5 EFFECT OF PERIODIC INHOMOGENEITY OF ELASTIC STRUCTURE ON INSTABILITY

Finalizing the paper, let us underline that all the models that were considered so far were homogeneous in the direction of motion of the vehicle. In reality, railway tracks are (quasi-) periodically inhomogeneous because of the sleepers. Could this inhomogeneity influence stability of a moving train? There is a good reason to answer this question positively. Indeed, imagine a vehicle that moves on a beam that is supported by a periodically inhomogeneous foundation. If this vehicle moves uniformly, then in the contact point between the beam and the vehicle, the stiffness of the latter varies periodically *in time*. This means, that the beam reaction can be equivalently represented by a spring with a stiffness that varies periodically in time. Vibrations of a mechanical system supported by such a spring are known to be subject of the parametric instability. Thinking by analogy, we could expect that the moving vehicle could also experience such instability. Verichev and Metrikine (in press) have recently proven that this is indeed the case and, at quite low velocities of the vehicle, there exist zones of parametric resonance. These zones, however, were found to be quite narrow so that, most probably, this kind of instability is of no practical importance.

REFERENCES

Achenbach, J.D. 1973. *Wave propagation in elastic solids.* Amsterdam: North-Holland PubliCo.

Bogacz, R. 1983. On dynamics and stability of continuous systems subjected to a distributed moving load. *Ingeneuor Archive* 53: 243-255.

Dean, G.D. 1990. The response of an infinite railroad track to a moving vibrating mass. *Transactions of ASME Journal of Applied Mechanics* 57: 66-73.

Denisov, G.G., Kugusheva, E.K. & Novikov, 1985 V.V. On the problem of the stability of one-dimensional unbounded elastic systems. *Journal of Applied Mathematics and Mechanics* 49(4): 533-537.

Dieterman, H.A. & Metrikine, A.V. 1996. The equivalent stiffness of a half-space interacting with a beam. Critical velocities of a moving load along the beam. *European Journal of Mechanics A/Solids* 15(1): 67-90.

Dieterman, H.A. & Metrikine, A.V. 1997a. Critical velocities of a harmonic load moving uniformly along an elastic layer. *Transactions of ASME Journal of Applied Mechanics* 64: 596-600.

Dieterman, H.A. & Metrikine, A.V. 1997b. Steady-state displacements of a beam on an elastic half-space due to a uniformly moving constant load. *European Journal of Mechanics A/Solids* 16(2): 295-306.

Filippov, A.P. 1961. Steady-state vibrations of an infinite beam on an elastic half-space under moving load. *Izvestia AN USSR OTN Mechanica and Mashinostroenie* 6: 97-105 (In Russian).

Fuchs, B.A., Shabat, B.V. & Berry, T.J. 1964. *Functions of Complex variables and some of their applications.* Oxford: Pergamon.

Ginzburg, V.L. 1979. *Theoretical Physics and Astrophysics.* Oxford: Pergamon.

Grundmann, H., Lieb, M. & Trommer, E. 1999. The response of a layered half-space to traffic loads moving along its surface. *Archive of Applied Mechanics* 69(1): 55-67.

Kaynia, A.M., Madhus, C. & Zackrisson P. 2000. Ground vibration from high-speed trains: prediction and countermeasure *Journal of Geotechnical and Geoenviromental Engineering* 126(6): 531-537.

Kruse, H. & Popp, K. 2001. A modular algorithm for linear, periodic train-track models. *Archive of Applied Mechanics* 71(6-7): 473-486.

Krylov, V.V. 1995. Generation of ground vibrations by superfast trains. *Applied. Acoustics* 44(2): 149-164.

Labra, J.J. 1975. An axially stressed railroad track on an elastic continuum subjected to a moving load. *Acta Mechanica* 22: 113-129.

Metrikine, A.V. 1994. Unstable lateral oscillations of an object moving uniformly along elastic guide as a result of anomalous Doppler effect. *Acoustical Physics* 40(1): 85-89.

Metrikine, A. V. & Dieterman, H. A. 1997. Instability of vibrations of a mass moving uniformly along an axially compressed beam on a viscoelastic foundation. *Journal of Sound and Vibration* 201(5): 567-576.

Metrikine, A. V. & Popp, K. 1999. Instability of vibrations of an oscillator moving along a beam on an elastic half-space. *European Journal of Mechanics A/Solids* 18(4): 679-701.

Metrikine, A.V. & Verichev S.N. 2001. Instability of vibration of a moving two-mass oscillator on a flexibly supported Timoshenko beam. *Archive of Applied Mechanics* 71(9): 613-624.

Metrikine, A.V., Vostrukhov, A.V. & Vrouwenvelder, A.C.W.M. 2001. Drag experienced by a high-speed train due to excitation of ground vibrations. *International Journal of Solids and Structures* 38(48-49): 8851-8868.

Neimark, Y.I. 1978. *Dynamic Systems and Controllable Processes.* Moscow: Nauka (in Russian).

Sheng, X., Jones, C.J.C. & Petyt, M. 1999. Ground vibration generated by a load moving along a railway track. *Journal of Sound and Vibration* 228(1): 129-156.

Suiker, A.S.J., de Borst, R. & Esveld, C. 1998. Critical behaviour of a Timoshenko beam half plane system under a moving load. *Archive of Applied Mechanics* 68 (3-4): 158-168.

Verichev, S.N. & Metrikine, A.V. 2002. Instability of a moving bogie on a flexibly supported Timoshenko beam. *Journal of Sound and Vibration* 253(3): 653-668.

Verichev, S.N. & Metrikine, A.V. Instability of vibrations of a mass that moves along a beam on a periodically inhomogeneous foundation. *J. of Sound and Vibration* (in press).

Vesnitskii, A.I. 2001. *Waves in systems with moving loads and boundaries.* Moscow: Nauka (in Russian).

Numerical and experimental investigation

Wave propagation – Moving load – Vibration reduction, Chouw & Schmid (eds.)
© 2003 Swets & Zeitlinger, Lisse, ISBN 90 5809 559 2

Hybrid experimental – numerical simulation of vibrating structures

U.E. Dorka
University of Kassel, Kassel, Germany

ABSTRACT: A general method to perform continuous substructure tests is discussed in the light of building and aerospace applications. The method is a hybrid experimental-numerical simulation of a vibrating system with tested substructures. Test results are presented and the issues of stability and accuracy are addressed. Future developments in substructure testing on shaking tables and multiple substructures on networked facilities are discussed.

1 INTRODUCTION

The simulation of vibrating systems, whether numerical or experimental, has become an important tool to assess the safety and performance of structures like bridges under wind or buildings under earthquakes or their effect on and interaction with subsystems like a vehicle (or vehicles) on a bridge or a satellite as payload in a carrier rocket.

In the latter case, time dependent non-linear damping may be encountered that cannot be identified correctly by the usual qualification procedures. Performed as sine-sweep tests, these procedures typically require payloads to be much stiffer than necessary and thus add mass and considerable costs. A procedure that can simulate the carrier-payload interaction correctly thus would increase payload efficiency.

Figure 1. The Normandy bridge (photo: Bouygues Co.).

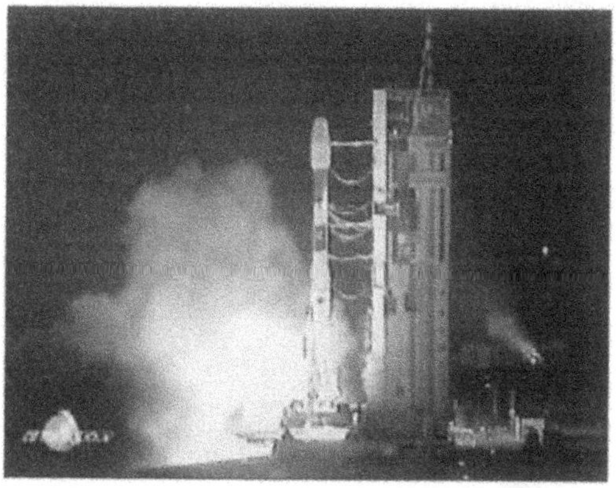

Figure 2. Ariane IV carrier rocket during a night launch (photo: Arianespace).

Often, we have a good mechanical model of such structures and can simulate them and their interaction with substructures numerically. But also very often, we cannot define a reasonable model, at least for some parts of a structure. This is usually the case when non-linearities are encountered, especially if they depend on time, load history or both. Such non-linearities can be found in buildings under earthquakes or satellites during launch.

And then there is the emerging field of structural control where active, passive or parameter controlled (semi–active) devices are used to enhance the performance of a structure under dynamic excitation. To verify a control device for a given application, e.g. a hybrid mass damper and its controls for a high rise under wind, an experiment must usually be per-

Figure 3. The Yokohama Landmark Tower uses an active system to reduce wind induced vibrations (photo: Japan Iron & Steel Federation).

formed. If this test can include the structure that the device will be applied to, complete knowledge of the combined system behavior is gained and the device and its controls can be optimized. Typically, such tests are performed after the completion of the building but not all important load cases can be simulated then.

If a good mechanical model of the structure is available (this is usually the case), a hybrid experimental – numerical simulation, with the device on a shaking table and the structure as an added numerical model can be performed. This will yield the required answers. It is this new method, often also referred to as dynamic substructure testing, that this paper is concerned with.

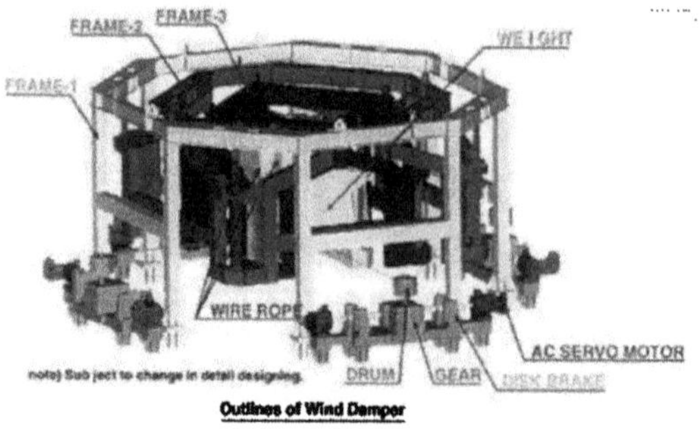

Figure 4. Tuned active damper (TAD) in the Yokohama Landmark Tower (drawing: Mitsubishi Heavy Industries Ltd.).

2 BASIC FORMULATION OF TESTING ALGORITHM

The basis for such tests is a discrete time integration algorithm. It is derived from a discrete formulation of the dynamic equilibrium:

$$M\tfrac{d^2}{dt^2}x + C\tfrac{d}{dt}x + Kx + f_r + f_s = p(t) \tag{1}$$

where f_r is the vector of numerical non-linear restoring forces and f_s is the vector of restoring forces measured at the interface between numerical model and tested substructure. M, C, K are the mass-, damping-, and stiffness matrices of the numerical model and $x, p(t)$ are the displacement vector and loading.

A general solution of this equation can be derived using a weighted residual formulation and a finite element discretisation in the time domain (Zienkiewicz, 1977):

$$\int_{-1}^{1} W \begin{bmatrix} M\left(u^{n-1}\tfrac{d^2}{dt^2}N_{n-1} + u^{n}\tfrac{d^2}{dt^2}N_{n} + u^{n+1}\tfrac{d^2}{dt^2}N_{n+1}\right) \\ + C\left(u^{n-1}\tfrac{d}{dt}N_{n-1} + u^{n}\tfrac{d}{dt}N_{n} + u^{n+1}\tfrac{d}{dt}N_{n+1}\right) \\ + K\left(u^{n-1}N_{n-1} + u^{n}N_{n} + u^{n+1}N_{n+1}\right) \\ + f_*^{n-1}N_{n-1} + f_*^{n}N_{n} + f_*^{n+1}N_{n+1} \end{bmatrix} d\xi \tag{2}$$

where u is the displacement vector at discrete time steps n, W a weighting function (compare Fig. 5), $\xi = t/\Delta t$, $f_* = f_r + f_s - p$ and:

$$\begin{aligned} N_{n-1} &= \xi(1+\xi)/2; \\ N_n &= (1-\xi)(1+\xi); \\ N_{n+1} &= -\xi(1-\xi)/2 \end{aligned} \tag{3}$$

are the shape functions of the displacements in time (Fig. 5). Note that the force vector f_* has been discretised in a consistent way using the same shape functions.

Performing the integration with various weighting functions yields all major time stepping algorithms that use three time steps:

$$u^{n+1} = \left[M + \gamma\Delta t C + \beta\Delta t^2 K\right]^{-1} \cdot \begin{Bmatrix} \left[2M - (1-2\gamma)\Delta t C - (\tfrac{1}{2} - 2\beta + \gamma)\Delta t^2 K\right]u^{n} \\ -\left[M - (1-\gamma)\Delta t C + (\tfrac{1}{2} + \beta - \gamma)\Delta t^2 K\right]u^{n-1} \\ + \beta\Delta t^2 f_*^{n+1} + (\tfrac{1}{2} - 2\beta + \gamma)\Delta t^2 f_*^{n} \\ + (\tfrac{1}{2} + \beta - \gamma)\Delta t^2 f_*^{n-1} \end{Bmatrix} \tag{4}$$

with:

$$\begin{aligned} \gamma &= \int_{-1}^{1} W(\xi + \tfrac{1}{2})d\xi \Big/ \int_{-1}^{1} W d\xi; \\ \beta &= \int_{-1}^{1} W \tfrac{1}{2}(1+\xi)\xi d\xi \Big/ \int_{-1}^{1} W d\xi; \end{aligned} \tag{5}$$

as constants defining a particular algorithm.

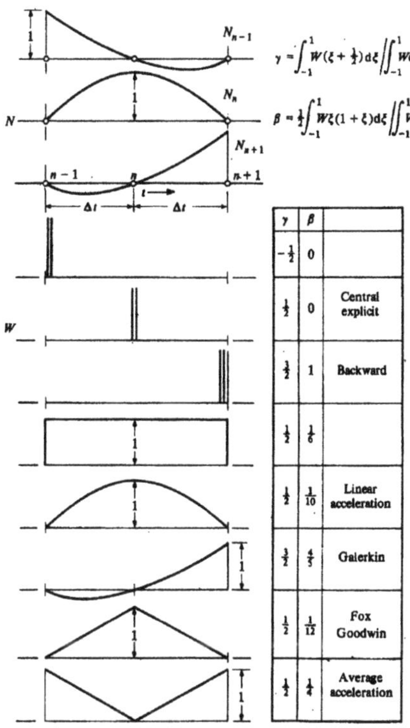

Figure 5. Shape functions to discretise the displacements in time and weighting functions that yield well known time stepping algorithms (Zienkiewicz, Fig. 21.7, 1977).

Historically, these algorithms have been derived by other reasoning. Zienkiewicz's formulation is the underlying fundamental approach and allows for a variety of weighting functions to fine-tune the algorithm's behavior.

The formulation can easily be extended to more then 3 time steps. A 4 step procedure that then comes naturally with the appropriate selection of a weighting function is the method suggested by Hilber, Hughes and Taylor (Hilber et al. 1977).

Different parameter combinations and thus different methods can now be compared numerically on the grounds of accuracy and numerical damping by studying the free response of a linear SDOF system. Considering that all linear systems can be separated into independent SDOF systems by mode decomposition, general statements can be made on accuracy and stability, at least in the context of linear systems.

The free vibration response of a SDOF system can be written in the form:

$$y(t) = Ye^{\upsilon t} \tag{6}$$

where Y is the amplitude and υ is the characteristic value. This equation may be expressed in a recurrence form as:

$$y_{n+1} = Ye^{\upsilon(t+\Delta t)} = \left(e^{\upsilon\Delta t}\right)Ye^{\upsilon t} = \lambda y_n \tag{7}$$

Substituting this into the general 3-step algorithm of equation (4) now expressed for a SDOF system yields a characteristic equation:

$$\begin{aligned}&\lambda^2\left[m + \gamma\Delta tc + \beta\Delta t^2 k\right]\\ &+ \lambda\left[-2m + (1-2\gamma)\Delta tc + \left(\tfrac{1}{2} - 2\beta + \gamma\right)\Delta t^2 k\right]\\ &+ \left[m - (1-\gamma)\Delta tc + \left(\tfrac{1}{2} + \beta - \gamma\right)\Delta t^2 k\right] = 0\end{aligned} \tag{8}$$

where in general, λ has complex roots that depend on γ, β and Δt. They characterize damped oscillatory results that are only bounded if $|\lambda| \leq 1$. It is self-evident that the same procedure can be applied to 4-step algorithms (see Zienkiewicz, pp. 589ff. 1977).

Studying now the case of an un-damped SDOF system where $|\lambda| = 1$ represents the exact solution, any recurrence scheme that has a $|\lambda| < 1$ produces a stable solution but introduces artificial numerical damping. Thus the value of $|\lambda|$ not only provides a necessary condition for an algorithm to be numerically stable, but also gives a measure of numerical damping (Fig. 6). Additionally, it provides information on the relative period error by comparing it to the exact value (Fig. 7).

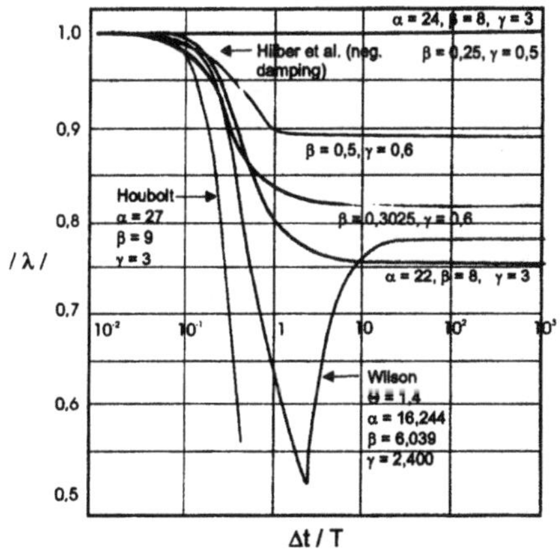

Figure 6. Numerical damping for 3-step and 4-step algorithms (Zienkiewicz, Figs 21.8, 21.11, 1977).

It is evident from Figures 6 & 7 that the well known implicit Newmark-β scheme (Newmark 1959) with parameters $\gamma = 0.5$ and $\beta = 0.25$ is the only unconditionally stable 3-step algorithm that has zero numerical damping and provides the least numerical softening. Furthermore, the figures show that, unless a distribution of numerical damping over the frequency range is desired, there is no considerable advantage in using a 4-step algorithm.

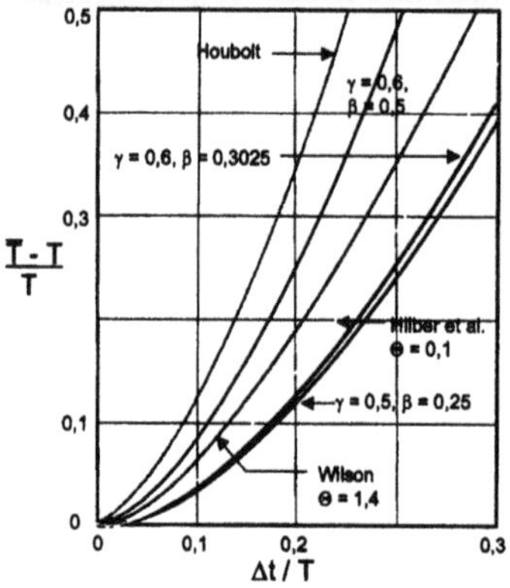

Figure 7. Numerical softening (expressed as period elongation) for 3-step and 4-step algorithms (Zienkiewicz, Figs 21.9, 21.12, 1977).

It also should be noted that only implicit algorithms, whether they are 3- or 4-point schemes, are able to provide unconditional stability. All explicit schemes have a cut-off frequency dependent on the selected time step which, if exceeded, will lead to unstable oscillations.

Implicit schemes require the knowledge of f_s^{n+1} for an exact solution which, in a test, is only available by measurement at the end of the time step. In purely numerical simulations with analytical non-linear restoring forces f_r, iterative procedures are used to minimize the error at the end of each time step. Such iterations cannot be used in substructure testing since they would produce high-frequency oscillations. Various methods have been suggested to deal with this problem like the operator splitting method (Nakashima et al., 1990), where the analytical part of the structure is solved by an implicit scheme and the tested substructure by an explicit one. Or f_s^{n+1} is estimated using the initial or some tangent stiffness of the substructure (contributions in: Donea & Jones, 1991).

None of these methods are unconditionally stable anymore and their stability depends highly on the structure to be tested and/or on the quality of the estimated stiffness. A more direct and concise way of dealing with this problem is viewing it as a linear control problem by noting that all implicit algorithms can be expressed in the general form:

$$u^{n+1} = u_0 + G\left(f_s^{n+1} + f_r^{n+1}\right) \tag{9}$$

which is a linear control equation during each time step with the initial vector u_0 (updated at the beginning of each time step), a gain matrix G (constant) and continuously varying force vectors f_s^{n+1} (measured) and f_r^{n+1} (calculated). Note the similarity in dealing with non-linear analytical restoring forces and the measured response of a substructure.

For a 3-step algorithm:

$$u_0 = \left[M + \gamma\Delta t C + \beta\Delta t^2 K\right]^{-1} \cdot \left\{ \begin{array}{l} \left[2M - (1-2\gamma)\Delta t C - \left(\tfrac{1}{2} - 2\beta + \gamma\right)\Delta t^2 K\right]u^n \\ - \left[M - (1-\gamma)\Delta t C + \left(\tfrac{1}{2} + \beta - \gamma\right)\Delta t^2 K\right]u^{n-1} \\ - \beta\Delta t^2 p^{n+1} + \left(\tfrac{1}{2} - 2\beta + \gamma\right)\Delta t^2 f_*^n \\ + \left(\tfrac{1}{2} + \beta - \gamma\right)\Delta t^2 f_*^{n-1} \end{array} \right\} \tag{10}$$

and:

$$G = \beta\Delta t^2 \left[M + \gamma\Delta t C + \beta\Delta t^2 K\right]^{-1} \tag{11}$$

Thewalt & Mahin (1987) applied equation (9) in a pseudo-dynamic test using the 4-step Hilber scheme (Hilber et al. 1977). They added the force response of the specimen in analog form directly to the displacement control signal modified by set amplifier gains that represented G.

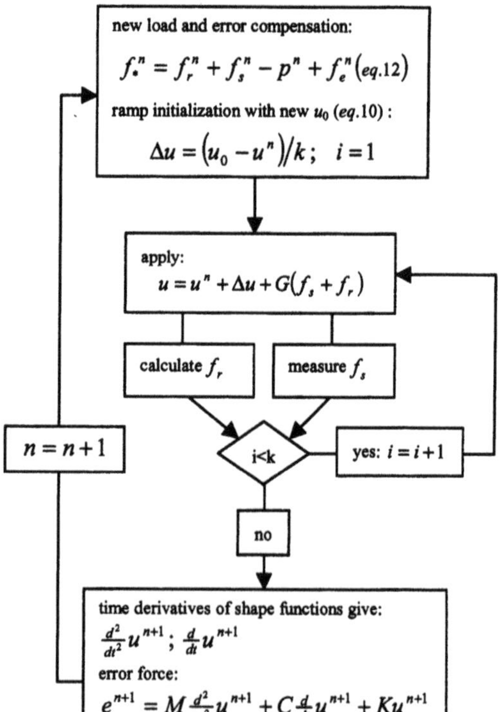

Figure 8. Flow chart of digital substructure algorithm with inner loop for sub stepping in the test and parallel treatment of analytical non-linearities.

In the digital world, a number of discrete sub-steps are required during each time step to perform the same operation (Dorka & Heiland, 1991). Here, the calculated change in u_0 at the beginning of each time step, expressed as $\Delta u_0 = u^n - u_0$, is applied as a linear ramp over all sub-steps. The forces f_s are measured at each sub step, multiplied by G and u is updated according to equation (9). If the number of sub steps k approaches infinity, equation (9) is correctly solved. If $k = 0$, the result is unstable. Thus, the number of sub steps is an important factor in performing a test. Figure 8 summarizes the basic steps of the algorithm.

Finally, it should be mentioned that the algorithm may easily be formulated for displacement, velocity or acceleration control.

3 STABILITY AND ACCURACY ISSUES

The number of sub steps k within each time step Δt obviously is an important factor for the stability and accuracy of a digital substructure test, but it is not the only one. Figure 9 gives the basic flow chart of a substructure test including the usually nonlinear

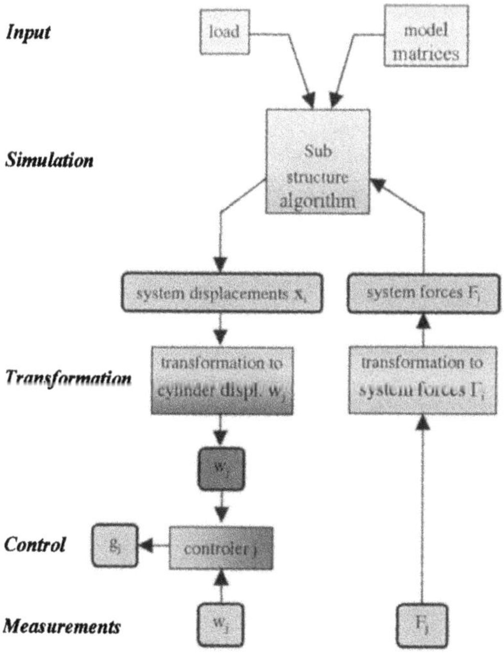

Figure 9. Flow chart of substructure test with the algorithm in its core. The transformations to cylinder displacements are geometrically non-linear in general. g_j may be a signal to adjust a hydraulic valve, a step motor or an electro-dynamic shaker. For simplicity, only one controller is shown.

geometric transformations of the input to the actuators and their controllers and the measurements and their transformations back to the algorithm.

Due to inaccuracies in the transformations and controllers and mechanical limitations of the actuators, positioning errors will occur especially in a continuously running test that can be minimized by adaptive controllers (e.g. Stoten & Benchoubane, 1990) but not completely avoided. They appear in a regular manner as overshoot or undershoot with a certain noise level that depends on the quality of the testing equipment.

Also, measurement errors may enter the algorithm. Again, regular errors may occur (incorrect amplification or insufficient resolution) accompanied by noise. Contrary to the positioning errors of the actuators, measurement errors can be avoided completely by a proper test setup.

In regular dynamic tests, filters are often used to clean measured signals of noise. This technique has only limited use in a feed back test since filters introduce delay time. If this delay is close to the duration of one sub step, proper feed back cannot occur and the test becomes unstable. This severely limits the application of filters to higher frequencies, even if one-sided recursive filters are used. With digital sensors, the measurement noise problem can be solved effectively and additional filters become obsolete.

Since not all errors can be avoided (especially the positioning error of the actuators can only be minimized) an equilibrium error occurs at the end of each time step. This error force, although small in a proper test setup and thus of no concern for accuracy, constitutes an additional excitation that tends to destabilize the test and therefore must be compensated. Bayer & Dorka (2000) suggested to minimize the error by a PID error minimization:

$$f_e^n = -P\left[e^n + I\Delta t\sum_i^n e^n + \frac{D}{\Delta t}\left(e^n - e^{n-1}\right)\right] \qquad (12)$$

where e^n is the equilibrium error force that can be calculated at the end of each time step (Fig. 8) and P, I and D are constants chosen in a suitable way. f_e^n is a compensation force vector added as additional loading to the system in the next time step (Fig. 8). Roik & Dorka (1989) used a simplified form of equation (12) with $P = 1$, $D = I = 0$ successfully in stabilizing rapid continuous pseudo-dynamic earthquake tests on stiff non-linear SDOF systems.

Dorka at al. (2002) investigated stability and accuracy issues using the sub stepping technique in combination with the error compensator of eq. (12) on the unconditionally stable implicit Newmark algorithm without numerical damping in a research project aimed at developing a real-time substructure algorithm for aerospace applications. It was a joint project by the University of Rostock and the German

Aerospace (DLR) financially supported by the German Research Society (DFG).

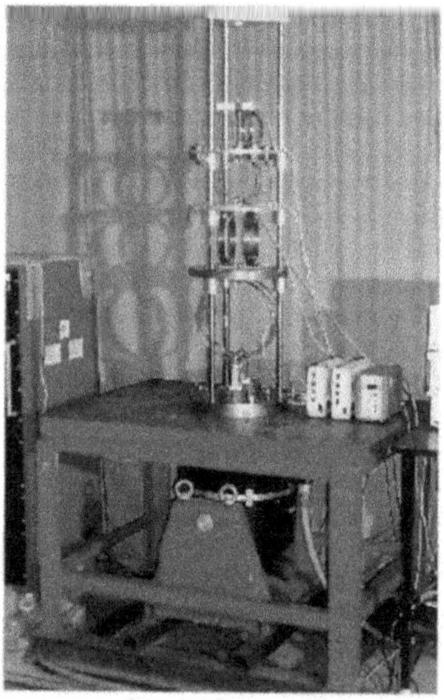

Figure 10. 4 DOF model structure used to develop a real-time substructure test algorithm for aerospace applications.

Besides numerical studies on the influence of the number of sub steps, choice of PID parameters and actuator over- and undershoot, tests were performed on a 4 DOF model structure (Fig. 10) representing typical eigenfrequencies of payload-launcher systems like the Ariane IV.

In the model, a pair of rings constitute a spring. Steel plates are placed between two pairs of rings to act as masses. Strain gauges attached to the rings were used to measure the spring forces and accelerometers to measure the accelerations of the masses. The model was excited vertically by an electrodynamic shaker which can be seen under the table in Fig. 10. Horizontal movements of the masses were prevented by three vertical guide bars clearly visible in Fig. 10.

The two top masses and springs represented the payload and were used as substructure whereas the two bottom masses and springs represented the launcher.

Sine-sweeps as well as a transient excitation taken from an Ariane IV stage separation (Fig. 11) were used in the tests. Only accelerations could be used for control with other signals (displacements, velocities) are being too small and lacking the necessary resolution.

The coupling force between substructure and main structure is the force feed back in the algorithm and therefore an important indicator of stability and accuracy. Figures 12 & 13 compare this force as

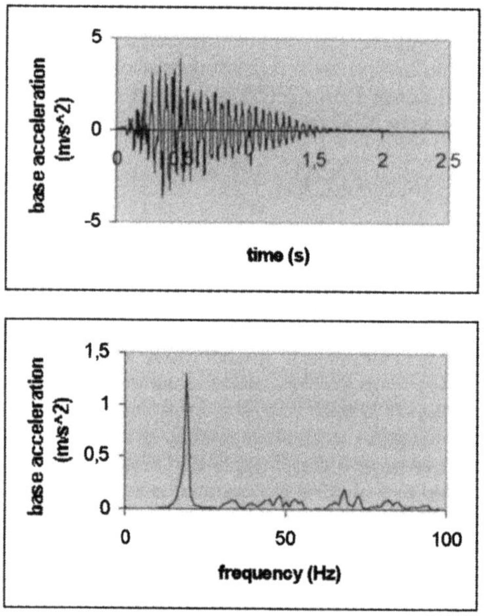

Figure 11. Measured accelerations from a 3rd stage Ariane IV separation used as signal in the tests.

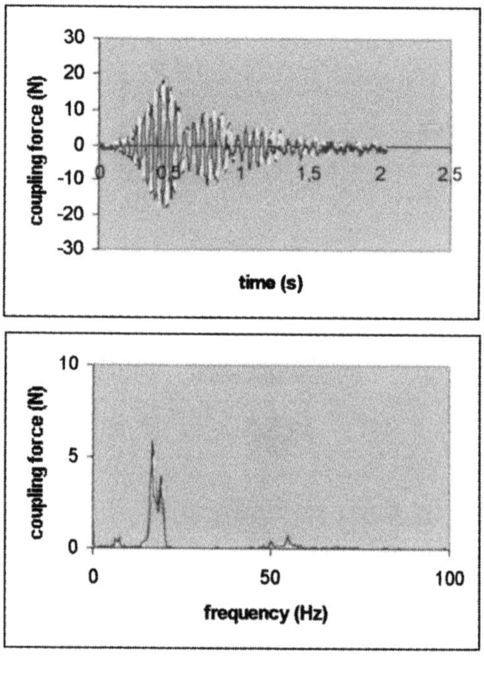

Figure 12. Comparison of coupling force between substructure and main structure for an algorithm with Δt=2ms, number of sub steps k=2 and P=0.9 (blue: exact, yellow: simulation, magenta: test).

measured in a substructure test to an exact solution for the total structure and a numerical simulation of the substructure test. The stage separation signal of Fig. 11 were used as base excitation.

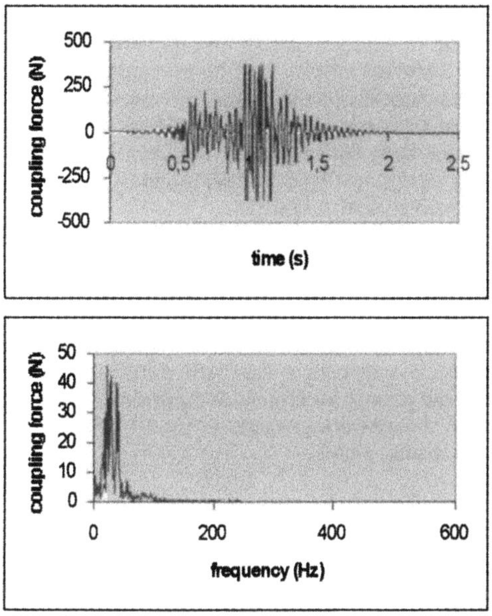

Figure 13. Comparison of coupling force between substructure and main structure for an algorithm without error force compensation and Δt=2ms, number of sub steps k=2 (blue: exact, yellow: simulation, magenta: test).

The exact solution for the complete system was obtained by Duhamel integral and mode superposition. The same technique was used to obtain the feed back response of the substructure in the numerical simulation of the test.

Whereas in Figure 12, the test reproduces almost exactly the results of the simulated test, the importance of the error compensation is seen in Figure 13 where a test with the same parameters, but without error force compensation, immediately becomes unstable.

As systematic studies of PID parameters on this model have shown that *I* and *D* will not improve the results here but even may add instability (especially *D*). The best value for *P* was 0.9. More work is needed to gain more insight into how an efficient error force compensator should look like since there is no question about its necessity.

As expected, increasing the number of sub steps increases accuracy and stability and so does a smaller time step. Both measures have the effect of reducing the un-avoidable errors due to cylinder positioning. A large number of sub steps and a small enough time step can even result in a stable test without the use of an error compensator. This could be shown by numerical studies that incorporated cylinder undershoot and overshoot. These studies could not be performed with the testing equipment because the time for one sub step would have been too small (only μs) to be applied in a test.

One should also note the differences between the exact solution and the substructure test or its simulation. The exact solution is the result of direct integration and modal superposition using the same mass and spring constants as in the numerical simulation of the test. The difference can be explained by the approximate nature of the time stepping scheme and is not introduced by the test setup which reproduces the simulated result very well (Fig. 12).

Aerospace applications require a rapid feed back control within a fraction of a millisecond which does not allow the application of hydraulic cylinders that have delay times in their valves and fluid of the same order. Also, acceleration control complicates the matter even further. Nevertheless, a real time substructure algorithm has been developed that fulfills the requirements for stability and accuracy in aerospace applications.

Other applications, like the qualification of a tuned mass damper in a high rise to mitigate wind induced vibrations are less demanding in this respect and time steps of 1/100 of a second or more produce enough accuracy in the results. Thus in building applications, hydraulic shaking tables with modern digital control can be used to run the substructure algorithm.

The discussion in the last two chapters shows that, for building applications (and similar ones), continuous (even real-time) substructure tests can be performed using the algorithm of Fig. 8 with the implicit Newmark-β scheme of $\gamma = 0.5$ and $\beta = 0.25$. Displacement control with 2 to 3 sub steps in the inner loop and an error force compensation based on *P* alone will yield a stable test with accurate and satisfactory results.

4 FUTURE DEVELOPMENTS AND DIRECTIONS

The substructure algorithm of Fig. 8 can be installed on any shaking table with modern digital control. The only additional requirement is the development of the geometrically non-linear transformations of the system displacements to cylinder displacements and in return, cylinder forces to system forces (Fig. 9). This must only be developed once for a particular table. Thus, sub-structuring on shaking tables will become a standard test in the near future.

In aerospace applications, where this would be of great advantage in qualifying subsystems as parts of mayor structures (like the modules of a space station) by including the interaction effects, the problem of time delay is very important because of the small sub step necessary (only several μs). To date, only electro-dynamic shakers are able to perform such tests and with larger forces, their size becomes excessive. Current work on superconductivity may reduce their size to a reasonable level which would open the door to lighter and more economic designs for payloads and space station modules.

Figure 14. Vehicle testing system that could be used to investigate vehicle-structure interactions in a substructure test with the vehicle as substructure including even a driver. (photo: IST)

For any testing system with fixed cylinder geometry, it is not particularly difficult to develop the non-linear geometric transformations necessary to run a substructure test. An example is a vehicle testing system (Fig. 14). Here, a vehicle could be used as a substructure driving over a bridge that is excited by traffic, wind or a combination of both for example. If a driver is included, even the response of the driver could be studied. The latter would of course require some additional audio-visual simulation.

It is another matter with variable test setups using cylinders in varying positions and directions. Here, the non-linear geometric transformations vary with each new test. Further studies are necessary to develop a concise method to safely arrive at the correct transformations quickly when such a test is planned.

Another future development is multiple substructures. Taking the example with the car, it is obvious that several such testing systems can be combined to simulate the interaction of several cars, drivers and a bridge. Various scenarios of bridge-vehicle-driver interaction could be tested in a very realistic simulation.

If an active mass damper is envisioned for the bridge, the device can be included in the simulation, if it is put on a shaking table that is networked to the system. Even soil-structure interaction can be included by adding dynamic tri-axial soil testing devices to the network that simulate the behavior of various soil substructures below the building.

The networking aspect of substructure testing facilities will be developed further in the near future since it makes economic sense to use existing facilities in multiple substructure testing. One can even envision facilities connected in a global network via satellite link.

But there are still many obstacles to overcome before this will be reality: Participating facilities must have the same standards beginning with the quality of the testing equipment all the way to data storage and exchange formats. Network links must be secure and able to transfer data in an un-interrupted manner. A software system must be developed that cannot only handle the algorithms on various substructure test beds but also provide efficient and secure communication between them, run continuous safety checks on the test's performance and allow remote monitoring of such an event.

But these problems can all be solved with today's technology. The US's NEES project aims at many of these tasks and with an 80 Million US$ funding, one can imagine how serious an effort this is. European large scale testing facilities are also going in this direction. The members of the CASCADE network are discussing plans how to include facilities networking in the 6. Framework program of the EU that is now in its initiation phase.

5 SUMMARY

Dynamic substructure testing is a hybrid experimental – numerical method to simulate the dynamic behavior of complex structures. Applications range from qualification tests on active mass drivers for high-rise buildings to simulations of the performance of new space station modules including the interaction effects with the main structure and other modules.

The heart of the test is a discrete time integration method. For complex systems, an implicit form is necessary for numerical stability. The best performance has the well known Newmark-β method with parameters $\gamma = 0.5$ and $\beta = 0.25$ which has no numerical damping and minimum period elongation.

All implicit methods require the knowledge of the restoring force at time $n+1$. Since iterations cannot be used in a test, estimator methods have been developed with their accuracy and stability depending on the quality of estimating certain parameters (like stiffness) and which are difficult to generalize.

Rewriting the time stepping equation leads to a general formulation for all algorithms as a linear control equation within each time step, where the measured restoring force is the feed back signal. This equation can be solved digitally by sub stepping. The number of sub steps is an important parameter for accuracy and stability of this general method.

In continuous substructure tests, positioning errors of the actuators lead to equilibrium errors in the equation of motion that need to be compensated least they destabilize the test. A PID type of error minimization known from standard control situations was used successfully in real-time substructure tests designed to develop a substructure algorithm for aerospace applications, but more work is needed to solve this problem completely.

In the future, substructure tests will be performed on shaking tables and in networks of substructure testing facilities allowing for an efficient and economic way of simulating large and complex structures with multiple substructures.

For aerospace applications, further developments in electro-dynamic actuators are important since hydraulic systems produce too large a delay time when the required sub step is in the order of μ seconds.

REFERENCES

Bayer, V., Dorka, U.E. 2000. Qualification of TMDs by real-time SubPSD-testing. *Proc. 2nd European Conf. Structural Control*, Champs sur Marne (in print).

Donea, J., Jones, P.M. (Eds.) 1991. *Experimental and Numerical Methods in Earthquake Engineering*. Ispra: Joint Research Centre.

Dorka, U.E., Heiland, D. 1991. Fast online earthquake simulation using a novel pc supported measurement and control concept. *Proc. 4th Int. Conf. Structural Dynamics*, Southhampton: 636-645.

Hilber, H.M., Hughes, T.J.R., Taylor, R.L. 1977. Improved numerical dissipation for time integration algorithms in structural mechanics. *Int. J. Earthquake Eng. Structural Dynamics* 5: 283-292.

Nakashima, M., Ishii, K., Ando, K. 1990. Integration techniques for substructure pseudo-dynamic test. *Proc. 4th National Conf. Earthquake Eng.* 2:515-524.

Newmark, N.M. 1959. A method for computation of structural dynamics. *Proc. American Society of Civil Engineers* 85: 67-94.

Roik, K., Dorka, U.E. 1989. *Fast online earthquake simulation of friction damped systems*. SFB151 Report no. 15, Ruhr-University Bochum.

Stoten, D.P., Benchoubane, H. 1990. Robustness of a minimal controller synthesis algorithm. *Int. Journal of Control* 51 No.4: 851-861.

Thewalt, C.R., Mahin, S.A. 1987. *Hybrid solution techniques for generalized pseudo-dynamic testing*. Report No. UCB/EERC-87/09, Univ. of California Berkeley.

Zienkiewicz, O.C. 1977. *The Finite Element Method*. Maidenhead: McGraw-Hill.

Wave propagation – Moving load – Vibration reduction, Chouw & Schmid (eds.)
© 2003 Swets & Zeitlinger, Lisse, ISBN 90 5809 559 2

Centrifuge simulations of wave propagation using a moving load system

K. Itoh, A. Takahashi, & O. Kusakabe
Department of Civil Engineering, Tokyo Institute of Technology, Tokyo, Japan

M. Koda & O. Murata
Railway Technical Research Institute, Tokyo, Japan

K.I. Lee
Department of Civil Engineering, Daejin University, Kyungki-do, Korea

ABSTRACT: This paper describes the behavior of wave generation and propagation using a multiple ball-dropping system which can simulate quasi-moving load caused by high-speed trains running through a viaduct. Characteristics of wave propagation were investigated using various combinations of distance of viaduct interval and EPS barrier. It was suggested that the superposition of the waves could be applicable to cases of moving load. A comparison between experiment results with the simple prediction method proposed by Yoshioka et al. (1980) showed good agreement. The information derived from the experiment is considered to be of use to verify the prediction methods that many researchers have proposed.

1 INTRODUCTION

Development of railway tracks for high-speed trains is growing rapidly throughout Europe, North America, and East Asia. Figure 1 shows European high-speed network in 2002 (International union of railway, 2002). The thick solid line shows in operation, the gray thick solid line under constructions, and gray thin solid line in project plan in 2020. If the European high-speed union carries out future projects, high-speed train network will be comfortable and convenient to the passenger's trip. In Japan, the Shinkansen network totals over 1850 km connecting Japan's major metropolitan areas and carries over 300 million passengers every year. Other Asian nations are now pursuing high-speed train systems of their own. A new high-speed rail system is under construction in Korea and in Taiwan.

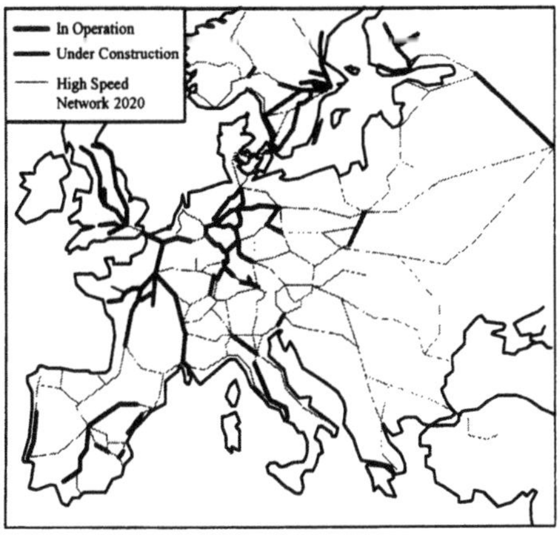

Figure 1 European high-speed network (UIC)

When constructing a high-speed railway system in urban areas, reduction of ground vibration of nearby structures generated by passages of high-speed train is of vital importance for environmental considerations. A typical prototype condition is the ground vibration transmitted through a series of viaduct foundations. This paper examines the characteristics of wave propagation from moving load such as high-speed train and proposes how to estimate the ground vibration at target locations using a centrifuge modelling test. In a centrifuge model testing, the scaling relationship agrees with the relevant effect of self weight induced stresses appropriate to the prototype earth structure and satisfies scaling laws for wavelength. The past researchers of examining the problem by using centrifuge test include Cheney et al. (1990), Siemer and Jessberger (1994), and Luong (1994). Their researches examined the observation of wave propagation generated by a single wave source. This paper describes that the behavior of the vibrations which were generated and propagated at the multiple wave sources such as a series of viaducts on which the high-speed train is running through.

2 EXPERIMENTAL SETUP

All the tests described here were conducted on Tokyo Tech Mark III centrifuge (Takemura et al., 1999). The container used was a steel cylindrical tube of 455 mm in diameter and 400 mm in height

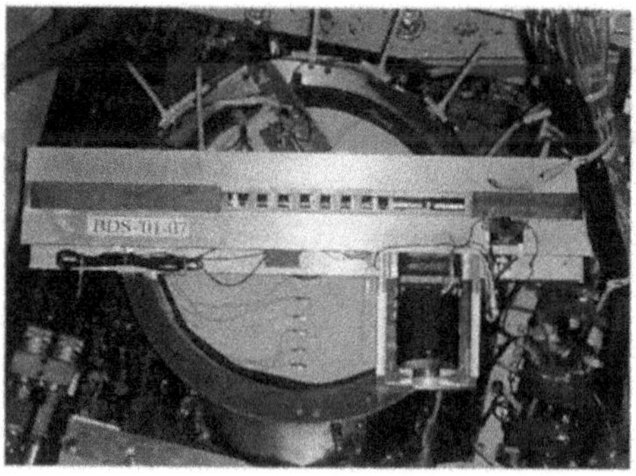

Figure 2 View of test container on centrifuge arm

Table 1 Material properties of Toyoura sand

Density of solid particles,	ρ_s (g/cm^3)	2.63
Maximum void ratio,	e_{max}	0.961
Minimum void ratio,	e_{min}	0.593
Coefficient of uniformity,	U_C	1.41
Mean particle diameter	D_{50} (mm)	0.18

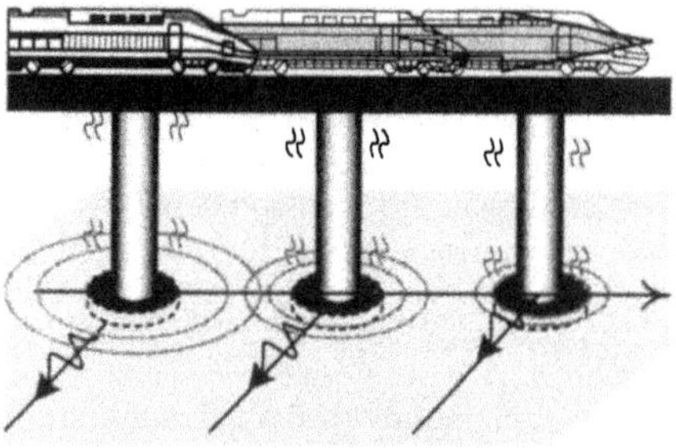

Figure 3 Concept of moving load

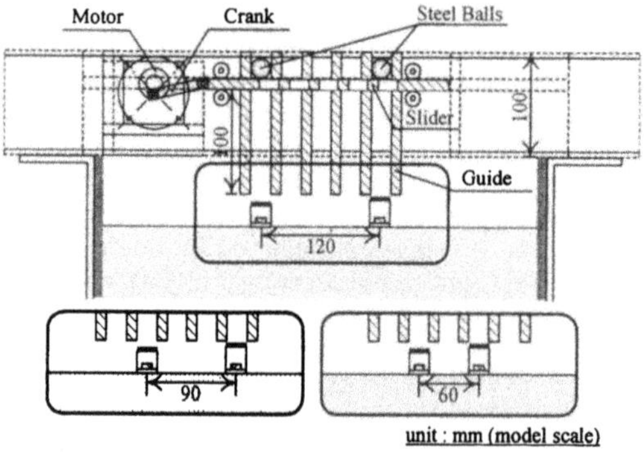

Figure 4 Experimental system

(Figure 2). One of the common problems encountered in physical modelling of vibration problems is that the walls of the container are rigid relative to the prototype ground systems and almost total reflection of the body waves would occur. In this study, a sheet of sponge rubber, with a thickness of 10 mm, was glued over the internal surfaces of the sidewalls and the bottom of the container to reduce the reflection waves. As a result, the maximum acceleration was

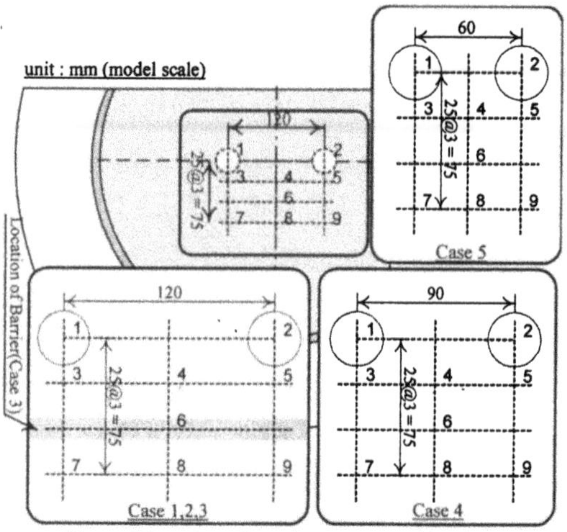

Figure 5 Location of accelerometers and barrier

Table 2 Experimental Programs

Test Case	1	2	3	4	5
Interval of viaduct (m)	6.0			4.5	3.0
Barrier	-	-	EPS	-	-
Train speed (km/h)	90.0	65.0	65.0	90.0	43.2
Time lag Δt (sec)	0.243	0.333	0.333	0.180	0.250

(in prototype scale)

reduced by 20-40 % in the case where the container was lined with the sponge rubber, compared to the case without the sponge rubber.

The soil used was air-dried Toyoura sand, of which the material properties are listed in Table 1. The sand was poured from a certain height for consistent production of uniform deposit with an average dry unit weight γ_d of 15.4 kN/m^3, and targeted relative density D_r of 80 %. The ground surface was leveled by vacuuming.

Moving load such as high-speed trains has been recognized as a potential source of ground vibrations. Figure 3 schematically illustrates the concept of moving load, where a moving load caused by a running high-speed train through a series of viaducts with shallow foundation is modelled. Figure 4 gives the experimental system of moving load simulation used in this study, called 'multiple ball-dropping system'. The mechanism of creating a moving load is as follows. Two steel balls are stored at the two extreme sides of this system with three different distances which are 6.0m, 4.5m and 3.0m in prototype scale (Figure 4). Two balls drop onto a corresponding foundation which type is circular shallow foundation. One foundation deliberately has a height slightly higher than the other foundation. The performance of this system is given by Itoh et al. (2002).

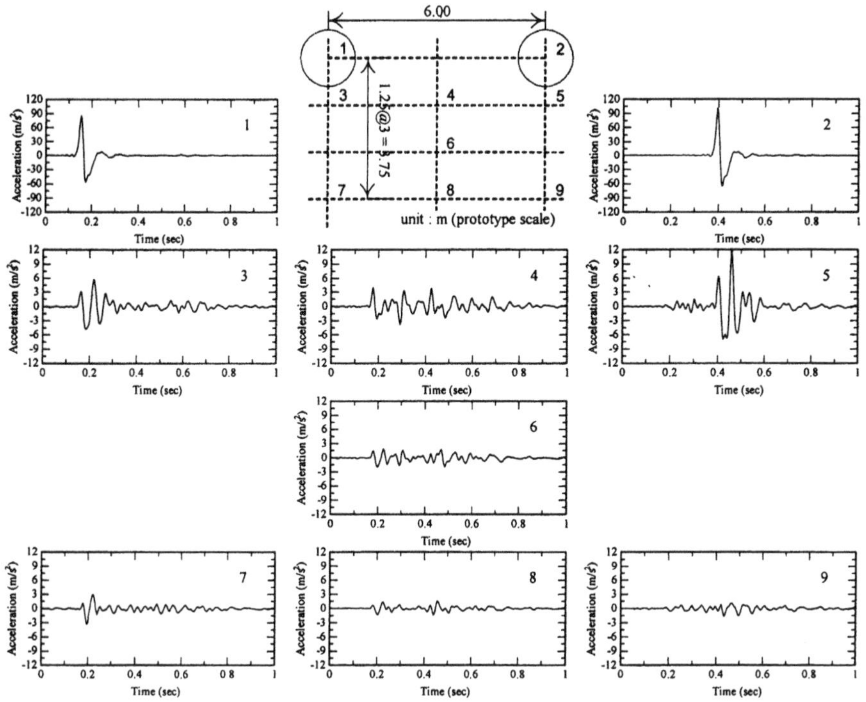

(a) Distance of interval viaduct =6.0m (Case-1)

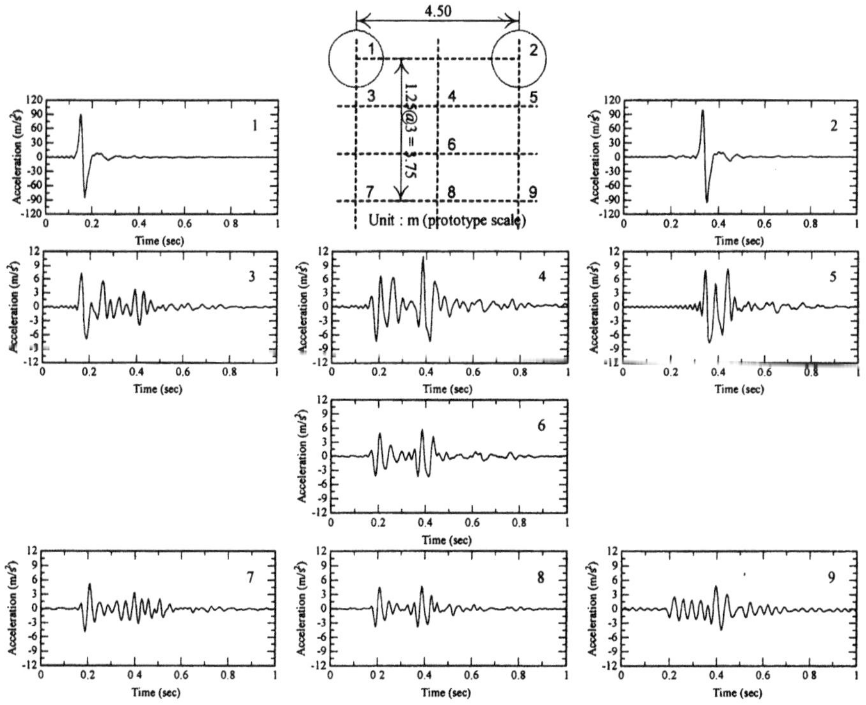

(b) Distance of interval viaduct =4.5m (Case-4)

Figure 6 Time histories in case of various viaduct spaces

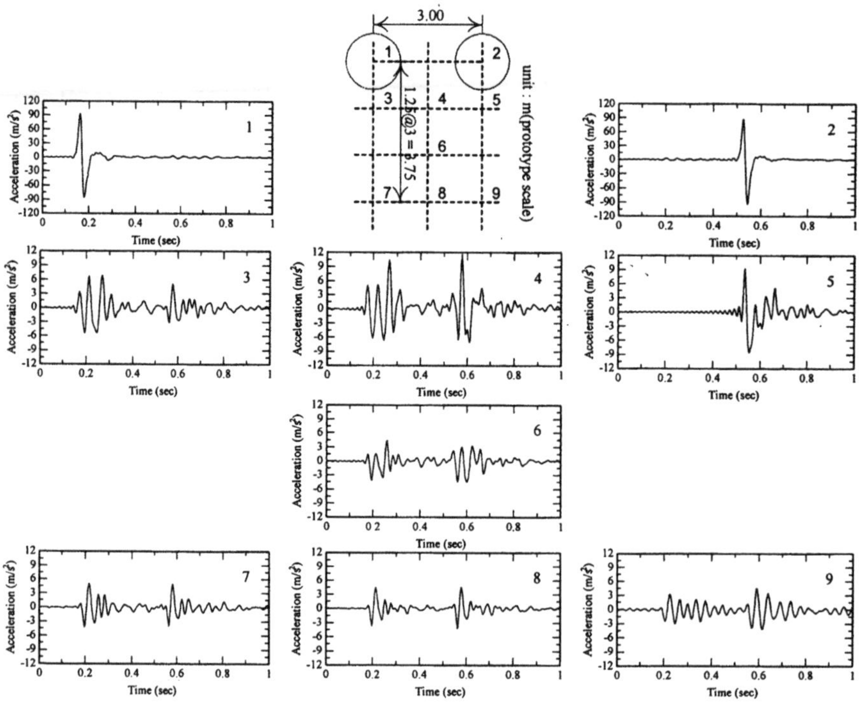

(c) Distance of interval viaduct =3.0m (Case-5)

Figure 6 Time histories in case of various viaduct spaces

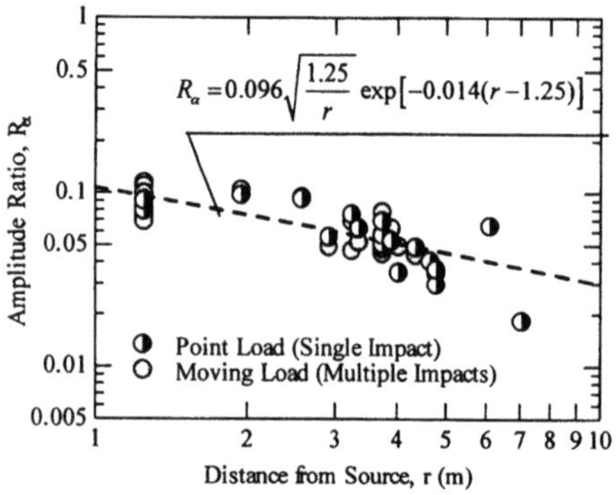

Figure 7 Attenuation of wave with distance from source

The motion was detected by piezo-electric accelerometers (CBC111BW and CBC107S, CBC Co., Ltd.) which provided data on the vertical accelerations at any positions on the ground surface. The experimental conditions are listed in Table 2. The effect of the vibration reduction method was examined by using Expanded Poly-Styrol (EPS) barrier in Case 3. The location of accelerometers and barrier is shown in Figure 5. All tests were carried out under the 50 G acceleration. Using the same model ground, three sets of the tests were carried out with a view to examine whether the superposition can hold or not. The first was to conduct a usual moving load test such as multiple impacts and in the second test only the No. 1 foundation was hit, and finally in the third test only the No. 2 foundation was hit such as single impact. On the ground surface, the first and the second dominate vibration frequencies were around 20 Hz and 80 Hz at prototype scale (Itoh et al., 2002). But the motion of the ground surface was detected by the piezo-electric accelerometer CBC111BW which has a resonance frequency of 4 kHz. The second dominating vibration frequency of 80 Hz is considered to be due to the resonance frequency of the accelerometer adopted (80 Hz × 50 G = 4 kHz). For the reason, it was decided that raw data were filtered by a low pass filter of 50 Hz to eliminate the influence of the resonance frequency of the accelerometer.

All the results of the centrifuge tests are presented at prototype scale, hereafter.

3 EXPERIMENTAL RESULTS AND DISCUSSIONS

3.1 *Characteristics of wave propagation*

The difference in foundation height was selected to be 9 mm in model scale, to provide the time lag as

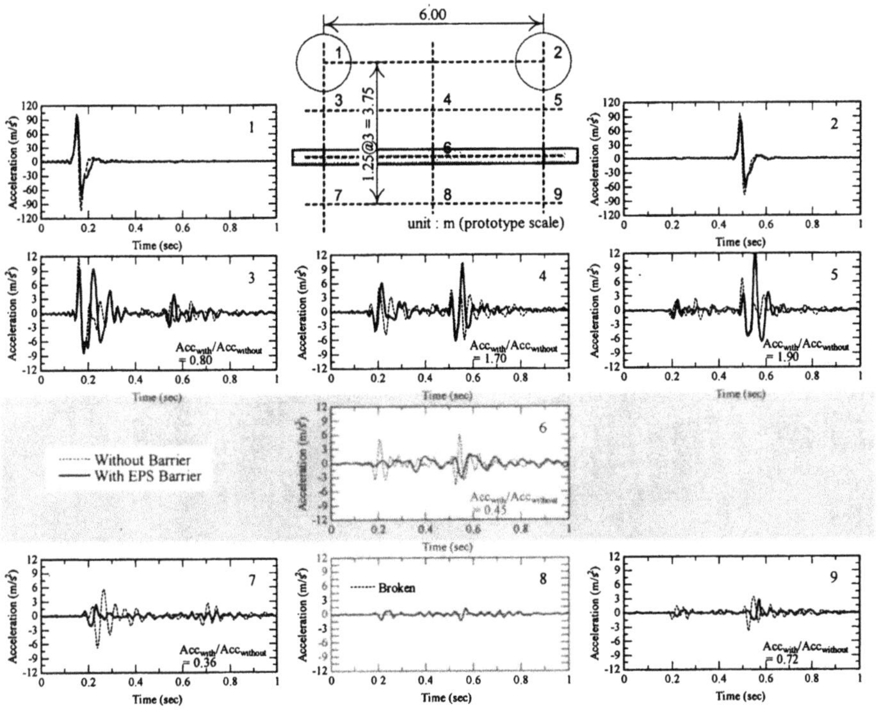

Figure 8 Comparisons of the wave between with and without EPS barrier

shown in Table 2. The time lag for the wave head from the one foundation to the next foundation differed on each case because all balls did not fall exactly at the same due to the limited precision of slider's slit.

Figure 6 presents a set of data of vertical acceleration at various points on the ground surface for each interval of viaduct. The amplitude of acceleration at the location 2.5 and 7.5 m from the foundation No. 1 (accelerometer No. 3 and 7) and the foundation No. 2 (accelerometer No. 5 and 9) was the same order, independent of the interval of viaduct. However, the accelerometer in the center of two viaducts indicates that amplitude of acceleration in viaduct interval of 3 m was larger than that of 6m, because the distances between the source and these positions were different for each case.

The geometrical and material damping in this ground was investigated. Figure 7 shows the ratio of the maximum acceleration measured on the ground surface to the input maximum acceleration measured in the model foundation, R_α, plotted against distance from source r in this test. The closed circles indicate the results by the moving load and the half closed circles the results by the point load. They represent the measured attenuation of R_α by the result of point load test, and the dashed line represents a theoretical attenuation predicted by Bornitz (1931). His equation includes geometrical and material damping, as follows;

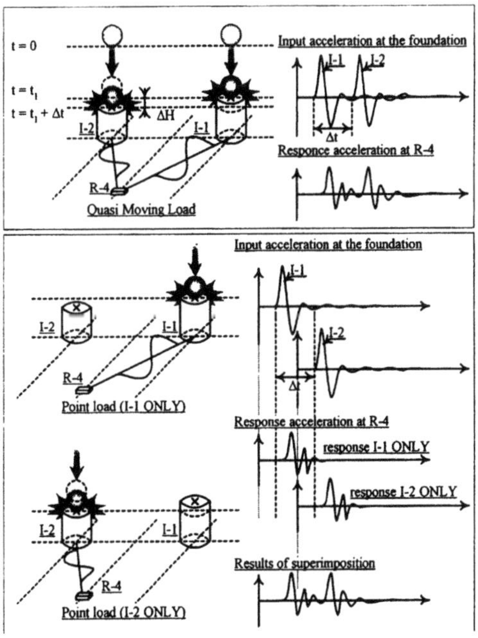

Figure 9 Experimental programs for verification of superimposition of waves from difference sources

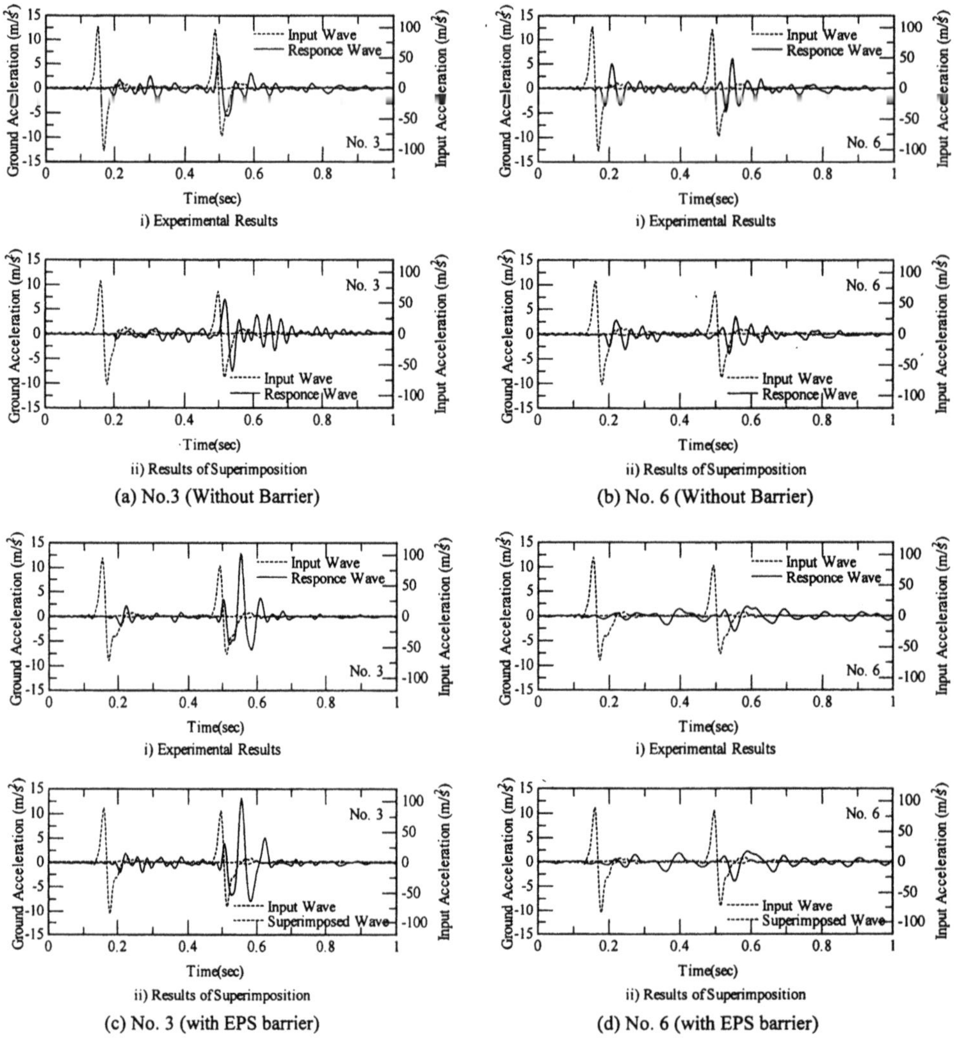

(a) No.3 (Without Barrier)

(b) No. 6 (Without Barrier)

(c) No. 3 (with EPS barrier)

(d) No. 6 (with EPS barrier)

Figure 10 Comparisons of the waves between calculated results of superimposition and experimental results (Case 2 &3)

$$R_\alpha = R_{\alpha 1}\sqrt{\frac{r_1}{r}}\,\exp\left[-\alpha(r-r_1)\right]$$

where,

r_1 = distance from source to point of known amplitude, r = distance from source to point in question, $R_{\alpha 1}$ = amplitude ratio of the component at distance r_1 from source, R_α = amplitude ratio of the component at distance r from source, and α = the coefficient of attenuation.

In this study, some parameters were decided as shown in Figure 7. Comparing to the results of point load test with the theoretical attenuation line, it is noted that the result of moving load test showed the same trend. The amplitude of acceleration is related to the distance from the nearest vibration source, independent of the input condition.

3.2 *Effect of EPS barrier*

An EPS barrier of rectangular shape (300 × 200 × 10 mm, in model scale) was selected for this test. During the model preparation, the barrier was set in advance and kept in a vertical position using the guide, when the sand was poured. When installing the EPS rectangular barrier of the embedded depth of 10 m, the motion of the ground changes as is shown in Figure 8, together with the case of without for comparison (Case 2 and 3). The velocity of the moving load can be readily calculated as about 65 km/h (= 18 m/s) at the two loading points 6 m apart. An apparent observation in this figure is the magni-

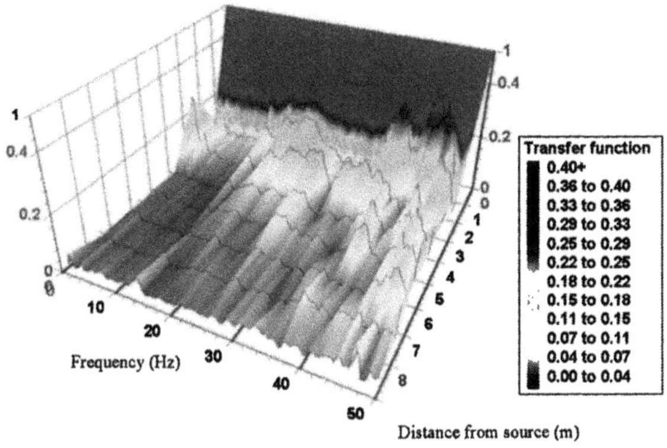

Figure 11 Transfer function H at multi-distance

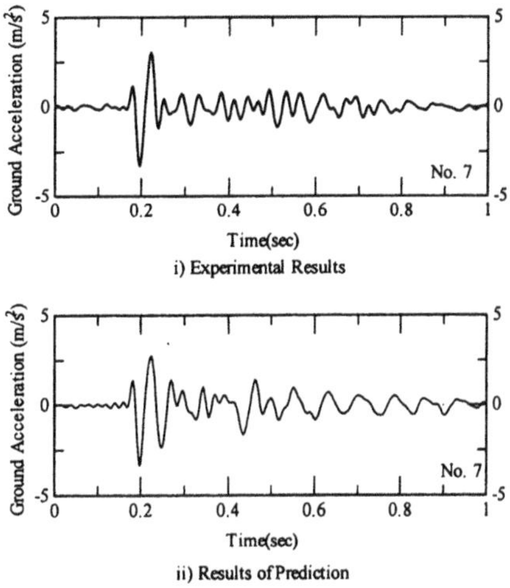

Figure 12 Comparison between the results by a prediction method and experiment results

fication of maximum acceleration in front of the barrier and the reduction on and behind the barrier. To be more precise, the peak amplitude of with EPS barrier was about twice as magnitude as without barrier in front of barrier. And the EPS barrier reduced 60 % amplitude on and behind the barrier.

3.3 *Comparisons of the waves between calculated results of superimposition and experimental results of moving load*

An attempt was made to examine whether the superposition of waves from difference source can hold or not, and in what circumstances, if it holes. Figure 9 explains what kind of experimental strategy was taken in this study. The data from the second test which hit only the foundation No. 1 was superimposed over the data from the third test which hit only the foundation No. 2, considering the time lag measured at the first test which was conduct a usual moving load. Thus calculated results are compared with the result of moving load test. Itoh et al. (2002) reported an investigation of the superposition of waves from different sources using air-dried Toyoura sand with unit weight γ_d of 15.4 kN/m^3 and 14.8 kN/m^3, and corresponding average relative densities of 80 % and 60 %, respectively. The results by them showed that accuracy of the superposition of the waves from different sources deteriorates, when a factor of safety for bearing failure decreases, as in the case of the looser ground D_r = 60%. This paper examines that the superimposition of waves with EPS barrier can be applicable or not in the case of D_r = 80 %.

In Figure 10, four cases are presented for the case of with and without EPS barrier at position of No. 3 and No. 6, together with the input waves. For the both cases (Case 2 and 3), the overall agreement between the experiment data of moving load and the calculation data of superposition is fairly well, both in magnitude and shape of the wave.

It was suggested that the superposition of the waves could be applicable to cases of moving load. Additionally, the superimposed result with EPS barrier has a yielded similar result without barrier.

3.4 *Prediction of the wave propagation by moving load*

Most prediction methods about the wave propagation on the ground vibration problems were judged by the maximum amplitude of acceleration/velocity. But there are some possibilities of magnifying the predicted value in the case of moving load such as supertonic condition. The vibrations caused by high-speed train generate and propagate from many vibration sources. Then the objective position records the wave which is superimposed from many sources. The prediction methods which simulate moving load need to be satisfied with the above condition. Yoshioka et al. (1980) originated a simple prediction method by a weight dropping experiment as follows;

$$O(\omega,P,j) \sim H(\omega,P,j)\cdot I(\omega,j) \qquad (3)$$

where,
$O(\omega,P,j)$ = the response acceleration spectra only by a j-th pier at the P point, $I(\omega,j)$ = the input acceleration spectra only by a j-th pier, $H(\omega,P,j)$ = the transfer function only by a j-th pier at the P point. Time domain records of input and response acceleration are already known from this experiment. This experiment can arrive at $H(\omega,P,j)$ by calculating the results of $O(\omega,P,j)$ and $I(\omega,j)$, as is seen in Figure 11. If the information of transfer function

$H(\omega, P, j)$ at many positions is obtained by this experiment, it is possible to extrapolate $H(\omega, P, j)$ at the objective distance from the source.

For example, this study predicts the response acceleration at the No. 7 location of the interval of viaduct of 6m (Case 2). The distances from two vibration sources to No. 7 location are 3.75 m and 7.08 m respectively. The transfer function $H(\omega, P, j)$ at these distances presumed that the relationship between the results of transfer function at the before location from the objective point and that at the behind location is linear. Figure 12 shows the comparison between the results by the prediction method and the experiment results. The overall agreement between the experiment results and the prediction results is good, both in magnitude and shape of the wave. More sophisticated prediction method than this method were proposed by many researchers (i. e. Lai et al. (2000) and Svinkin (2001) etc). The information derived from this experiment is considered to be of use to verify the methods that many researchers have proposed.

4 CONCLUSIONS

This paper describes an experimental investigation of wave propagation using a multiple ball-dropping system, which can simulate a moving load. The following conclusions were drawn from a series of centrifuge tests.

1. The amplitude of acceleration was related to the distance from the nearest vibration source, independent of the input condition.
2. It was suggested that the superposition of the waves could be applicable to cases of moving load not only without barrier but also with EPS barrier.
3. A comparison between experiment results with the simple prediction method proposed by Yoshioka et al. (1980) showed good agreement. The information presented in this paper is of use to verify similar prediction methods.

ACKNOWLEDGMENTS

This research was partially funded by the Grant-in-Aid for Scientific Research, Japan (Scientific Research (B) No. 14350253). This support is gratefully acknowledged.

REFERENCES

Bornitz, G. (1931): *Uber die Ausbreitung der von Grozklolbenmaschinen erzeugten Bodenschwingungen in die Tiefe*, J. Springer (Berlin).

Cheney, J. A., Brown, R. K., Dhat, N.R. and Hor, O. Y. Z. (1990): "Modeling free-field conditions in centrifuge models," *Journal of the Geotechnical Engineering, ASCE*, Vol. 116(9), pp. 1347-1367.

International union of railway (2002): Map Europe 2020, http://www.uic.asso.fr/home/home_en.html

Itoh, K., Koda, M., Lee, K. I., Murata, O. and Kusakabe, O. (2002): "Centrifugal simulations of wave propagation using a multiple ball dropping system," *International Journal of Physical Modelling in Geotechnics*, Vol. 2, No. 2, pp.33-51.

Lai, C. G., Callerio, A., Faccioli, E. and Martino, A. (2000): "Mathematical modelling of railway-induced ground vibrations," *Proceedings of International Workshop WAVE2000*, Balkema, Bochum, pp. 99-110.

Luong, M. P. (1994): "Efficiency of a stress wave mitigation barrier," *Proceeding of International Conference on Centrifuge Modelling-Centrifuge 94*, Balkema, Singapore, pp. 283-288.

Siemer, Th. and Jessberger, H.L. (1994): "Wave propagation and active vibration control in sand," *Proceeding of International Conference on Centrifuge Modelling-Centrifuge 94*, Balkema, Singapore, pp. 307-312.

Svinkin, M. R. (2001): "An impulse response function approach for predicting ground and structure vibrations," *Proceedings of the fifteenth international conference on soil mechanics and geotechnical engineering*, Balkema, Istanbul, pp. 785-787.

Takemura, J., Kondoh, M., Esaki, T., Kouda, M. and Kusakabe, O. (1999): "Centrifuge model tests on double propped wall excavation in soft clay," *Soils and Foundations*, Vol. 39, No. 3, pp. 75-87.

Yoshioka, O., Nagai, M., Kanema, T. and Mitsuzuka, T. (1980): "On prediction method of train-induced ground vibration by a weight dropping experiment and a microbus running test," *Butsuri-Tanko (Geophysical Exploration)*, Vol. 33, No. 6, pp. 333-351 (in Japanese).

Wave propagation – Moving load – Vibration reduction, Chouw & Schmid (eds.)
© 2003 Swets & Zeitlinger, Lisse, ISBN 90 5809 559 2

Dynamic response of a plate element of a steel girder caused by high-speed train

M. Okamura
Yamanashi University, Kofu, Japan

ABSTRACT: On the thin-walled steel members of railway bridges, it has been clarified by measuring the vibration of a railway bridge that out-of-plane flexural vibration of the plate elements of the member increases suddenly with speedup of a train. This paper aims to elucidate the mechanism of the sudden increase of the out-of-plane flexural vibration. Firstly, the phase velocity and group velocity of the flexural waves in plate elements of thin-walled steel members are computed. Secondly, the dynamic responses of the plates under high speed moving loads are analyzed. The results show that it is necessary to investigate not only in terms of vibration of the plate element related to periodicity of axle alignment of the train but also in terms of propagation of flexural wave in the plate element in order to elucidate the dynamic behavior of the plate element under a high-speed train.

1 INTRODUCTION

The dynamic behavior of railway bridges caused by a traveling train has attracted the attention of researchers for many years. The dynamic behavior of bridge girders was numerically analyzed on the basis of beam theory, and it was shown that the dynamic behavior of the girders is strongly affected by the traveling speed and axle alignment of the train.

These dynamic effects of a traveling train on railway bridges have been accounted for by using an impact factor in the design of railway bridges. The impact factor depends on the train speed, the span and the fundamental frequency of the bridge. For instance, the maximum speed of the Shinkansen bullet train for the design of railway bridges is about 300 km/h. However, it is recommended to elucidate the dynamic behavior of railway bridges caused by a traveling train for speeds over 300 km/h because train speeds tend to increase over the years.

Regarding the dynamic behavior of a steel railway box girder under a high-speed train, the following were clarified by field measurements on the Shinkansen: 1) Out-of-plane flexural vibrations of web plates increased with speedup of a train. 2) The acceleration amplitude of the vibration especially increases when the train is traveling at a certain speed. Nevertheless, the mechanism of these phenomena and their relation with the out-of-plane flexural vibrations have not been clarified.

Therefore, in order to elucidate the mechanism of the phenomenon, dynamic behavior of a plate element of a steel box girder caused by high speed traveling loads is numerically analyzed. The dynamic behavior of the plate is investigated not only in terms of the vibration problem but also in terms of the wave motion problem.

2 OUTLINE OF ANALYSYS

2.1 *Plate and load characteristics*

The plate element of a steel girder and traveling loads used in this study are illustrated in Figures 1 and 2.

The dimensions of the plate were determined with reference to the steel box girder that was used for the field measurements. It is assumed that the longitudinal ends of the plate are simply supported and other ends are free. The material properties of the plate are given as follows: Young's modulus E = 206 Gpa, Poisson's ratio, ν = 0.3, and density ρ = 7850 kg/m^3.

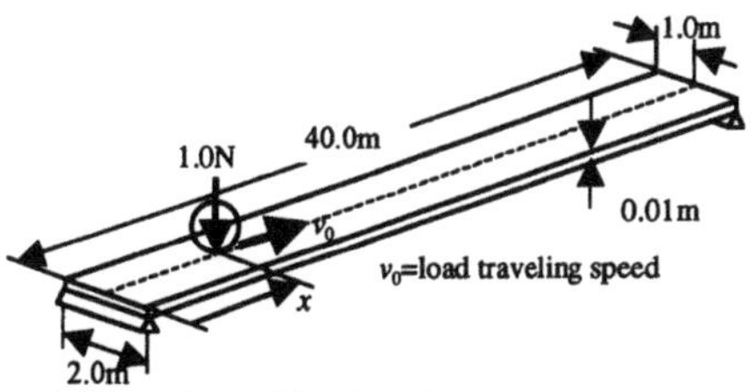

Figure 1. Dimensions of the plate element

Figure 2 shows a coach with four wheel axles. A concentrated load denotes an axle force of a train. Two loading cases were numerically analyzed; 1) an axle load is loaded, 2) twelve axle loads which denote a train of three coaches are loaded.

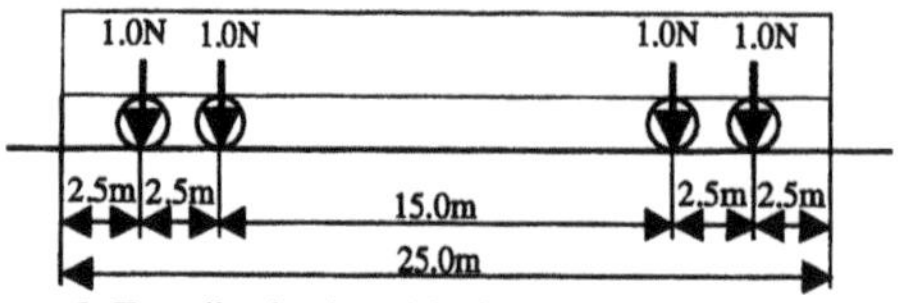

Figure 2. Traveling load model of a coach

2.2 *Analytical Method*

The plate element was idealized as finite strip elements. A typical finite strip i with nodal lines 1 and 2 is shown in Figure 3.

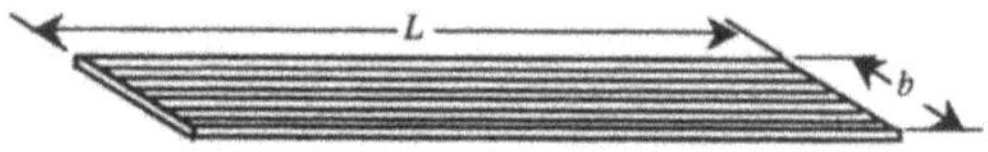

(1) Division into finite strips

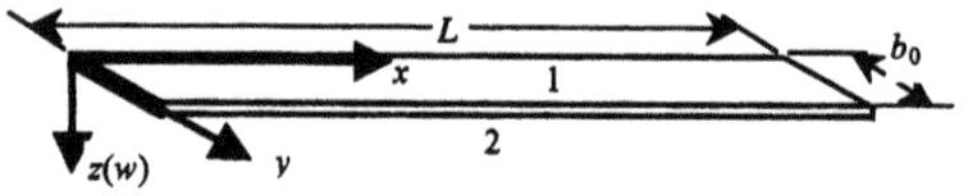

(2) Typical finite strip element

Figure 3. Finite strip model

If it is assumed that the longitudinal ends of the strip are simply supported, the displacement function of w in directions z is written as

$$w = \sum_{m=1}^{r} \langle N \rangle \{W_m\} \sin\left(\frac{m\pi}{L}x\right) \tag{1}$$

where r = number of terms of series in the longitudinal direction; and $\{W_m\}$ = vector of nodal line displacement corresponding to the m-th term of the longitudinal series. $\langle N \rangle$ is shape function in the transverse direction, which has the form

$$\langle N \rangle = \langle 1-3\eta^2+2\eta^3 \quad y(1-2\eta+\eta^2) \quad 3\eta^2-2\eta^3 \quad y(\eta^2-\eta) \rangle \tag{2}$$

where $\eta = y/b_0$.

Substituting the displacement functions of Equation (1) into the equation of virtual work, the equations of motion for a strip are derived. Transforming the matrices in the local coordinate system for strips into the global coordinate system and rearranging them according to the compatibility of displacements and the equilibrium of forces, the equation of motion for the plate is obtained as

$$[M]\frac{d^2\{\delta\}}{dt^2} + [K]\{\delta\} = \{F\} \tag{3}$$

where $[M]$ = mass matrix; $[K]$ = stiffness matrix; $\{\delta\}$ = displacement vector; and $\{F\}$ = force vector.

If $\{\delta\}$ is sinusoidal wave propagated in the longitudinal direction of the plate, it takes the form of

$$\{\delta\} = \{\delta_0\}\sin\frac{2\pi}{l}\left(x - c_p t\right) \tag{4}$$

where $\{\delta_0\}$ = amplitude; l = half-wavelength and c_p = phase velocity.

Substituting Equation (4) into Equation (3), an eigen equation is obtained as

$$\det\left|[K] - \left(\frac{\pi}{l}c_p\right)^2 [M]\right| = 0 \tag{5}$$

Group velocity c_g of the waves which are propagated in the longitudinal direction of the plate can be derived that

$$c_g = c_p - l\frac{dc_p}{dl} \tag{6}$$

The dynamic responses of the girder were simulated by applying a modal analysis and Newmark's generalized acceleration method. It was assumed that the loads started to travel from the left-hand end of the plate which was standing still at time $t = 0$.

The plate was divided into 10 strip elements (see Fig. 3(1)). The numerical calculations were performed by taking 250 terms of harmonic functions in the longitudinal direction and 8 modes of cross-sectional deformation for each term of harmonic function in the longitudinal direction. Since the loads move on the center line of the plate, the cross-sectional modes deforming symmetrically were adopted.

3 CHARACTERISTICS OF WAVE PROPAGATION IN THE PLATE

Phase velocity and group velocity of the plate element are obtained by solving Equation (5) and (6). The dispersion curves of phase velocity and group velocity of the plate are shown in Figure 4. The abscissa b/l denotes the wavelength ratio, where b is the width of the plate and l is the longitudinal half-wavelength.

The minimums of secondary and tertiary phase velocity dispersion curves are 305.7 km/h and 657.6 km/h, respectively. These points also mean intersections of phase velocity and group velocity curves. Inequality between phase velocity curves and group velocity curves are reversed at these points, and the secondary and tertiary group velocities are faster than the phase velocities for shorter wavelengths. In addition, since the phase velocities of the waves around the minimum points are constant with respect to wavelength ratio, it is predicted that the wave groups propagating at about 305 or 660 km/h are nondispersive and maintain their shapes.

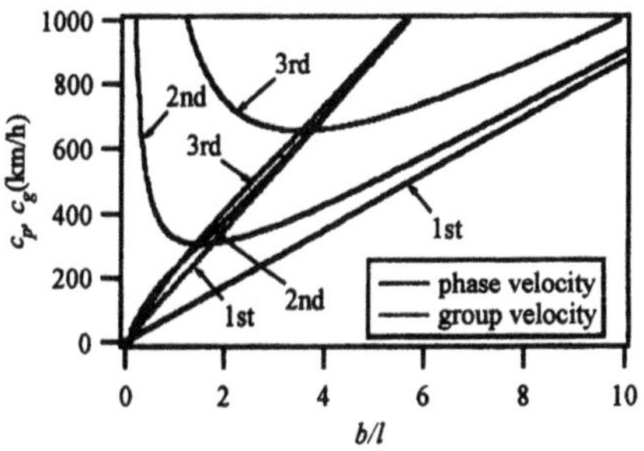

Figure 4. Phase velocity and group velocity dispersion curve

4 DYNAMIC BEHAVIOR CAUSED BY THE HIGH-SPEED TRAVELING LOADS

4.1 *Wave motion under an axle load*

The longitudinal distributions of vertical acceleration in the plate for varying load speed were obtained when the load arrived at the midspan.

Figure 5 shows the wave shapes of the vertical acceleration a_w at the loading line of the plate when the load arrives at the midspan. The vertical and horizontal arrows respectively denote the longitudinal position and the forward direction of the load.

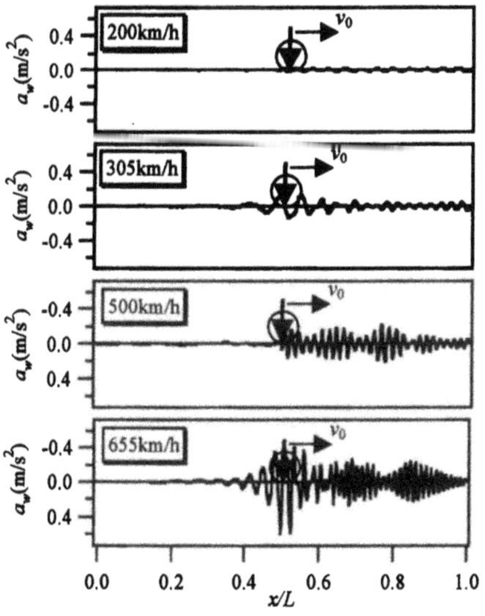

Figure 5. Wave shape of vertical acceleration at loading line

The wave shape around the loading point at v_0 = 305 km/h and 655 km/h show large amplification. The waveforms are asymmetric with respect to the loading point. The reason for the asymmetrical wave shape is explained as follows; since the secondary and tertiary group velocities are faster than the phase velocities for short wavelengths, the wave group of short wavelength propagates both forward and backward of the load, and the wave group of long wavelength propagates only behind the load.

The relation between the maximum amplitude $a_{w,max}$ taken from these wave shapes of the vertical acceleration and the load speed v_0 is shown in Figure 6. The maximum amplitude of the acceleration increase greatly at v_0 = 305 km/h and 650 km/h. The same phenomenon was observed in the field measurements, although there was a little difference in load speed. In addition, Figures 4 and 6 show that these values coincide with the minimums of the phase velocity dispersion curves. This means that the vertical acceleration of the plate increase greatly when the load speed is almost equal to the minimums of phase velocity dispersion curves.

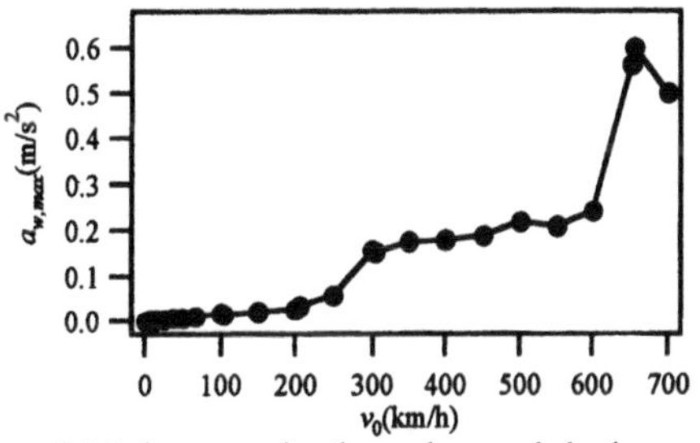

Figure 6. Maximum acceleration under an axle load versus load traveling speed

The longitudinal distributions of vertical acceleration in the plate for varying load speed were obtained when the loads arrived at a certain position. The results are shown in Figure 7.

They are contour maps of the vertical acceleration in the plate with the passage of the load at v_0 = 305 km/h, where t_0 is the time required for loads to travel the span of the plate. The characters ($0.25t_0$, $0.50t_0$, and so on) on the upper left corner of the contour maps denote the elapsed time required since the load starts to moves from the left-side end of the plate. The load leaves from left-side end of the plate at t = 0 and arrives at right-side end at $t = 1.00t_0$.

When the load is on the plate, the wave group with large amplitude generated around the load moves with the traveling load. When the load arrives at the right-side end of the plate, the wave group is reflected and propagates in reverse direction. After the load passes over the plate, the wave group propagates with maintaining its shape and speed.

These figures indicate the characteristics of nondispersive waves.

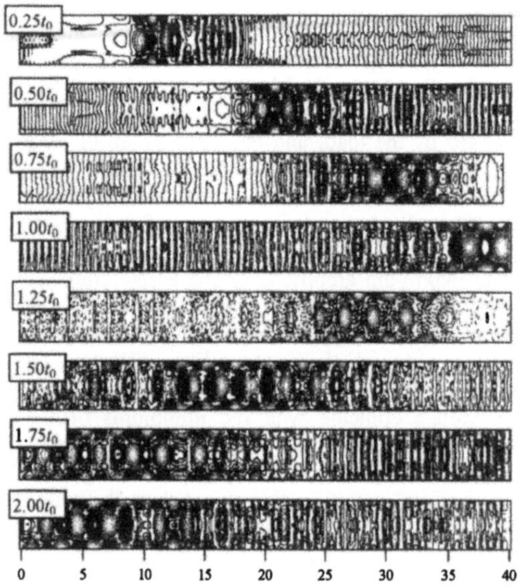

Figure 7. Change of contour of vertical acceleration with the passage of the load at v_0=305km/h

Furthermore, the response histories of vertical acceleration a_w at the central point of midspan section for varying load were obtained. The results are shown in Figure 8. The abscissa t/t_0 denotes the dimensionless time.

When the load travels at v_0 = 305 km/h or 650 km/h, the amplitudes increase at t/t_0 = 0.5 and 1.5. This means that the wave group traveling at v_0 = 305 km/h or 655 km/h is nondispersive and moves with the load speed.

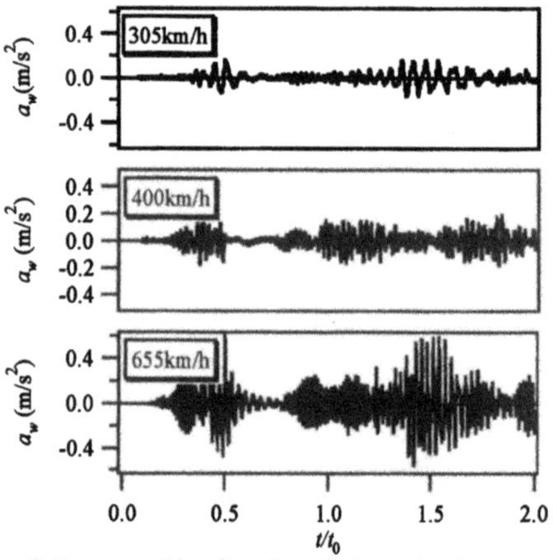

Figure 8. Response histories of vertical acceleration a_w at the central point of midspan section

The Fourier spectra of vertical acceleration a_w of the plate at the midspan at v_0 = 305, 400 and 655 km/h are shown in Figure 9. Moreover, in order to investigate dominant frequencies of these spectra, the relation between phase velocity and frequency is shown in Figure 10. This figure was obtained by transforming the wavelength ratio of the horizontal axis of Figure 4 into frequency.

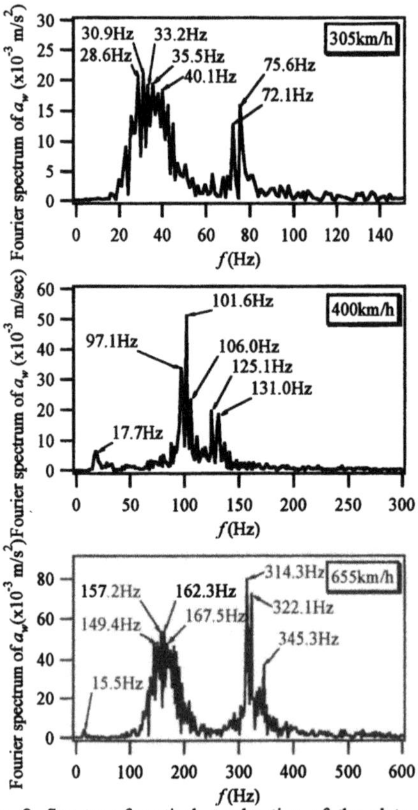

Figure 9. Spectra of vertical acceleration of the plate at the central point of midspan section

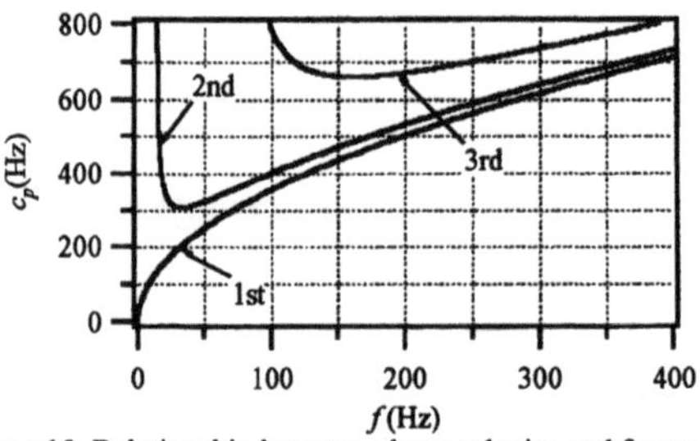

Figure 10. Relationship between phase velocity and frequency

According to Figure 9 and 10, it is found that the dominant frequencies coincide with the frequencies of intersection of phase velocity dispersion curves and a horizontal line for the load traveling speed. It is confirmed that the increase of vertical acceleration amplitude of the plate with speedup of the load is caused by the superposition of nondispersive waves because there are many peaks around the frequencies of the minimum points of the phase velocity dispersion curves in Figure 9.

4.2 *Wave motion under a train*

It is known that the dynamic response of a girder caused by a traveling train is influenced by frequency characteristics of the girder, train speed, periodicity of axle arrangement, and so on. However, the influence of nondispersive flexural waves on the dynamic response of the girder under a traveling train have not been clarified. Consequently, the dynamic responses of the plate element caused by the twelve traveling loads modelling a train of three coaches were simulated.

The response histories of vertical acceleration a_w at the central point of midspan section of the plate are shown in Figure 11. The abscissa t/t_0 denote dimensionless time, where t_0 is the time required for all axle loads of the train pass over the plate.

The response history curves show periodic characteristics due to axle alignment especially at v_0 = 400 km/h and 655 km/h. The amplitude of each response history curve increases as the dimensionless time t/t_0 increases. The characteristics of nondispersive wave generated at v_0 = 305 km/h and 655 km/h as mentioned above do not appear on these figures.

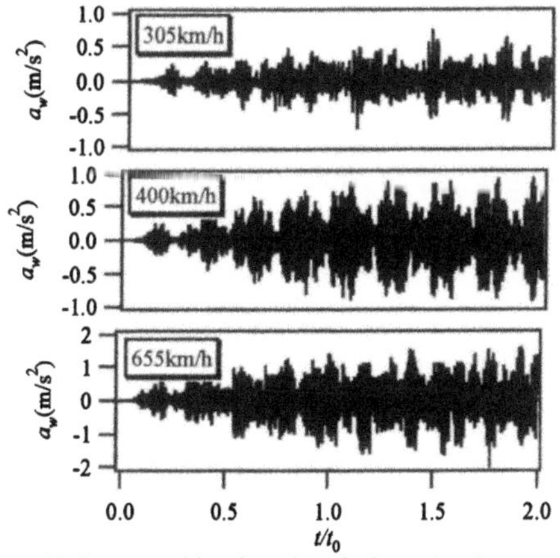

Figure 11. Response histories of vertical acceleration a_w at the central point of midspan section

The Fourier spectra of vertical acceleration a_w at the central point of midspan section of the plate are shown in Figure 12. The number of peaks become smaller compared to Figure 9 although dominant frequencies are almost the same.

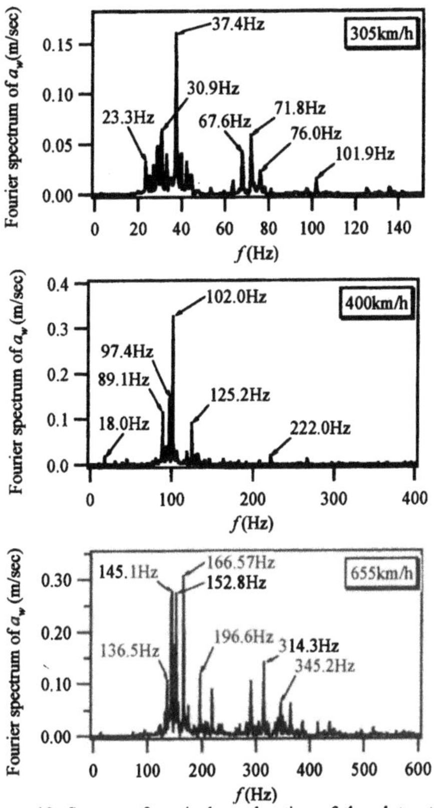

Figure 12. Spectra of vertical acceleration of the plate at midspan

The relationship between maximum amplitude of the vertical acceleration of the plate and the train speed is shown in Figure 13. The maximum amplitude of the acceleration caused by the train increases as the train speed increases. There are some peaks other than those for the minimums of the phase velocity dispersion curves of the plate.

To investigate these frequencies for the peaks, the relationship among the dominant frequencies of the vertical acceleration caused by the train, phase velocity of the plate and train speed is shown in Figure 14. Solid lines denote natural frequency-phase velocity curves, and dotted lines denote the relation between frequency and train speed which depend on axle intervals of the train (see the equation written in the Figure 14). The circles denote the dominant frequency taken from the Fourier spectra of vertical acceleration (Figure 12).

According to Figures 13 and 14, the train speeds for the maximum amplitude of the vertical acceleration of the plate are almost equal to the minimums of phase velocity dispersion curve or the speeds determined by natural frequency of the plate and axle alignment of the train.

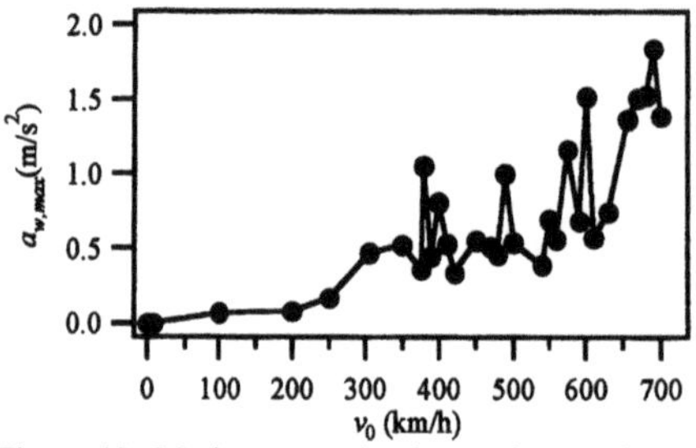

Figure 13. Maximum acceleration under a train versus load traveling speed

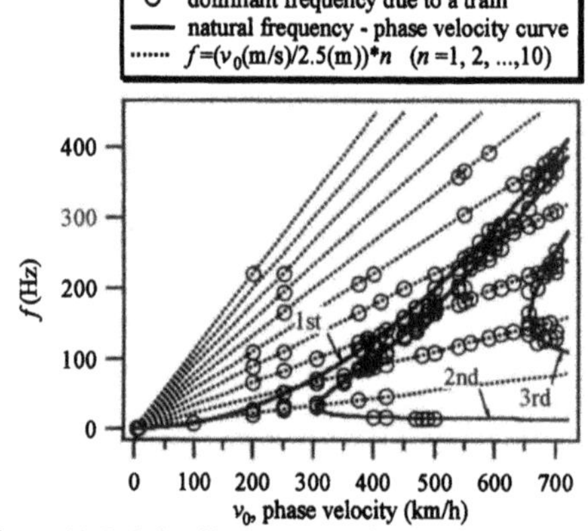

Figure 14. Relationship among dominant frequency, phase velocity and load traveling speed

5 CONCLUSION

The mechanism of the phenomenon that out-of-plane flexural vibration of plate elements of a steel railway girder increases with speed of a train was numerically investigated, noting the relationship between train speed and propagation velocity of flexural waves in the plate.

Results obtained from these analyses are as follows:

1) The wave shape around the traveling load especially increase when the speed of the traveling load approach the minimums of the phase velocity dispersion curves.

2) The wave group, which is generated by the traveling load at the speed of minimum of the phase velocity dispersion curve, reflects at the longitudinal end of the plate element and propagates maintaining its shape and speed after the passage of load on the plate element.

3) It is necessary to investigate not only in terms of vibration of the plate element related to periodicity of axle alignment of the train but also in term of propagation of flexural waves in the plate element in order to elucidate the dynamic behavior of plate element under a high-speed train.

REFERENCES

Dyle, J. F. 1997. *Wave propagation in structures*, New York: Springer.

Fung, Y. C. 1965. *Foundations of solid mechanics*, New Jersey: Prenntice-Hall, Inc.

Matsuura, A. 1976. Study of dynamic behavior of bridge girder for high speed railway. *Proc. of JSCE*, 256, 35-47.

Okamura, M. and Fukasawa, Y. (1994), "A fundamental study on the impulsive behavior of thin-walled steel I-section beams," *Journal of Structural Eng.*, 40A: 749-758.

Sugimoto, I., Miki, C., Ichikawa, A. and Itoh, Y. 1997. Dynamic behavior and stress features at the end of vertical stiffener of steel railway box girder under high-speed train. *Journal of Structural Eng.*, 43A: 1003-1012.

Wave propagation – Moving load – Vibration reduction, Chouw & Schmid (eds.)
© 2003 Swets & Zeitlinger, Lisse, ISBN 90 5809 559 2

Experimental validation of a numerical prediction model for traffic induced vibrations by in situ experiments

L. Pyl, G. Degrande & G. Lombaert
K.U. Leuven, Department of Civil Engineering, Heverlee, Belgium

W. Haegeman
Ghent University, Soil Mechanics Laboratory, Zwijnaarde, Belgium

ABSTRACT: A 3D numerical prediction model for road traffic induced vibrations, accounting for dynamic soil-structure interaction at the source and the structure, is developed. First, this paper deals with the essential elements of the numerical prediction model. Second, the experimental setup is discussed. Vibration measurements are performed in and near a single family dwelling during the passage of a truck on an uneven concrete road with joints for vehicle speeds between 25 km/h and 51 km/h. Additional measurements are performed for truck passages on an artificial profile installed on the uneven road for vehicle speeds between 23 km/h and 58 km/h. Third, the preliminary results of the validation based on the vibration measurements are presented. Although a further refinement of the model is needed and the input parameters have to be optimized, a good agreement is obtained.

1 INTRODUCTION

Road traffic induced vibrations in buildings are a matter of growing environmental concern as they may cause malfunctioning of sensitive equipment, discomfort to people and structural damage. Regarding the cost of vibration isolation measures, there is a clear demand to develop accurate numerical tools.

An existing source model for the prediction of free field traffic induced vibrations is used for the calculation of the incident wave field due to the passage of a truck on an uneven road (Lombaert et al. 2000). Dynamic interaction between the road and the layered soil is accounted for by means of a substructuring technique (Aubry and Clouteau 1992). An analytical beam model is used for the road and the soil is modelled using boundary elements. This model has been validated by means of in-situ experiments (Lombaert and Degrande 2001).

The response of the structure due to this incident wave field is computed with the program MISS, using a subdomain formulation for dynamic soil-structure interaction (Aubry and Clouteau 1992). A finite element method is applied to the structure, while the unbounded soil domain is calculated with a boundary element method using the Green's functions of a layered halfspace.

The experimental results of an in situ measurement campaign, that has been performed in and near a structure located at a distance of about 15.0 m from a busy road, are used to validate the numerical prediction model (Pyl et al. 2002).

The objective of this paper is to present the preliminary results of the experimental validation of the numerical prediction model. First, the essential elements of the model are briefly recapitulated. Second, the experimental setup is discussed and finally the results of the numerical prediction are compared with the experimental results.

2 THE NUMERICAL MODEL

2.1 *The subdomain formulation*

To calcultate the dynamic response of a structure subjected to an incident wave field, the problem is decomposed into two subdomains: the structure and the soil. It is assumed that the incident wave field is not influenced by the presence of the structure. The dynamic soil-structure interaction problem is solved by enforcing continuity of displacements and equilibrium of stresses on the interface Σ between both subdomains.

The displacement vector in the structure is equal to $\mathbf{u}_{st} = \mathbf{\Phi}\boldsymbol{\alpha}$ where the matrix $\mathbf{\Phi}$ collects the eigenmodes $\boldsymbol{\phi}_j$ of the structure ($j = 1, ..., q$) with q a set of linearly independant eigenmodes that is usually much smaller than the total number of degrees of freedom n. The vector $\boldsymbol{\alpha}$ contains the modal coordinates.

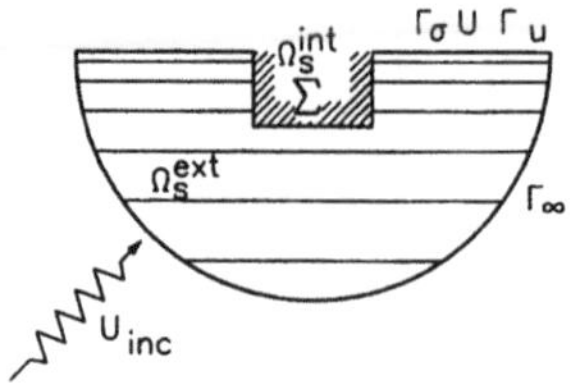

a. $\mathbf{u}_0$.

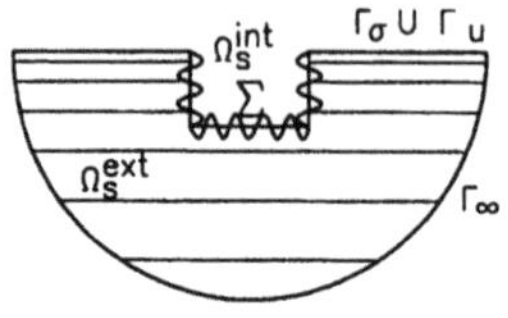

b. $\mathbf{u}_{sc}$.

Figure 1: Decomposition of the displacement field $\mathbf{u}_s$ into (a) a displacement field $\mathbf{u}_0$, and (b) a scattered wave field $\mathbf{u}_{sc}$.

The displacement vector in the soil $\mathbf{u}_s$ is decomposed into a displacement field $\mathbf{u}_0$ and a scattered wave field $\mathbf{u}_{sc}$ (figure 1). The scattered wave field $\mathbf{u}_{sc}$ in the soil is due to the motion of the interface Σ. This motion is reradiated into the semi-infinite soil domain Ω_s^{ext}. The scattered wave field $\mathbf{u}_{sc}$ can generally be written as the superposition of displacement modes $\boldsymbol{\Psi}$ with unknown modal coordinates $\boldsymbol{\alpha}_{sc}$.

The displacement field $\mathbf{u}_0$ is decomposed into an incident wave field $\mathbf{u}_i$ and a locally diffracted wave field $\mathbf{u}_{d0}$ (Aubry and Clouteau 1992) (figure 2):

$$\mathbf{u}_s = \mathbf{u}_0 + \mathbf{u}_{sc} = \mathbf{u}_i + \mathbf{u}_{d0} + \mathbf{u}_{sc} \quad (1)$$

The incident wave field $\mathbf{u}_i$ is defined on the soil domain without excavation. The locally diffracted wave field is equal to $-\mathbf{u}_i$ on the interface Σ.

2.2 *Displacement continuity*

Displacement continuity on Σ is expressed as:

$$\mathbf{u}_{st} = \mathbf{u}_i + \mathbf{u}_{d0} + \mathbf{u}_{sc} \quad \text{on} \quad \Sigma \quad (2)$$

As $\mathbf{u}_i + \mathbf{u}_{d0} = \mathbf{u}_0$ is equal to zero on the interface Σ, this equation reduces to:

$$\mathbf{u}_{st} = \mathbf{u}_{sc} \quad \text{or} \quad \boldsymbol{\Phi}_\Sigma \boldsymbol{\alpha} = \boldsymbol{\Psi}_\Sigma \boldsymbol{\alpha}_{sc} \quad \text{on} \quad \Sigma \, (3)$$

If the projections of the displacement modes $\boldsymbol{\Phi}_\Sigma$ are equal to $\boldsymbol{\Psi}_\Sigma$, the unknown modal coordinates $\boldsymbol{\alpha}_{sc}$ are equal to the unknown modal coordinates $\boldsymbol{\alpha}$.

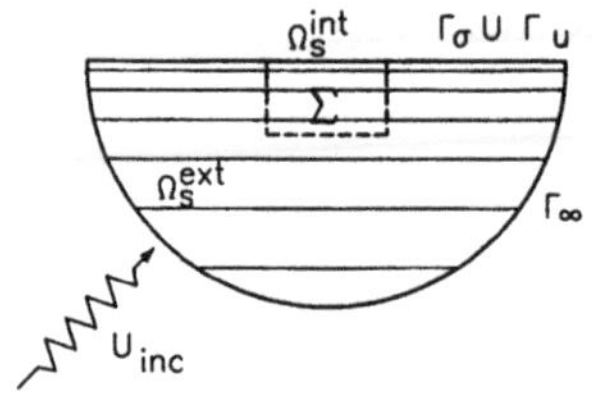

a. $\mathbf{u}_i$.

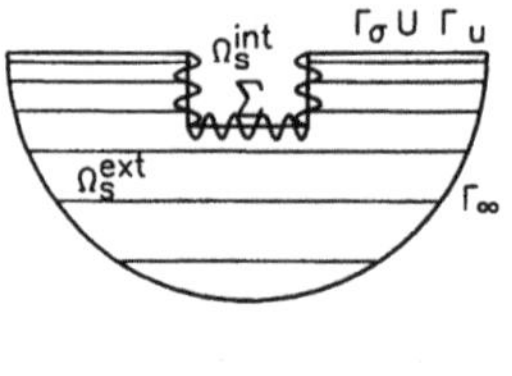

b. $\mathbf{u}_{d0}$.

Figure 2: Decomposition of the displacement field $\mathbf{u}_0$ into (a) an incident wave field $\mathbf{u}_i$, and (b) a locally diffracted wave field $\mathbf{u}_{d0}$.

2.3 *Stress equilibrium*

The stress equilibrium on the interface Σ is imposed in the weak sense expressing that the virtual work of the sum of the stresses $\mathbf{t}_{st}(\mathbf{u}_{st})$ and $\mathbf{t}_s(\mathbf{u}_s)$ on the interface Σ, related to a virtual displacement field $\mathbf{v}$, should be zero. Using the decomposition of the soil displacement vector $\mathbf{u}_s$, the virtual work equation becomes:

$$\int_\Sigma \mathbf{v} \cdot \mathbf{t}_{st}(\mathbf{u}_{st}) \, d\Sigma + \int_\Sigma \mathbf{v} \cdot \mathbf{t}_s^{\mathbf{n}^{\text{ext}}}(\mathbf{u}_{sc}) \, d\Sigma =$$

$$- \int_\Sigma \mathbf{v} \cdot \mathbf{t}_s^{\mathbf{n}^{\text{ext}}}(\mathbf{u}_i) \, d\Sigma - \int_\Sigma \mathbf{v} \cdot \mathbf{t}_s^{\mathbf{n}^{\text{ext}}}(\mathbf{u}_{d0}) \, d\Sigma \quad (4)$$

where $\mathbf{n}^{\text{ext}}$ is the unit outward normal vector on the boundary Σ of the semi-infinite layered halfspace Ω_s^{ext}. The unknown loading terms are collected on the right hand side of equation (4).

The first term on the left hand side of equation (4) represents the virtual work of the stresses $\mathbf{t}_{st}(\mathbf{u}_{st})$ on the interface Σ. Using the principle of virtual work, this term is elaborated as the sum of the virtual work of the internal forces and the inertial forces in the structure:

$$\int_\Sigma \mathbf{v} \cdot \mathbf{t}_{st}(\mathbf{u}_{st}) \, d\Sigma =$$

$$\int_{\Omega_{st}} \boldsymbol{\varepsilon}(\mathbf{v}) : \boldsymbol{\sigma}(\mathbf{u}_{st}) \, d\Omega + \int_{\Omega_{st}} \mathbf{v} \cdot \rho \ddot{\mathbf{u}}_{st} \, d\Omega \quad (5)$$

The first term on the right hand side of equation (4) denotes the virtual work of the stresses $\mathbf{t}_s^{\mathbf{n}^{\text{ext}}}(\mathbf{u}_i)$ on the interface Σ. As this term is defined on the bounded interior soil domain Ω_s^{int} with unit outward normal vector $\mathbf{n}^{\text{int}}$, an analogous expression as in equation (5) is obtained:

$$-\int_{\Sigma} \mathbf{v} \cdot \mathbf{t}_s^{\mathbf{n}^{\text{ext}}}(\mathbf{u}_i)\, d\Sigma = +\int_{\Sigma} \mathbf{v} \cdot \mathbf{t}_s^{\mathbf{n}^{\text{int}}}(\mathbf{u}_i)\, d\Sigma =$$

$$\int_{\Omega_s^{\text{int}}} \varepsilon(\mathbf{v}) : \sigma(\mathbf{u}_i)\, d\Omega + \int_{\Omega_s^{\text{int}}} \mathbf{v} \cdot \rho \ddot{\mathbf{u}}_i\, d\Omega \quad (6)$$

As the first term on both sides of equation (4) is defined on a bounded domain, a finite element method is used for the calculation.

The second term on the left hand side of equation (4) represents the virtual work of the stresses $\mathbf{t}_s(\mathbf{u}_{sc})$ on the interface Σ. The semi-infinite soil domain is analysed to calculate the unknown tractions $\mathbf{t}_s(\mathbf{u}_{sc})$.

Using dynamic reciprocity and $\mathbf{u}_{d0} = -\mathbf{u}_i$ on Σ, the second term on the right hand side of equation (4) can be elaborated as:

$$\begin{aligned} -\int_{\Sigma} \mathbf{v} \cdot \mathbf{t}_s^{\mathbf{n}^{\text{ext}}}(\mathbf{u}_{d0})\, d\Sigma &= -\int_{\Sigma} \mathbf{t}_s^{\mathbf{n}^{\text{ext}}}(\mathbf{v}) \cdot \mathbf{u}_{d0}\, d\Sigma \\ &= +\int_{\Sigma} \mathbf{t}_s^{\mathbf{n}^{\text{ext}}}(\mathbf{v}) \cdot \mathbf{u}_i\, d\Sigma \quad (7) \end{aligned}$$

As the second term on both sides of equation (4) is related to a semi-infinite soil domain Ω_s^{ext}, a boundary element method based on the Green's functions for a layered halfspace is used.

2.4 *Discretized equations*

The discretized form of the scalar equilibrium equation (4) in the frequency domain is:

$$\begin{aligned} &\underline{\mathbf{v}}^{\mathrm{T}} \left(\int_{\Omega_{st}} \mathbf{B}^{\mathrm{T}}\mathbf{D}\mathbf{B}\, d\Omega - \omega^2 \int_{\Omega_{st}} \mathbf{N}^{\mathrm{T}}\rho\mathbf{N}\, d\Omega \right) \underline{\mathbf{u}}_{st} \\ &+\underline{\mathbf{v}}^{\mathrm{T}} \int_{\Sigma} \mathbf{N}^{\mathrm{T}}\mathbf{N}\mathbf{G}^{-1}\mathbf{H}\, d\Sigma\, \underline{\mathbf{u}}_{sc} \\ &= \underline{\mathbf{v}}^{\mathrm{T}} \left(\int_{\Omega_s^{\text{int}}} \mathbf{B}^{\mathrm{T}}\mathbf{D}\mathbf{B}\, d\Omega - \omega^2 \int_{\Omega_s^{\text{int}}} \mathbf{N}^{\mathrm{T}}\rho\mathbf{N}\, d\Omega \right) \underline{\mathbf{u}}_i \\ &+\underline{\mathbf{v}}^{\mathrm{T}} \int_{\Sigma} (\mathbf{G}^{-1}\mathbf{H})^{\mathrm{T}}\mathbf{N}^{\mathrm{T}}\mathbf{N}\, d\Sigma\, \underline{\mathbf{u}}_i \qquad (8) \end{aligned}$$

The displacement vector $\mathbf{u}_{st}$ in the first term on the left hand side of equation (4), is approximated as $\mathbf{N}\underline{\mathbf{u}}_{st}$, where $\mathbf{N}$ are the shape functions. In a Galerkin formulation, a similar decomposition is used for the virtual displacements $\mathbf{v} = \mathbf{N}\underline{\mathbf{v}}$. The strain-displacement relation $\varepsilon = \mathbf{L}\mathbf{u}_{st}$ and the constitutive equation $\sigma = \mathbf{D}\varepsilon$ are introduced. The matrix $\mathbf{B}$ represents the matrix product $\mathbf{LN}$. The first term on the right hand side of equation (4) is calculated analogously.

To calculate the unknown tractions $\mathbf{t}_s(\mathbf{u}_{sc})$ due to the scattered wave field in the second term on the left hand side of equation (4), the interface Σ is discretized into boundary elements and the tractions and displacements on the interface are defined by their nodal values and element based shape functions. The boundary integral equation is evaluated numerically and a fully-populated non-symmetrical system of equations $\mathbf{H}\underline{\mathbf{u}}_{sc} = \mathbf{G}\mathbf{t}(\underline{\mathbf{u}}_{sc})$ is obtained, with the unknown tractions $\mathbf{t}(\underline{\mathbf{u}}_{sc})$. The second term on the right hand side of equation (4) is calculated analogously.

As equation (8) must hold for any virtual displacement field $\underline{\mathbf{v}}$, the resulting equilibrium equation in matrix-vector form is equal to:

$$\begin{bmatrix} \mathbf{S}_{ss} & \mathbf{S}_{sb} \\ \mathbf{S}_{bs} & \mathbf{S}_{bb} + \mathbf{S}_{bb}^g \end{bmatrix} \begin{Bmatrix} \mathbf{u}_s^t \\ \mathbf{u}_b^t \end{Bmatrix} = \begin{Bmatrix} \mathbf{0} \\ \mathbf{S}_{bb}^f \mathbf{u}_b^f \end{Bmatrix} \quad (9)$$

The total structural displacement vector $\underline{\mathbf{u}}_{st}$ in equation (8) contains the degrees of freedom of the discretized model and is split into a vector $\mathbf{u}_b^t$ and $\mathbf{u}_s^t$ corresponding to the degrees of freedom of the soil-structure interface and the remaining degrees of freedom of the structure, respectively.

The term between brackets on the left hand side of equation (8) is the impedance of the structure and is decomposed into the block matrices $\mathbf{S}_{ss}$, $\mathbf{S}_{bb}$, $\mathbf{S}_{bs}$ and $\mathbf{S}_{sb}$.

The second term on the left hand side of equation (8) results in a force $\mathbf{S}_{bb}^g\underline{\mathbf{u}}_{sc}$ with $\mathbf{S}_{bb}^g$ the complex, frequency dependent dynamic stiffness matrix of the unbounded soil domain with excavation. The displacement field $\underline{\mathbf{u}}_{sc}$ is replaced by $\mathbf{u}_b^t$ as continuity of displacements is enforced on the interface Σ.

The term between brackets on the right hand side of equation (8) is the impedance $\mathbf{S}_{bb}^e$ of the excavated part of the soil Ω_s^{int} and the second term corresponds to the impedance $\mathbf{S}_{bb}^g$ of the unbounded soil domain Ω_s^{ext}.

As the sum of the impedances $\mathbf{S}_{bb}^e$ and $\mathbf{S}_{bb}^g$ equals the impedance $\mathbf{S}_{bb}^f$ of the soil domain Ω_s without excavation, the term on the right hand side of equation (8) results in a force $\mathbf{S}_{bb}^f\mathbf{u}_b^f$, where $\mathbf{u}_b^f$ denotes the projection of $\underline{\mathbf{u}}_i$ on the interface Σ.

Introducing modal decomposition in equation (9) results into:

$$\begin{Bmatrix} \underline{\mathbf{\Phi}}_s^{\mathrm{T}} \underline{\mathbf{\Phi}}_b^{\mathrm{T}} \end{Bmatrix} \begin{bmatrix} \mathbf{S}_{ss} & \mathbf{S}_{sb} \\ \mathbf{S}_{bs} & \mathbf{S}_{bb} + \mathbf{S}_{bb}^g \end{bmatrix} \begin{Bmatrix} \underline{\mathbf{\Phi}}_s \\ \underline{\mathbf{\Phi}}_b \end{Bmatrix} \alpha$$

$$= \begin{Bmatrix} \mathbf{0} \\ \underline{\mathbf{\Phi}}_b^{\mathrm{T}} \mathbf{S}_{bb}^f \mathbf{u}_b^f \end{Bmatrix} \qquad (10)$$

where $\underline{\mathbf{\Phi}}_b$ and $\underline{\mathbf{\Phi}}_s$ are the eigenmodes corresponding to the degrees of freedom of the soil-structure interface and the remaining degrees of freedom of the structure, respectively.

3 THE EXPERIMENTAL SETUP

3.1 *The structure*

Figure 3: Front and side view of the single family dwelling.

The structure is a two storey single family dwelling with a rectangular plan (figure 3) and an embedded concrete box foundation. The house is located at a distance of 15.0 m from the busy road N18 connecting the 2 Belgian municipalities of Mol and Retie.

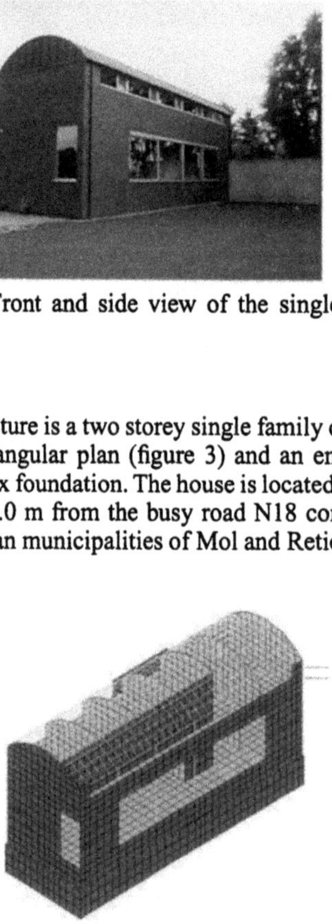

Figure 4: Finite element model of the structure.

The structure is modelled using shell and beam elements (Talloen 2002) (figure 4). The floors are made of Stalton beams with intermediate members of burnt clay. An orthotropic plate model is developed and an updating to the measured data is applied to the model. The updating parameters are used to construct an equivalent plate that is implemented in the finite element model of the structure. The slightly different boundary conditions of the plate in the complete finite element model result in a first bending mode of 13.5 Hz of the plate versus 14.5 Hz from measured data.

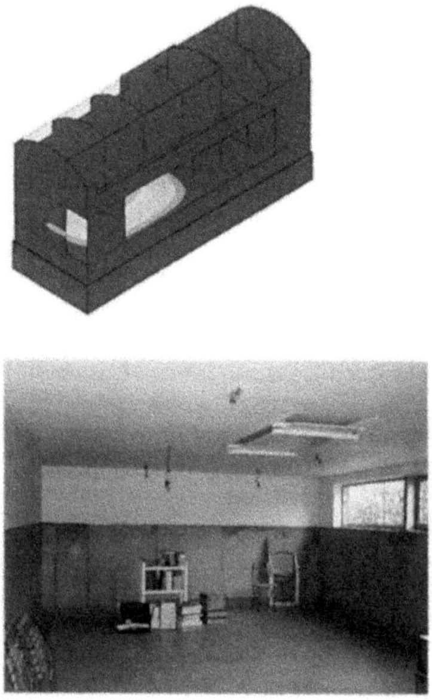

Figure 5: Bending mode of the slab on the first floor at 13.5 Hz and constrained-layer damping system.

Previously, a lot of vibration nuisance was experienced on the first floor during the passage of heavy trucks and buses. Therefore, a constrained-layer damping system has been attached to the slab on the first floor (Dewulf and De Roeck 2000a; Dewulf and De Roeck 2000b) (figure 5). During the measurements this constrained-layer damping system has been removed.

Figure 6a shows the torsion mode of the structure at 16.3 Hz and figure 6b shows the second bending mode of the slab on the first floor at 20.9 Hz.

For validation purposes, a complete finite element model of the structure in Retie is introduced in the numerical prediction model. However, a model of the slab on the first floor could be sufficient as the plate's first bending mode causes most problems.

3.2 *The dynamic excitation*

The road is constructed with concrete slabs with an average length of 12.0 m and a width of 3.5 m. The transverse joints between the concrete slabs are filled with a bitumen. Especially on weekdays, the N18 is a very busy road with a lot of trucks. The present measurements were performed with a two-axle Volvo FL6 truck, however, while all other traffic was deviated. The truck has a wheel base of 5.20 m. The total mass

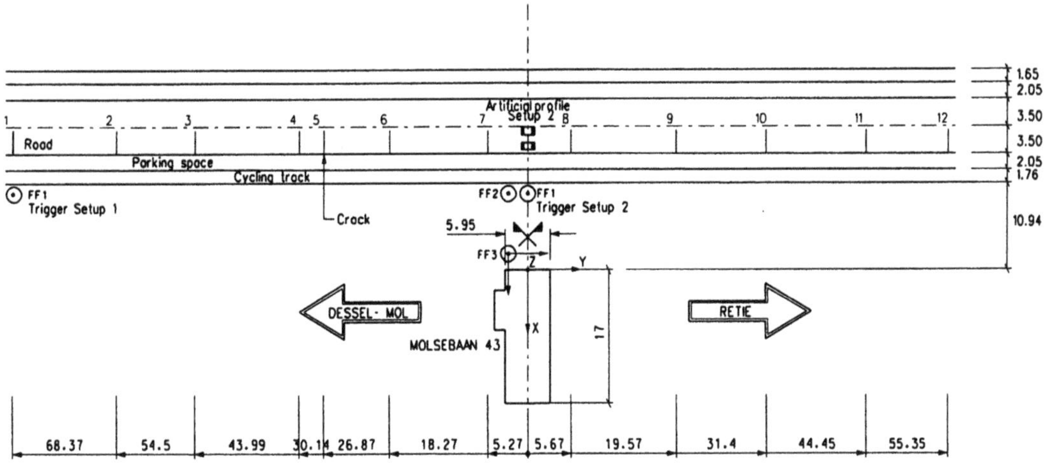

Figure 7: Plan view.

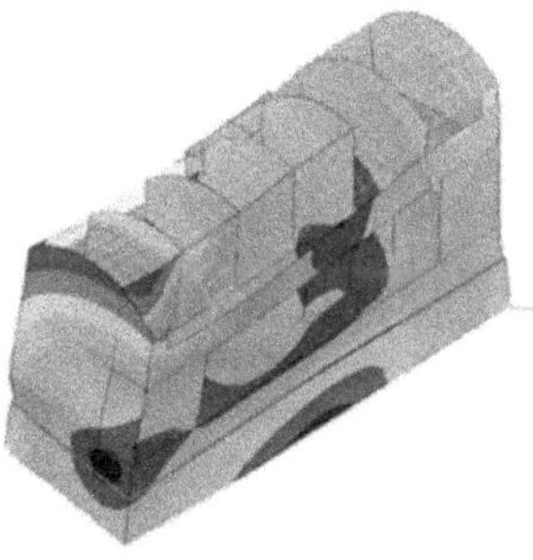

a. 16.3 Hz.

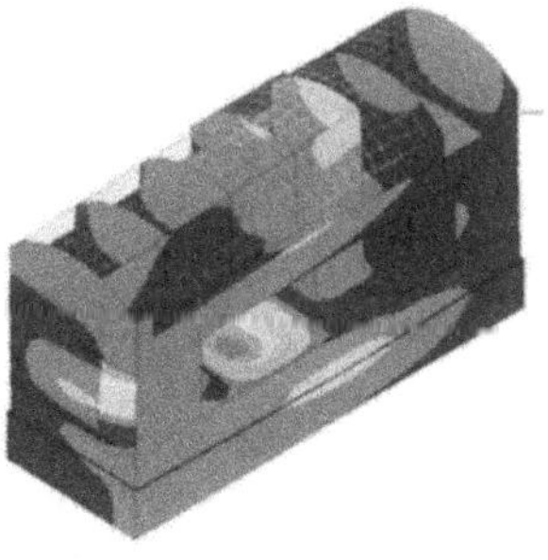

b. 20.9 Hz.

Figure 6: Torsion mode of the structure and second bending mode of the slab on the first floor.

of the vehicle is 8620 kg. The front and rear axle mass are equal to 2800 kg and 5820 kg, respectively.

Two measurement setups are used in this campaign and will be referred to as setup 1 and 2.

In setup 1, 21 passages of the truck on the uneven road have been recorded for a wide range of vehicle speeds between 25 km/h and 51 km/h. The major unevennesses (both joints and cracks) in the road surface in front of the house are numbered from 1 to 12 (figure 7). Unevenness 5 is a crack in the concrete plate.

The Belgian Road Research Center has measured the height of the discontinuities between the concrete plates, using a ZEISS DiNi 10 digital altimeter and an invar measuring staff, with a precision of 0.3 mm (figure 8a) (N.N. 2001b). Additionally, the relative vertical motion of the concrete plates during the passage of the vehicle's rear axle at low speed has been measured using a faultimeter (figure 8b) (N.N. 2001a). In joint 6, the relative motion is small with respect to the height of the joint, and will be disregarded. The joints can consequently be modelled as a step function.

In setup 2, 39 passages of the truck have been recorded for a range of vehicle speeds between 23 km/h and 58 km/h on an artificial profile installed in front of the house on the uneven road (figure 7). The artificial plywood unevenness has a total height of $H = 0.054$ m, a flat top part with a length $L = 1.30$ m and slopes with a length $l = 0.30$ m. It has been designed as to control the frequency content of the dynamic axle loads within a vehicle speed range between 25 km/h and 75 km/h (Lombaert 2001).

3.3 *The road*

In the source model, a two layer road model with a concrete plate with thickness $d = 0.20\,\text{m}$, Young's modulus $E = 400 \times 10^8\,\text{N/m}^2$, material density $\rho = 2500\,\text{kg/m}^3$ and Poisson's ratio $\nu = 0.2$ on top of a granular subbase with thickness $d = 0.30\,\text{m}$, Young's

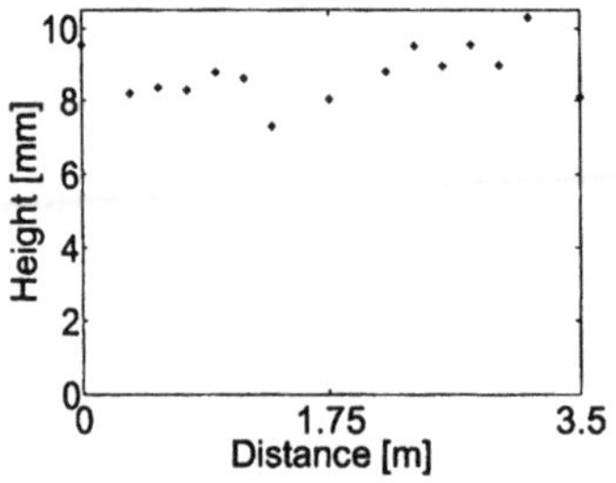

a. The height.

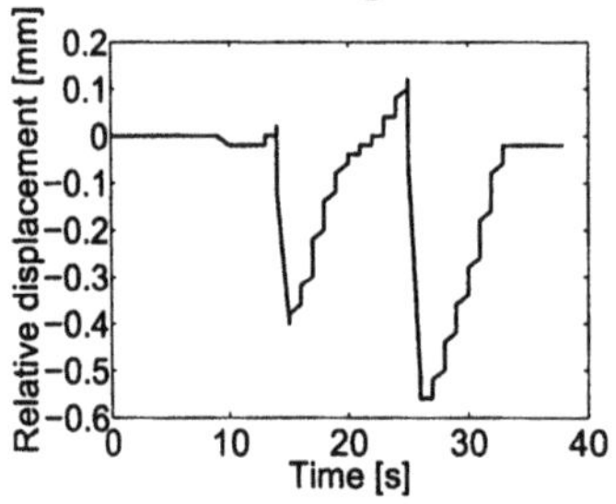

b. The relative motion of the plate's ends.

Figure 8: The height and relative motion of joint 6.

modulus $E = 5 \times 10^8\,\text{N/m}^2$, material density $\rho = 2000\,\text{kg/m}^3$ and Poisson's ratio $\nu = 0.5$ is considered. The material properties are determined from the values reported in literature (N.N. 1983). Equivalent bending and torsional characteristics are used for the calculation of the road's impedance as the wavelength in the road is much larger than the thickness of the road. The Poisson's ratio ν_i of each layer is used to calculate the shear modulus G_i in the torsional stiffness (Lombaert 2001). The joints are disregarded in the calculation.

3.4 *The dynamic soil characteristics.*

Table 1: The parameters of the soil model

d	ν	ρ	E	C_s	C_p	β^s
[m]	[-]	kg/m^3	[$\times 10^6$N/m^2]	[m/s]	[m/s]	[-]
4.0	0.42	1800	204	200	529	0.0500
1.0	0.49	2000	48	90	876	0.0375
∞	0.48	2000	369	250	1148	0.0250

The dynamic soil characteristics of the site are determined by means of four seismic cone penetration tests (SCPT), two spectral analysis of surface wave tests (SASW) and borehole experiments with undisturbed sampling. Undisturbed sampling was impossible due to the non plastic behavior of the material. Borehole experiments reveal the presence of a sand of Mol upto a large depth. A loose sand layer is present at a depth of 4 m. Table 1 summarizes the layer thickness d, the Poisson's ratio ν, the density ρ, the Young's modulus E, the S-wave velocity C_s, the P-wave velocity C_p and the hysteretic material damping in the soil represented by a material damping ratio β^s. In this frequency domain analysis, material damping in the elastic medium is modelled replacing the Lamé coefficients λ and μ by their complex counterpart $\lambda(1 + 2\beta^s i)$ and $\mu(1 + 2\beta^s i)$. As the ground water table is located near the free surface, a high Poisson's ratio ν is chosen to account for the saturation of the soil. The soil density is estimated as $\rho = 1800\,\text{kg/m}^3$ for dry sand and $\rho = 2000\,\text{kg/m}^3$ for saturated sand. The determination of the material damping ratio from geophysical tests as the SASW and the SCPT is presently studied. In anticipation of these results, the values of the damping ratio are estimated as 0.05 for the first layer, 0.0375 for the second layer and 0.025 for the halfspace. This agrees with the previous experimental observation that material damping decreases with depth.

3.5 *Data acquisition*

In both setups, simultaneous measurements are performed in the vehicle, the free field and the structure, using a mobile data acquisition system in the vehicle and a fixed data acquisition system in the free field and in the structure.

The accelerations of the truck have been measured on both sides of the vehicle's front and rear axle.

In the free field and the structure, measurements have been performed in 16 channels with seismic piezoelectric accelerometers.

Each channel in the free field is represented by a label FFij where the number i denotes the number of the measurement point and the character j denotes the direction in the Cartesian frame of reference. Five accelerometers in the front yard register the free field vibrations (figure 7). In setup 1, measurement point FF1 is located at a distance of 68.37 m from the center of the house (figure 7), close to a wooden board that is used as a trigger to synchronize the measurements in the truck, the free field and the structure. In setup 2, measurement point FF1, located in front of the house at a distance of 10.9 m (figure 7), is used as a trigger.

Each channel in the structure is represented by a label Fijk where the number i denotes the floor number (basement, first floor and second floor), while the characters j and k indicate the number of the measurement point and the direction in the Cartesian frame of reference. Eleven accelerometers in the building register the structural vibrations (figure 9).

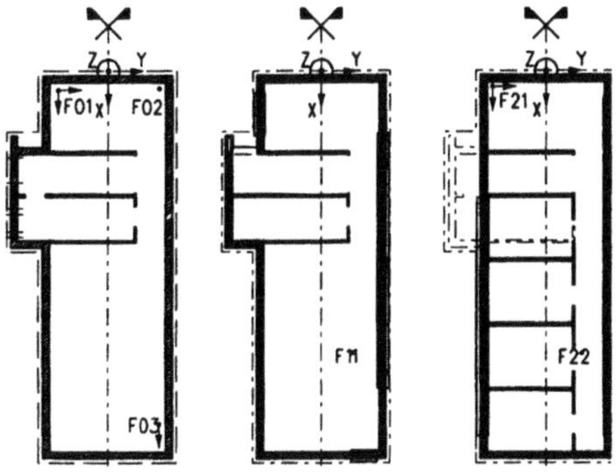

a. Basement. b. First floor. c. Second floor.
Figure 9: Location of the measurement points.

4 THE EXPERIMENTAL RESULTS

The discussion of the experimental results is limited to setup 2 as the free field and structural response are predicted for the passage of the truck on the artificial profile only.

4.1 *The response of the truck*

Figure 10 shows the time history and frequency content of the acceleration at the left hand side of the truck's front axle for a vehicle speed $v = 25$ km/h, $v = 48$ km/h and $v = 58$ km/h.

For the lowest vehicle speed, the impacts at the ascending and the descending of the front axle on the profile are well separated, while this is no longer the case for higher vehicle speeds. The duration of the acceleration of the front axle is shorter for higher vehicle speeds.

The dominant frequencies for the axle vibrations are situated between 10 Hz and 20 Hz, which corresponds to the eigenfrequencies of the axle hop modes.

Previous numerical predictions (Lombaert and Degrande 2001) have shown that, for the passage of a vehicle on a trapezoidal unevenness, the distance between the lobes in the frequency content of the vehicle response is equal to the ratio $v/(L+l)$ of the vehicle speed v and the mean length $L+l$ of the profile. This is confirmed experimentally by the frequency content of the front axle's acceleration. This phenomenon is not observed for the acceleration of the rear axle, which is due to the loss of contact between the rear axle and the road.

For increasing vehicle speeds, the frequency content of the acceleration of the front axle shifts to higher frequencies. A similar trend will be observed for the frequency content of the free field and structural vibrations.

Figures 11a and 11b show that the peak acceleration increases for increasing vehicle speed. This increase is almost linear for the peak acceleration of the front axle. Figure 11c demonstrates that the vehicle speeds do not have a large influence on the dominant frequencies f_d, which correspond to the eigenfrequencies of the vehicle.

4.2 *The free field and structural response*

Figure 12 shows the time history and frequency content of the velocity in the vertical direction for measurement points FF3 in the free field, point F01 in the stiff edge of the basement, point F11 at the first floor and point F21 located in the stiff corner of the second floor. The time history and the frequency content of the velocity are shown on the same vertical scale for all channels. The vertical peak particle velocity (PPV) for measurement points F01 and F21 are almost the same, due to the high structural stiffness in the vertical direction. The frequency content of the velocity is mainly situated below 20 Hz. The dominant frequencies for all channels are situated between 10.0 Hz and 15.0 Hz, which corresponds to the dominant frequencies for the axle's accelerations. The peak at 14.5 Hz in the frequency content for F11z corresponds to the floor's first bending mode.

The results of passages at vehicle speeds between 23 km/h and 58 km/h show that the PPV in the free field and the structure increase for higher vehicle speeds (figure 13). The frequency content shifts to higher frequencies, while the dominant frequencies remain almost the same. The observed trend is very similar as the axle's peak acceleration.

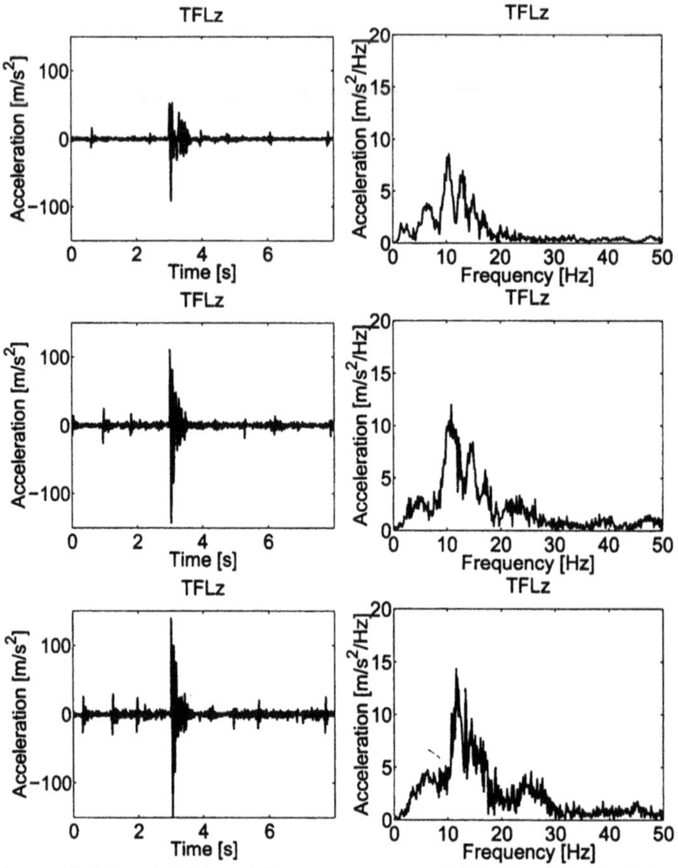

Figure 10: Time history (left hand side) and frequency content (right hand side) of the acceleration at the left hand side of the truck's front axle for a vehicle speed of $v = 25$ km/h (top), $v = 48$ km/h (middle) and $v = 58$ km/h (bottom).

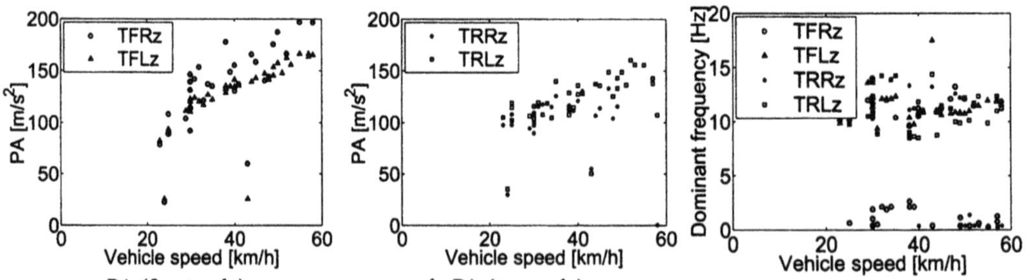

a. PA (front axle). b. PA (rear axle). c. f_d.

Figure 11: Peak acceleration for the (a) front axle, (b) rear axle and (c) dominant frequency in all channels as a function of the vehicle speed.

5 VALIDATION

This paper reports on the free field and structural response assuming a flexible surface foundation for the structure. The dynamic characteristics of the soil, the road and the structural components are implemented in the numerical prediction model. The first 20 eigen-

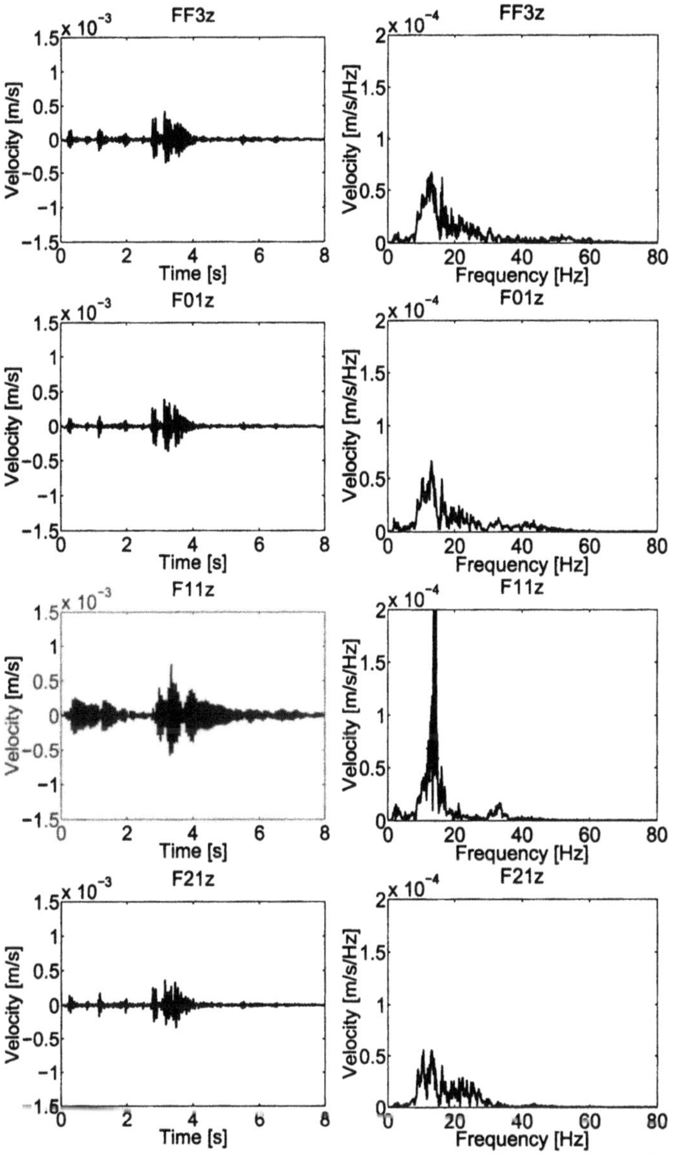

Figure 12: Time history (left hand side) and frequency content (right hand side) of the velocity in measurement points FF3z, F01z, F11z and F21z for the passage on the artificial profile for a vehicle speed $v = 48$ km/h.

modes of the structure are accounted for; 1 % of critical damping is assumed in all modes ($\xi_j = 1.0\%$).

Figure 14 shows the time history and frequency content of the velocity in points FF3z, F01z and F11z and F21z for the passage on the artificial profile for a vehicle speed $v = 50$ km/h. The amplitudes of the predicted free field and structural response are overestimated, especially in F21z. The PPV's in FF3z are equal to 0.41 mm/s versus 0.85 mm/s for the measured and predicted data, respectively. The ratio of the predicted and the measured response is equal to 2.06. As the predicted free field amplitude is twice the measured amplitude, this affects the predicted structural response. The comparison of the time history and frequency content in the structure (figures 12 and 14) illustrates that a good agreement between the mea-

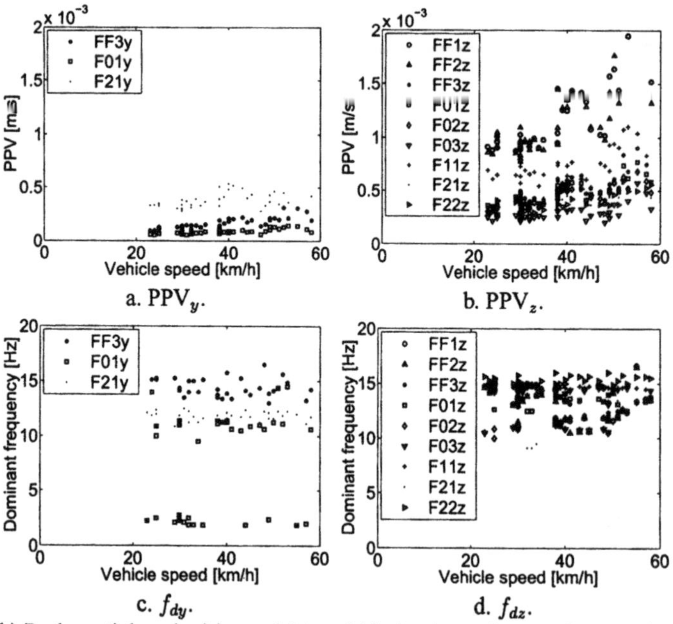

a. PPV_y. b. PPV_z. c. f_{dy}. d. f_{dz}.

Figure 13: (a) and (b) Peak particle velocities and (c) and (d) dominant frequencies as a function of the vehicle speed for the free field and the structural response.

surements and the predictions is obtained. The PPV is overestimated by a factor between 1.7 and 3.4. A large overestimation is observed in F21z with PPV's equal to 0.36 mm/s versus 1.2 mm/s for the measured and predicted data, respectively.

The dominant frequencies are situated between 10 Hz and 20 Hz with a peak at 13.04 Hz for F11z, except for F21z with dominant frequencies in the higher frequency range between 20 Hz and 30 Hz.

6 CONCLUSIONS

A numerical prediction model for traffic induced vibrations in buildings has been developed. The validation relies on measured data. The vibration measurements are performed in and near a single family dwelling during truck passages on an artificial profile installed on an uneven road for vehicle speeds between 23 km/h and 58 km/h. Although the amplitudes are overestimated, comparable time history and frequency content in the free field and the structure are obtained. A thorough study of the input parameters in the numerical prediction model and a further refinement of the model is needed to obtain a better agreement.

7 ACKNOWLEDGEMENTS

The results presented in this paper have been obtained within the frame of the STWW-project IWT000152 'Traffic induced vibrations in buildings'. The financial support of the Ministry of the Flemish Community is gratefully acknowledged.

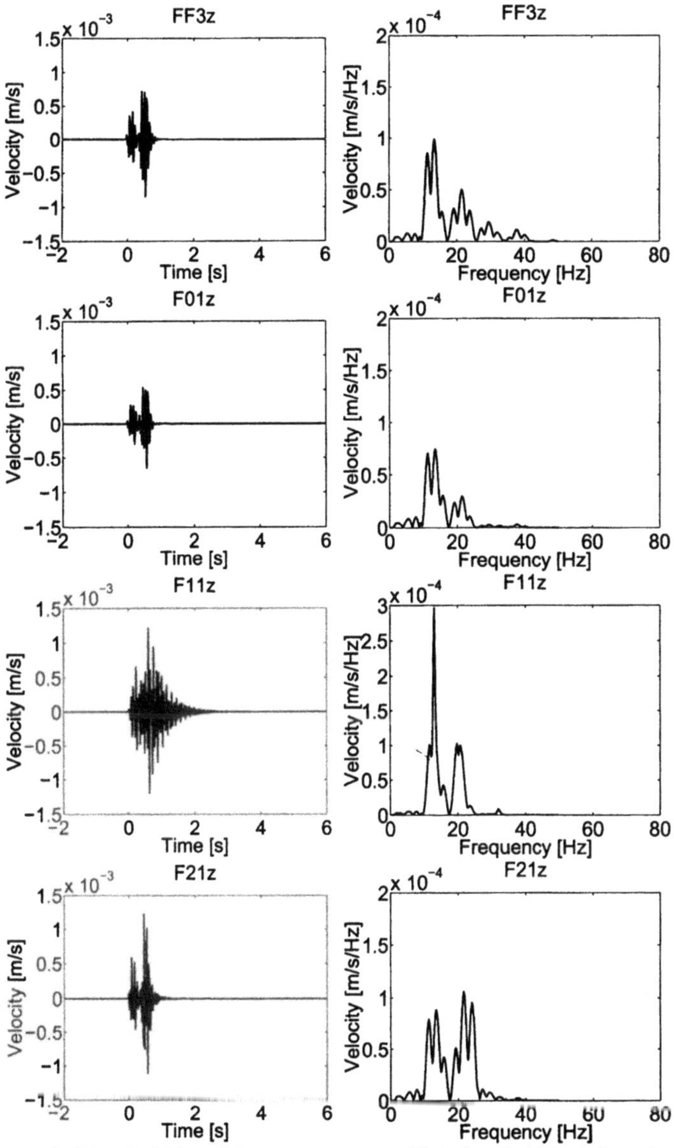

Figure 14: Time history (left hand side) and frequency content (right hand side) of the velocity in measurement points FF3z, F01z, F11z and F21z for the passage on the artificial profile for a vehicle speed $v = 50$ km/h.

REFERENCES

Aubry, D. and D. Clouteau (1992). A subdomain approach to dynamic soil-structure interaction. In V. Davidovici and R. Clough (Eds.), *Recent advances in Earthquake Engineering and Structural Dynamics*, pp. 251–272. Nantes: Ouest Editions/AFPS.

Dewulf, W. and G. De Roeck (2000a, January). Design of a PED system: Molsebaan 43, 2470 Retie. Internal report BWM-2000-04, Department of Civil Engineering, Katholieke Universiteit Leuven.

Dewulf, W. and G. De Roeck (2000b, March). Generalization of the combination test floor + CLD system. Internal report BWM-2000-05, Department of Civil Engineering, Katholieke

Universiteit Leuven. I.W.T. project 970175: Passive energy dissipation systems in civil engineering.

Lombaert, G. (2001). *Development and experimental validation of a numerical model for the free field vibrations induced by road traffic.* Ph. D. thesis, Department of Civil Engineering, Katholieke Universiteit Leuven.

Lombaert, G. and G. Degrande (2001). Experimental validation of a numerical prediction model for free field traffic induced vibrations by in situ experiments. *Soil Dynamics and Earthquake Engineering 21*(6), 485–497.

Lombaert, G., G. Degrande, and D. Clouteau (2000). Numerical modelling of free field traffic induced vibrations. *Soil Dynamics and Earthquake Engineering 19*(7), 473–488.

N.N. (1983). *Handleiding voor het dimensioneren van wegen met een bitumineuze verharding*, Volume O.C.W.-A 49/83.

N.N. (2001a, August). Faultimeter metingen. Technical Report EP 61211, Opzoekingscentrum voor de Wegenbouw.

N.N. (2001b, August). Topografische opmeting van de trapvorming van dwarsvoegen. Technical Report EP 61210, Opzoekingscentrum voor de Wegenbouw.

Pyl, L., G. Degrande, G. Lombaert, and W. Haegeman (2002, September). Building response measurements as a validation tool for a numerical prediction model for traffic induced vibrations in buildings. In H. Grundmann (Ed.), *Proceedings of the 5th European Conference on Structural Dynamics: Eurodyn 2002*, Munich, Germany, pp. 979–984.

Talloen, K. (2002). Eindige elementenmodellering van een eensgezinswoning voor de berekening van trillingen ten gevolge van wegverkeer. Master's thesis, Department of Civil Engineering, Katholieke Universiteit Leuven.

Wave propagation – Moving load – Vibration reduction, Chouw & Schmid (eds.)
© 2003 Swets & Zeitlinger, Lisse, ISBN 90 5809 559 2

Properties of train-induced vibration at railway tunnel lining

K. Tsuno & S. Konishi
Railway Technical Research Institute, Tokyo, Japan

M. Furuta
Construction Div, Tokyo Metropolitan Subway Construction Corporation, Tokyo, Japan

ABSTRACT: Train-induced vibration is considered to be one of the factors that cause the spalling of tunnel concrete lining. This paper presents and discusses the train- induced vibrations measured near a track and on a side wall and upper slab of a subway open-cut tunnel. At the same time, simulation by 2-D FEM analysis is conducted to simulate the propagation of vibration. The analytical results are found to have the similar tendency to the measurement data.

1 INTRODUCTION

Recently, a serious problem is posed by the spalling of lining concrete and repair materials in railway tunnels. The vibration induced by trains for a long period is considered to be one of the factors that cause this phenomenon. However, only a few studies investigated the influence of train-induced vibration on tunnel linings. Further studies are deemed necessary. Therefore, this paper presents and discusses both measured and analytical results on the vibration induced by trains at tunnel linings. In our research, the acceleration induced by train is measured and simulated by 2-D FEM analysis.

2 MEASUREMENT OF VIBRATION IN OPEN CUT TUNNEL

2.1 *Outline of measurement*

Vibration acceleration is measured inside a subway tunnel constructed by the open-cut method. The configuration of the measurement point is shown in Table 1. The measurement points are located near a track and on a side wall and upper slab of tunnel as shown in Figure 1. The distance between the point near the track and the rail where trains get through is 1.0 m. At each measurement point, vibration acceleration is measured in the three directions, namely in the longitudinal (axial) direction of track (X-direction), lateral direction (Y-direction) and vertical direction (Z-direction). The vibration is measured by piezoelectric pickups and amplified by amplifier in consideration of the frequency contents of vibration acceleration ranging from several Hz to several hundred Hz. The piezoelectric pickups are set up at each measurement point as shown in Figure 2(a), and then covered with the vessels made from expanded polystyrene in order to prevent the influences of air pressure from running trains as shown in Figure 2(b). Moreover, the cords connecting the piezoelectric pickups with the accelerometer are fixed by adhesive tapes in order not to be moved by air pressure. The specific characteristics of train-induced vibration inside the tunnel lining are investigated by 1/3 octave band analysis in this paper.

Table 1 Configuration of the measurement point

Tunnel structure	Open cut tunnel (Double decks)
Shape of line	Straight line
Track structure	Fastened to concrete track
Sort of rails	50N, Long rail
Outline of Cars	20m/car, 8 Cars
Train speed	Local:40km/h, Express:62km/h
Number of trains	Local:13 trains, Express:3 trains

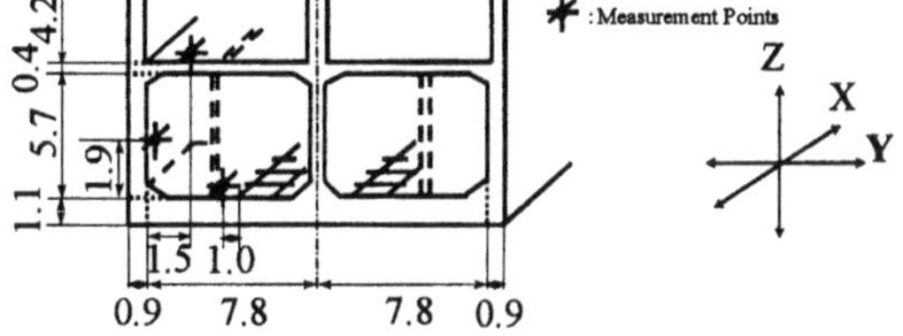

Figure 1 Measurement point in the open cut tunnel

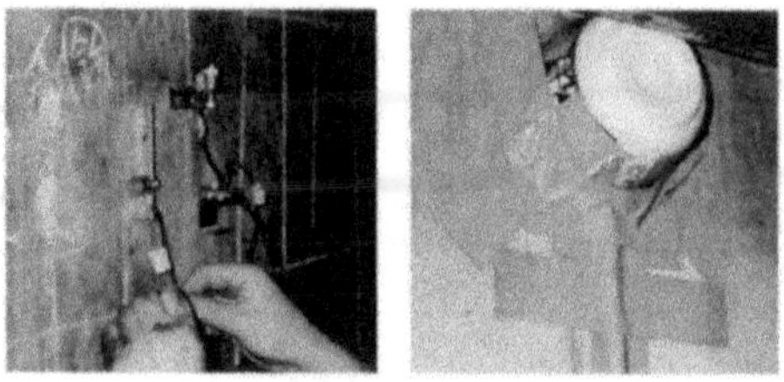

(a)Setup of pickup (b) Cover for pickup

Figure 2 Setup of piezoelectric pickups

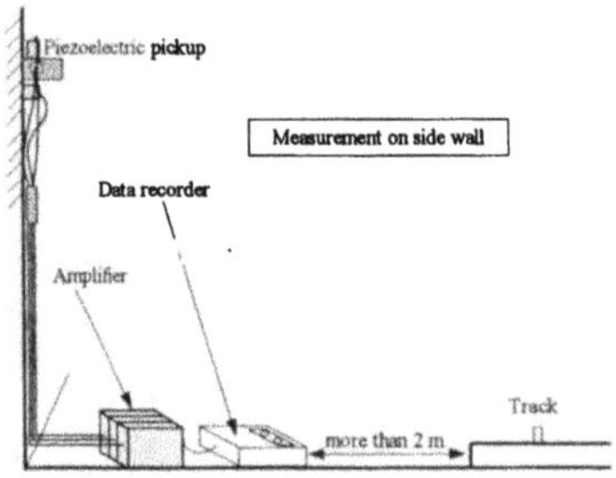

Figure 3 Measurement equipment

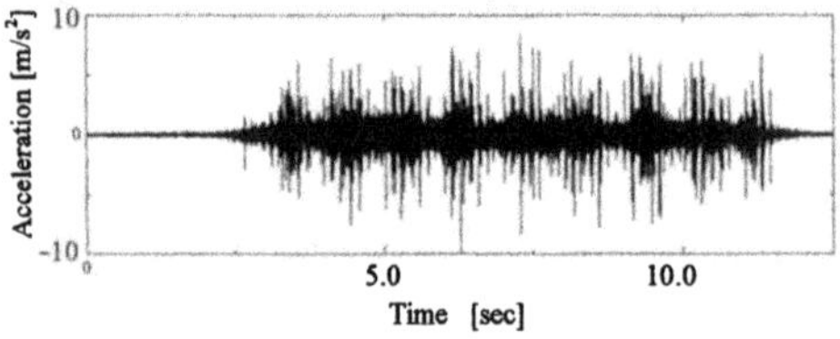

(a) Near the track in the Z direction

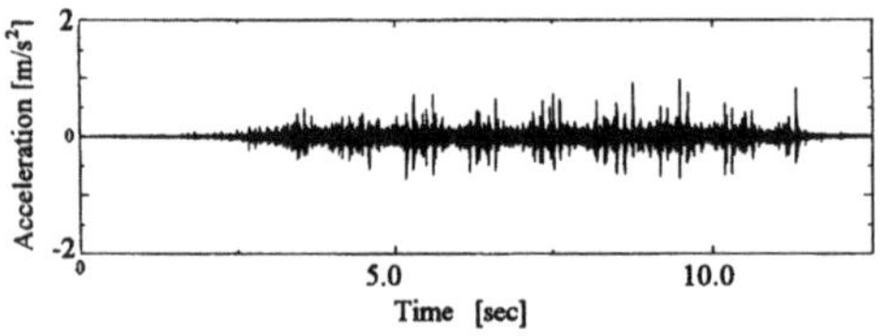

(b) Side wall in the Y direction

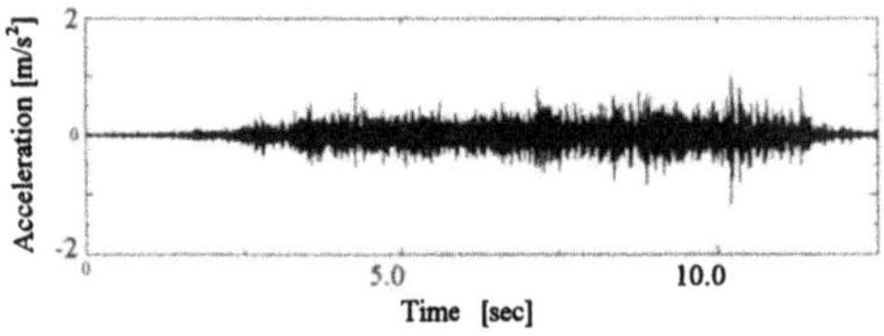

(c) Upper slab in the Z direction

Figure 4 Vibration wave induced by the train

2.2 *Amplitude of vibration acceleration*

Figure 4 shows the vibration waves induced by one of the express trains which is gained by analog-digital conversion. At a point near the track, the amplitude becomes large while trains are passing the point. The peak value of acceleration is about 10 m/s^2. On the other hand, at the upper slab, the peak value of acceleration is about 1 m/s^2. It is believed that the flat of wheels causes this irregular shape of vibration wave.

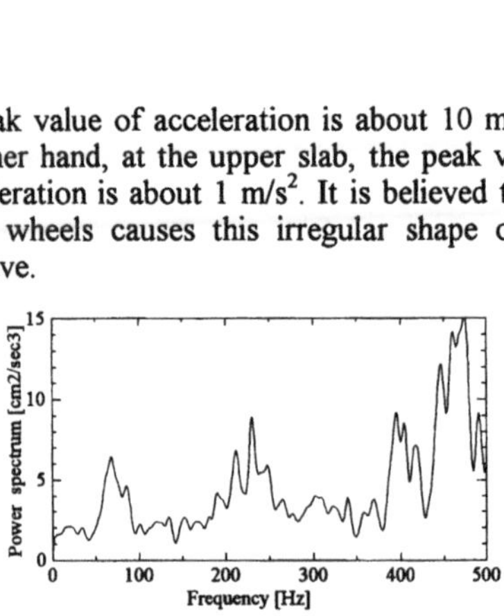

a) Near the track in the Z direction

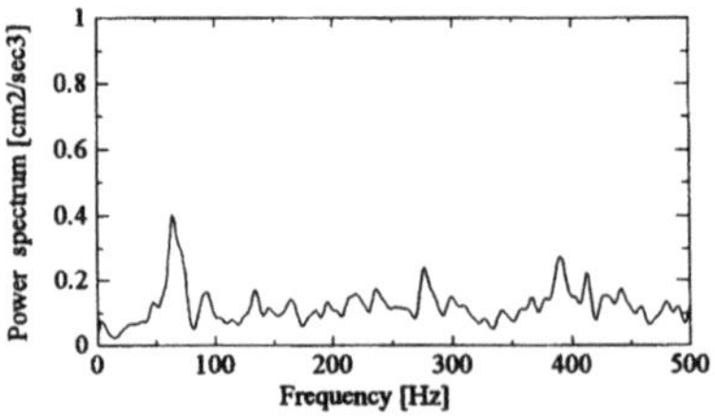

b) On the side wall in the Y direction

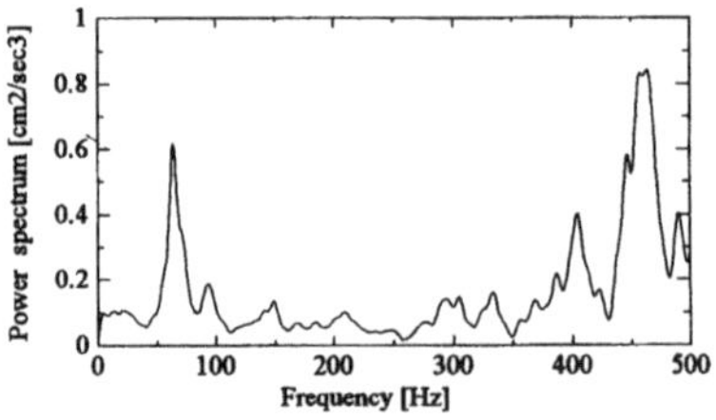

c) On the side wall in the Z direction

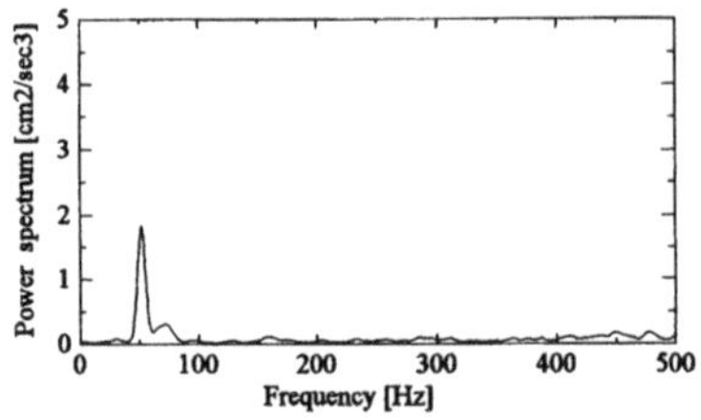

d) On the upper slab in the Y direction

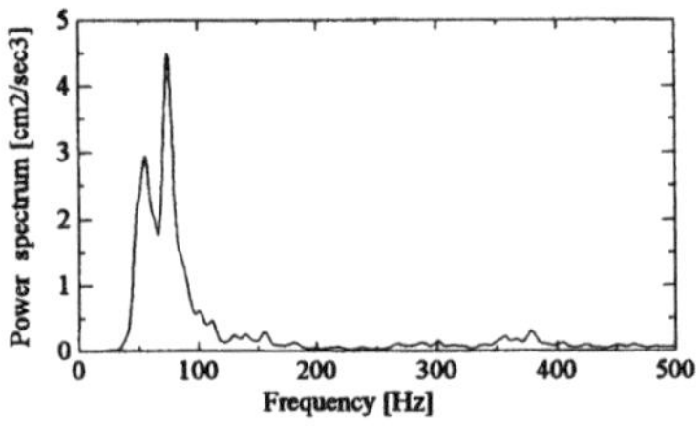

e) On the upper slab in the Z direction

Figure 5 Power spectrum of each measurement point

2.3 *Power spectrum*

Figure 5 shows the power spectrum of the same express train mentioned at Section 2.2 calculated by FFT analysis. The measured analog data are converted into digital data to calculate the power spectrum. The smoothing by the Parzen window is carried out, and the bandwidth of smoothing is 10.0 Hz.

Concerning the spectrum in the Z- direction, the spectrum near the track has the clear peaks at about 65, 240, 400 and 470Hz, while the one on the side wall has the peaks at about 65, 400 and 470Hz. The spectrum on the upper slab has the peaks at 55 and 80 Hz and there is no peak at more than 100 Hz. The ingredient of high frequencies near 400 Hz is thought to be caused by the contact between wheels and a rail. This ingredient reduces according to the distance from track and doesn't exist on the upper slab.

2.4 *Vibration acceleration level (VAL)*

The 1/3 octave analysis (JIS C1513, ISO266-1975) is usually applied to estimate train-induced vibration in Japan. Then, in this paper, the measurement results are evaluated by using the 1/3 octave band analysis. Table 3 shows the central frequencies of each octave band.

Table 2 and Figure 6 shows the results of 1/3 octave band analysis for 16 passing trains. The maximum and minimum values corresponding to the 16 trains and the AP are drawn in the Figure 6. The AP indicates the All Pass from 1Hz to 500Hz. The following are confirmed from the 1/3 octave band analysis.

- At a point near the track, the all pass vibration acceleration levels (abbreviated as VAL (AP)) in the Z-direction are 90 to 100 dB, while those in the X- and Y- directions are almost 100 dB.
- On the side wall, the VAL (AP) are less than 80 dB in all three directions.
- On the upper slab, the VAL (AP) in the Z- direction are 80 to 90 dB, which is 5 to 10 dB larger than those in the X- and Y- directions.

2.5 *Predominant frequency*

Predominant frequencies evaluated by the 1/3 octave band analysis have the following features.

- The vibration acceleration levels (VAL) tend to increase gradually as the frequency becomes higher from 10 to 500 Hz. This tendency is observed in another measurement results in the subway shield tunnel where the same type of cars pass and the rail is fastened to concrete track (Furuta, M. & Nagashima, F. 1991).
- Near the track, the VAL at high frequency levels (more than 200 Hz) are large, when compared with those at the side wall and upper slab. Peaks are observed at 63 Hz in the Z- direction.
- A peak exists at 63 or 80Hz on the side wall.
- On the upper slab, peaks are observed at 63 to 80 Hz in the Z- direction. The VAL is about 75 dB near the peaks. These values are at the same levels as those near the track. On the other hand, peaks are observed at 50 Hz in the Y- direction.

2.6 *Results of the measurement*

It can be said that this measurement site is under typical conditions as a subway tunnel; train speed is 40 to 70 km/h; the long rails and concrete tracks are used and the track is maintained satisfactorily. The following is obtained by the 1/3 octave band analysis of this measurement.

- The VAL (AP) and peak values of VAL in the normal direction to free surface or in the Z- direction are not always the largest among those in the three directions.
- It can be said that the general values of VAL (AP), which are different to some extent among the three directions (X, Y and Z), are 100dB near the track and 70 to 80 dB on the side wall and upper slab.
- The peaks of VAL appear at 50 to 80 Hz in the Z-direction near the track and in all directions at the side wall and upper slab.

Table 2 The average, minimum and maximum values of VAL (AP) of 16 trains

Location	Direction	Average	Min	Max
Near the track	X-direction	100.4dB	97.6dB	105.7dB
	Y-direction	97.4dB	92.3dB	108.7dB
	Z-direction	91.1dB	87.3dB	97.1dB
On the side wall	X-direction	69.2dB	67.2dB	72.1dB
	Y-direction	72.7dB	68.1dB	78.1dB
	Z-direction	73.3dB	69.2dB	78.7dB
On the upper slab	X-direction	73.0dB	70.3dB	76.5dB
	Y-direction	74.9dB	71.5dB	79.3dB
	Z-direction	81.8dB	77.4dB	89.3dB

Note: $1 cm/sec^2 = 60 dB$, $1 m/sec^2 = 100 dB$

Table 3 Central frequencies of 1/3 each octave band

1	0.8 Hz	11	8 Hz	21	80 Hz
2	1 Hz	12	10 Hz	22	100 Hz
3	1.25 Hz	13	12.5 Hz	23	125 Hz
4	1.68 Hz	14	16 Hz	24	160 Hz
5	2 Hz	15	20 Hz	25	200 Hz
6	2.58 Hz	16	25 Hz	26	250 Hz
7	3.15 Hz	17	31.5 Hz	27	315 Hz
8	4 Hz	18	40 Hz	28	400 Hz
9	5 Hz	19	50 Hz	29	500 Hz
10	6.3 Hz	20	63 Hz	-	---

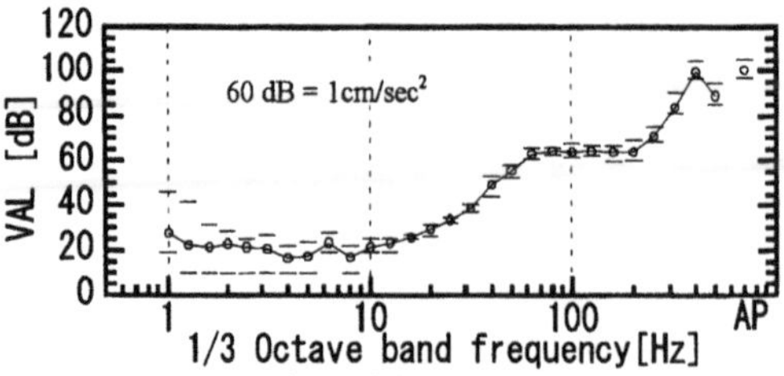

a) Near the track in the X direction

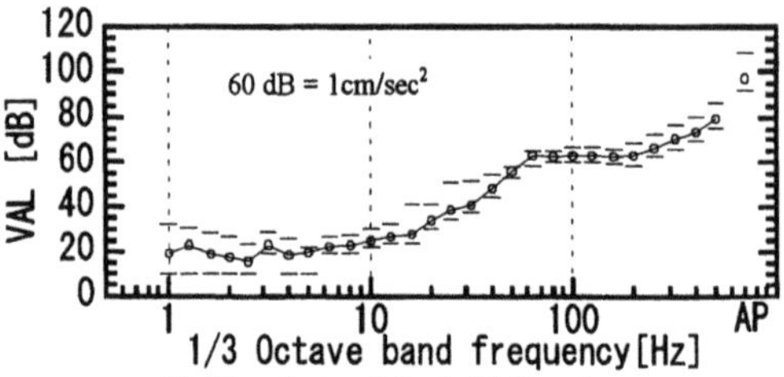

b) Near the track in the Y direction

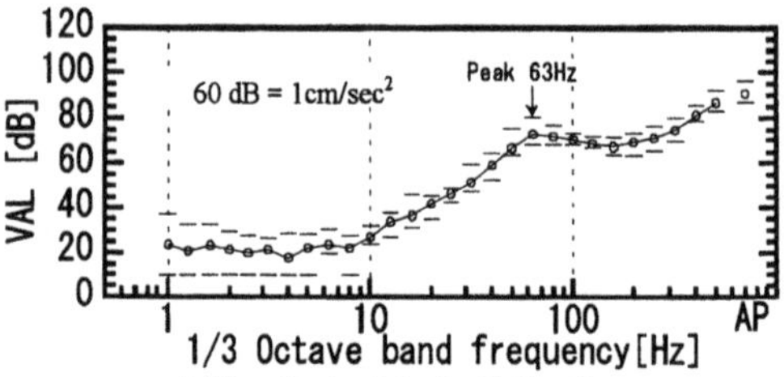

c) Near the track in the Z direction

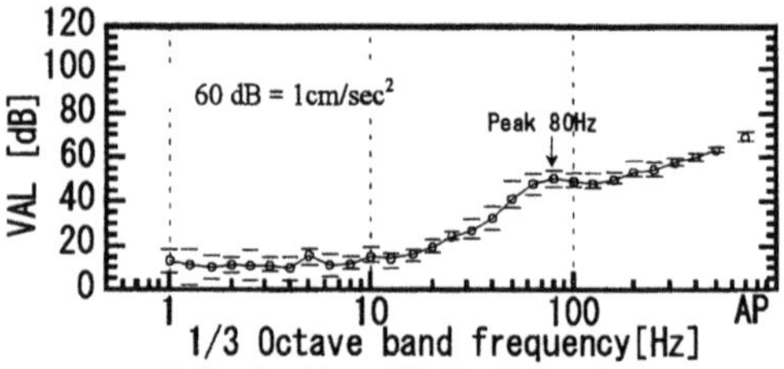

d) On the side wall in the X direction

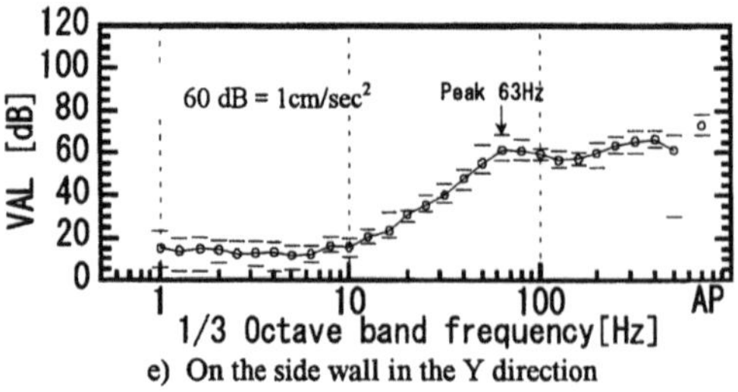

e) On the side wall in the Y direction

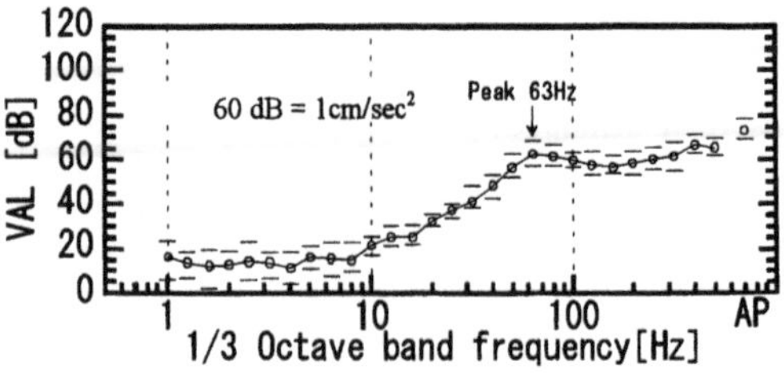

f) On the side wall in the Z direction

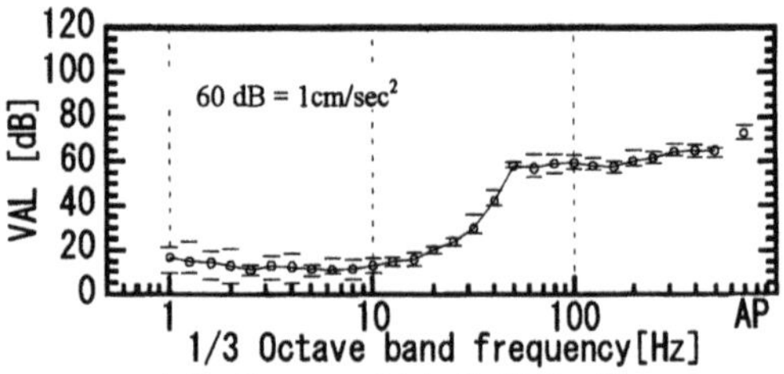

g) On the upper slab in the X direction

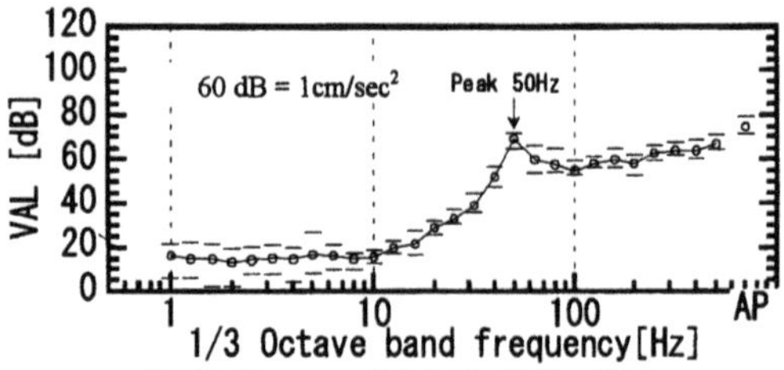

h) On the upper slab in the Y direction

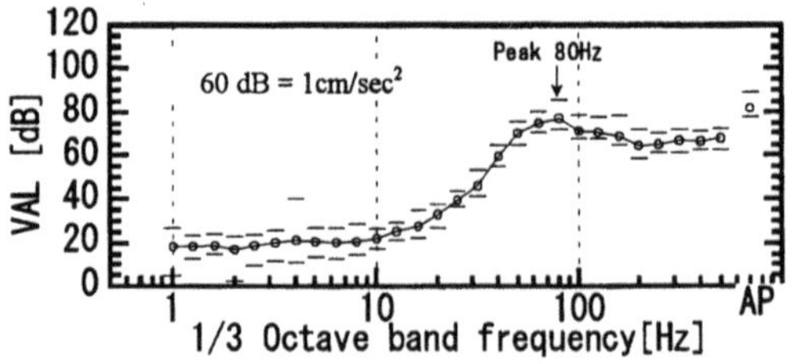

i) On the upper slab in the Z direction

Figure 6 Results obtained by 1/3 octave band analysis

3 SIMULATION BY FEM ANAYISIS

3.1 *Outline of FEM analysis*

The numerical results are obtained by a 2-D finite element method (FEM). Triangular elements and quadrilateral elements are generally applied in the FEM analysis. These elements are easy to formulate, but difficult to correctly model the curve boundary. Therefore, it is said that this feature disturbs the accuracy of analysis. Iso- parametric elements are applied to overcome this problem in this simulation. A consistent mass matrix is applied.

The damping force is generally assumed to be an approximate value of C x du/dt. However, both viscous damping and structural damping are considered in this simulation to express the damping at different frequencies. In other words, the viscous damping, which is similar to the Mass matrix (M), is assumed to be αM, while the structural damping, similar to the Stiffness matrix (K), is assumed to be βK. The damping matrix (C) is defined by the following equation, in consideration of Rayleigh damping.

$$C = \alpha M + \beta K \tag{1}$$

Where

α, β: Constants
M: Mass matrix
K: Stiffness matrix

The Newmark's beta method, which is one of the direct integral analyses, are applied in this simulation.

3.2 *FEM Mesh*

Figure 7 shows the FEM mesh used in this simulation. This model includes 959 iso- parametric elements and 3,112 nodes. The iso-parametric elements have 8 nodes as shown in Figure 8. The length of each side of elements are less than 0.8 m by referring the past research (Nagashima, F. et al. 1988).

The standard viscous boundary condition shown in Figure 8 is arranged on the perimeter of analysis area to prevent the influence of boundary condition.

3.3 *Input data*

The measurement waves near the track which is induced by the express trains mentioned in Section 2.2 is used as an input wave. The interval of input wave data is 1/2,000 sec. The dynamic modulus of elasticity and the modulus of structure and soil are shown in Table 4.

The contents of α and β of are set to satisfy the condition that the damping coefficients (h) are 0.05 at 5 Hz and 0.1 at 200 Hz. The values of α and β ,in this simulation, are shown in Table 5. Figure 9 describes the damping curve (relationship between the damping coefficients (h) and frequencies).

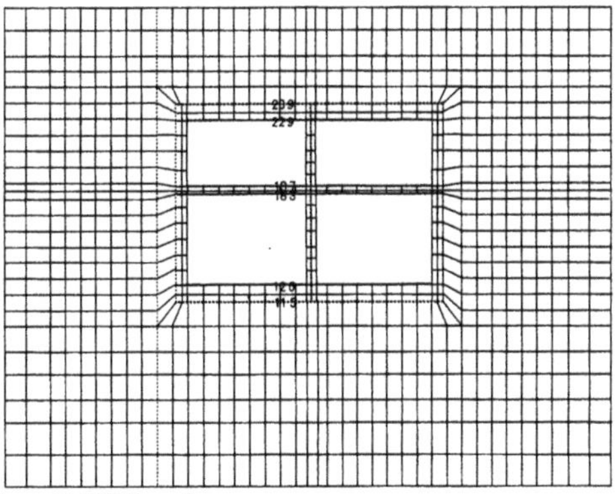

Figure 7 FEM mesh

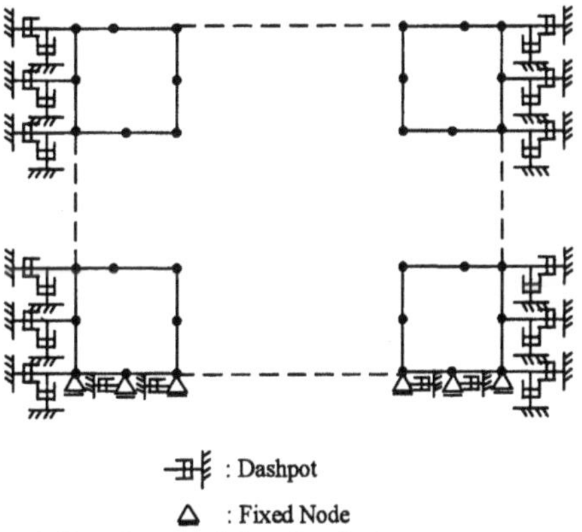

Figure 8 Boundary condition of analysis area

Table 4 Modulus of structure and ground

	Ground	Tunnel concrete
Dynamic modulus of elasticity E [kN/m²]	5.0×10^4	2.1×10^7
Poisson's ratio ν	0.45	0.167
Density ρ [kg /m³]	1600	2300

Table 5 Constants α and β

α	2.9864
β	0.0002

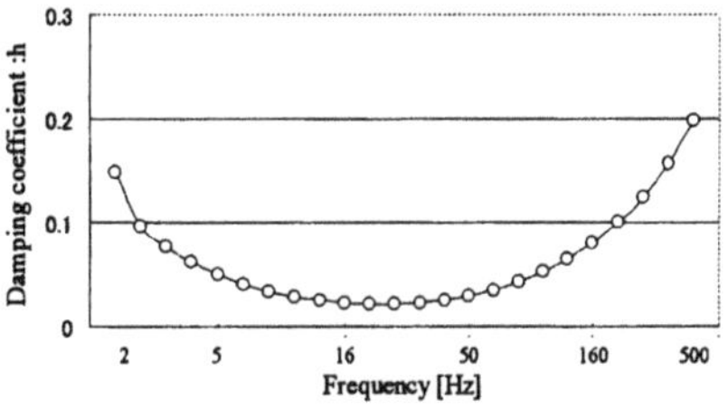

Figure 9 Damping coefficient (h) as a function of frequency

3.4 *Results*

3.4.1 *Acceleration waves*

Figure 10 shows the waves on the side wall in the lateral direction (Y- direction in chapter 2) and that at the upper slab in the vertical direction (Z- direction in Chapter 2) gained by the measurement and the 2-D FEM simulation. The analytical waves are roughly similar to the measured waves.

3.4.2 *Power spectrum*

Figure 11 shows the power spectrum of measurement and analytical waves. Both spectra have a peak at near 60 Hz, but analytical spectra have more peaks than measurement spectra from 0 to 250 Hz. Then the power spectra of analytical waves don't match those of measurement waves.

3.4.3 *1/3 octave band analysis*

In Chapter 2, a 1/3 octave band analyzer is used so as to estimate the measurement results. On the other hand, there is a method to calculate the vibration acceleration level (VAL) corresponding to each 1/3 octave band from the power spectrum, which is calculated by Fourier analysis (FFT) of the obtained dispersion data of vibration acceleration.

The VAL of each 1/3 octave band of analytical and measured waves are calculated by the FFT method, as shown in Figure 12. Figure 12 also includes the VAL of measurement wave by a 1/3 octave band analyzer. In this case, the VAL by FFT method in the frequency range lower than 40 Hz is larger than that by a 1/3 octave band analyzer for the same measurement data. These differences are attributable to problem of spectrum resolution. The VAL of analytical waves shows higher levels in the frequency bands lower than 40 Hz, and almost the same levels in the frequency bands from 50Hz to 100Hz, as compared with that of measured waves.

The VAL (1 to 90 Hz) of analytical waves are compared with that of measured waves in Table 6. The differences of the values between the analytical and measured waves are maximum 3.3 dB at the side wall and maximum 5.3 dB at the upper slab. It can be said that the analytical VAL (1 to 90 Hz) roughly corresponds to the measured values.

3.5 *Discussion*

The analytical power spectra don't match the measurement power spectra. Then, it is difficult to estimate the power spectrum by the simulation method included in this paper. However, the analytical VAL of each 1/3 octave band from 50 to 100 Hz and the VAL (1 to 90 Hz) roughly correspond to measured values, as shown in Figure 12. Therefore, the simulation method included in this paper can be used to roughly estimate the VAL from 50 to 100 Hz and the VAL (1-90Hz) in subway tunnels. On the other hand,

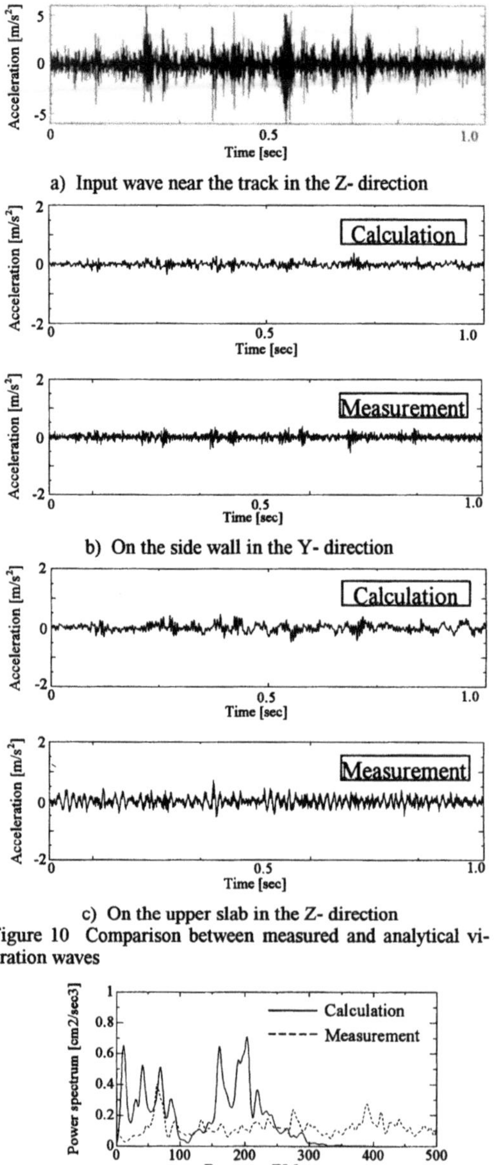

a) Input wave near the track in the Z- direction

b) On the side wall in the Y- direction

c) On the upper slab in the Z- direction

Figure 10 Comparison between measured and analytical vibration waves

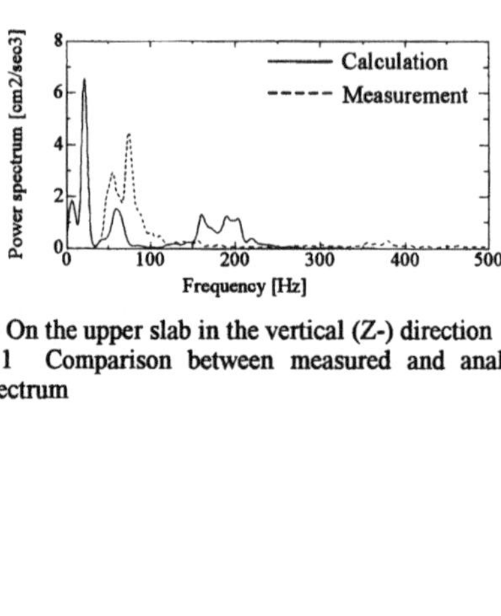

a) On the side wall in the lateral (Y-) direction

b) On the upper slab in the vertical (Z-) direction

Figure 11 Comparison between measured and analytical power spectrum

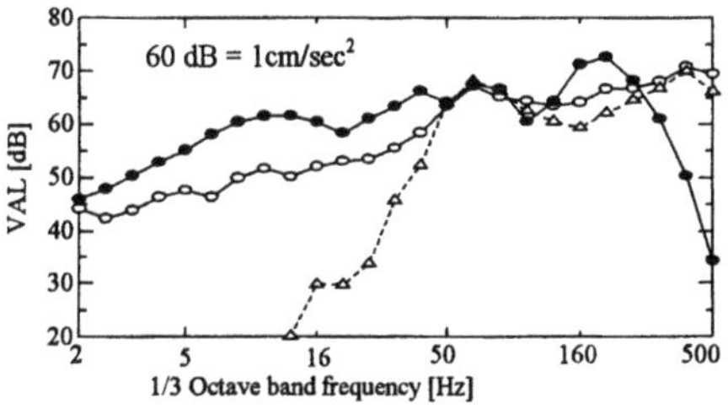

a) On the side wall in the lateral (Y-) direction

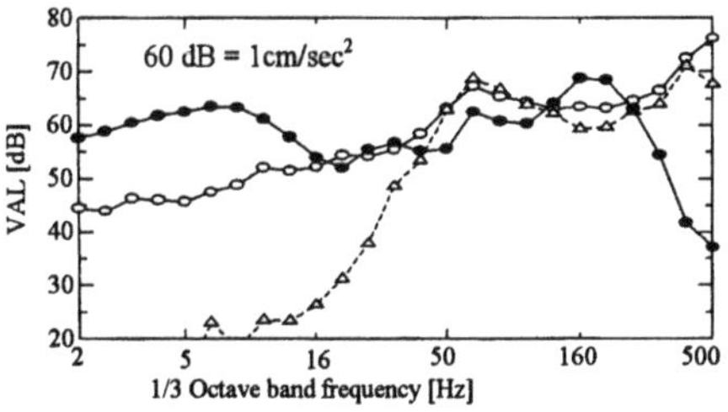

b) On the side wall in the vertical (Z-) direction

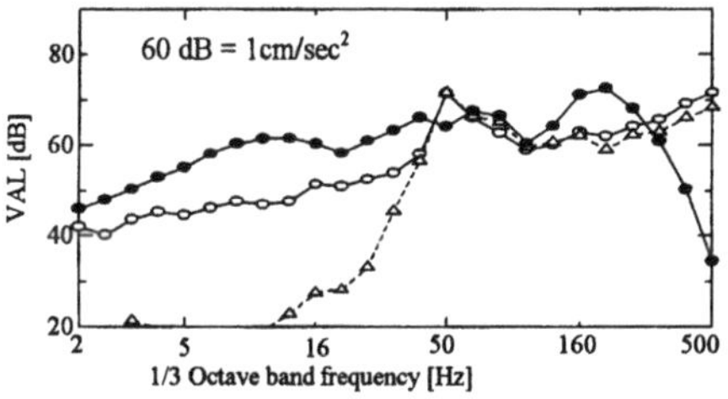

c) On the upper slab in the lateral (Y-) direction

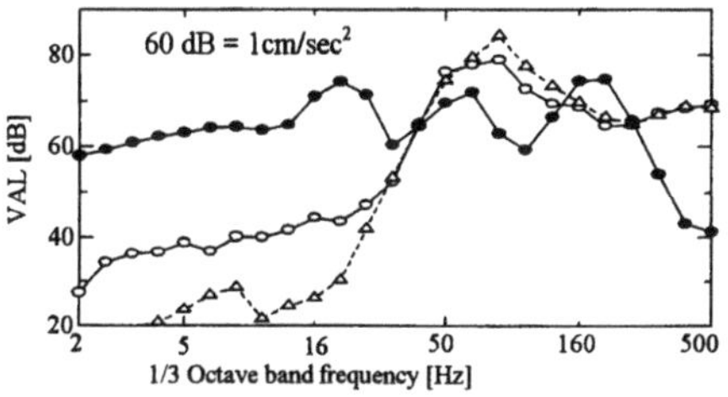

d) On the upper slab in the vertical (Z-) direction

-●- : Analytical results (1/3 octave band by FFT method)
-○- : Measurement results (1/3 octave band by FFT method)
-△- : Measurement results
(1/3 octave band by 1/3 octave band analyzer)

Figure 12 Comparison between measured and analytical VAL by 1/3 octave band analysis

Table 6 Measured and analytical VAL (1 to 90Hz) [dB]

Location and direction		Simulation (FFT)	Measure-ment (FFT)	Differ-ence
Side wall	Lateral (Y)	74.5	71.2	+3.3
	Vertical (Z)	72.7	71.3	+1.4
Upper slab	Lateral 1 (Y)	68.2	73.5	-5.3
	Vertical (Z)	80.1	82.8	-2.7

Note: 60 dB = 1cm/sec^2

the analytical VAL except for 50 to 100 Hz doesn't match that of measured waves and there is room for modification in our analytical method.

4 CONCLUSION

Vibration acceleration is measured near the track and on the side wall and upper slab of a subway open-cut tunnel in three directions at the same time. The measured VAL (AP) is found to be 100dB near the track and 80 dB on the side wall and upper slab. Moreover, it can be confirmed that the peak values of VAL in the normal direction to free surface are not always the largest among those in the three directions. The peaks of VAL appear at 50 to 80 Hz.

Numerical results are obtained by the 2-D finite element method (FEM). The power spectra aren't estimate satisfactorily. However, it is confirmed that the VAL of each 1/3 octave band and the VAL (1-90Hz) is roughly estimated by our simulation method.

We will continue research into the measurement and simulation of train-induced vibration in subway tunnels and on the ground surface, and also the influence of train- induced vibration on the tunnel lining concrete.

REFERENCES

Furuta, M. & Nagashima, F. 1991. An investigation and analysis of subway-induced vibration, *Proc. of tunnel engineering, Vol 1,* PP.101-106

Furuta, M. & Nagashima, F. 1994. Dynamic response analysis of subway–induced vibration on shield tunnel-ground systems, and calculation examples), *Proc. of tunnel engineering, Vol 4,* PP.93-100

Nagashima, F. et al. 1988. Vibration transmissibility of the subway shield tunnel and the alluvial soil peripheral to it, *Journal of structural engineering Vol 34A,* PP.837-846

Noise pollution

Wave propagation – Moving load – Vibration reduction, Chouw & Schmid (eds.)
© 2003 Swets & Zeitlinger, Lisse, ISBN 90 5809 559 2

Numerical study on the effect of the near pressure field around a train on the railside environment

T. Doi
Kobayashi Institute of Physical Research, Tokyo, Japan

T. Ogawa
Dept. of Mechanical Engineering, Seikei University, Tokyo, Japan

ABSTRACT: The near pressure field around a train which causes the environmental problem such as rattling of a building window nearby the railroad is numerically studied. The panel method, the BEM for the potential flow theory, is utilised. The numerical result is validated with the field measurement data and the comparison shows good agreement with each other. Computation of a flow field where a train runs on various types of a track is carried out and the result shows the track structure is important for the railside environment nearby the track. The effect of the conventional measures against noise on the near pressure field is examined and found to be small. Influence of the existence of a building near the track is also discussed. The result shows that a building affects the pressure field, and that it is necessary to take the existence of a building into consideration for accurate assessment of the pressure variation.

1 INTRODUCTION

A high-speed train forms a near pressure field where strong pressure distribution is observed around the nose of the train. Since this near pressure field moves with the train, a pressure variation is observed along side the railway. This variation in pressure may cause environmental disturbances at the railside such as annoying sounds and window vibration.

In past studies, there have been computational and experimental researches in two dimensions [1][2]. In addition, the three dimensional investigation has been carried out [3]. In these studies, ground condition is assumed to be the level ground. The near pressure field is affected by the type of track structure a train is traveling on such as an elevated structure, a bank structure, or a cut (depressed) structure. Therefore, it is important to investigate the near pressure field to assess the railside environment. In this paper, the near pressure field created around a train running on various track structures is numerically investigated by using the panel method. The computational result is compared with field measurement data, and the effect of the track structure on the pressure field, such as the pressure variation strength, and the distance attenuation is discussed.

2 NUMERICAL METHOD

In this study, the flow around a train nose is assumed to be an inviscid incompressible flow. The modern high-speed train runs at about Mach number from 0.2 to 0.4 and the effect of compressibility on the near pressure field is negligible, particularly when traveling outside a tunnel. The nose of a high-speed trains is aerodynamically shaped and the flow at the nose of the train is rarely separated. Thus, the flow around the nose of a train can be approximated as a potential flow.

The panel method [4] is a boundary element method that is based on the potential flow theory and is often used to compute the flow field around an object. The basic equation is the Laplace equation of the velocity potential (eq.(1)) and the velocity componet normal to the solid wall can be obtained with the surface integral of σ (eq.(2)).

$$\Delta\Phi = 0 \qquad (1)$$

$$q(x) \equiv \frac{\partial\Phi(x)}{\partial n(x)} = -\frac{\sigma(x)}{2} - \int_\Gamma \sigma(y)\,\frac{\partial\Psi^*(y,x)}{\partial n(x)}\,d\Gamma(y) \qquad (2)$$

where $q(x)$ is the velocity component normal to the solid wall, Γ is the domain boundary and σ is the source density.

Eq.(3) is the discretized form of eq.(2). The surface of the domain boundary Γ is decomposed into panels and the integral is carried out on the panels.

$$q(x_i) = -\frac{\sigma(x_i)}{2} - \sum_{j=1}^{N}\left\{\int_{Panelj} \frac{\partial \Psi^*(y,x_i)}{\partial n(x)} d\Gamma(y)\right\}\sigma(y_j) \tag{3}$$

The source density σ is determined so that the velocity component normal to the solid wall should be 0. Then, the flow velocity at the observation point is calculated from the following equation:

$$u(x) = u_0 - \sum_{j=1}^{N}\left\{\int_{Panelj} \frac{(x-y)}{4\pi r^3} d\Gamma(y)\right\}\sigma(y_j) \tag{4}$$

where $u(x)$ is the velocity at an observation point; u_0 is the uniform flow velocity.

Since the source density is computed by the surface integral as shown these expressions, this method needs far less computer resources than those methods using computational fluid dynamics, such as the finite difference method, and is quite effective for evaluating pressure fields.

3 RESULTS

3.1 *Validation of the numerical result*

Figure 1 shows the panels of a high-speed train running at 270km/h on a bank structure and the pressure contour plots. The density of the air is 1.226kg/m^3. The effects on the railside of positive pressure at the nose and negative pressure at the shoulder part of the body can be seen.

This computation took only 35 sec. on a PC (Pentium3, 600 MHz).The computational result was validated using field measurement data[5]. Figure2 (a) shows the cross section of the bank structure and an observation point. Figure 2(b) shows the pressure histories obtained at 2.6 m from the center of the track. The results show good agreement with each other and it turns out that the near pressure field around a high-speed train can be well predicted using panel method.

3.2 *Influence by the track structure*

Computations were carried out for the various types of rail structures mentioned above. Figure 3 shows the pressure histories for a train running on a cut track, a bank track, and an elevated track. The positive pressure variation was due to the stagnation point at the train nose and negative pressure was observed when the shoulder of the train passed the measurement point. As seen from the figure, there was a significant difference in pressure variation when observation point is near the track.

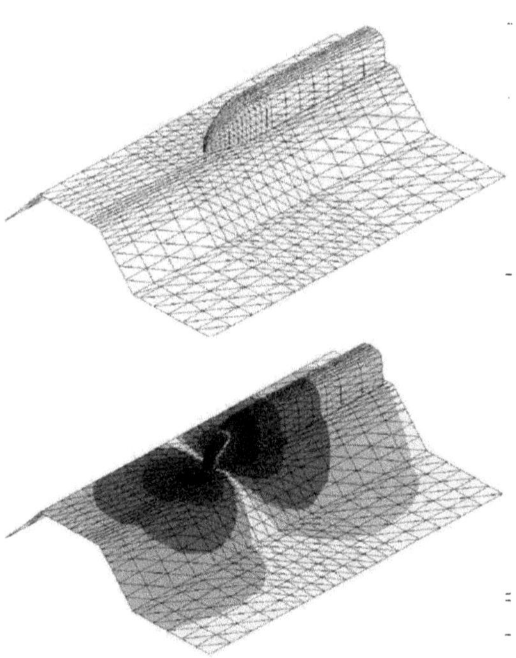

Figure 1. The panels of a train running on a bank structure and the pressure contour plots.

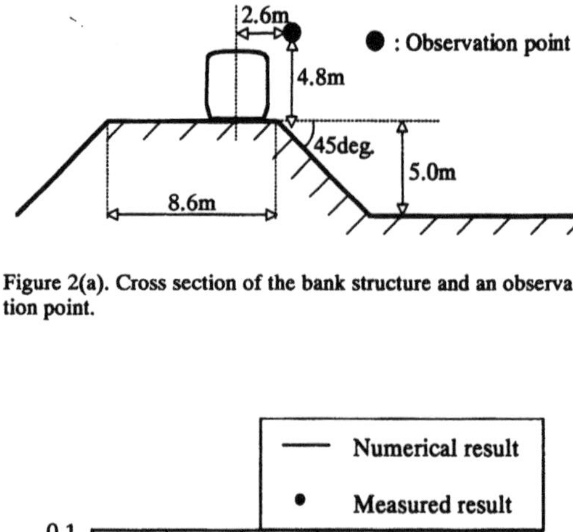

Figure 2(a). Cross section of the bank structure and an observation point.

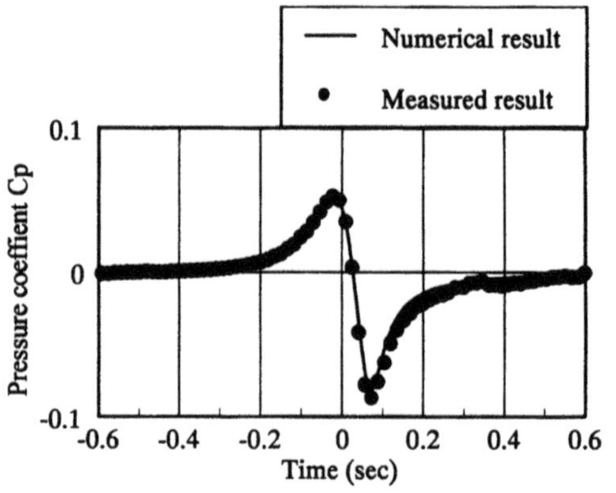

Figure 2(b). Pressure variation with time (2.6m from track center, R.L.:4.8m).

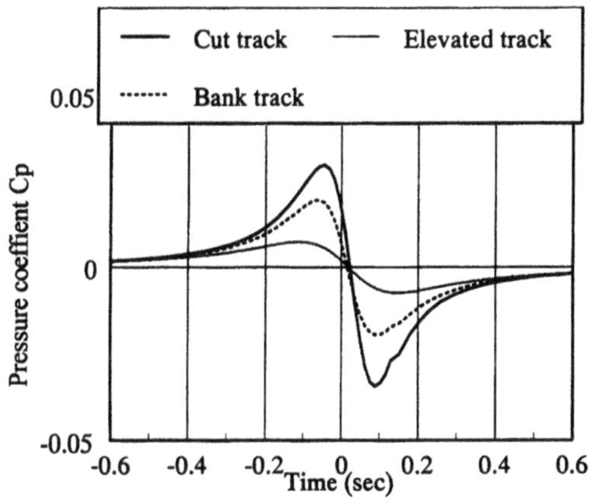

Figure 3. Difference in the pressure variation waveform by the track structure (6 m from track center, ground level).

The pressure variation for the cut track was much larger than that for the elevated track. Thus, the type of the track is very important in an assessment of railside environment conditions.

Figures 4(a) - 4(g) show the distributions of the maximum pressure variation (the peak-to-peak value) for each track structure. As seen from the figure, the pressure variation depends on the type of a structure. For example, In the elevated structure, the pressure variation below the elevated track was smaller than that above the elevated track because of the shielding effect of the base plate.

In order to investigate the difference of pressure fields on the track structure, the velocity, and pressure distribution were studied. Figure 5 shows the velocity and pressure contour plots on level ground and for a 120-degree space. This figure shows that a high-pressure region was observed at the train nose where the flow stagnates, and a decrease in pressure was observed at the shoulder of the train body where the flow was accelerating. In order to fill the narrow space on the slope side in a 120-degree space, the flow velocity changed largely from a uniform flow, due to the narrow gap between the train and the slope in the 120-degree space, and the pressure variation became large.

The difference in distance attenuation caused by the presence of ground is shown in Fig. 6. The pressure variation decayed in proportion to the approximately second power of distance for all the cases. However, the pressure amplitude with the ground was twice as strong as that without the ground. The rise in pressure resulted from the increase in the effective train cross-section area caused by the mirror image.

Figure 7 shows the distance attenuation in pressure variation when a train runs along a slope. The results obtained for different angle slopes, the bank, and the cut structure are compared. It was found that the pressure variation became large as the slope became steeper. The distance attenuation also depended on the geometry of a structure, since the pressure fluctuated wherever the velocity accelerates or decelerates.

These results led to the conclusion that the track structure greatly affects the pressure field when assessing the railside environment. Although the effect of the track structure on the pressure field was large near the slope, the effect became weaker where the slope became flat ground, as can be seen the distance attenuation characteristic of the cut and the bank in Figure 7. Therefore, the effect of the track structure is continued during the slope. Although the variations of pressure become about half that of the level ground condition, as the results in Fig. 7 show, if the trend of the downward slope continues like that of the 240-degree space shown in Fig. 4 (g), a slope like this case is physically impossible. Even though the downward slope is an effective track geometry for reducing the pressure variations, it is only effective while a slope continues, and the pressure variation is not reduced if the slope changes to the flat ground.

3.3 *The countermeasures of the pressure variation*

The reduction method for the pressure field at level ground is considered below. There are several methods used to reduce pressure variation, for example, the appropriate design of a train shape. In this study, the effects of several countermeasures applied to railsides, for example a conventional noise barrier, are examined.

Generally, the countermeasures under consideration can be applied only near the track, and a wall constructed at the railside must be short so that view from passing trains is not obstructed. Therefore, the measures examined in this study are applied within the region shown in Fig. 8(a); the maximum height is 2 m and the maximum distance from the track is 12.5 m. Pressure variation was evaluated 20 m from a track where houses and buildings exist.

The countermeasures to be examined are shown in Figs. 8(b)- 8(e). Figures 8(b) and 8(c) show the conventional sound-proof wall which is 2 m tall and installed at 3 and 12.5 m away from a track, respectively. In the countermeasures shown in Figs. 8(d) and 8(e), a 2 m tall bank and a 2 m deep cut is created on the ground. The left figures in Fig. 8 show the strength of the pressure variation, and the right figures show the ratio of the pressure variation with a measure to that without a measure; the ratio became smaller than 1 when a measure was effective.

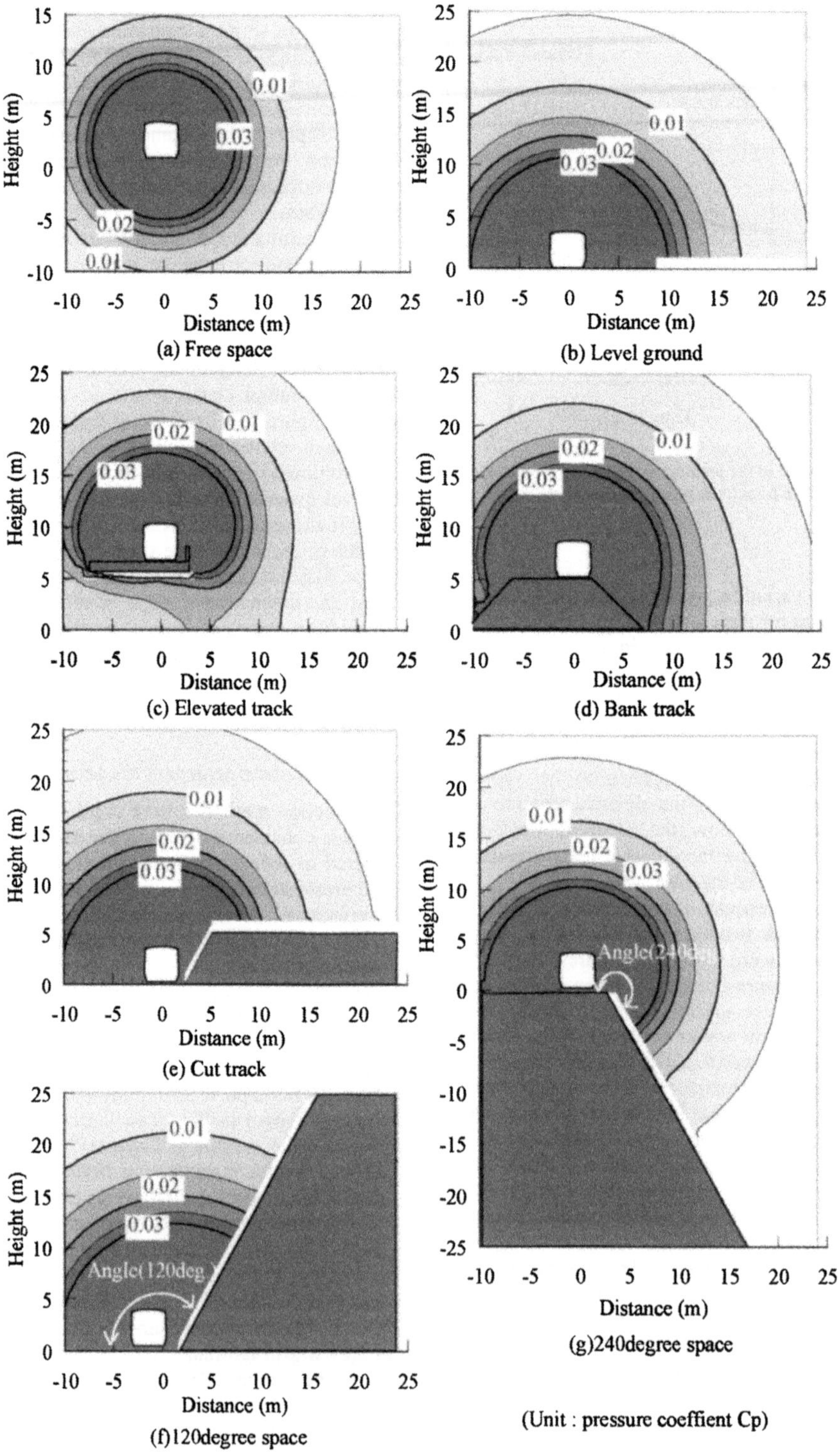

Figure 4. Distributions of the maximum pressure variation.

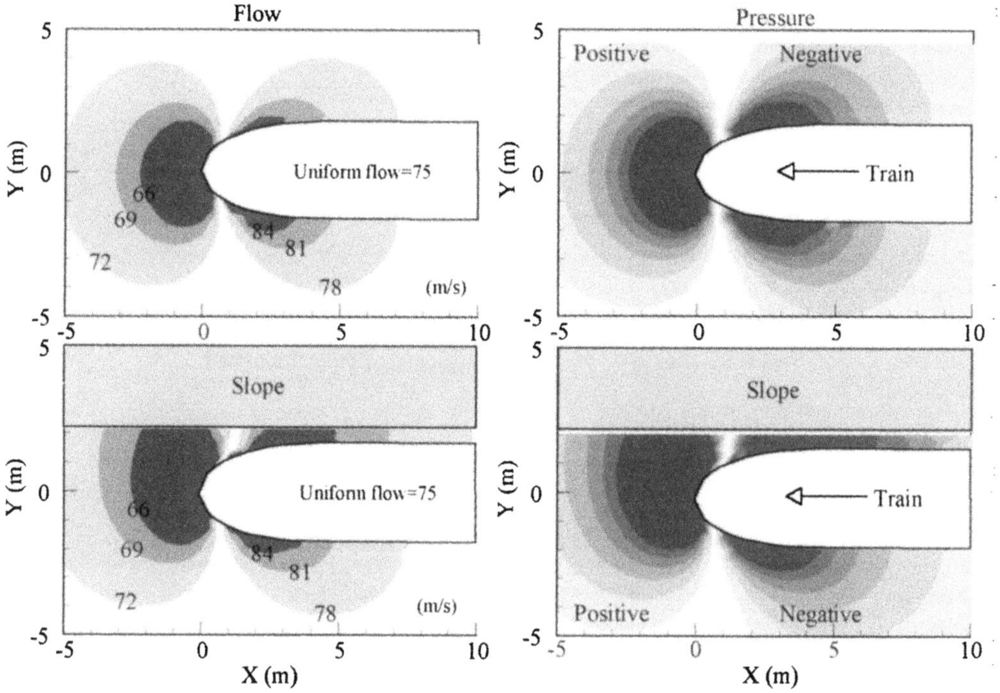

Figure 5. Flow contour plots and pressure contour plots in X-Y section (Z=0),
(Upper : Level ground, Lower : 120-degree space).

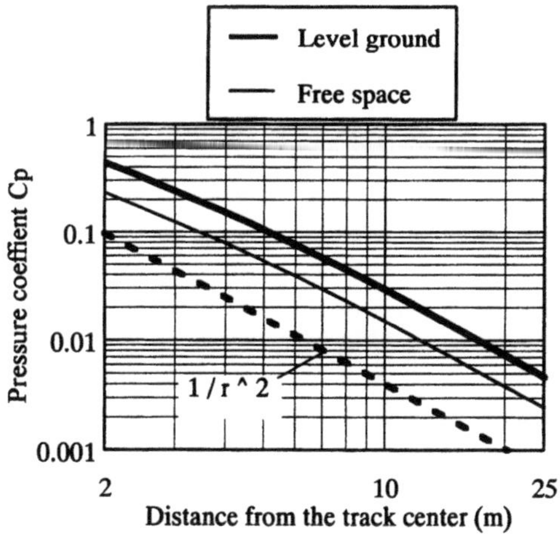

Figure 6. The difference in distance attenuation caused by presence of ground.

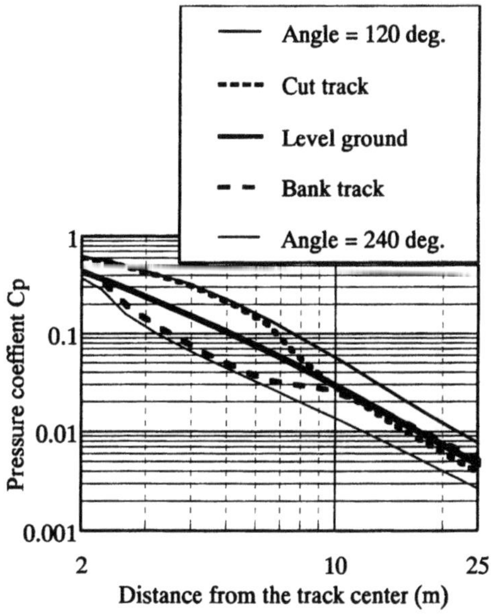

Figure 7. The effect of the gradient angle on the near pressure field.

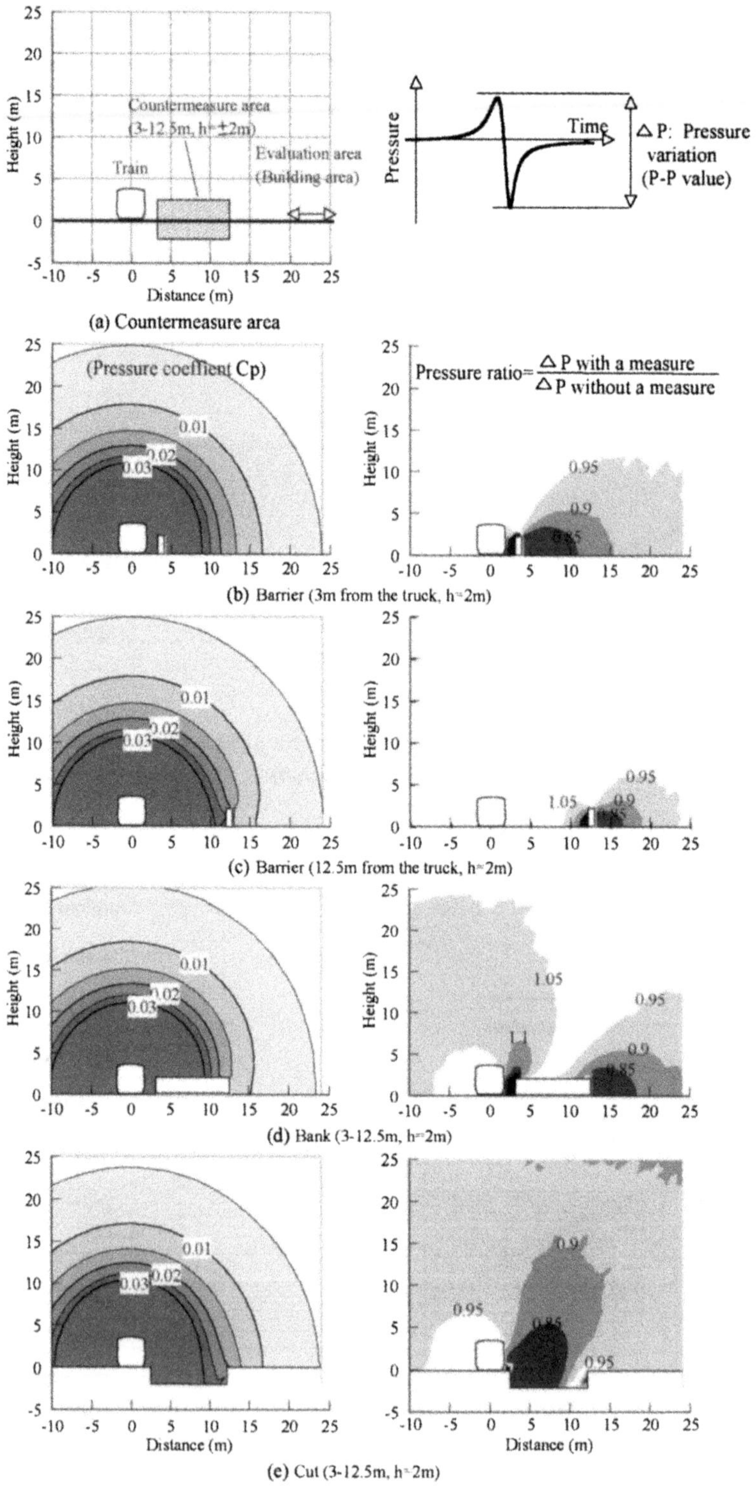

Figure 8. Comparison of reduction method (Left : pressure contour maps, Right : relative pressure(Ref. level ground)).

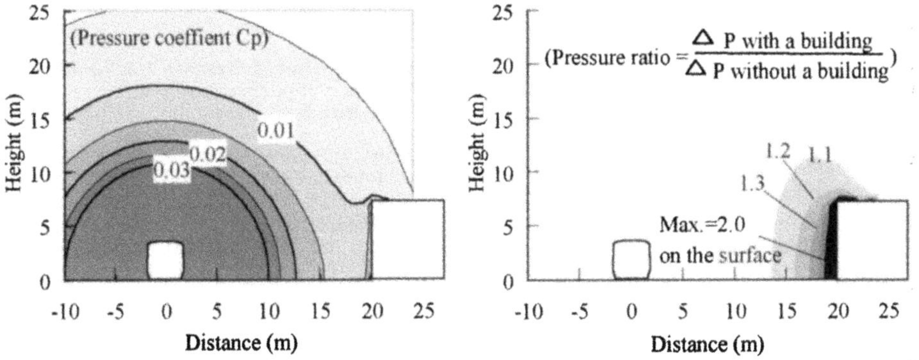

Figure 9. Influence of a building (Left : distribution of the maximum pressure variation, Right : pressure ratio(Ref. level ground)).

Figures 8(b) and 8(c) shows that the pressure variation is reduced about 10% at the rear side of the wall although the variation became large at the front of the wall. The evaluation of the pressure variation was quantitative, since the panel method is based on the potential flow theory and the flow separation that usually occurs at the rear side of an object was not taken into account. Although it is effective to install a wall adjacent to the railside, If the wall is 10 m or more away, it is almost ineffective. Although the pressure variation near the train increased when a 2 m bank was installed (fig. 8(d)), after 12.5 m the bank changed to the flat ground, there was a reduction effect of 10 % or more. On the other hand, the magnitude of the reduction effect beyond 12.5 m was only 10 % or less although the pressure variation was decreased in a wide area around the cut section when a 2 m deep cut was installed (fig. 8(e)).

From these results, it is thought that a countermeasure height of about 2 m is effective in only about 10 % of the case and the reduction effect is not enough to stop the vibration of fittings. To create an effective countermeasure, a higher wall, optimization of shape of the train nose, etc., are required.

3.4 *Influence by the existence of a building*

The influence on a building has not yet been discussed. However, since it is assumed that the pressure field is changed by the presence of a building, it is still necessary to grasp an understanding of this influence. Figure 9 shows the change of the pressure field that occurs around a building when the building height is 7 m and it is 20-27 m from the track under level ground conditions. The left figure expresses the distributions of the maximum pressure variation and the right figure expresses the ratio that broke the strength of the pressure variation, in the case of a building present, by the strength of the pressure variation in case where there was no building. Thus the pressure ratio is larger, when it is larger than 1. The pressure change became larger as the train approached the front of the building, and it doubled on the surface of the front building although pressure change was only 10% compared with that of the level ground conditions measured about 5 m from the building. Probably, the pressure increase for an actual building becomes smaller so that the building is infinitely long in the running direction for this calculation condition. However, it turns out that the pressure on the front of a building may reach a maximum of about twice that of the case where no building was present.

In the past, the pressure variation that affects a building was estimated in the no building condition. Since the pressure increase caused by a building cannot be ignored compared with the effect on the pressure field of a countermeasure and the influence of the track structure, must take into consideration the existence of the building.

4 SUMMARY

The effect of track structures on the pressure field was numerically investigated using the panel method. It was confirmed that a solution can be obtained in a very short time and the numerical results agree well with field measurement results. These results indicate that the panel method is effective for evaluation of the near pressure field around a train nose.

The result investigated about influence of the track structure indicated that there is significant difference in pressure variation when measurement point is nearby the track and influence of the slope appears with in a slope.

Next, the work on the countermeasures for reducing the near pressure field made it clear that the conventional measure equipment like a noise barrier of 2m height is ineffective. Therefore, a different countermeasure against noise is necessary to deal with the near pressure field.

Furthermore, the calculation in the case of a building near the railside made it clear that the pressure increases up to about double compared with the case where a building is not present. In the past, although the pressure variation that affects buildings was estimated in the no building condition, this calculation result implied that pressure variation might be underestimated using the conventional method.

5 REFERENCES

1. Teranishi, M., Tkano, J., Kitamura, T. & Yamada, S. 1997.9. Measurement of pressure distribution around model train; Proc. Autumn Meet. Acoust. Soc. Jpn: 665-666 (In Japanese).
2. Kitamura, T., Matubayashi, K., Kosaka, T. & Yamada, S. 2002. Estimation of low frequency noise of high-speed train with potential flow model; J. Acoust. Soc. Jpn. 58(7): 379-385 (In Japanese).
3. Kikuchi, K., Uchida, K., Nakatani, K., Yoshida, Y., Maeda, T. & Ynagizawa, M. 1996. Numerical analysis of pressure variation due to train passage using the boundary element method; Railway Technical Research Report, Vol.37, No.4.
4. Jaswon, M.A. & Symm, G.T. 1977. Integral Equation Methods in Potential Theory and Elastostatics. London: Academic press.
5. Saito, O.& Doi, T. 2002. Measurement of the pressure distribution formed around a high-speed train; Proceeding of Inter-noise 2002(in press).

Wave propagation – Moving load – Vibration reduction, Chouw & Schmid (eds.)
© 2003 Swets & Zeitlinger, Lisse, ISBN 90 5809 559 2

Prediction model of wayside noise level of Shinkansen

K.Nagakura & Y.Zenda
Railway Technical Research Institute, Tokyo, Japan

ABSTRACT: This paper introduces a prediction model of the wayside noise level of Shinkansen. The noise indices used in the model are the A-weighted sound exposure level of a train set pass L_{AE}, equivalent continuous A-weighted sound pressure level $L_{Aeq,T}$ and maximum A-weighted sound pressure level (slow) of a train set pass $L_{pA,Smax}$. In this model, Shinkansen noise sources are divided into four components, namely, the noise from the lower parts of cars, concrete bridge structure noise, aerodynamic noise from the upper parts of cars and pantograph noise. Each noise component is regarded as a row of discrete point sources, the positions and power levels of which are decided according to the noise component, train velocity, car type, track and structure. As a result, the wayside noise level of Shinkansen can be predicted under various conditions. The average and standard deviation of the difference between the measured value L_{meas} and the predicted value L_{calc} are 0.7 dB and 1.5 dB, respectively.

1 INTRODUCTION

The environmental quality standards for Shinkansen superexpress railway noise prescribes that the maximum A-weighted sound pressure level (slow) of a train set pass $L_{pA,Smax}$ at the wayside of the track shall be 70 dB(A) or less in the area for mainly residential used and 75 dB(A) or less in other areas, including commercial and industrial areas, where normal living conditions should be preserved. When Shinkansen commenced operation, all Shinkansen cars are of the same type (Series 0) running at the maximum velocity of about 200km/h. Currently, there are several types of cars running at various maximum velocities, from 220 to 300 km/h (see Table 1), and the noise level depends on the car type. Therefore, it is required to predict the noise levels of Shinkansen under different conditions of construction, car type, velocity and other factors.

No prediction models have been published so far for Shinkansen noise, although detailed studies for analysis and control of the Shinkansen noise have progressed (Moritoh et al. 1996, Nagakura 1996). In Europe, several prediction models have been developed by either the national railway companies, local research institutes, or the national Ministries of Environment or Traffic. Reviews of these European prediction models are presented by Leeuwen (2000). However, these models cannot be applied to the Shinkansen noise because the noise source characteristics of European railways differ from those of Shinkansen. Under the circumstances, the Railway Technical Research Institute proposed a prediction model of wayside noise level of Shinkansen to the requirement of Environment Agency.

Table 1. Dimensions of Shinkansen cars in operation

Series	u_{max}	N	l	n
Series 0	220	4	100	2,4
		6	150	2,4,6
		12	300	2,4,6,8,10,12
		16	400	2,4,6,8,10,12,14,16
Series 100	220	16	400	2,6,12
Series 100N	230	16	400	4,12,14
Series 200	240	8	200	4,6
		10	250	4,6
		12	300	4,10
		16	400	4,8,14
Series 300	270	16	400	6,12
Series 500	300	16	400	5,13
Series 700	285	8	200	2,7
		16	400	5,12
Series E1	240	12	300	6,10
Series E2	275	8	200	4,6
Series E4	240	8	200	4,6
		16	400	4,6,12,14
Series 400	240	7	143.5	2,4
Series E3	275	6	123	2,5

u_{max} (km/h): Maximum velocity, N: Number of cars,
l (m): Length of a train set,
Pantographs are located at the n th car.

2 SURVEY OF PREDICTION MODEL

2.1 *Flow chart of prediction model*

This model predicts the A-weighted sound exposure level of a train set pass L_{AE}, equivalent continuous A-weighted sound pressure level $L_{Aeq,T}$ and maximum A-weighted sound pressure level (slow) of a train set pass $L_{pA,Smax}$. Figure 1 shows a flow chart of the prediction model. First, the conditions of construction, car type, velocity, track and measuring point are introduced, according to which a noise source model is defined. In the prediction model, the noise sources of Shinkansen are regarded as rows of discrete point sources, the position and power levels of which are decided according to the noise component, train velocity, car type, track and structure. Next, the time history of the instantaneous A-weighted sound pressure level of one point source moving on the track (which we call a "unit pattern") is calculated. Once the unit pattern is obtained, L_{AE} is obtained by integrating the unit pattern and summing them for all point sources contained in a train set. $L_{Aeq,T}$ can be calculated if the number of train passes during the time length T is known. $L_{pA,Smax}$ is calculated as a definition or obtained directly from L_{AE} by using a conversion equation.

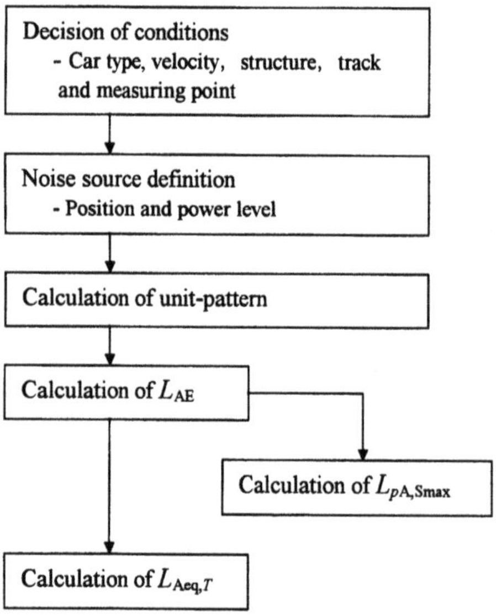

Figure 1. Flow chart of prediction model

2.2 *Extent of application*

This prediction model can be applied under the conditions shown in Table 2. The extent of velocity has a lower limit because the contributions of noise components such as noise from air conditioners and inverters, which are not considered in the model, cannot be neglected at the velocity below 150 km/h. The restriction for a measuring point can be removed if the effects of the attenuation due to air absorption and ground absorption are considered in the model.

Table 2. Extent of application of the prediction model

Type of car	All Shinkansen cars in operation
Velocity	150km/h-maximum velocity in operation
Track	Ballast track, slab track and vibration-reducing track
Construction	Concrete bridge structure and enbankment
Sound barrier	Straight type with or without absorbing materials
Measuring point	At the point at a horizontal distance of 12.5 to 50 m from the track and at a height lower than the upper end of the sound barrier

3 NOISE SOURCE DEFINITION

3.1 *Classification of noise sources*

In the prediction model, Shinkansen noise sources are divided into four components, (1) the noise generated by the lower parts of cars, which consists of the rolling noise, aerodynamic noise, and gear noise, (2) concrete bridge structure noise, (3) aerodynamic noise generated by the upper parts of cars and (4) pantograph noise, which consists of the aerodynamic noise and spark noise. Each noise source is regarded as a row of discrete non-directive point sources (monopole sources). The noises (1) to (3) are radiated to a half space and the noise (4) is radiated to a full space.

3.2 *Position of source*

The positions of the noise sources are shown in Figure 2. The coordinates of the noise sources differ in detail according to the type of cars. Table 3 shows the coordinates of the noise sources for all types of cars in operation in the case of the standard concrete bridge structure.

3.3 *Sound radiation characteristics*

Figure 3 shows the time history of the A-weighted sound pressure level (fast) of Shinkansen cars measured at the point 25 m away from the track. The calculated time histories of discrete monopole sources and dipole sources whose axes are parallel to sleepers are also shown in the same Figure. It is found that the radiation characteristics of the Shinkansen noise are similar to those of monopole rather than those of dipole. Thus, all sources are assumed to be monopole sources in the prediction model.

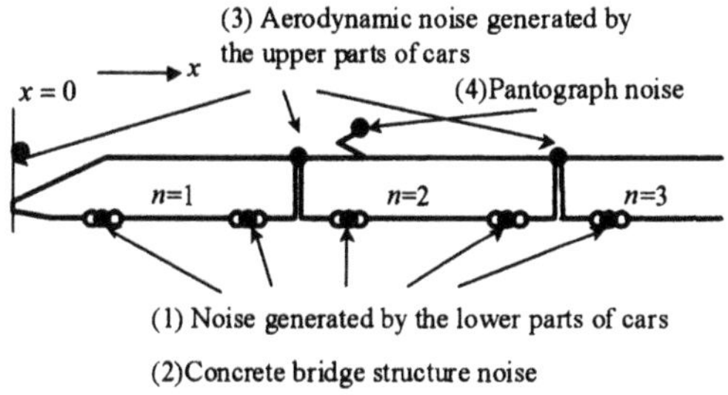

(a) Side view

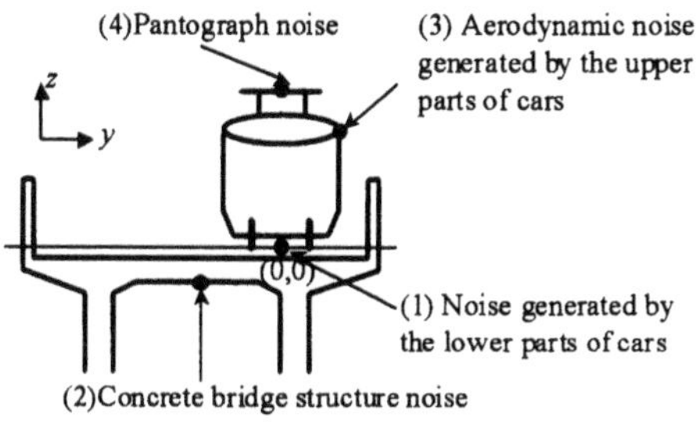

(b) Cross section

Figure 2. Position of noise sources

Table 3. Coordinates of noise sources

	Coordinate x(m)	
(1)* (2)*	$25(n-1)+3.75$ $25(n-1)+21.25$	All series except series 400,E3
	$20.5(n-1)+3.175$ $20.5(n-1)+17.325$	Series 400,E3
(3)*	$25(n-1)$	All series except series 400,E3
	$20.5(n-1)$	Series 400, E3
(4)*	$25(n-1)+1.25$	Series 0, 100, 100N, 200,300,E1, E2, E4
	$25(n-1)+6.6$	Series 500
	$25(n-1)+18.65$ $25(n-1)+6.35$	Series 700 (n =2,5) Series 700 (n =7,12)
	$25(n-1)+6.35$	Series 400,E3
	Coordinates (y, z)(m)	
(1)*	(0.0, 0.0)	All series
(2)*	(-2.15, -0.9)	All series
(3)*	(1.6, 3.7)	Series 0,100,100N, 200
	(1.6, 3.5)	Series 300,500,700,E2
	(1.6, 4.0)	Series E1, E4
	(1.4, 3.7)	Series 400,E3
(4)*	(0.0, 5.0)	All series

* (1) Noise generated by the lower parts of cars, (2) Concrete bridge structure noise, (3) Aerodynamic noise generated by the upper parts of cars, (4) Pantograph noise

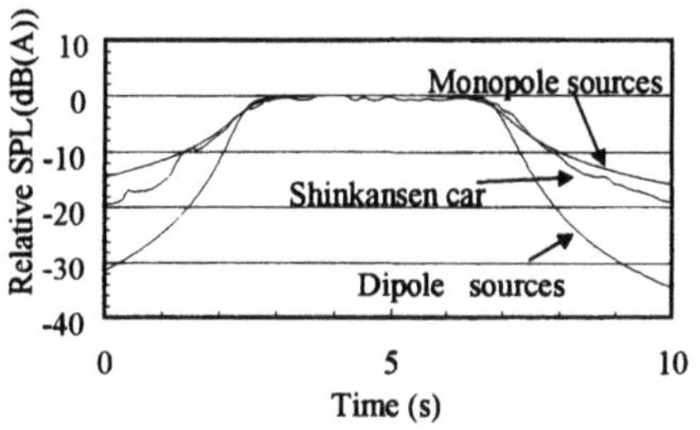

Figure 3. Time histories of A-weighted sound pressure level (time constant: fast, length of a train set = 300m, velocity = 240km/h)

3.4 *Power level of noise source*

3.4.1 *Dependency of power level on velocity*

The dependency of the power level on velocity is defined by the factor "n", where the power level is proportional to the nth power of velocity. Here the factor "n" for each noise component is decided on the basis of experimental data of field tests and wind tunnel tests.

(1) Noise generated by the lower parts of cars

Figure 4 shows the relation between the velocity and the A-weighted sound pressure level near the rail (at a point 2 m away from the rail and at the height of a wheel axle). The A-weighted sound pressure level increases in proportion to the 2-4th power of the velocity. Thus the factor "n" is decided to be 3 for the noise generated by the lower parts of cars in the prediction model.

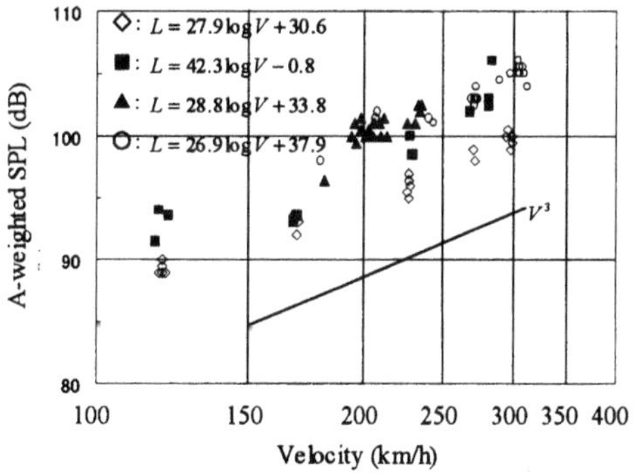

Figure 4. Relation between the velocity and the A-weighted sound pressure level near the rail

(2) Concrete bridge structure noise

Figure 5 shows the relation between the velocity and the A-weighted sound pressure level under a concrete bridge structure. The A-weighted sound pressure level increases in proportion to the 2-5th power of the velocity. The factor "n" is roughly decided to be 3 for the concrete bridge structure noise in the prediction model.

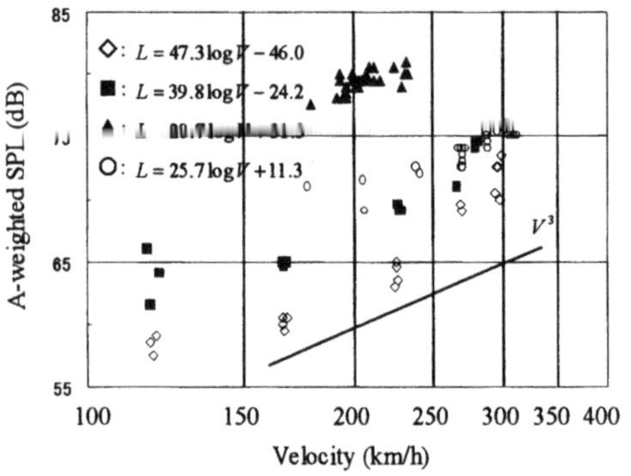

Figure 5. Relation between the velocity and the A-weighted sound pressure level under a concrete bridge structure

(3) Aerodynamic noise generated by the upper parts of cars
There are no enough data of field tests which show the dependency on velocity of the aerodynamic noise generated by the upper parts of cars. However, the theoretical analysis on aerodynamic noise by Curle (1955) and results of wind tunnel experiments show that the aerodynamic noise generated from trains increases in proportion to the 6th power of velocity in most cases. Thus, the factor "n" is decided to be 6 for the aerodynamic noise generated by the upper parts of cars in the prediction model.
(4) Pantograph noise
Figure 6 shows the relation between the velocity and the A-weighted sound pressure level of the pantograph noise measured with a microphone array. In most cases, the A-weighted sound pressure level increases proportional to the 5-7th power of the velocity. This is because the main component of the pantograph noise is aerodynamic noise at the velocity above 200 km/h. Thus, the factor "n" is decided to be 6 for the pantograph noise in the prediction model.

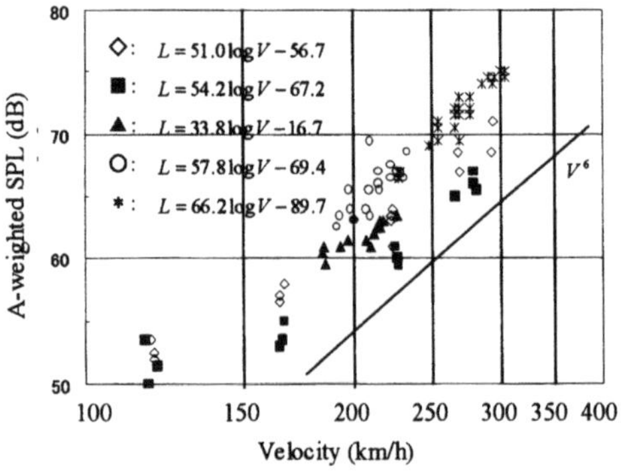

Figure 6. Relation between the velocity and the A-weighted sound pressure level of pantograph noise measured with a microphone array

3.4.2 *Decision of power level*
Once the factor "n" is decided in section 3.4.1, the power level of the noise source $L_{wi}(u)$ can be given by the equations (1) to (4) where the subscripts i = 1, 2, 3 and 4 correspond to (1) the noise generated by the lower parts of cars, (2) concrete bridge structure noise, (3) aerodynamic noise generated by the upper parts of cars and (4) pantograph noise, respectively.

$$L_{w1}(u) = L_{w1}(200) + 30\log(u/200) - \Delta L_1 \quad (1)$$

$$L_{w2}(u) = L_{w2}(200) + 30\log(u/200) - \Delta L_2 \quad (2)$$

$$L_{w3}(u) = L_{w3}(200) + 60\log(u/200) \quad (3)$$

$$L_{w4}(u) = L_{w4}(200) + 60\log(u/200) \quad (4)$$

In the above equations, $L_{wi}(200)$ denotes the power level at 200 km/h and u is the train velocity (km/h). ΔL_1 is the correction factor for the noise generated by the lower parts of cars, which is 0 dB for slab tracks and 5 dB for ballast tracks. ΔL_2 is the correction factor for the concrete bridge structure noise, which is 5 dB for slab tracks with a rubber isolator, 8 dB for ballast tracks with a ballast mat, and 10 dB for tracks with sleepers covered with a resilient material. The values of $L_{wi}(200)$ depend on the type of cars. In order to decide $L_{wi}(200)$, the A-weighted sound pressure level of each noise component at the wayside is estimated first on the basis of the measured data (Nagakura 1996, Kitagawa & Nagakura 2000). $L_{wi}(200)$ can be calculated inversely by using equations shown in Chapter 4. The values of $L_{wi}(200)$ are shown in Table 4.

Table 4. Values of $L_{wi}(200)$ (dB)

Series	L_{w1} (200)	L_{w2} (200)	$L_{w3}(200)$ Leading car	$L_{w3}(200)$ Other cars	L_{w4} (200)
0	114.5	90	100	99	104
100	114.5	90	100	94.5	103
100N	114.5	90	100	94.5	103
300	113.5	88	97	91	101
500	112.5	88	94	87	99.5
700	113.5	88	95	88	97
200	114.5	90.5	100	96	104
E1	115	90.5	100	96	103
E2	113.5	89	97	88	100
E4	114	90	98	95	103
400	114.5	89	97	95	103
E3	113.5	88.5	97	92	102

4 PROCESS OF CALCULATION

4.1 *Calculation of unit pattern*

When a point source moves along a track, the time history of the instantaneous A-weighted sound pressure level $L_{pA,p}(t)$ is given by the equations (5) to (6).

$$L_{pA,\,pn}(t) = L_w - \alpha - 20\log R_n(t) - \Delta L_{pn}(t) \quad (5)$$

$$L_{pA,\,p}(t) = 10\log(10^{L_{pA,p1}(t)/10} + 10^{L_{pA,p2}(t)/10}) \quad (6)$$

Here L_w is the power level of the point source; $R_n(t)$ is the distance between the noise source and the measuring point; and $\Delta L_{pn}(t)$ is the screening attenuation for a point source by a sound barrier. The subscript "n=1" denotes the values for the direct sound and the subscript "n=2" denotes the values for the sound reflected by the ground. The value of α is 8 for half space radiation and 11 for full space radiation.

4.2 *Calculation of L_{AE}*

The A-weighted sound exposure level of a point source pass $L_{AE,p}$ is calculated by the equation (7).

$$L_{AE,\,p} = 10\log\int_{-\infty}^{\infty} 10^{L_{pA,p}(\tau)/10} d\tau \quad (7)$$

If the cross section of the construction is constant along the track, which is often the case in railways, the equations (5) to (7) are reduced to the equations (8) to (9).

$$L_{AE,\,pn} = L_w - 10\log \beta v r_n - \Delta L_{ln} \quad (8)$$

$$L_{AE,\,p} = 10\log(10^{L_{AE,p1}/10} + 10^{L_{AE,p2}/10}) \quad (9)$$

where the value of β is 2 for half space radiation and 4 for full space radiation. v(m/s) is the train velocity; r_n(m) is the distance between the measuring point and the row of point sources; and ΔL_{ln} is the screening attenuation for a line source by a sound barrier.

The A-weighted sound exposure level of a train set pass L_{AE} is calculated as the summation of $L_{AE,p}$ for all point sources contained in a train set.

$$L_{AE} = 10\log\sum_n 10^{L_{AE,p,n}/10} \quad (10)$$

where $L_{AE,p,n}$ denotes the A-weighted sound exposure level of the nth point source.

4.3 *Calculation of $L_{Aeq,T}$*

The equivalent continuous A-weighted sound pressure level during the time length T ($L_{Aeq,T}$) is defined by the equation (11).

$$L_{Aeq,T} = 10\log\left[(T_0/T)\sum_n 10^{L_{AE,n}/10}\right] \quad (11)$$

where $L_{AE,n}$ denotes the A-weighted sound exposure level of the nth train set during the time length T and T_0=1(s).

4.4 *Calculation of $L_{pA,Smax}$*

$L_{pA,Smax}$ is defined as the maximum level of the time history of the sound pressure level (slow) $L_{pA,S}(t)$, which is calculated by the equations (12) to (13).

$$L_{pA}(t) = 10\log\sum_n 10^{L_{pA,p,n}(t)/10} \quad (12)$$

$$L_{pA,\,S}(t) = 10\log\int_{-\infty}^{t} 10^{L_{pA}(\tau)/10} e^{-(t-\tau)/T_c} d\tau \quad (13)$$

where the summation is done for all point sources contained in a train set; $L_{pA,p,n}(t)$ is the unit pattern of nth point source; and T_c=1 (s) is a time constant. If the cross section of the construction is constant along the track, $L_{pA,Smax}$ is approximated by the equation (14) to significantly simplify the calculation.

$$L_{pA,\,Smax} = L_{AE} - 10\log(l/v) - \Delta L \quad (14)$$

where l (m) is the length of a train set; v (m/s) is the velocity; r(m) is the distance between the measuring point and the row of point sources; and ΔL is a correction term which depends on l/v and l/r. At the limit of $l/v \to \infty$ or $l/r \to \infty$, ΔL converges to 0. The values of ΔL are given in advance by a numerical simulation of line source of length l moving at the velocity v. The results of the simulation are shown in Table 5. Figure 7 compares the left side of the Equation (14) with the right side by using measured data of $L_{pA,Smax}$ and L_{AE}. It is found that Equation (14) gives a good approximation.

5 SCREENING ATTENUATION BY SOUND BARRIER

5.1 *Screening attenuation for point source*

The screening effect by a sound barrier for a point source $\Delta L_{pn}(t)$ in the equation (5) is estimated from the path length difference δ defined in Figure 8 by applying the typical spectrum of each noise source to the Maekawa chart. The spectra of noise sources depend on the train velocity. However, the difference of $\Delta L_p(t)$ due to the velocity, which is at most 1 dB, is not significant within the velocity range of 150 to 300km/h. Figures 9 and 10 show the charts for calculating the screening attenuation for a point source. If the inner walls of the barrier are not covered with an absorbing material and the car runs on the track nearer the sound barrier, the effect of the barrier is weakened because of the multiple reflection between the inner wall of the barrier and car body. In this case, the dotted line shall be used in Figure 9.

5.2 *Screening Attenuation for Line Source*

The screening effect by a sound barrier for a line source ΔL_{ln} in the equation (8) is calculated by the equation (15).

Table 5. Values of ΔL obtained by numerical simulation (dB)

		u							
		0	112.5	150	225	300	450	600	900
r	0	0.00	0.00	0.00	0.00	0.00	0.00	0.00	0.00
	9.4	0.00	0.08	0.14	0.25	0.39	0.73	1.11	1.85
	12.5	0.00	0.15	0.21	0.33	0.47	0.82	1.20	1.95
	18.8	0.00	0.24	0.32	0.46	0.61	0.97	1.36	2.12
	25	0.00	0.32	0.41	0.58	0.75	1.12	1.51	2.27
	37.5	0.00	0.48	0.61	0.81	1.00	1.39	1.79	2.54
	50	0.00	0.64	0.81	1.04	1.25	1.65	2.05	2.78
	75	0.00	0.96	1.20	1.48	1.71	2.13	2.52	3.23
	100	0.00	1.26	1.57	1.90	2.15	2.57	2.95	3.63
	150	0.00	1.80	2.22	2.65	2.93	3.35	3.71	4.33
	200	0.00	2.24	2.76	3.28	3.59	4.02	4.36	4.94
	300	0.00	2.85	3.54	4.24	4.62	5.09	5.41	5.93

l=300(m), r(m), u(km/h), v=u/3.6(m/s)

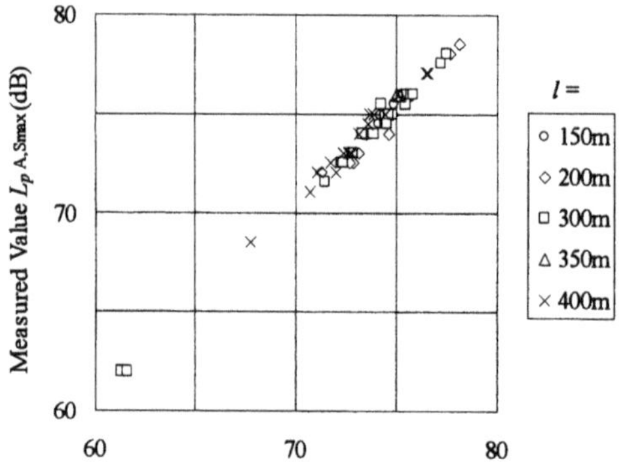

Figure 7. Relation between the left side and right side of the Equation (14)

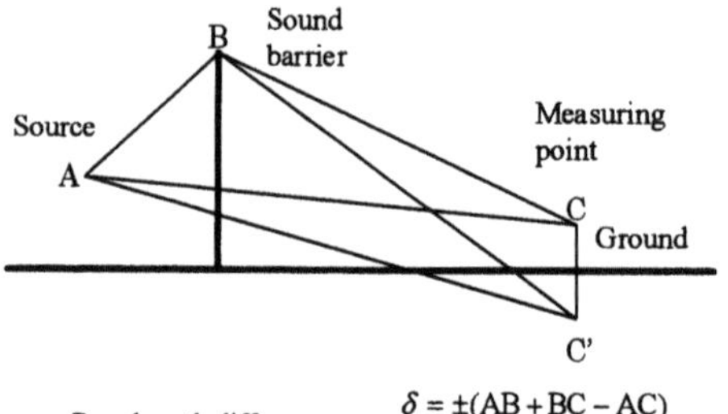

Figure 8. Definition of pass length difference

$$\Delta L_l = -10\log\left(\int_{-\infty}^{\infty} 10^{-\Delta L_P(x)/10} / (x^2 + r^2)dx\right) \qquad (15)$$

where the x-axis coincides with the line source, the origin of which is in front of the measuring point; r is the distance between the measuring point and the line source; and $\Delta L_P(x)$ is the screening effect by a sound barrier for the point source located at the position x (see Figure 11). The values obtained from the equation (15) are shown in Figure 12 for the noise generated by lower parts of cars, and in Figure 13 for the aerodynamic noise generated by the upper parts of cars and pantograph noise. The doted line shall be used again in Figure 12 if the inner walls of the barrier are not covered with an absorbing material and the car runs on the track nearer the sound barrier.

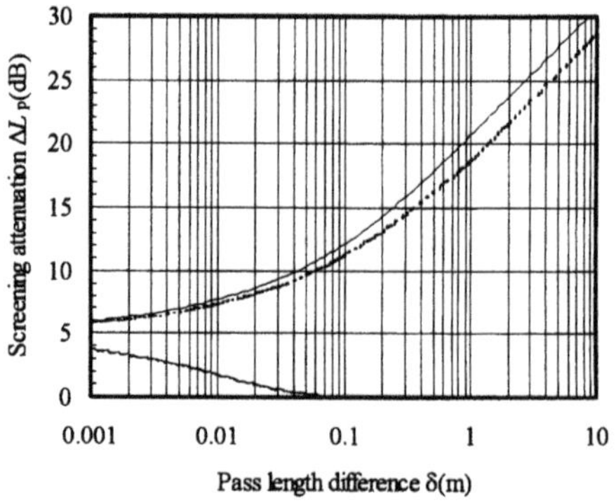

Figure 9. Screening attenuation for a point source (Noise generated by the lower parts of cars)

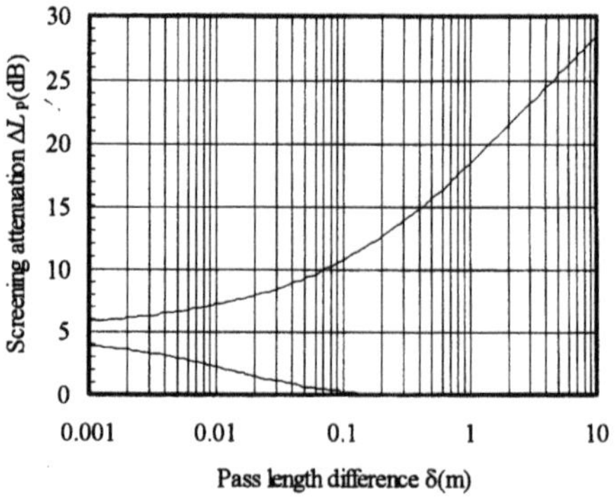

Figure 10. Screening attenuation for a point source (Aerodynamic noise generated by the upper parts of cars or pantograph noise)

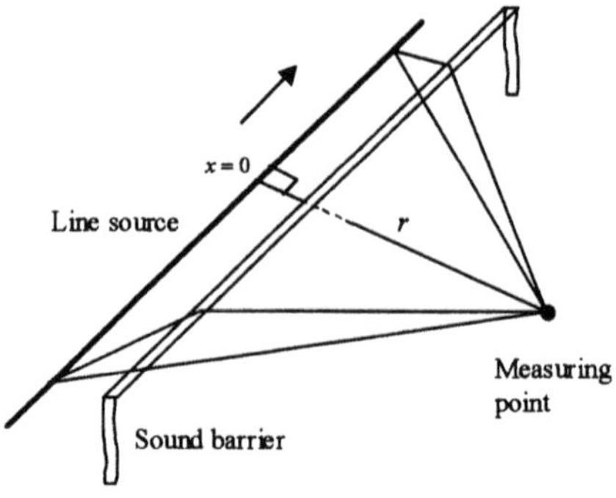

Figure 11. Geometry of line source, measuring point and sound barrier

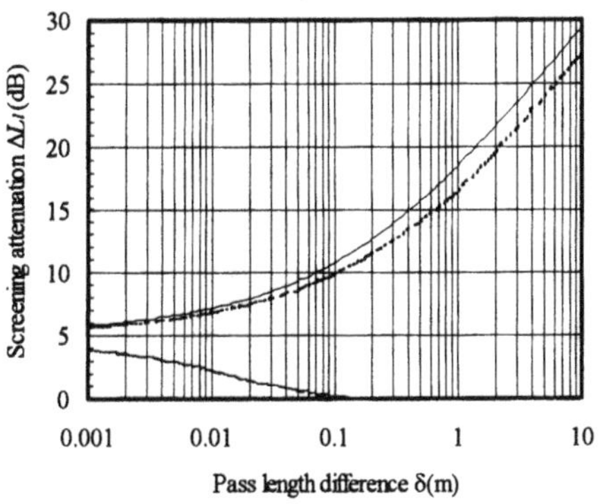

Figure 12. Screening attenuation for a line source (Noise generated by the lower parts of cars)

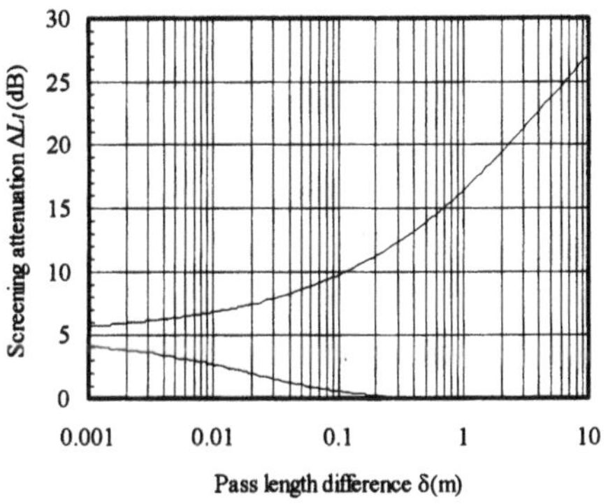

Figure 13. Screening attenuation for a line source (Aerodynamic noise generated by the upper parts of cars or pantograph noise)

6 CASE STUDY

6.1 *Condition of case study*

In order to confirm the validity of the prediction model, the measured values L_{meas} were compared with the predicted values L_{calc}. A case study was carried out by using data of a total of 2,595 train passes running at 22 sections which contain eight ballast track sections, 10 slab track sections and four vibration-reducing track sections. The construction of all sections is a concrete bridge structure with a sound barrier, which is the standard construction of Shinkansen. All types of cars except Series E4 and Series 400 are included in the data. The measuring point is 12.5 to 50 m away from the track and 1.2 m above the ground. $L_{pA,Smax}$ is used as a noise index.

6.2 *Results*

The result of case study for all data is shown in Figure 14. The average and standard deviation of the difference between L_{meas} and L_{calc} are 0.7 dB and 1.5 dB, respectively. This result shows that the prediction model is valid enough.

In order to clarify the cause of the variance, the standard deviation of the difference between L_{meas} and L_{calc} was calculated when one of the conditions was fixed, such as the section, type of cars and distance from the track to the measuring point. The results are shown in Table 6. If the section is fixed, the standard deviation of the difference between L_{meas} and L_{calc} is lowered to 1.1 dB, which suggests that the difference of the section causes the variance. This is because the power levels of rolling noise and concrete bridge structure noise strongly depend on the conditions of the rail and wheel surface which vary according to the section. If the type of cars is fixed, the standard deviation of the difference between L_{meas} and L_{calc} is lowered to 1.3 dB, thus the type of cars could have some effect on the variance. The distance from the track to the measuring point is considered to have very little effect on the variance.

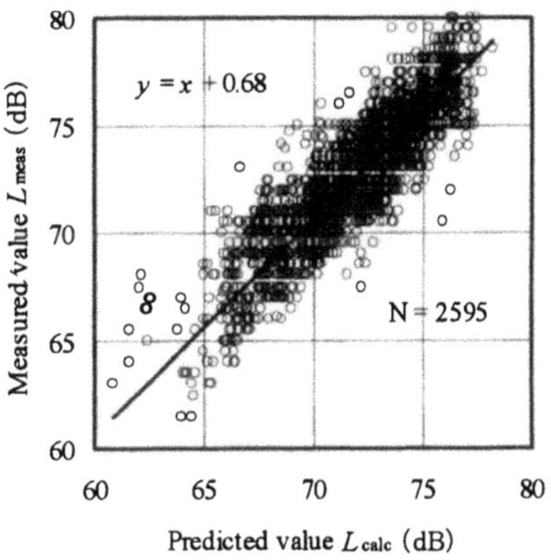

Figure 14. Result of case study

Table 6. Standard deviation of (L_{meas}-L_{calc})

Fixed condition	Standard deviation of (L_{meas}-L_{calc})
Section	1.1 dB
Type of cars	1.3 dB
Distance from track to measuring point	1.5 dB

Next, the data were divided into three categories according to the velocity, namely, from 150 km/h to 200 km/h, from 200 km/h to 250 km/h and above 250 km/h and case studies were carried out separately. Furthermore another case study was carried out using the data below 150 km/h, which are out of application of the model. The results are shown in Figure 15 and Table 7. The prediction model is valid above 150 km/h although its precision becomes worse as the velocity decreases.

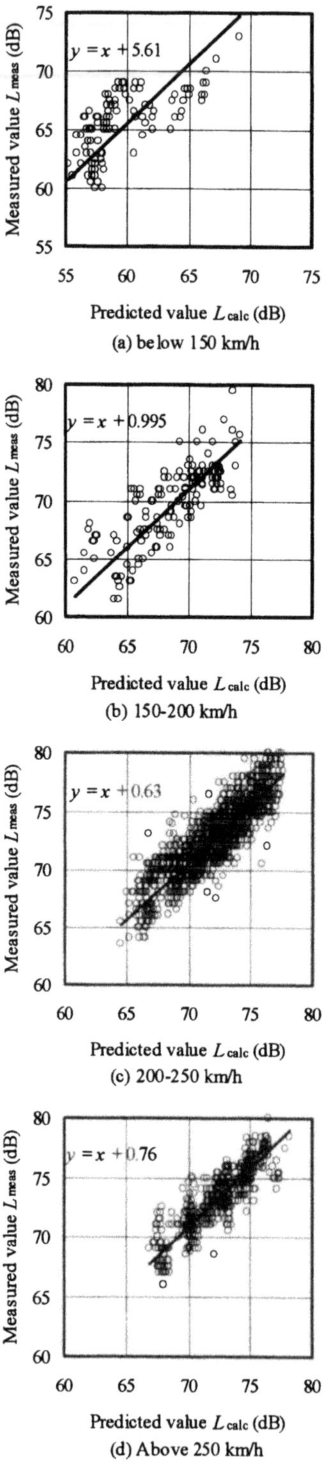

Figure 15. Results of case study for each velocity range

Below 150 km/h, L_{meas} is greater than L_{calc} by 5 dB, which is due to the effects of the noise from air conditioners, inverters, etc.

Table 7. Standard deviation of (L_{meas}-L_{calc}) for different velocity categories

Velocity	Number of data	Average of (L_{meas}-L_{calc})	Standard deviation of (L_{meas}-L_{calc})
Below 150km/h	116	5.61 dB	2.39 dB
150-200km/h	158	1.00 dB	2.06 dB
200-250km	1809	0.63 dB	1.44 dB
Above 250km/h	628	0.76 dB	1.30 dB

7 CONCLUSIONS

This paper prescribes the prediction model of the wayside noise level of Shinkansen (L_{AE}, $L_{Aeq,T}$, $L_{pA,Smax}$). In this model, Shinkansen noise sources are divided into four components and regarded as rows of discrete point sources, the positions and power levels of which are decided according to the noise component, train velocity, car type, track and structure. Thus, the wayside noise level of Shinkansen can be predicted under various conditions. As the result of a case study, the average and standard deviation of the difference between the measured value L_{meas} and the predicted value L_{calc} are 0.7 dB and 1.5 dB, respectively. The variance of the difference between L_{meas} and L_{calc} is mainly due to the variance of the power levels of rolling noise and concrete bridge structure noise which strongly depend on the conditions of the rail and wheel surface.

8 REFERENCES

Curle, N. 1955. Influence of solid boundaries upon aerodynamic sound. *Proc. Roy. Soc.*, A231: 505

Kitagawa, T. & Nagakura, K. 2000. Aerodynamic noise generated by Shinkansen cars. *Journal of Sound and Vibration*, 231(3): 913-927

Leeuwen, H.J.A. 2000. Railway noise prediction models: A comparison. *Journal of Sound and Vibration*, 231(3): 975-987

Moritoh, Y. et al. 1996. Noise control of high-speed Shinkansen. *Journal of Sound and Vibration*, 193(1): 319-334

Nagakura, K. 1996. The Method of Analyzing Shinkansen Noise. *Quarterly Report of RTRI*, Vol.37, No.4

Wave propagation – Moving load – Vibration reduction, Chouw & Schmid (eds.)
© 2003 Swets & Zeitlinger, Lisse, ISBN 90 5809 559 2

Suppression of an acoustic shock wave and damping of pressure waves in a tunnel by a double array of Helmholtz resonators

N. Sugimoto
Department of Mechanical Science, Graduate School of Engineering Science
University of Osaka, Toyonaka, Osaka 560-8531, Japan

ABSTRACT: This paper describes a method to suppress emergence of an acoustic shock wave and also to damp pressure waves in a tunnel generated by entry of a high-speed train. It exploits both action of wave dispersion and damping by connecting a double array of Helmholtz resonators to the tunnel axially. Emphasis is placed on difference in mechanisms of suppression of shock and of damping of pressure waves. Using the nonlinear wave equations derived previously, the numerical calculations are carried out to examine the effects of the double array. There are two important parameters called the coupling and tuning parameters characterizing each array. It is shown that the double array acts very effectively when the tuning parameter of one array is set to be 10 or higher than it for suppression of shock and the one of the other array is set around unity for damping of pressure waves, while the coupling parameters of both arrays are chosen to be unity.

1 INTRODUCTION

When a train enters a long tunnel at high speed, an acoustic shock wave (simply called a shock hereafter) emerges in the tunnel. Radiation of the shock from the tunnel exit gives rise to bursting sound and low-frequency pressure waves (infrasound), which cause an environmental noise and vibration problem (Ozawa 1979). Because the train speed is well subsonic, the shock does not emerge immediately on entry. It tends to emerge from pressure disturbances generated by entry in the course of propagation at a speed almost equal to the sound speed. Such a process is illustrated in Figure 1.

In the case of tunnels for the Shinkansen entering at speed 300 km/h, the noise and vibration problem is solved to be within allowable level. To clear this level, hoods of suitable length and with windows are installed at the tunnel portal as the countermeasures on the ground side (Ozawa 1979, Maeda 2001). The hoods act to make milder the pressure gradient at the wave front so that the shock-formation point may be beyond the tunnel exit. But it is noted that the action of the hoods cannot suppress shock formation essentially.

As a train speed is increased so that the magnitude of the pressure disturbances is increased, the shock tends to emerge in such a tunnel that no shock has emerged so far. For the magnetically levitated trains traveling at speed 500 km/h or higher, it is not certain whether or not the conventional countermeasures are still applicable, so a new alternative method should be devised. The present author has proposed a concept of shock-free propagation by ex-

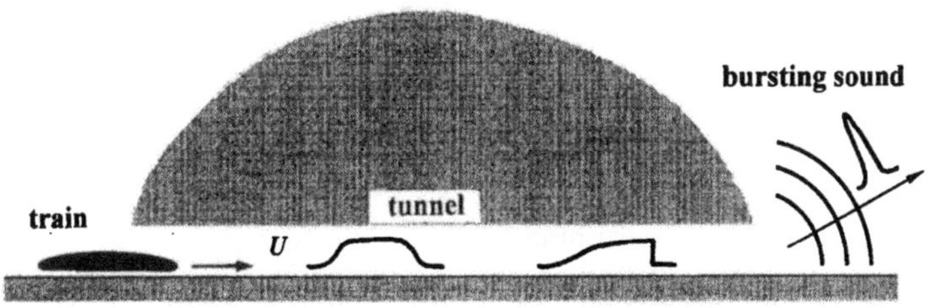

Figure 1. Train entering a tunnel and the resulting radiation of busting sound and pressure waves from the tunnel exit.

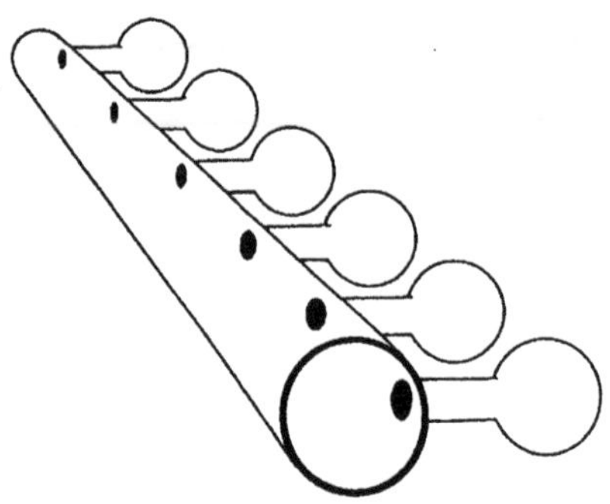

Figure 2. Conceptual figure of a tunnel with an array of Helmholtz resonators.

ploiting wave dispersion, which hinders shock formation naturally by its own physical mechanisms. But because the dispersion is absent in the propagation of pressure waves in the tunnel, an array of Helmholtz resonators is proposed to be connected to the tunnel in the axial direction (Sugimoto 1992). The conceptual figure is depicted in Figure 2. Each resonator's effect on wave propagation is small but their cumulative effects in the array can counteract shock formation, which also results from cumulative effects of weak nonlinearity in the propagation of pressure waves. It has already been shown not only in theory (Sugimoto 1992, 1993, 2001) but also by experiment (Sugimoto *et al.* 1999) that the array can indeed inhibit the shock formation.

Propagation of the pressure waves in the tunnel with the array is controlled by two parameters, one being called the "coupling parameter" of the array $K\ (=\kappa/2\varepsilon)$, and the other the "tuning parameter" $\Omega\ (=(\omega_0/\omega)^2)$. The former parameter is defined by the ratio of the "size parameter" of the array κ ($= V/Ad \ll 1$) to the acoustic Mach number ε ($= u_0/a_0 \ll 1$), where V, A, d denote, respectively, cavity's volume of the resonator, the cross-sectional area of the tunnel, and the axial spacing between neighbouring resonators, while u_0 and a_0 denote, respectively, a typical axial speed of air and the linear sound speed. The tuning parameter is defined as the ratio squared of resonator's natural angular frequency ω_0 to a typical angular frequency of the pressure waves ω. It is revealed that suppression of shock can be achieved by choosing Ω much higher than unity while K is chosen to be of order unity.

It should be emphasized that the mechanism to suppress the shock formation is different from the one to damp the pressure waves. While the magnitude of pressure disturbances remains to be small, the shock may be inhibited by damping of the pressure waves. But as the magnitude is large, the damping becomes weak relatively with the nonlinearity so that it alone cannot suppress the shock formation. In reality, it is required not only to suppress the shock but also to damp the pressure waves. To this end, a double array is proposed to be connected to the tunnel. This paper examines propagation of the pressure wave in the tunnel with the double array by solving numerically the nonlinear wave equations derived previously to demonstrate its effectiveness.

2 BASIC EQUATIONS

First the nonlinear wave equations are presented for unidirectional propagation of pressure waves in a tunnel with a double array of Helmholtz resonators. For the details of derivation, see Sugimoto (2001). Each array, designated by the array 1 or 2, is assumed to consist of identical resonators connected with equal axial spacing to the tunnel. Then each array constitutes spatially periodic structure, but the tunnel itself is not necessarily of periodic structure because the ratio of the axial spacing in each array is irrational in general.

The quantities pertaining to each array m (m = 1, 2) are distinguished by the subscript m. Unless the subscript is attached, it is the case with a single array. The size parameter of each array and the natural angular frequency of the resonator are denoted by κ_m and ω_m, respectively, while the volume of cavity, the cross-sectional area of the throat, its length and the axial spacing between neighbouring resonators in each array are denoted by V_m, B_m, L_m and d_m, respectively. In the same fashion, the coupling parameters and the tuning parameters are designated, respectively, by $K_m\ (=\kappa_m/2\varepsilon)$ and $\Omega_m\ (=(\omega_m/\omega)^2)$ with $\kappa_m (=V_m/Ad_m)$ and $\omega_m\ (=(B_m a_0^2/L_{em}V_m)^{1/2})$ where L_{em} is the effective length of the throat with end corrections and is given by $L_m + 2\times 0.82 r_m$, r_m being the hydraulic radius of the throat.

The unidirectional propagation in the tunnel is described by

$$\frac{\partial f}{\partial X} - f\frac{\partial f}{\partial \theta} = -\delta_R \frac{\partial^{1/2} f}{\partial \theta^{1/2}} + \beta\frac{\partial^2 f}{\partial \theta^2} - K_1\frac{\partial g_1}{\partial \theta} - K_2\frac{\partial g_2}{\partial \theta}, \tag{1}$$

while the response of each array is described by

$$\frac{\partial^2 g_m}{\partial \theta^2} + \delta_{rm}\frac{\partial^{3/2} g_m}{\partial \theta^{3/2}} + \Omega_m g_m = \Omega_m f$$
$$+\varepsilon\left[\left(\frac{\gamma-1}{\gamma+1}\right)\frac{\partial^2 g_m^2}{\partial \theta^2} - \frac{2V_m}{(\gamma+1)B_m L_{em}}\left|\frac{\partial g_m}{\partial \theta}\right|\frac{\partial g_m}{\partial \theta}\right], \tag{2}$$

for m = 1 and 2 with the definition of the fractional derivative of order 1/2 given by

$$\frac{\partial^{1/2} f}{\partial \theta^{1/2}} = \frac{1}{\sqrt{\pi}} \int_{-\infty}^{\theta} \frac{1}{\sqrt{\theta - \theta'}} \frac{\partial f(\theta', X)}{\partial \theta'} \mathrm{d}\theta', \quad (3)$$

and $\partial^{3/2} g / \partial \theta^{3/2} = \partial / \partial \theta \cdot (\partial^{1/2} g / \partial \theta^{1/2})$ where εf and εg_m denote, respectively, the dimensionless axial velocity of air in the tunnel, $[(\gamma+1)/2] u / a_0$, and excess pressure in the cavity, $[(\gamma+1)/2\gamma] p'_{cm} / p_0$, u, p'_{cm}, and γ being, respectively, the axial velocity of air, the excess pressure in the cavity of the resonator in the array m, and the ratio of specific heats; εf is equal, to the lowest order, to the dimensionless excess pressure in the tunnel, $[(\gamma+1)/2\gamma]\, p'/p_0$, p' being the excess pressure in the tunnel; the independent variables θ and X denote, respectively, the retarded time $\omega(t - x/a_0)$ and the far-field coordinate $\varepsilon \omega x / a_0$, x and t being the axial coordinate and the time; the coefficients δ_R, δ_{rm} and β are defined as follows:

$$\left.\begin{aligned} \delta_R = C\frac{\sqrt{\nu/\omega}}{\varepsilon R^*}, \quad \delta_{rm} = \frac{2C_{Lm}\sqrt{\nu/\omega}}{r_m} \\ \beta = \frac{\nu_d \omega}{2\varepsilon a_0^2} \end{aligned}\right\}, \quad (4)$$

with $C = 1 + (\gamma - 1)/\sqrt{Pr}$, $R^* = R/(1 - RB_1/2Ad_1 - RB_2/2Ad_2)$ and $C_{Lm} = L'_m / L_{em}$ where ν is the kinematic viscosity, Pr is the Prandtl number, and $L'_m (= L_m + 2r_m)$ is the length with the viscous end corrections. Here it is noted that δ_R and δ_{rm} are proportional to the ratio of a typical thickness of the boundary layer $(\nu/\omega)^{1/2}$ to the hydraulic radius of the tunnel R and the throat r_m, respectively, and the terms with them in 1 and 2 designate the hereditary effects due to the boundary layer on the tunnel wall and the throat wall, respectively; β represents the effect of diffusivity of sound, ν_d $(= \nu\,(4/3 + \mu_v/\mu + (\gamma - 1)/Pr))$, μ_v and μ being the bulk and shear viscosities, respectively. The value of β is usually very small so that its effect is negligible in the outside of a shock layer.

3 EMERGENCE OF SHOCKS

Let us first demonstrate that an acoustic shock wave emerges in a tunnel without an array of resonators. In this case, we have only to solve Equation 1 with $K_1 = K_2 = 0$. Supposing a tunnel of semi-infinite length $(x \geq 0)$, we prescribe an initial condition of f at $X = 0$ as

$$f(\theta, X = 0) = \exp(-\theta^2), \quad (-\infty < \theta < +\infty). \quad (5)$$

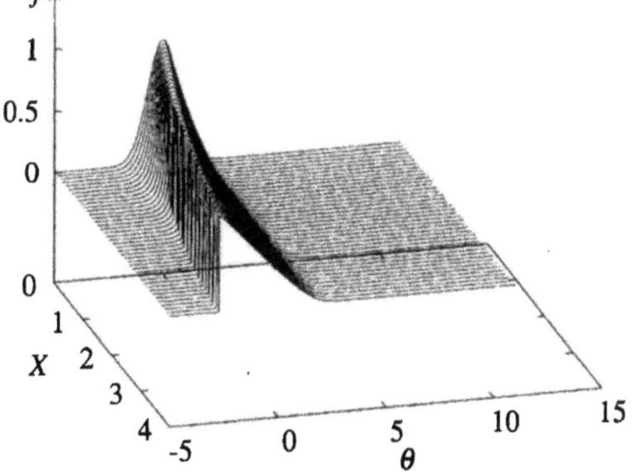

Figure 3. Spatial evolution of the Gaussian-shaped pulse in the tunnel without the array and emergence of shock at $X = 1.17$.

In a case of a tunnel with an array, initial conditions of g_m are to be given as solutions to Equations 2 with f prescribed.

Here the initial condition is remarked. Equations 1 and 2 describe a far-field behaviour in the tunnel. The initial condition is given as a matching condition between a near field and a far field in the tunnel. The matching is executed in a matching region in $R \ll x \ll a_0/\varepsilon\omega$ where the tunnel entrance is located at $x = 0$ (Sugimoto 2001). The matching condition is derived from a behaviour in a distant field of the near field, which is obtained by solving the near field together with the open space outside of the tunnel. The matching condition is prescribed at $X = 0$ but note that the point $X = 0$ does not mean the entrance.

The pressure waves in the matching region take a square (trapezoidal) pulse whose temporal width is determined roughly by the time l/U, l and U being, respectively, the train length and the train speed on entry. When the blockage ratio χ, i.e. ratio of the train's cross-sectional area to the tunnel's one, is small enough, the maximum pressure of the square pulse, $\Delta p'$, is given by the simple formula (Sugimoto 1994, Sugimoto & Ogawa 1998):

$$\frac{\Delta p'}{p_0} = \frac{\gamma \chi M^2}{1 - M^2}, \quad (6)$$

for $\chi (\ll 1)$ where $M (= U/a_0 < 1)$ denotes the Mach number of the train. For example, when U is 130m/s (468 km/h) and $a_0 = 340\,\text{m/s}$, we have M=0.38. Taking $\gamma = 1.4$ for air and $\chi = 0.1$ in the case of the magnetically levitated trains, it follows from the relation 6 that $\Delta p'/p_0$ takes the value 0.024. From the definition of εf, ε becomes to be $[(\gamma+1)/2\gamma]\, \Delta p'/p_0$, if the maximum value of f at $X = 0$ is

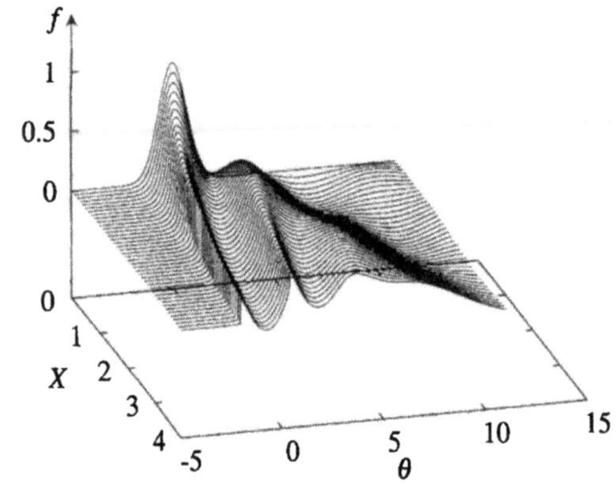

Figure 4. Spatial evolution of the Gaussian-shaped pulse in the tunnel with the single array of Helmholtz resonators having K = 1 and $\Omega = 1$, and emergence of two shocks at X = 1.55 and X = 3.87.

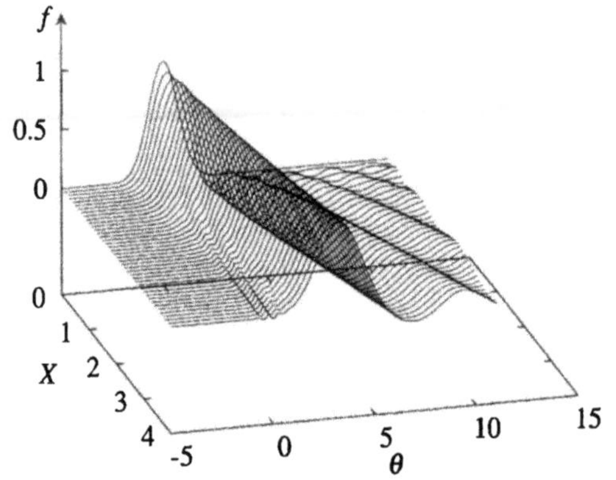

Figure 5. Spatial evolution of the Gaussian-shaped pulse in the tunnel with the single array of Helmholtz resonators having K = 1 and $\Omega = 10$, and suppression of emergence of shock.

chosen to be unity. Thus it follows that $\varepsilon = 0.02$ in this case.

In usual cases, the pulse has a very wide plateau in comparison with the step-up region because the train length is very long. In order to examine a far-field behaviour produced by a long train, we must impose a wide pulse rather than the Gaussian-shaped one. However, to see effects of the array on the pressure waves generated not only by entry of the nose but also of the tail into the tunnel, it is simple to consider the condition 5 modelling a short train.

As for the temporal gradient of the pressure profile, a typical time is determined by D/U, i.e. the ratio of the tunnel diameter D (= $2R$) divided by the train speed. In fact, Howe (1998) obtained the temporal gradient explicitly for the train of semi-infinite length (without taking account of entry of the tail of the train), and the result is consistent with this estimation. The typical angular frequency ω introduced vaguely in the section 2 is related to the inverse of the typical time. For the Gaussian-shaped pulse in the form of $p'/p_0 = (\Delta p'/p_0)\exp[-(\omega t)^2]$, the typical "angular frequency" $\omega/2\pi$ is assumed to be 5 s^{-1}, although $U/2\pi D$ takes the value $2.1\,s^{-1}$ for U = 130 m/s and D =10 m. Incidentally, the pressure gradient $\partial/\partial t\cdot(p'/p_0)$ in the case of the Gaussian-shaped pulse takes the maximum $\Delta p'/p_0\,(2/e)^{1/2}\omega$. In view of Howe's result, ω is given by $(2e)^{1/2}\,(1-M^2)\,H_oU\,/D \approx 3.3\,s^{-1}$, where H_o is the maximum value shown in Figure 3.6.8 (Howe 1998) and is about 0.8 for M = 0.4. But note that this result holds in the case of the tunnel without the array. Yet with the array connected, the result is expected to be valid in the distant near field because the cumulative effects of the array remain small in this field.

Figure 3 shows the spatial evolution from the condition 5 for $\varepsilon = 0.02$ and $\delta_R = 0.01$ on assuming $\omega/2\pi = 5\ s^{-1}$ and $R = 5\,m$. For air at room temperature, we use the following data: Pr=0.72, μ_v/μ = 0.60, C = 1.47, $\nu = 1.45\times10^{-5}\,m^2/s$ and $\beta = 2.5\times10^{-7}$. It is clearly seen that the discontinuity in the pressure profile, i.e. shock, appears at X = 1.17. This corresponds to the distance 633 m in x. In passing, since β is very small, the shock fitting is made by setting it to be zero (i.e. the discontinuity is introduced without taking account of the structure of shock layer).

4 SUPPRESSION OF SHOCK FORMATION

The effect of the array of resonators and the difference in mechanisms of suppression of shock and of damping of pressure waves are first demonstrated. To suppress the shock, one is apt to consider that the pressure waves should be damped. So the array is tuned so that Ω may be set around unity. As an example, we consider a tunnel of the cross-sectional area $A = 25\pi\,m^2$ (R = 5m) with a single array of resonators having $V = 16\ m^3$, $B = \pi/16\,m^2$ (r = 0.25 m), L = 1 m connected with the spacing d = 5 m. Then the size parameter takes the value 0.04 while the natural frequency of the resonator becomes 5.0 Hz

In Figure 4, we show the evolution of the pulse in the tunnel with K = 1 and $\Omega = 1$. As is seen, the pressure waves are damped in comparison with the profiles in Figure 3. But the shock emerges at X = 1.55, though the shock formation is delayed. In addition, a new shock emerges behind the leading one at X = 3.87. From a viewpoint of suppression of shock, this situation has become worse than the one in the tunnel without the array. For the purpose of suppression of shocks, it has already been shown that the tuning

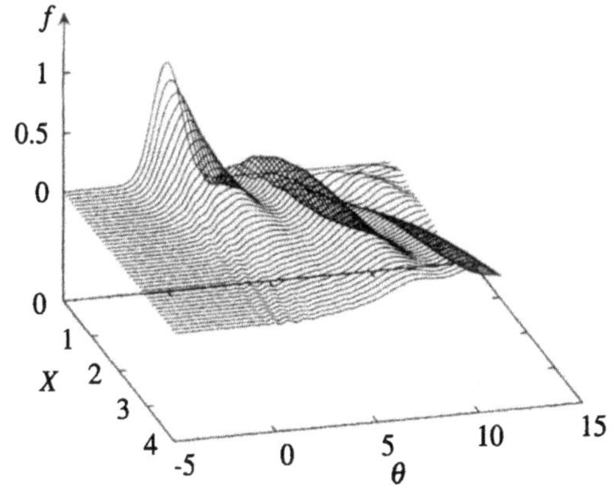

Figure 6. Spatial evolution of the Gaussian-shaped pulse in the tunnel with the double array of Helmholtz resonators having $K_1 = 1$ and $\Omega_1 = 10$, and $K_2=1$ and $\Omega_2=1$, and simultaneous action of suppression of shock and damping of pressure wave.

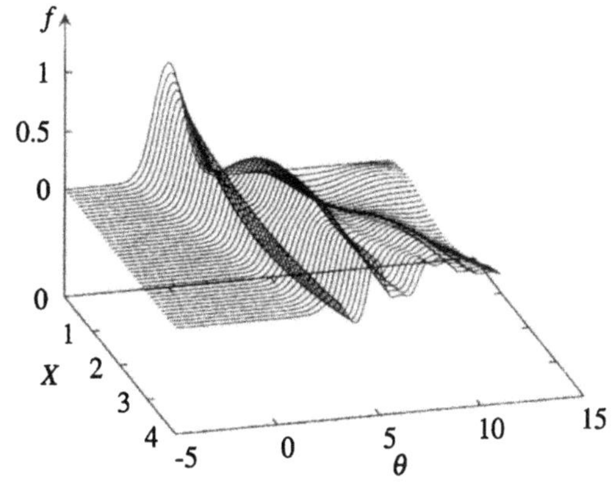

Figure 7. Spatial evolution of the Gaussian-shaped pulse in the tunnel with the double array of Helmholtz resonators having $K_1 = 1$ and $\Omega_1=100$, and $K_2=1$ and $\Omega_2=1$, and escape of the pressure waves from the damping action of the array 2.

parameter should be set to be much higher than unity (Sugimoto 1992). In this case, the dispersion due to the array can compete with the nonlinear steepening of the waveform to suppress emergence of shock. For Ω smaller than unity, on the contrary, shocks always appear. To set Ω much higher than unity, a smaller resonator is required. The case with $K=1$ and $\Omega=10$ may be realized by choosing the resonator having $V=4\,\mathrm{m}^3$, $B=\pi/16\,\mathrm{m}^2$, $L_e = 0.6$ m and d = 2.5 m. Then the size parameter takes the value 0.02. For $K=1$, the axial spacing may either be halved to 1.25 m or the double array of the resonators of the same size may be connected per axial spacing 2.5 m. The natural frequency of the resonator becomes 15.5 Hz.

Figure 5 shows the shock-free propagation in this case. The initial profile is neither steepened nor damped significantly and therefore emergence of shock is suppressed admittedly. Although the value of the coupling parameter is the same in Figures 4 and 5, i.e. the volume of cavity per unit axial length, V/d, is equal in both cases, the difference in the values of the tuning parameter controls emergence of shock. As the value of Ω is decreased from 10 to approach unity, shocks tend to emerge. On the other hand, as the value of Ω is increased larger than 10, shocks will never appear. This is because the array yields the higher-order dispersion. In the limit as $\Omega \to \infty$, the initial pulse will eventually split into a series of acoustic solitons (Sugimoto 1992).

When the pressure waves free from the shock are radiated from the tunnel exit, no bursting sound will be heard. The sound pressure radiated from the tunnel exit is proportional to the time-derivative of the pressure profile at the tunnel exit (Sugimoto 2001). Therefore no discontinuity in the profile produces no high-frequency (audible) sound. But when the pressure waves shown in Figure 5 are radiated from the tunnel exit, their typical frequency is as low as 5 Hz. Since such an infrasound is far-reaching, it is now required to suppress propagation of these waves.

5 DAMPING OF PRESSURE WAVES

To reduce the shock-free pressure waves, damping action of the array must be invoked by choosing the tuning parameter to be a value near unity. Both requirements are fulfilled if a double array of resonators is considered, one having the tuning parameter much higher than unity for suppression of shock, and the other having the tuning parameter near unity for damping of pressure waves.

To demonstrate the effectiveness of the double array, we solve the case with one array having $K_1=1$ and $\Omega_1=10$, and the other having $K_2=1$ and $\Omega_2=1$. The respective arrays are the ones used previously: the former consists of the resonators with $V_1=4\,\mathrm{m}^3$, $B_1=\pi/16\,\mathrm{m}^2$, $L_{e1}=0.6\,\mathrm{m}$, and $\omega_1/2\pi$ = 15.5 Hz connected with $d_1=2.5\,\mathrm{m}$, while the latter consist of the resonators with $V_2=16\,\mathrm{m}^3$, $B_2 = B_1 = \pi/16\,\mathrm{m}^2$, $L_{e2}=1.41\,\mathrm{m}$ and $\omega_2/2\pi=5.0\,\mathrm{Hz}$ connected with $d_2=5\,\mathrm{m}$. The evolution from the initial pulse is shown in Figure 6. Of course, no shock is formed and simultaneously the pressure waves are damped significantly at $X = 4$.

Next we see qualitative change in evolution by varying the values of the parameters. Since the array 2 is introduced for damping, Ω_2 is fixed at unity and only one parameter out of K_1, Ω_1 and K_2 is changed from the values used in Figure 6. Firstly as K_2 is increased, it is obvious that the pressure waves are damped further. As K_2 is decreased, the damping is not expected but emergence of shock is suppressed, as shown in Figure 5, which is the degenerate case with $K_2=0$. On the contrary, when K_1 is decreased to 0.5, no shock emerges. But when it is decreased to 0.1, two shocks emerge.

On the other hand, when Ω_1 is decreased to 5, no shock still emerges for $K_1 = K_2 = 1$, and the situation is similar to the one shown in Figure 6. If Ω_1 is set larger than 10, of course, the shock is suppressed. In a case with extremely large value of Ω_1, however, the initial pulse is split into a few pulses, which escape from the damping action of the array 1, because a typical frequency of the pulses split is much higher than 5 Hz. Then the array 1 becomes idle. Figure 7 shows the evolution when Ω_1 is set equal to 100 with $K_1 = K_2 = 1$ and $\Omega_2 = 1$. Such a new array may consist of six resonators, each of which has $V_1 = 0.53\text{m}^3$, $B_1 = \pi/16\text{m}^2$, L_{e1} = 0.43 m, and $\omega_1/2\pi = 50.2\,\text{Hz}$, and is connected with $d_1 = 1\text{m}$. It is found from this result that Ω_1 should not be set extremely higher than 10.

6 CONCLUSIONS

In order to satisfy two requirements of suppression of emergence of shock and simultaneously of damping of propagation of pressure waves produced by entry of a high-speed train, the present paper has proposed to connect the double array of Helmholtz resonators with the tunnel axially. Since the pressure disturbances generated originally are propagated in the form of nonlinear infrasound, it is very difficult to make them damped by local action only. This is also the case of suppression of shock because it results from small but cumulative effects of weak nonlinearity. To control such a slow phenomenon over a long distance, therefore, the continual action along the tunnel, even if small locally, is essential. In this sense, the array of resonators is suitable to it.

Effects of the array are controlled by two parameters, one being the coupling parameter and the other the tuning parameter. For the present purpose, it is concluded that such a double array is very effective that the tuning parameter of one array is set equal to unity, while the other is set to be 10 or higher than it, with both coupling parameters fixed at unity. From a technical point of view, each resonator in the double array may be installed on the ceiling of the tunnel as shown in Figure 8. The cavity's shape needs not to be spherical and no special requirements are necessary. Because such resonators are easily realized from an economical point of view, the array is fully feasible in real tunnels

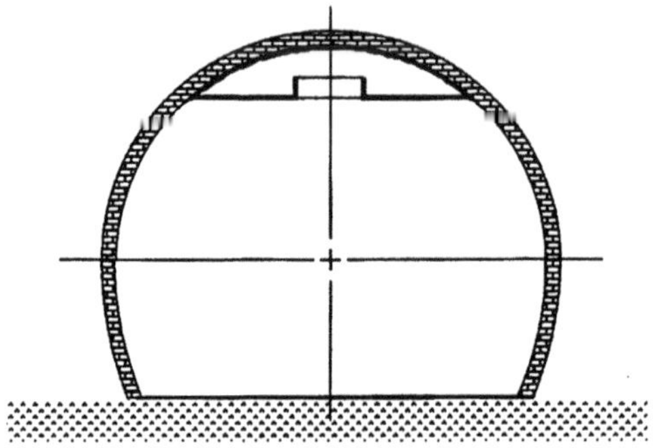

Figure 8. Example of installation of Helmholtz resonators on the ceiling of a real tunnel.

7 REFERENCES

Howe, M. 1998. *Acoustics of Fluid-Structure Interactions*. Cambridge: Cambridge University Press.

Maeda, T. 2001. Micropressure waves radiating from a Shinkansen tunnel portal. In V. V. Krylov (ed.), *Noise and vibration from high-speed trains*: 187-211. London: Thomas Telford.

Ozawa, S. 1979. Studies of micropressure wave radiated from a tunnel exit. (in Japanese) *Railway Technical Research Report* No.1121: 1-92. Tokyo: The Japan National Railways.

Sugimoto, N. 1992. Propagation of nonlinear acoustic waves in a tunnel with an array of Helmholtz resonators. *J. Fluid Mech.* 244: 55-78.

Sugimoto, N. 1993. 'Shock-free tunnel' for future high-speed trains. In *Speed-up technology for railway and maglev vehicles; Proc. international conference, Yokohama, 22-26 November 1993.* Vol. 2: 284-292. Tokyo: Japan Society of Mechanical Engineers.

Sugimoto, N. 1994. Sound field in a tunnel generated by traveling of a high-speed train. J. E. Ffowcs-Williams, D. Lee & A. D. Pierce (eds.), *Structural acoustics, scattering and propagation, theoretical and computational acoustics* Vol.1: 45-56. Singapore: World scientific.

Sugimoto, N. & Ogawa, T. 1998. Acoustic analysis of the pressure field in a tunnel, generated by entry of a train. *Proc. R. Soc. Lond.* A 454: 2083-2112.

Sugimoto, N., Masuda, M., Ohno, J. & Motoi, D. 1999. Experimental demonstration of generation and propagation of acoustic solitary waves in an air-filled tube. *Phys. Rev. Lett.* 83(20): 4053-4056.

Sugimoto, N. 2001. Emergence of an acoustic shock wave in a tunnel and a concept of shock-free propagation. In V. V. Krylov (ed.), *Noise and vibration from high-speed trains*: 213-247. London: Thomas Telford.

Wave propagation – Moving load – Vibration reduction, Chouw & Schmid (eds.)
© 2003 Swets & Zeitlinger, Lisse, ISBN 90 5809 559 2

Road traffic noise prediction model "ASJ Model 1998" proposed by the Acoustical Society of Japan

K. Yamamoto
Kobayasi Institute of Physical Research

H. Tachibana
Institute of Industrial Science, University of Tokyo

ABSTRACT: The prediction model of road traffic noise "ASJ Model 1998" is presented, which is proposed by the Acoustical Society of Japan. The model includes general prediction procedure, source description, calculation methods for sound propagation and application procedures to all types of road structures. In this paper, the outline of the model is described and the accuracy of the prediction is shown.

1 INTRODUCTION

In September 1998, the "Environmental Quality Standards for Noise" in Japan was revised and the noise index for road traffic noise was changed to L_{Aeq} from L_{50}. To correspond to this change, the Research Committee in Acoustical Society of Japan just completed the substantial revision of the former prediction model "ASJ Model 1993" (Tachibana 1994) and published a new one "ASJ Model 1998" on April 1st in 1999. This model has been adopted as the standard prediction method in the Environmental Impact Assessment for road traffic noise.

In deriving the new model, existing knowledge was widely taken into account and various research projects were carried out to develop a reasonable model. To improve the accuracy of the prediction model, efforts have been made to supplement a large amount of field data with a support of public corporations related to Ministry of Land, Infrastructure and Transport.

This paper describes the outline of the procedures for the calculation of road traffic noise, which is based upon ASJ Model 1998.

2 GENERAL METHOD

2.1 *Field of application*

The calculation conditions and the field of application of the ASJ Model 1998 are as follows.

(1) Type of road : general roads (flat, bank, cut, and viaduct), special parts (interchange, depressed or semi-underground, vicinity of a tunnel mouth, parts where viaduct and flat road exist together, and double-deck viaduct)
(2) Traffic volume: unlimited
(3) Vehicle running speed : 40 to 140 km/h for freeways and urban roads under steady running condition, 10 to 60 km/h for urban roads under unsteady running condition including acceleration and deceleration, and 0 to 80 km/h for the special parts like interchange.
(4) Prediction area : as far as 200 m in horizontal distance from the road under consideration and up to the height of 12 m from the ground. (Although there is no restriction in the calculation principle, the validity of the model has been examined for the limited areas mentioned above.)

Meteorological condition : the condition of no wind and no strong temperature inversion is assumed as the standard condition.

2.2 *Principle of calculation*

In the calculation of road traffic noise based on L_{Aeq}, it is the point to obtain "unit pattern", a time history of sound pressure, observed at a prediction point when a road vehicle runs on the road under consideration. By squaring and integrating the unit pattern, the sound pressure exposure is obtained and then by considering the traffic volume and by averaging the total sound exposure, L_{Aeq} can be calculated. The concrete procedure is as follows.

Firstly, the traffic-lane of the road under consideration is properly divided into finite segments [see Fig.1 (a)]. For the i-th segment, the sound propagation from the center point of the segment to the prediction point is calculated. In this calculation, the sound power P_i (or sound power level $L_{W,i}$) of the

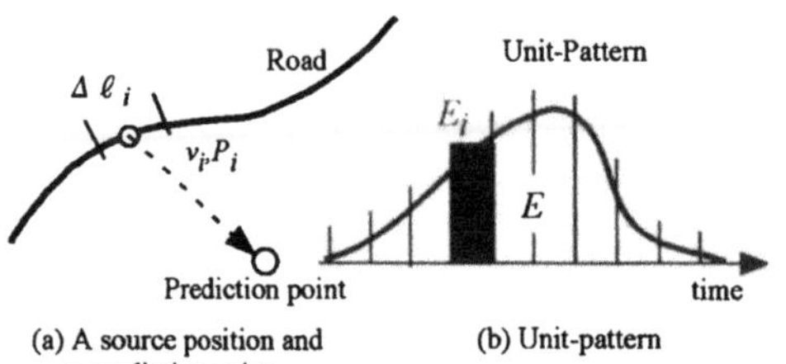

Figure 1 Calculation principle of road traffic noise in "ASJ Model 1998"

sound source (vehicle) which is assumed to be an omni-directional point source on a reflecting plane is provided for each vehicle type and running condition and the sound pressure p_i (or sound pressure level $L_{p,i}$) at the prediction point is calculated.

Next, the sound pressure exposure E_i [Pa²s] at the prediction point over the time interval Δt_i [s] in which the sound source exists within the i-th segment is calculated as follows:

$$E_i = p_i^2 \cdot \Delta t_i = p_i^2 \frac{\Delta \ell_i}{v_i} \qquad (1)$$

where, $\Delta \ell_i$ is the length [m] of the i-th segment, and v_i is the running speed [m/s] of the sound source in the segment. The calculation is made for each segment of the traffic-lane and by adding up these results the total sound pressure exposure E [Pa²s] over the time interval during which the sound source passes the lane under consideration is obtained as follows [see Fig.1 (b)]:

$$E = \sum_i E_i = \sum_i p_i^2 \cdot \Delta t_i = \sum_i p_i^2 \cdot \frac{\Delta \ell_i}{v_i} = \sum_i p_i^2 \cdot \frac{3.6 \Delta \ell_i}{V_i} \qquad (2)$$

where, V_i is the running speed [km/h] of the sound source in the i-th segment ($v_i = V_i / 3.6$).

As expressed by the following equation, the quantity expressed in level (dB) of the total sound pressure exposure E is the "sound pressure exposure level" L_{pE}. (When considering A-weighted sound pressure, L_{pE} corresponds to "sound exposure level" L_{AE}.)

$$L_{pE} = 10\log_{10} \frac{E}{E_0} \qquad (3)$$

where, $E_0 = 4 \times 10^{-10}$ Pa²s (the reference value of sound pressure exposure).

From the value of L_{pE} calculated as mentioned above and by considering the traffic volume N [vehicles/h], the "equivalent continuous sound pressure level" L_{peq} for 1 hour (3600 s) is obtained as follows: (When considering the A-weighted sound pressure, L_{peq} corresponds to L_{Aeq}.)

$$L_{peq} = 10\log_{10}(10^{L_{pE}/10} \frac{N}{3600}) = L_{pE} + 10\log_{10} N - 35.6 \qquad (4)$$

By making the calculation for all lanes of the road under consideration and for all vehicle types, and by adding up these results on energy-base, L_{peq} in each octave or 1/3 octave band at the prediction point is obtained. Finally, L_{Aeq} is obtained by adding up the values of L_{peq} in all frequency bands with A-weighting.

3 SOURCE DESCRIPTION

The ASJ model 1998 provides sound power levels as a sound emission level of road vehicles. The power level depends upon the types of vehicles and running speed in different flow conditions. It is also given by the running mode, i.e., steady running condition and transient running condition as shown in Fig.2.

3.1 *Types of road vehicles*

Road vehicles are classified into four types as shown in Table 1. This classification corresponds to that of the motor vehicle noise regulations that is based on vehicle weight and engine output. For simplicity of prediction procedure, two types classification is available, i.e., heavy vehicle and light vehicle.

3.2 *Sound power levels*

3.2.1 *Sound power level under transient running condition*

For general urban roads, transient running condition including acceleration and deceleration has to be considered. The A-weighted sound power level is

Table 1 Motor vehicle classification

Vehicle weight and engine output	Four-type classification	Two-type classification
Motor vehicles with GVW (gross vehicle weight) of over 3.5 t, and maximum engine output of over 150 kW	Large-sized vehicles	Heavy vehicles
Motor vehicles with GVW of over 3.5 t, and maximum engine output of 150 kW or less	Medium-sized vehicles	
Motor vehicles with GVW of 3.5 t or less	Small-sized vehicles	Light vehicles
Motor vehicles used exclusively for carrying passengers, with capacity of 10 or fewer passengers	Passenger cars	

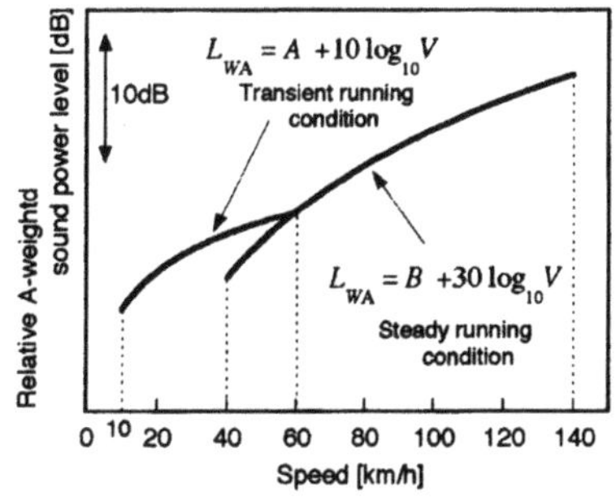

Figure 2 Calculation model for sound power level.

expressed by $10\log_{10}V$ in the speed within the range of 10km/h and 60km/h. The formula is as follows:

$$L_{WA} = A + 10\log_{10} V, \quad 10\text{km/h} \le V \le 60\text{km/h} \quad (5)$$

L_{WA} : A-weighted sound power level [dB]
V : running speed [km/h]
A : coefficient

3.2.2 *Sound power level under steady running condition*

On freeways and urban roads under steady running condition, vehicles run at the top gear position in the speed range greater than 40km/h. For this condition, the expression of $30\log_{10}V$ is adopted as the speed dependence. The A-weighted sound power level L_{WA} is given by the next formula:

$$L_{WA} = B + 30\log_{10} V, \quad 40\text{km/h} \le V \le 140\text{km/h} \quad (6)$$

B : coefficient

3.2.3 *Coefficient A and B*

Based on field measurements, coefficients A and B are determined by regression analysis. The coefficients are shown in Table 2 and Table 3 for four-type and two-type classification, respectively.

Table 2 Coefficients A and B for four type classification

Classification	Coefficient A	Coefficient B
Large-sized	90.0	54.4
Medium-sized	87.1	51.5
Small-sized	83.2	47.6
Passenger	82.0	46.4

Table 3 Coefficients A and B for two type classification

Classification	Coefficient A	Coefficient B
Heavy vehicles	88.8	53.2
Light vehicles	82.3	46.7

1.3 *Sound power spectrum*

Sound power spectrum is expressed by the next simple formula:

$$\Delta L(f) = -10\log_{10}\left[1 + \left(\frac{f}{2000}\right)^2\right] \quad (7)$$

$\Delta L(f)$: relative sound power level for the frequency band centered at f (Hz)

It is an approximation to the average of field measurement data. The spectrum is available to both of steady and transient running conditions. A-weighted sound power spectra in octave bands and in 1/3 octave bands are shown in Fig. 3, where the overall level is assumed to be 0 dB.

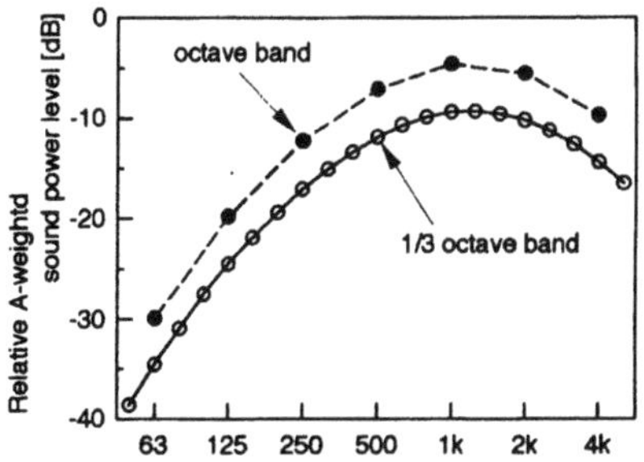

Figure 3 A-weighted power spectrum.

1.4 *Correction at drainage asphalt pavement*

A consideration of the noise reduction by drainage asphalt pavement is provided. The power level of vehicles running on drainage asphalt pavement is calculated as follows:

$$L_{WA,dr} = L_{WA} + \Delta L_{WA,dr} \quad (8)$$

where $L_{WA,dr}$ is sound power level on drainage asphalt pavement and L_{WA} is sound power level on normal asphalt pavement. The correction term $\Delta L_{WA,dr}$ is given by the next formula:

$$\Delta L_{WA,dr} = -3.5\log_{10}V + 3.2 \quad (9)$$

This formula is applicable in the speed range of 40 to 140 km/h for light vehicles and 40 to 120 km/h for heavy vehicles in two-type classification.

1.5 *Power level in acceleration, deceleration and idling mode*

When a vehicle start moving and increasing its speed up to a constant speed, the power level L_{WA} is expressed by next expressions with the use of regression coefficients B and C.

$$L_{WA} = \begin{cases} B + 3 & 0\text{km/h} \le V < 10\text{km/h} \\ C + 10\log_{10}V & 10\text{km/h} \le V < 80\text{km/h} \\ B + 30\log_{10}V & 80\text{km/h} \le V < 140\text{km/h} \end{cases} \quad (10)$$

When a vehicle reduces the speed down to 0km/h, the sound power level is given by the next expressions:

$$L_{WA} = \begin{cases} B + 30 & 0\text{km/h} \le V < 10\text{km/h} \\ B + 30\log_{10}V & 10\text{km/h} \le V \le 140\text{km/h} \end{cases} \quad (11)$$

When a vehicle stop moving and the engine is on idling mode, the sound power level is specified to a constant value.

$$L_{WA} = B + 30 \quad V = 0\text{km/h} \tag{12}$$

Table 4 gives coefficient C for four-type and two-type classification.

Table 4 Coefficient C for four and two type classification

Classification	C	Classification	C
Large-sized	92.5	Heavy vehicles	91.3
Medium-sized	89.6		
Small-sized	85.7	Light vehicles	84.8
Passenger	84.5		

4 GENERAL PROCEDURES FOR THE CALCULATION OF SOUND PROPAGATION

This section contains basic calculation method for sound propagation from road vehicles to prediction points. The factors included are diffraction, ground effects, meteorological effects and sound reflection. Two approaches are considered. One is a precise version, denoted by "A-method (precision method)" and it is derived from wave theory. The other is a simple version, denoted by "B-method (engineering method)" which is based on geometrical acoustics. The propagation factors are separated into a series of correction terms, each of which has physical significance to road traffic noise engineers.

4.1 *A-method (precision method)*

A-method is derived from theory of sound diffraction and that of sound propagation over a finite impedance boundary. The calculation procedure seems complicated, however, one can deal with cases having an arbitrary noise spectrum and acoustical properties of ground and barrier surfaces.

4.1.1 *Basic equation*

For a certain position of a source, A-weighted sound pressure level at a prediction point is calculated by the following equation,

$$L_{pA} = 10\log_{10} \sum_{i=1}^{n} 10^{L_{pA,i}/10} \tag{13}$$

where L_{pA} is the sound pressure level (in A-weighting) at i-th frequency. It is calculated by the next equation,

$$L_{pA,i} = L_{WA,i} - 8 - 20\log_{10} r + 10\log_{10}\left|\frac{\phi_i}{\phi_{g,i}}\right|^2 \tag{14}$$

where $L_{WA,i}$: sound power level (in A-weighting) of a source at i-th frequency [dB],

r : distance between a source and a prediction point [m],

ϕ_i : complex quantities of sound pressure at i-th frequency for the field that includes sound diffraction and reflection,

$\phi_{g,i}$: complex quantities of sound pressure at i-th frequency for free field.

4.1.2 *Equations for diffraction and ground effects*

For the complex quantity of sound pressure at a distance of R from a point source in free field, the next expression is to be used:

$$\phi_g = \frac{e^{ikR}}{R} \tag{15}$$

By using this expression, one can express the total sound pressure as the sum of contributions corresponding to sound path at geometry of a source and receiver.

$$\phi = \phi_1 + \phi_2 + \cdots + \phi_n = \sum_{j=1}^{n} \phi_j \tag{16}$$

$$\text{where } \phi_j = Q_j \cdot D_j \cdot \frac{e^{ikR_j}}{R_j} \tag{17}$$

The terms Q and D in the equation above are complex reflection coefficient (Kouyoumjian et al. 1974) and diffraction coefficient (Kawai 1983), respectively. An example of sound paths configuration and related expressions is shown in Fig. 4.

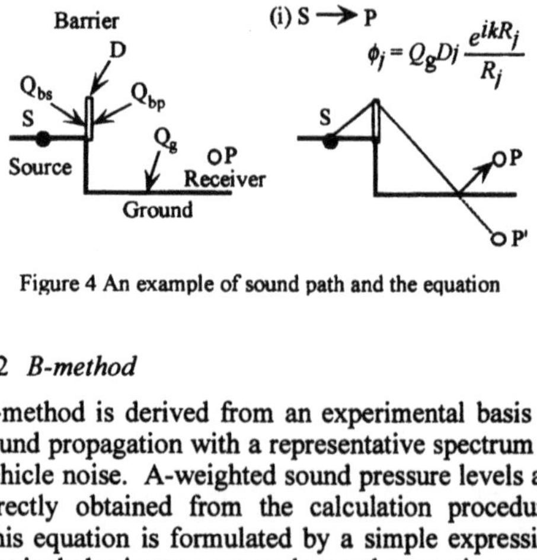

Figure 4 An example of sound path and the equation

4.2 *B-method*

B-method is derived from an experimental basis of sound propagation with a representative spectrum of vehicle noise. A-weighted sound pressure levels are directly obtained from the calculation procedure. This equation is formulated by a simple expression that includes inverse square law and correction terms for diffraction and ground effects.

4.2.1 *Basic equation*

A-weighted sound pressure level L_{pA} [dB] is given by the next equation:

$$L_{pA} = L_{WA} - 8 - 20\log_{10} r + \Delta L_d + \Delta L_g \tag{18}$$

where L_{WA} is A-weighted sound power level, ΔL_d and ΔL_g are the correction terms of diffraction and ground effect, respectively.

4.2.2 *The correction term for diffraction effect*

A road shoulder or a noise-shielding barrier provides diffraction effect. The correction term due to diffraction is given by the numerical expression shown as follows:

$$\Delta L_d = \begin{cases} -20 - 10\log_{10}(\delta) & \delta \geq 1 \\ -5 \pm \dfrac{-15}{\ln(1+\sqrt{2})} \cdot \sinh^{-1}\left(|\delta|^{0.414}\right), & -0.0537 \leq \delta < 1 \\ 0, & else \end{cases} \quad (19)$$

The symbol δ denotes the path length. Plus and minus signs are used for $\delta>0$ and $\delta<0$, respectively. The plus sign in δ is given for a receiver in a barrier shadow zone. The curve for the correction ΔL_d is shown in Fig 5.

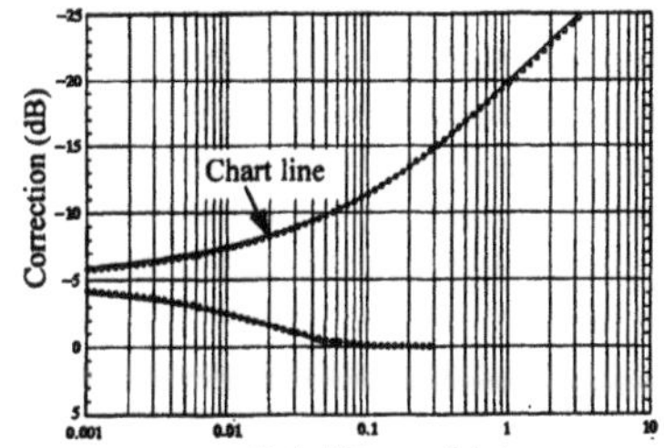

Figure 5 Correction chart for single edge diffraction.

4.2.3 *The correction term for ground effect*

Sound that propagates along a ground surface receives attenuation over inverse square law. It depends upon the ground impedance, heights of source and receiver above the ground, the distance between the two. In B-method, the correction term is expressed by,

$$\Delta L_g = \sum_{i=1}^{n} \Delta L_{g,i} \quad (20)$$

where $\Delta L_{g,i}$ is the correction for the *i*-th ground and is given by the next empirical equation:

$$\Delta L_{g,i} = \begin{cases} -K_i \log_{10}\left(\dfrac{r_i}{r_{o,i}}\right), & r_i \geq r_{o,i} \\ 0, & else \end{cases} \quad (21)$$

Symbol K is a coefficient that characterizes the attenuation rate per doubling of distance, r is a distance, and r_o is a specific distance where the ground effect starts increasing. K and r_o are obtained by numerical simulation of sound propagation and given by numerical expressions for three types of absorptive ground (Yamamoto 1994).

4.3 *Multi-edged diffraction*

Embankments, hills and rows of buildings may form a multi-edge diffraction. For double diffraction as shown in Fig.6, correction term $\Delta L_{d,dd}$ is simply expressed by the next formula (Takagi 1996):

$$\Delta L_{d,dd} = \begin{cases} \Delta L_{SXP}, & \text{zone I \& II} \\ \Delta L_{SYP} + \Delta L_{SXY} + 5, & \text{zone III} \end{cases} \quad (22)$$

where ΔL is a correction term calculated by Eq.(19) and the suffix denotes the sound path.

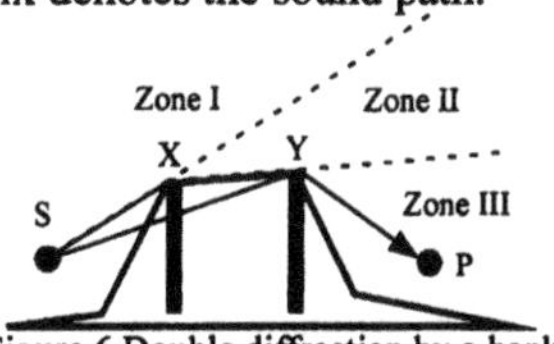

Figure 6 Double diffraction by a bank

4.4 *Sound reflection*

Noise barriers, retaining walls and buildings are usually shielding obstacles against noise, but they often become significant sound reflection objects. In this model, treatment of specular reflection is employed, however, the size of the reflection surface is considered. As is shown in Fig.6, the reflected wave from a finite size surface (a stripe of wall) is regarded as a transmitted wave through the slit that has the same size as the reflection wall. The calculation method is derived on the assumption that Babinet's principle holds in sound energy field.

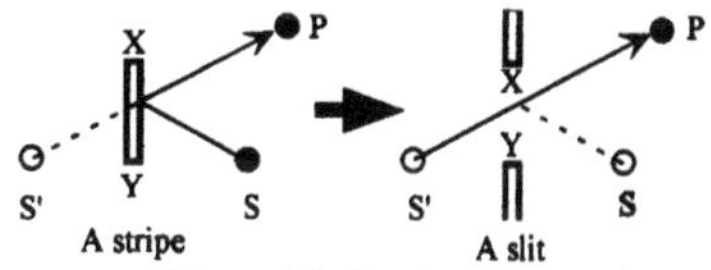

Figure 6 Reflection from a stripe

The correction $\Delta L_{d,slit}$ due to the diffraction of slit opened between the edge X and Y is specified as

$$\Delta L_{d,slit} = 10\log_{10}\left|10^{\Delta L_{S'XP}/10} - 10^{\Delta L_{S'YP}/10}\right| \quad (23)$$

where ΔL is a correction of a single diffraction and the suffix denotes the sound path. On calculation the barrier direction must be carefully specified as shown in Fig.7.

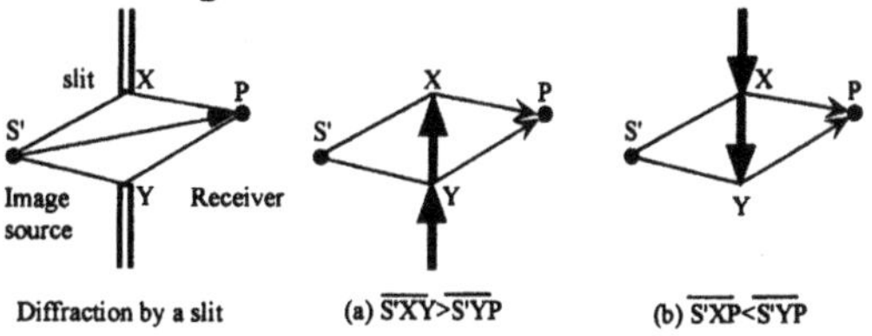

(Note:the top of the arrow denotes the edge of a semi-infinite barrier)

Figure 7 Barrier edge and the direction

4.5 *Air absorption and meteorological effect*

Correction terms due to air absorption is specified on the basis of the standard atmospheric condition (temperature 20°C, humidity 60%) and given by the next expression:

$$\Delta L_{A,air}(r) = -0.3452(r/1000)^3 - 2.011(r/1000)^2 - 6.840(r/1000) \qquad (24)$$

where r [m] is the distance between a point source and a prediction point. The correction is obtained from ISO 9613-part2.

For meteorological effect, deviation due to vector wind is provided. It is obtained by field measurement and given by the next empirical model:

$$\Delta L_{m,\ell ine} = 0.88 \cdot U_{vec} \cdot \log_{10}(\ell/15), \quad \ell > 15 \qquad (25)$$

where U_{vec} is vector wind [m/s] and ℓ is horizontal distance [m] from the center line of a road. This term is applied to correct the finial predicted values of $L_{Aeq,T}$.

5 PREDICTION METHOD FOR SPECIAL ROAD

The ASJ Model 1998 essentially covers the procedures of noise prediction for roads in all types. Here is some more information that is provided in the calculation stages for special cases.

5.1 *Interchange*

Interchange section generally consists of main road, ramp, tollgate, fork, junction and branch road. Geometrical configuration of interchange and traffic flow are quite complex. In particular, speed of vehicles is not constant and thereby sound power level is variable with regard to the speed in operation. Figure 8 shows a schematic configuration of an interchange showing a behavior of a vehicle.

The ASJ Model 1998 provides acceleration and deceleration of speed of vehicles as shown in Table 5 to calculate speed profile in computer programming step. The service time for paying highway charge is specified for noise exposure time and it is shown in Table 6.

Table 5 Acceleration and deceleration of speed of vehicles

	Light vehicles	Heavy vehicles
Acceleration (m s^{-2})	0.7	0.6
Deceleration (m s^{-2})	-1.0	-0.8

Table 6 Service time at tollgate

Receiving a card at an entrance (s)	6
Toll collection by cash at an exit (s)	14
Toll collection by ticket at an entrance (s)	8

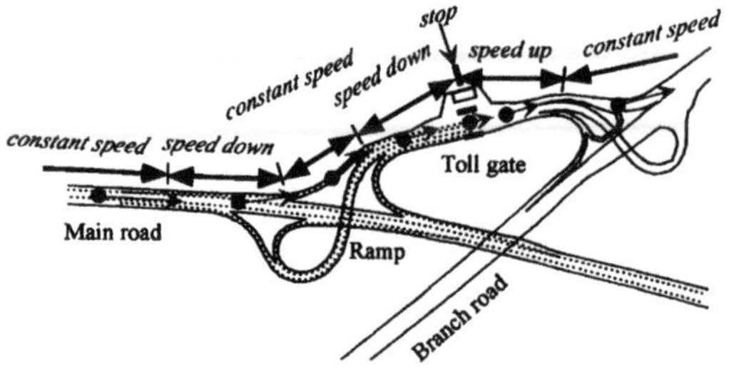

Figure 8 Schematic configuration showing an interchange and a behavior of a vehicle

5.2 *Depressed and semi-underground road*

For the noise mitigation strategy, depressed road and semi-underground road (the same as depressed road but it has a large overhang portion supported at the top edge) are coming into use nowadays in suburbs. The problems in prediction are the treatment of multiple sound reflection between the retained walls. In programming, image sources must be specified in the walls as shown in Fig.9. The contributions from these sources to prediction point are calculated as the transmitted sound through the walls (see 4.4). The contribution of i-th image source $L_{pA,i}$ is expressed by the next regular formula:

$$L_{pA,i} = L_{WA} - 8 - 20\log_{10} r_i + \Delta L_{d,i} + \Delta L_{d,s\ell it,i} + \Delta L_{g,i} \qquad (26)$$

where $\Delta L_{d,s\ell it,i}$ is the correction due to slit diffraction. The slit corresponds to the stripe of the wall that reflects sound from the real source. Total sound pressure level L_{pA} is computed by the energy summing of the contributions from real source and image sources.

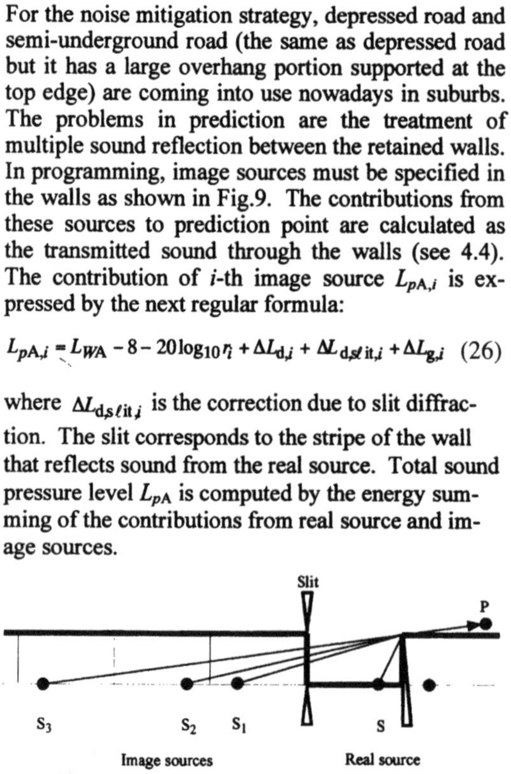

Figure 9 Real and image sources at depressed road

5.3 *Road tunnel*

Tunnel is a good noise shielding measure against road traffic noise, but sometimes it causes serious noise problem when residences are located close to the tunnel mouth. In prediction, two imaginary sources are assumed. One is a point source that represents a direct contribution of sound from a vehicle in tunnel. The other is a surface source that represents residual sound with multiple reflections on the walls inside the tunnel. The model is developed on

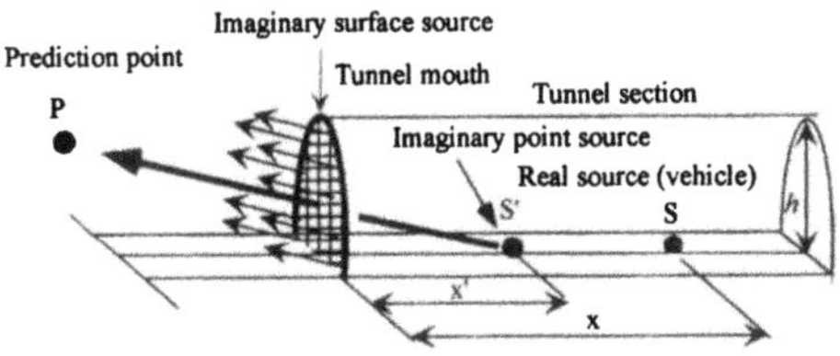

Figure 10 Schematic configurations of imaginary sources

the basis of sound energy balance inside the tunnel (Takagi 2000). A schematic configuration of sound sources and a prediction point is shown in Fig.10. A-weighted sound pressure level L_{pA} at the prediction point P is given by,

$$L_{pA} = 10\log_{10}\left(10^{L_{TD}/10} + 10^{L_{TR}/10}\right) \tag{27}$$

where L_{TD} is a sound pressure level due to the imaginary point source S' that relates to the direct sound, and L_{TR} is a sound pressure level due to the imaginary surface source that relates to reflected sound.

5.4 *Overhead road and double deck viaduct*

Noise reflection from the underside of an overhead roadway and a double deck viaduct is provided in the ASJ model. The influence of reflection causes build-up of noise to the side areas when noise barrier is applied to the lower road (see Fig.11). There are three types of viaduct structure, i.e., structures supported on flat girder, box girder and I girder (see Fig.12).

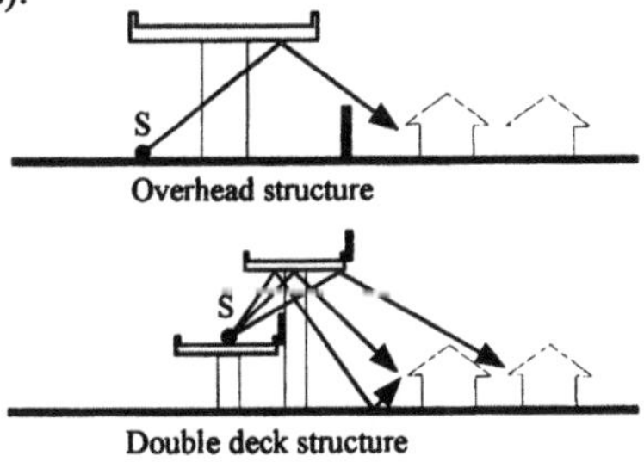

Figure 11 Noise reflection from the underside of viaduct

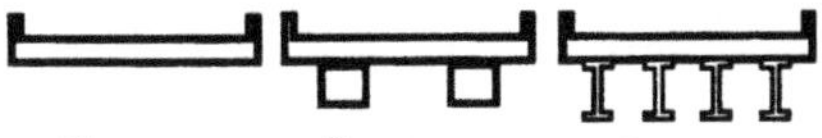

Figure 12 Types of girder

The treatment of sound reflection depends upon the roughness of the surface, however all cases are treated as flat surface and specular reflection is applied instead of scattered reflection for simplicity.

6 ADDITIONAL PROCEDURES

There are two more important issues in road traffic noise prediction. One is structure noise that is radiated from viaduct. The other is noise propagation into built-up area. These two are the recent topic being watched and the prediction procedure is required by public organizations involved in noise control and monitoring.

6.1 *Structure noise of viaduct*

Structure of viaducts generally vibrates when vehicles are passing on it. The force of vehicles generates mechanical vibrations on the structure. The vibrations of the slab and the girder produce noise in audible frequency, called "structure noise of viaduct", and the noise propagates to the side areas.

The calculation model is available when the next condition meets.

1) Type of structure: Metal viaduct (truss and arch structures are excluded) and concrete viaduct.
2) Vehicles to be considered: Heavy vehicles
3) Running speed: greater than 60 km/h

Assuming that a point source is moving on the underside of the viaduct, A-weighted sound pressure level $L_{pA,str}$ is simply given by the next equation:

$$L_{pA,str} = L_{WA,str} - 8 - 20\log_{10} r \tag{28}$$

where $L_{WA,str}$ is A-weighted sound power level. A power level of 95.3 dB is specified for both concrete and metal structure viaducts at the present stage.

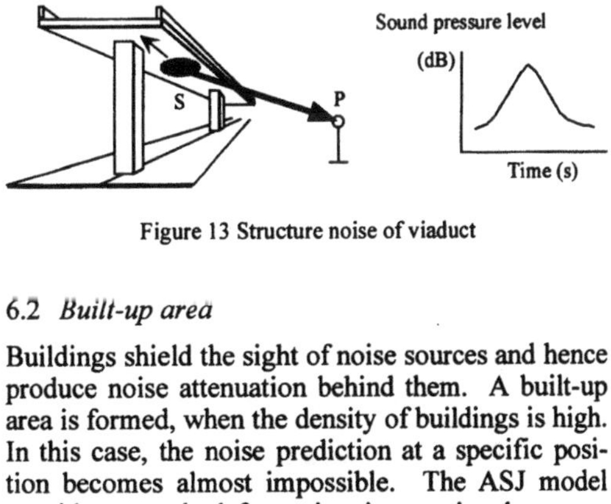

Figure 13 Structure noise of viaduct

6.2 *Built-up area*

Buildings shield the sight of noise sources and hence produce noise attenuation behind them. A built-up area is formed, when the density of buildings is high. In this case, the noise prediction at a specific position becomes almost impossible. The ASJ model provides a method for estimating sectional energy-averaged level $\overline{L_{Aeq}}$ (see Fig.14).

The sectional energy-averaged level is defined as a spatial average of L_{Aeq}, which represents a noise level at a certain range of distance parallel to a road. It is given by the next formula (K.Uesaka 2000, inter-noise 2000):

$$\overline{L_{Aeq,T}} = L_{Aeq,T} + \overline{\Delta L_{builds.}} \tag{29}$$

where $\overline{L_{Aeq,T}}$ is the predicted value without buildings and $\overline{\Delta L_{builds.}}$ is the correction of shielding effect by buildings. The ground is assumed to be reflective. The correction is expressed as a function of the density β_{all} for all buildings between road and prediction point. A simple expression is as follows:

$$\overline{\Delta L_{builds.}} = 10\log_{10}(1-\sqrt{\beta_{all}}) - 0.775\{\beta_{all}/(1-\beta_{all})\}^{0.630}\times(d_{road}-w_1)^{0.859} \quad (30)$$

where w_1 is the average width of first row of buildings and d_{road} is the distance from the road edge.

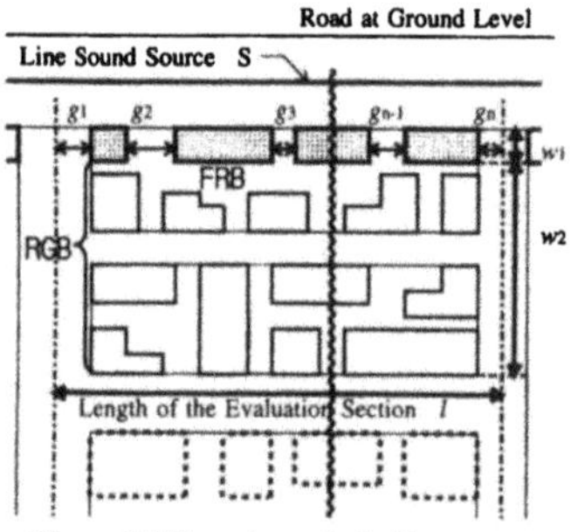

Figure 14 Plan view of a built-up area

7 ACCURACY

The accuracy of the ASJ Model 1998 has been tested up to now by comparing calculated noise levels with measured ones. The analysis was made for the mean difference between the predicted (estimated) and the measured noise level (x: [measured]-[predicted]), the standard deviation of the difference (s) and the correlation coefficient (r). Table 7 shows the accuracy based on the data collected around national highways at ground level, which includes impervious and pervious road surface, and daytime and nighttime data. Table 8 shows the accuracy for expressways in various road constructions, where background noises obtained at individual measuring sites are added in the calculated noise level.

Table 7 Accuracy based on the data collected at national highways at ground level

Type	Sample size	x	s	r
Impervious road surface				
Daytime (6:00 – 22:00)	951	-0.6	1.8	0.84
Nighttime (22:00 – 6:00)	951	1.4	2.1	0.92
Pervious road surface (drainage asphalt)				
Daytime (6:00 – 22:00)	285	-0.4	2.0	0.73
Nighttime (22:00 – 6:00)	285	1.5	2.1	0.86

Table 8 Accuracy based on the data collected at express highways (Note: back ground noises at individual sites were taken into account for the noise level calculation)

Type	Sample size	x	s	r
Bank roads	864	1.3	2.5	0.93
Elevated roads	814	1.5	2.2	0.91
Flat roads	833	0.5	2.9	0.87
Depressed roads	833	0.7	3.0	0.83

The relationships between the calculated and the measured noise level ($L_{Aeq,10min.}$) are shown in Fig.15 for national highway and in Fig.16 for bank road and elevated road (viaduct).

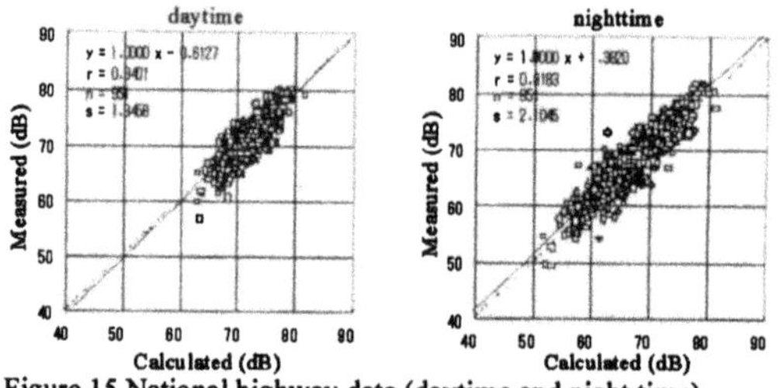

Figure 15 National highway data (daytime and night time)

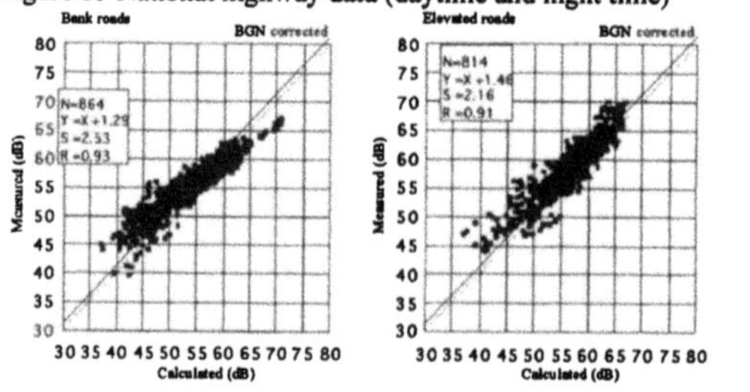

Figure 16 Expressway data (bank roads and elevated roads)

8 CONCLUDING REMARKS

The outline of the ASJ Model 1998 has been presented. The model is available to monitoring of noise at roadside areas as well as road traffic noise prediction. However, there still exist various kinds of problems that should be solved in the future. The Research Committee of Road Traffic Noise in the Acoustical Society of Japan is continuing research works for a well-revised model "ASJ RTN-Model 2003".

REFERENCE

Tachibana H., et. al. 1994. ASJ prediction methods of road traffic noise. Proc. inter-noise 94, 283-288.

Kouyoumjian R. G. et al. 1974. A uniform geometrical theory of diffraction for an edge in a perfectly conducting surface. Proc. IEEE 62, 1448-1461.

Kawai T. 1983. On sound propagation above a locally reacting boundary. J. Acoust. Soc. Jpn., 39, 374-379, (in Japanese).

Yamamoto K. 1994. Revised expression of vehicle noise propagation over ground. J. Acoust. Soc. Jpn(E), 15, 233-241.

Takagi K., et al. 1996. Comparison of simple prediction methods for noise reduction by double barriers. Proc. inter-noise 96, 779-782.

Takagi K., et al. 2000. Prediction of road traffic noise around tunnel mouth. Proc. inter-noise 2000.

Uesaka K., et al. 2000. Prediction and Evaluation Method for Road Traffic Noise in Built-up Areas. Proc. inter-noise 2000.

Final discussion

Final discussion

The discussion note is prepared by N. Chouw and G. Schmid based on the note provided by N. Chouw, O. v. Estorff, S. Hirose, and G. Schmid.

In the final discussion the following topics have been addressed:

Wave barrier

Massarsch: Air cushions surrounded by solid material have been developed as a wave barrier. These wave barriers have been patented. They are effective in the frequency range between 12 to 18Hz. This approach has been applied to isolate a house near to a rail track in Düsseldorf, Germany, with a floor resonant frequency of 15Hz. The method has also been applied in marine environment in order to shield off pressure waves. Wave barriers with chopped rubber as filling material of a trench did not proof to be effective.

For more information Dr. R. Massarsch is a good contact person.

Chouw: Wave impeding barrier (wib) based on cut-off frequency of a soil layer over bedrock is already used in more than 30 projects in Japan. This reduction approach has been patented.

For more information please contact Profs. G. Schmid and N. Chouw.

Damping

In the mechanical formulation damping is often not clearly defined in many publications. However, a clear definition is necessary, and the limitation of the assumption should be understood.

For example, *damping ratio* in case of viscous damping is valid only for a single mass-spring-damper system, i.e. a single-degree-of-freedom (SDOF) system, or a generalized SDOF system, where the eigenmode shapes are the modal coordinates. *Lost factor* is valid for a cyclic excitation. *Geometrical damping* is related to geometrical wave spreading in an open system. However, it is not related to energy dissipation.

It was noted that all damping is of hysterical nature. The so-called frequency independent *hysteretic damping* violates causality. This incorrect damping model will cause the system to respond to the load before the load is applied, and thus leads to non-causal behaviour in the time domain. Frequency independent damping cannot be correct, since at zero frequency the damping value has to be zero.

In reality the energy dissipation can be highly non-linear, where energy dissipation depends on strain level and frequency.

Distinction of *energy dissipation* and *attenuation of vibrations* should be made for a clear mechanical formulation of a system. In current literature for both phenomena the word damping is used.

For soil the material damping is usually determined by using resonant-column tests where shear modulus, density and Poisson's ratio are significant, and the Poisson's ratio is difficult to determine.

In addition, in numerical analysis numerical damping may be present.

Dimensionality

Question: In which cases a 2- or 2.5-dimensional model is sufficient compared to the full 3-dimensional model?

Answer: Generally, a simplification from 3-dimensional to 2-dimensional is not possible. It depends on the source and observation points. A possible answer can be found by comparing the 2- and 3-dimensional results. For a validation of the model field experimental result should be considered. In this context it is also interesting how the waves travel along the track (wave guide/ leaky waves). A 3-dimensional model can help to understand the situation.

Field and numerical benchmark tests are available, for example:

Karl Popp and Werner Schiehlen (eds.): *System Dynamics and Long-Term Behaviour of Railway Vehicles, Track and Subgrade*, Lecture Notes in Applied Mechanics Vol. 6, Springer Verlag, 2003, 488pp., ISBN 3-540-43892-0

and

Yoshihiko Sato: New Railroad Track Mechanics, Chinese Railroad Publisher, November 2001, 504 pp., ISBN 7-113-04313-5 (in Chinese and Japanese)

In the given publications one can also find information on *which data* and *at which locations* to be measured. Both depend on the experience in the specific country.
Vertical and horizontal component of ground motions

In many design codes mainly the vertical component of the ground motions is considered. However, experience shows that horizontal and vertical components can be of the same magnitude.

Miscellaneous

Wave propagation – Moving load – Vibration reduction, Chouw & Schmid (eds.)
© 2003 Swets & Zeitlinger, Lisse, ISBN 90 5809 559 2

Addresses of scientific committee members

Prof. Dr.-Ing. Nawawi Chouw
Okayama University
Department of Environmental and Civil Engineering
3-1-1, Tsushima Naka
Okayama 700-2530
Japan
chouw@cc.okayama-u.ac.jp

Prof. Dr. Eng. Sohichi Hirose
Tokyo Institute of Technology
Department of Mechanical and Environmental Informatics
2-12-1 Ookayama, Meguro-ku
Tokyo 152-8552
Japan
shirose@cv.titech.ac.jp

Prof. Günther Schmid, Ph.D.
Ruhr University Bochum,
Geb. IA 6/37
Universitaetsstr. 150
D-44780 Bochum
Germany
Guenther.Schmid@ruhr-uni-bochum.de

Prof. Dr. Eng. Takeo Taniguchi
Okayama University
Department of Environmental and Civil Engineering
3-1-1, Tsushima Naka
Okayama 700-2530
Japan
taniguti@cc.okayama-u.ac.jp

Addresses of invited speakers

Prof. Dr.-Ing. Nawawi Chouw
Okayama University
Department of Environmental and Civil Engineering
3-1-1, Tsushima Naka
Okayama 700-2530
Japan
chouw@cc.okayama-u.ac.jp

Dr. Didier Clouteau
LMSSM
Ecole Centrale de Paris-CNRS/URA 850
92295 Chatenay-Malabry
France
clouteau@mss.ecp.fr

Prof. Dr. Geert Degrande
Katholieke Universiteit Leuven
Faculteit Toegepaste Wetenschappen
Departement Burgerlijke Bouwkunde
Afdeling Bouwmechanica
Kasteelpark Arenberg 40
B-3001 Heverlee
Belgium
Geert.Degrande@bwk.kuleuven.ac.be

Prof. Dr.-Ing. Uwe E. Dorka
University of Kassel
Department of Civil Engineering
Kurt-Wolters-Straße 3
D-34125 Kassel
Germany
uwe.dorka@uni-kassel.de

Prof. Dr.-Ing. Otto von Estorff
University Hamburg-Harburg
Eissendorferstrasse 42
D-21073 Hamburg
Germany
estorff@tuhh.de

Prof. Hong Hao, Ph.D.
University of Western Australia
Department of Civil and Resource Engineering
35 Stirling Highway
Crawley, Western Australia 6009
Australia
hao@civil.uwa.edu.au

Prof. Dr. Eng. Sohichi Hirose
Tokyo Institute of Technology
Department of Mechanical and Environmental Informatics
2-12-1 Ookayama, Meguro-ku
Tokyo 152-8552
Japan
shirose@cv.titech.ac.jp

Dr. K.R. Massarsch
Geo Engineering AB
Ferievagen 25
S-168 41 Bromma
Sweden
rainer.massarsch@chello.se

Prof. Dr.-Ing. Stavros A. Savidis
Technical University Berlin
Gustav-Meyer-Allee 25
D-13355 Berlin
Savidis@tu-berlin.de

Prof. Günther Schmid, Ph.D.
Ruhr University Bochum
Geb. IA 6/37
Universitaetsstr. 150
D-44780 Bochum
Germany
Guenther.Schmid@ruhr-uni-bochum.de

Prof. Dr. Eng. Nobumasa Sugimoto
Osaka University
Department of Mechanical Science
Toyonaka
Osaka 560-8531
Japan
sugimoto@me.es.osaka-u.ac.jp

Prof. Yeong-Bin Yang, Ph.D.
National Taiwan University
Department of Civil Engineering
No. 1 Sec 4 Roosevelt Road
Taipei 10617
Taiwan
ybyang@ntu.edu.tw

Dr. Eng. Osamu Yoshioka
Central Japan Railway Company
Nakagawa-ku, Chioryou machi 1-1
Nagoya 454-0815
Japan
yosi@jr-central.co.jp

Addresses of speakers selected from the "Call for Papers"

Prof. Dr. Eng. Kazuhisa Abe
Niigata University
Department of Civil Engineering and Architecture
8050 Igarashi 2-Nocho
Niigata 950-2181
Japan
abe@eng.niigata-u.ac.jp

Prof. Dr. Eng. Maher Adam
Zagazig University
El-Hosy St.
Kaha 13743
Egypt
maheradam@yalla.com

Prof. Sofia W. Alisjahbana, Ph.D.
School of Graduate Studies
Civil Engineering Program
Tarumanagara University
Jln. Letjen. S. Parman, No. 1
11440 Jakarta
Indonesia
wangsadi@indosat.net.id

Dr. Shigeru Aoki
Metropolitan College of Technology
Department of Mechanical Engineering
Shinagawa-ku
Higashiooi 1-10-40
Tokyo
Japan
aoki@tokyo-tmct.ac.jp

Prof. Dr. Ömer Aydan
Department of Marine Civil Engineering
Tokai University
Shimizu
Japan
aydan@scc.u-tokai.ac.jp

Dr.-Ing. Christopher Bode
Technical University Berlin
Gustav-Meyer-Allee 25
D-13355 Berlin
Germany
christopher.bode@grundbau.tu-berlin.de

Dr. Eng. Tetsuya Doi
Kobayashi Institute of Physical Research
Kokubunji-shi
Higashimoto-cho 3-20-41
Tokyo 185-0022
Japan
doi@kobayasi-riken.or.jp

Dr. Eng. Masaru Furuta
Tokyo Metropolitan Construction Ltd.
Shinjuku-ku
Nishi Shinjuku 2-8-1
Tokyo 163-8001
Japan
furuta@tokyo.email.ne.jp

Prof. Dr. Eng. Kiyoshi Hayakawa
Rtsumeikan University
Department of Civil Engineering
Japan
kiyoshi@se.ritsumei.ac.jp

Kazuya Itoh
Tokyo Institute of Technology
Faculty of Engineering
2-12-1, O-OkayamaUniversity Meguro-ku
Tokyo 152-8552
Japan
k-ito@cv.titech.ac.jp

Dr. Eng. Shinji Konishi
Railway Technical Research Institute
Tunnel group
2-8-38, Hikari-cho
Kokubunji-shi
Tokyo 185-8540
Japan
konishi@rtri.or.jp

Kang-Il Lee, Ph.D.
Daejin University
Korea
lpk1007@ianyang.ac.kr

PYL Lincy
Katholieke Universiteit Leuven
Faculteit Toegepaste Wetenschappen
Departement Burgerlijke Bouwkunde
Afdeling Bouwmechanica
Kasteelpark Arenberg 40
B-3001 Heverlee
Belgium
lincy.pyl@bwk.kuleuven.ac.be

Dr. Andrei V. Metrikine
Delft University of Technology
Faculty of Civil Engineering and Geosciences
Netherlands
A.Metrikine@citg.tudelft.nl

Imad Mualla, Ph.D.
Technical University of Denmark
Brovej, Building 118
DK-2800 Kgs. Lungby
Denmark
ihm@damptech.com

Dr. Eng. Kiyoshi Nagakura
Railway Technical Research Institute
2-8-38, Hikari-cho
Kokubunji-shi
Tokyo 185-8540
Japan
naga@rtri.or.jp

Dr. Eng. Ali Niousha
Kajima Technical Research Institute
Tokyo
Japan
niousha@katri.kajima.co.jp

Dr. Miyoshi Okamura
Yamanashi University
Department of Civil and Environmental Engineering
Yamanashi-ken
Takeda 4-3-11
Koufu-shi 400-8511
Japan
miyoshi@mail.yamanashi.ac.jp

Masatsugu Otsuki
Keio University
Department of System Design Engineering
3-14-1, Hiyoshi, Kohoku-ku
Yokohama-shi 223-8522
Japan
otsuki@sd.keio.ac.jp

Dr.-Ing. Gero Pflanz
BMW AG
EK-212
D-80788 Munich
Germany
Gero.Pflanz@bmw.de

Prof. Dr. Eng. Yoshihiko Sato
Railway Track System Research Institute
Inage-ku
Kurosadai 1-11-8
Chibaken
Chibashi
Japan
satoy@sa2.so-net.ne.jp

Dr. Eng. Kiwamu Tsuno
Railway Technical Research Institute
Tunnel group
2-8-38, Hikari-cho
Kokubunji-shi
Tokyo 185-8540
Japan
tsuno@rtri.or.jp

Dr. Zhongqi Wang
Nanyang Technological University
School of Civil and Structural Engineering
Nanyang Avenue
Singapore 639798
czqwang@ntu.edu.sg

Dr. Eng. Kohei Yamamoto
Kobayashi Institute of Physical Research
Kokubunji-shi
Higashimoto-cho 3-20-41
Tokyo 185-0022
Japan
yamamoto@kobayasi-riken.or.jp

Prof. Jong- Dar Yau, Ph.D.
Tamkang University
Department of Architecture and Building Technology
5, Lane 199, Kinghua Street,
Taipei 10620
Taiwan
jdyau@mail.tku.edu.tw

Wave propagation – Moving load – Vibration reduction, Chouw & Schmid (eds.)
© 2003 Swets & Zeitlinger, Lisse, ISBN 90 5809 559 2

Author index

Milton Keynes UK
Ingram Content Group UK Ltd.
UKHW020822141024
449569UK00008B/523

9 789058 095596